Universitext

Universitext

Universitext is a series of textbooks that presents material from a wide variety of mathematical disciplines at master's level and beyond. The books, often well class-tested by their author, may have an informal, personal, even experimental approach to their subject matter. Some of the most successful and established books in the series have evolved through several editions, always following the evolution of teaching curricula, into very polished texts.

Thus as research topics trickle down into graduate-level teaching, first textbooks written for new, cutting-edge courses may make their way into *Universitext*.

For further volumes:
www.springer.com/series/223

Karl-G. Grosse-Erdmann · Alfred Peris Manguillot

Linear Chaos

Springer

Karl-G. Grosse-Erdmann
Institut de Mathématique
Université de Mons
Le Pentagone, 7000 Mons
Belgium
kg.grosse-erdmann@umons.ac.be

Alfred Peris Manguillot
Institut Universitari de Matemàtica
Pura i Aplicada
Universitat Politècnica de València
Edifici 7A, 46022 València
Spain
aperis@mat.upv.es

ISSN 0172-5939 e-ISSN 2191-6675
Universitext
ISBN 978-1-4471-2169-5 e-ISBN 978-1-4471-2170-1
DOI 10.1007/978-1-4471-2170-1
Springer London Dordrecht Heidelberg New York

British Library Cataloguing in Publication Data
A catalogue record for this book is available from the British Library

Library of Congress Control Number: 2011936377

Mathematics Subject Classification: 47-01, 37B05, 47A16, 47D06

Cover design: VTeX UAB, Lithuania

Printed on acid-free paper

Springer is part of Springer Science+Business Media (www.springer.com)

Preface

According to a widely held view, chaos is intimately linked to nonlinearity. It is usually taken to be self-evident that a linear system behaves in a predictable manner.

However, as early as 1929, G.D. Birkhoff obtained an example of a linear operator that possesses an important ingredient of chaos: the existence of a dense orbit. Later, G.R. MacLane (1952) found the same phenomenon for the differentiation operator, which, after all, is the fundamental operator in analysis. And S. Rolewicz (1969) showed that not only nonlinear shifts but also linear shifts can have dense orbits. Motivated by these sporadic examples, researchers began in the nineteen-eighties to study the dynamical properties of general linear operators; henceforth, operators with a dense orbit were called hypercyclic. As a first important result a useful condition for hypercyclicity, the so-called Hypercyclicity Criterion, was obtained.

A further decisive step was taken by G. Godefroy and J.H. Shapiro (1991). Not only did they come up with whole new classes of hypercyclic operators, they also proposed to accept Devaney's definition of (nonlinear) chaos as the right definition for linear chaos: a linear operator is chaotic if it has a dense orbit, it has a dense set of periodic points and it has sensitive dependence on initial conditions. They then showed that many linear operators are chaotic, including the three classical operators of Birkhoff, MacLane and Rolewicz.

The fact that chaos for linear systems has only been discovered recently is easily explained: as Rolewicz showed, hypercyclicity, and hence also linear chaos, requires an infinite-dimensional setting.

Over the last quarter of a century, the study of hypercyclic and chaotic operators has turned into a fascinating and very active research area. It has produced an astounding number of deep and beautiful results. As representative examples we mention here only Ansari's theorem that every power of a hypercyclic operator is hypercyclic, the Ansari–Bernal theorem that every infinite-dimensional separable Banach space supports a hypercyclic operator, Grivaux's theorem that every Hilbert space operator is the sum of two chaotic operators, and the Bourdon–Feldman theorem that any orbit that is somewhere dense is (everywhere) dense.

It seems fair to say that while research in linear dynamics is still expanding in both depth and breadth, the foundations have reached a certain stage of maturity. At the same time the basic ideas as well as the applications of the field have a broad appeal also to nonspecialists.

It is therefore our aim to make the theory of hypercyclic operators and linear chaos accessible to a wider audience. The book is aimed at advanced undergraduate or beginning graduate students, both as a basis for a lecture course and for self-study. We have strived at a self-contained exposition. Each chapter contains a large number of exercises and ends with a section that gives references and directs the reader to further literature.

We have tried to keep the necessary prerequisites for reading this book to a minimum. Since the concept of a hypercyclic operator requires both a topological and a linear structure, the reader is supposed to be familiar with metric spaces (up to the Baire category theorem) and with the basic theory of Hilbert and Banach spaces, as it is often presented in advanced undergraduate courses on analysis. Moreover, since many examples in the theory are given by operators on spaces of holomorphic functions the reader is also expected to have had an introductory course on complex analysis. Additional, more advanced tools that are only needed occasionally will be provided in the two appendices.

The book is divided into two parts. Part I presents an introduction to the dynamics of linear operators. Its chapters form a unity and are best studied in that order. In contrast, Part II covers selected topics from linear dynamics. Its chapters are largely independent so that they can be read in an arbitrary order. An occasional cross reference should pose no problem.

More specifically, Chapter 1 introduces the reader to the fundamental concepts of the theory of (not necessarily linear) dynamical systems. Its highlights are the Birkhoff transitivity theorem, which is of fundamental importance for all that follows, and a close study of the various concepts of maps with complicated behaviour, including chaotic maps. In Chapter 2, the notions and results from the first chapter are revisited in the context of linearity. Among other things it is proved that the operators of Birkhoff, MacLane and Rolewicz are chaotic, and that linear dynamics can be as complicated as nonlinear dynamics. We begin the chapter with an introduction to a straightforward generalization of Banach spaces, the so-called Fréchet spaces; they provide the setting for some important chaotic operators. Chapter 3 presents several criteria for hypercyclicity and chaos, in increasing order of sophistication. It culminates in the Hypercyclicity Criterion, which is discussed in detail. In Chapter 4, some important classes of hypercyclic and chaotic operators are described: weighted shift operators on sequence spaces, differential and composition operators on spaces of holomorphic functions, and adjoint multiplication operators. In addition to the shift operators, which are studied throughout the book, the reader may want to concentrate on one or two additional classes, depending on his or her own personal preference. In Chapter 5 we discuss the spectral properties of hypercyclic and chaotic operators. As an application we derive properties that preclude hypercyclicity or chaos. Finally, Chapter 6 presents some of the deepest, most beautiful and most useful results from linear dynamics. It contains, among other things, Ansari's theorem on the powers of hypercyclic operators, the Bourdon–Feldman theorem on somewhere dense orbits and the León–Müller theorem on the hypercyclicity of unimodular multiples of hypercyclic operators.

In the second part, Chapter 7 discusses the continuous analogue of hypercyclic and chaotic operators in the form of semigroups. While the theories run parallel in great parts, hypercyclic and chaotic semigroups have important applications to partial differential equations. In Chapter 8 we obtain, among other things, the Ansari–Bernal theorem on the existence of hypercyclic operators. We also discuss here the richness of the set of hypercyclic operators. The contents of Chapter 9 are motivated by recent work on the application of ergodic theory to linear dynamics. While the technical difficulties involved prevent us from studying these tools here, we will discuss a new concept that has come out of these investigations, the frequently hypercyclic operators. Chapter 10 is devoted to the question of whether there is, for a given operator, an infinite-dimensional closed subspace all of whose nonzero vectors are hypercyclic, while Chapter 11 studies the existence of common hypercyclic vectors for (uncountable) families of operators. The final Chapter 12 treats hypercyclicity and linear chaos in their most natural (and most general) setting, the topological vector spaces. After a brief introduction to such spaces we revisit many of the results previously obtained in the book and show that they hold in great generality.

At this point it seems important to add a disclaimer concerning our strategy for attaching names to theorems. In keeping with the usual practice in mathematics we have attributed results to the author(s) who first proved them. But the form in which the result is presented in the book may well be due to additional contributions from further authors. The reader is advised to consult the relevant sources and comments section for complete references.

We would like to say a few words about the differences with a recent monograph by F. Bayart and É. Matheron [44] on the topic of linear dynamics. While their book is intended to be accessible to readers with a reasonable background in functional analysis at the graduate level, we have tried to make our text more basic and self-contained, so that students should be able to follow it at an earlier stage of their studies without additional material. In a certain sense, the two books are complementary. While we cover the foundations and the main body of linear dynamics in detail, Bayart and Matheron proceed to present some technically demanding topics like the counterexamples to Herrero's problem due to De la Rosa–Read and Bayart–Matheron, the applications of ergodic theory to linear dynamics, or Read-type operators for which every nonzero vector is hypercyclic.

The starting point of this book was a mini-workshop on hypercyclicity and linear chaos at Oberwolfach in August 2006, where we agreed that a book on this topic ought to be written. Since then, stays at the Berlin Mathematical School/Technische Universität Berlin (2007), the Centre International de Rencontres Mathématiques in Marseille (2008, 2009, 2010), and the Universitat Politècnica de València/Universitat de València (2008), made it possible, with their support and their perfect working conditions, that the book was written. We are, in particular, grateful to Günter M. Ziegler (Berlin), Pascal Chossat (CIRM) and Manuel Maestre (València) for the invitations to their institutions.

It is with great pleasure that we thank Richard Aron, Salud Bartoll, Luis Bernal-González, Juan Bès, Manuel De la Rosa, Elisabetta Mangino, Étienne Matheron, Quentin Menet, Raymond Mortini, Joel Shapiro, Dirk Werner, and the referees; they have helped us, in various ways and at different stages of the project, by providing many valuable suggestions and interesting discussions, by detecting embarrassing mistakes and by offering constructive remarks. We are much indebted to Klaus Cloppenburg for providing us with the index for the book. Our thanks also go to the editors and their staff at Springer-Verlag, London, especially Karen Borthwick, Joerg Sixt, and Lauren Stoney, for their valuable assistance.

We acknowledge the support of MICINN and FEDER, Project MTM2007-64222, and Generalitat Valenciana, Project PROMETEO/2008/101.

Above all we thank our partners, Klaus and Olga, for their love, encouragement and support. The first author wants to dedicate the book to the memory of his father who could not live to see its completion. The second author dedicates the book to Olga, his parents, and his family.

Mons, València
June 2011

Karl-G. Grosse-Erdmann
Alfred Peris Manguillot

Contents

Part I Introduction to linear dynamics

Part I
Introduction to linear dynamics

Chapter 1
Topological dynamics

This chapter provides an introduction to the theory of (not necessarily linear) dynamical systems. Fundamental concepts such as topologically transitive, chaotic and (weakly) mixing maps are defined and illustrated with typical examples. The Birkhoff transitivity theorem is derived as a crucial tool for showing that a map has a dense orbit. Moreover we obtain several characterizations of weakly mixing maps that will be of great significance later on.

The interest in this chapter is that it derives results on dynamical systems that do not require linearity. From Chapter 2 onwards, all our systems will be linear.

1.1 Dynamical systems

The theory of dynamical systems studies the long-term behaviour of evolving systems.

As a motivating example we consider the size of a population, which we assume to be given by the value N_n at discrete times $n = 0, 1, 2, \ldots$. In a simple model the size at time $n + 1$ will only depend on the size at time n. The population is then described by a law

$$N_{n+1} = T(N_n), \quad n = 0, 1, 2, \ldots,$$

where T is a suitable map. It follows that

$$N_n = (T \circ \ldots \circ T)(N_0), \quad n = 1, 2, \ldots$$

with n applications of the map T. Thus the behaviour of the population is completely determined by the initial population N_0 and the map T.

More generally, we assume that the possible states of a (physical, biological, economic, ..., or abstract) system are described by the elements from a set X and that evolution of the system is described by a map $T : X \to X$;

K.-G. Grosse-Erdmann, A. Peris Manguillot, *Linear Chaos*, Universitext,

DOI 10.1007/978-1-4471-2170-1_1,

that is, if $x_n \in X$ is the state of the system at time $n \geq 0$, then

$$x_{n+1} = T(x_n), \quad n = 0, 1, 2, \ldots.$$

Since we want to measure changes in the values x_n, we require the underlying space to be a metric space. And since we want small changes in x_n only to result in small changes in x_{n+1} we require continuity of T.

Definition 1.1. A *(discrete) dynamical system* is a pair (X, T) consisting of a metric space X and a continuous map $T : X \to X$.

Often we will simply call T (when the underlying space X is taken for granted) or $T : X \to X$ a dynamical system. Moreover we adopt the notation used in operator theory to write Tx for $T(x)$.

What we are interested in is the evolution of the system that starts with a certain state x_0. For this we define the *iterates* $T^n : X \to X$, $n \geq 0$, by the n-fold iteration of T,

$$T^n = T \circ \ldots \circ T \quad (n \text{ times})$$

with

$$T^0 = I,$$

the identity on X.

Definition 1.2. Let $T : X \to X$ be a dynamical system. For $x \in X$ we call

$$\mathrm{orb}(x, T) = \{x, Tx, T^2x, \ldots\}$$

the *orbit* of x under T.

Returning to the previous discussion, suppose that the size N_n of a population changes proportionally to its actual size, that is, it follows the law

$$\frac{N_{n+1} - N_n}{N_n} = \gamma, \quad n \geq 0,$$

with some constant $\gamma > -1$. One may write this equivalently as

$$N_{n+1} = (1 + \gamma) N_n,$$

so that the corresponding dynamical system is given by

$$T : \mathbb{R}_+ \to \mathbb{R}_+, \quad Tx = (1 + \gamma)x.$$

The orbit of $x \in \mathbb{R}_+$ can be calculated explicitly as

$$\mathrm{orb}(x, T) = \{(1 + \gamma)^n x \ ; \ n \geq 0\}.$$

Thus, the orbit tends to 0, x and ∞ for $-1 < \gamma < 0$, $\gamma = 0$ and $\gamma > 0$, respectively.

As a more realistic model for the evolution of a population the following has been suggested. If we assume that the environment limits the size of the population by a certain number $L > 0$ then we might assume the law to be

$$\frac{N_{n+1} - N_n}{N_n} = \gamma(L - N_n), \quad \gamma > 0.$$

Rescaling by $M_n = (L + \gamma^{-1})^{-1} N_n$ and setting $\mu = \gamma L + 1$ we obtain that

$$M_{n+1} = \mu M_n (1 - M_n), \quad n \geq 0.$$

This refined model leads to the following dynamical system.

Example 1.3. **(Logistic map)** Let $\mu \in \mathbb{R}$. The *logistic map* $L_\mu : \mathbb{R} \to \mathbb{R}$ is given by $L_\mu x = \mu x(1-x)$, $x \in \mathbb{R}$. Figure 1.1 shows the graph of L_μ for $\mu = 3$.

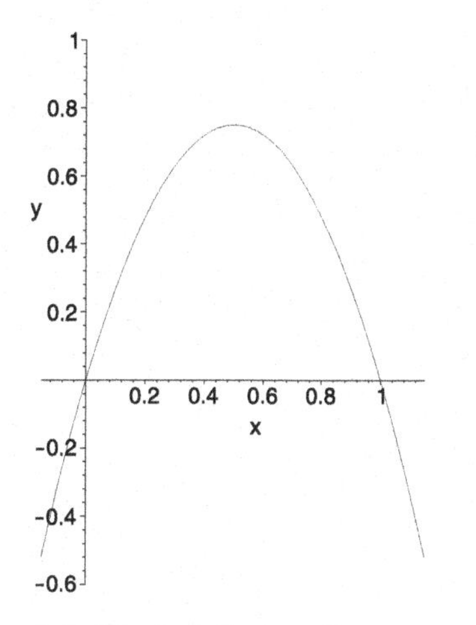

Fig. 1.1 The logistic map L_3

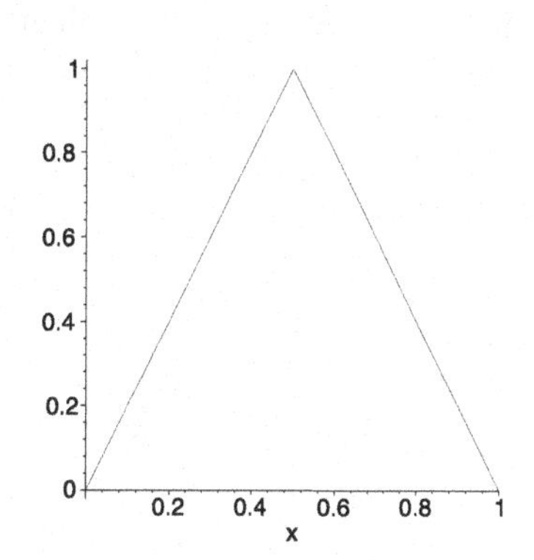

Fig. 1.2 The tent map T

We introduce several other popular dynamical systems.

Example 1.4. (a) **(Quadratic map)** A quadratic map is defined by the real dynamical system $Q_c : \mathbb{R} \to \mathbb{R}$, $x \to x^2 + c$, with a parameter $c \in \mathbb{R}$, or by the corresponding complex dynamical system $Q_c : \mathbb{C} \to \mathbb{C}$, $z \to z^2 + c$, with $c \in \mathbb{C}$.

(b) **(Doubling map on the circle)** Let $T : \mathbb{C} \to \mathbb{C}$ denote the square function $Tz = z^2$. Its iterates are $T^n z = z^{2^n}$. It follows that the orbits for points z with $|z| < 1$ tend to 0, while for $|z| > 1$ the orbits tend to infinity. As we will see later, the dynamics of T for points on the unit circle $\mathbb{T} = \{z \in \mathbb{C}\,;\, |z| = 1\}$ are much more interesting. Since $T(\mathbb{T}) \subset \mathbb{T}$, we usually consider the dynamical system $T : \mathbb{T} \to \mathbb{T}$, $z \to z^2$, the so-called *doubling map*. The name refers to the fact that T doubles the argument of the complex number z.

(c) **(Circle rotation)** The system $T : \mathbb{T} \to \mathbb{T}$, $z \to e^{i\alpha}z$, $\alpha \in [0, 2\pi[$, describes the rotation of the point z on the unit circle by the angle α. We will see that its dynamical behaviour depends to a large extent on the question of whether the rotation is *rational* ($\alpha \in \pi\mathbb{Q}$) or *irrational* ($\alpha \notin \pi\mathbb{Q}$).

(d) **(Tent map)** The tent map is given by $T : [0, 1] \to [0, 1]$, $Tx = 2x$, if $x \in [0, \frac{1}{2}]$, and $Tx = 2 - 2x$, if $x \in]\frac{1}{2}, 1]$. The name derives from the shape of its graph; see Figure 1.2.

(e) **(Doubling map on the interval)** We consider the interval $[0, 1]$ in which we identify 0 and 1; the metric on this space is given by $d(x, y) = \min(|x - y|, 1 - |x - y|)$. Then $T : [0, 1] \to [0, 1]$, $x \to 2x \pmod 1$ describes a dynamical system.

(f) **(Shift on the interval)** When we identify again 0 and 1, the map $T : [0, 1] \to [0, 1]$, $x \to x + \alpha \pmod 1$ with $\alpha \in [0, 1[$ describes the shift by α, modulo 1, of any point on the unit interval.

Every mathematical theory has its notion of isomorphism. When do we want to consider two dynamical systems $S : Y \to Y$ and $T : X \to X$ as equal? There should be a homeomorphism $\phi : Y \to X$ such that, when $x \in X$ corresponds to $y \in Y$ via ϕ then Tx should correspond to Sy via ϕ. In other words, if $x = \phi(y)$ then $Tx = \phi(Sy)$. This is equivalent to saying that $T \circ \phi = \phi \circ S$.

We recall that a homeomorphism is a bijective continuous map whose inverse is also continuous. In many applications, however, it is already enough to demand that ϕ is continuous with dense range.

Definition 1.5. Let $S : Y \to Y$ and $T : X \to X$ be dynamical systems.

(a) Then T is called *quasiconjugate* to S if there exists a continuous map $\phi : Y \to X$ with dense range such that $T \circ \phi = \phi \circ S$, that is, the diagram

$$\begin{array}{ccc} Y & \xrightarrow{\ S\ } & Y \\ {\scriptstyle\phi}\downarrow & & \downarrow{\scriptstyle\phi} \\ X & \xrightarrow{\ T\ } & X \end{array}$$

commutes.

(b) If ϕ can be chosen to be a homeomorphism then S and T are called *conjugate*.

Conjugacy is clearly an equivalence relation between dynamical systems, and conjugate dynamical systems have the same dynamical behaviour. What makes this notion even more interesting is the fact that it is by no means always obvious if two systems are conjugate or not.

Example 1.6. We refer to the various dynamical systems introduced above.

(a) For any $\mu \neq 0, 2$, the logistic maps L_μ and $L_{2-\mu}$ are conjugate; one can take an affine function $x \to ax + b$ for ϕ, as is easily verified. It therefore suffices to study these maps for $\mu \geq 1$.

(b) Any logistic map L_μ, $\mu \neq 0$, is conjugate to a suitable real quadratic map Q_c; again, the conjugacy can be given by an affine function.

(c) The logistic map L_4, when restricted to the interval $[0,1]$, is conjugate to the tent map. In fact, an easy calculation shows that one may take $\phi(x) = \sin^2(\frac{\pi}{2}x)$.

(d) When we identify the points 0 and 1, the map $\phi : [0,1] \to \mathbb{T}$, $x \to e^{2\pi i x}$ is a homeomorphism. It clearly defines a conjugacy between the doubling map on $[0,1]$ and the doubling map on the circle; and it defines a conjugacy between the shift by α on $[0,1]$ and the circle rotation by the angle $2\pi\alpha$.

Definition 1.7. We say that a property $\mathcal{P}$ for dynamical systems is *preserved under (quasi)conjugacy* if the following holds: if a dynamical system $S : Y \to Y$ has property $\mathcal{P}$ then every dynamical system $T : X \to X$ that is (quasi) conjugate to S also has property $\mathcal{P}$.

For example, the property of having dense range is clearly preserved under quasiconjugacy, while the property of being surjective is preserved under conjugacy but not under quasiconjugacy.

1.2 Topologically transitive maps

One way of defining a new dynamical system from a given dynamical system T is by restricting it to a subset. However, one has to ensure that T maps this subset into itself.

Definition 1.8. Let $T : X \to X$ be a dynamical system. Then a subset $Y \subset X$ is called *T-invariant* or *invariant under T* if $T(Y) \subset Y$.

Thus, if $Y \subset X$ is T-invariant, then $T|_Y : Y \to Y$ is also a dynamical system.

Example 1.9. The interval $[0,1]$ is invariant under the logistic map L_μ for $0 \leq \mu \leq 4$.

The study of a mathematical object is often simplified by breaking it up into smaller parts and by studying these separately. If such a splitting is not possible then one usually says that the object is irreducible. In the case of dynamical systems, we might regard $T : X \to X$ as irreducible if X cannot be divided into two T-invariant subsets with nonempty interior. In that direction we have the following result.

Proposition 1.10. *Let $T : X \to X$ be a dynamical system. Then we have the implications* (i) $\Longleftarrow$ (ii) $\Longleftrightarrow$ (iii) $\Longleftrightarrow$ (iv) $\Longleftrightarrow$ (v), *where*

(i) *X cannot be written as $X = A \cup B$ with disjoint T-invariant subsets A, B such that A and B have nonempty interior;*

(ii) *X cannot be written as $X = A \cup B$ with disjoint subsets A, B such that A is T-invariant and A and B have nonempty interior;*
(iii) *for any pair U, V of nonempty open subsets of X there exists some $n \geq 0$ such that $T^n(U) \cap V \neq \varnothing$;*
(iv) *for any nonempty open subset U of X the set $\bigcup_{n=0}^{\infty} T^n(U)$ is dense in X;*
(v) *for any nonempty open subset U of X the set $\bigcup_{n=0}^{\infty} T^{-n}(U)$ is dense in X.*

Proof. (ii)$\Longrightarrow$(i) is trivial.

(ii)$\Longrightarrow$(iv). Let $A = \bigcup_{n=0}^{\infty} T^n(U)$ and $B = X \setminus A$. Then A is T-invariant, and it has nonempty interior since it contains U. By (ii), B must have empty interior, which implies that A is dense.

(iii)$\Longrightarrow$(ii). Suppose that $X = A \cup B$, $A \cap B = \varnothing$ and $T(A) \subset A$. Then $\mathrm{int}(A)$ and $\mathrm{int}(B)$ are open sets with $T^n(\mathrm{int}(A)) \cap \mathrm{int}(B) \subset A \cap B = \varnothing$ for all $n \geq 0$. By (iii) this can only be the case if either A or B has empty interior.

We clearly have that (iii)$\Longleftrightarrow$(iv). For (iii)$\Longleftrightarrow$(v) one need only note that $T^n(U) \cap V \neq \varnothing$ is equivalent to $U \cap T^{-n}(V) \neq \varnothing$. $\square$

We see that condition (iii) is slightly stronger than the irreducibility of a dynamical system; see also Exercise 1.2.1. Since this condition will turn out to be of fundamental importance for the theory it is given its own name.

Definition 1.11. A dynamical system $T : X \to X$ is called *topologically transitive* if, for any pair U, V of nonempty open subsets of X, there exists some $n \geq 0$ such that $T^n(U) \cap V \neq \varnothing$.

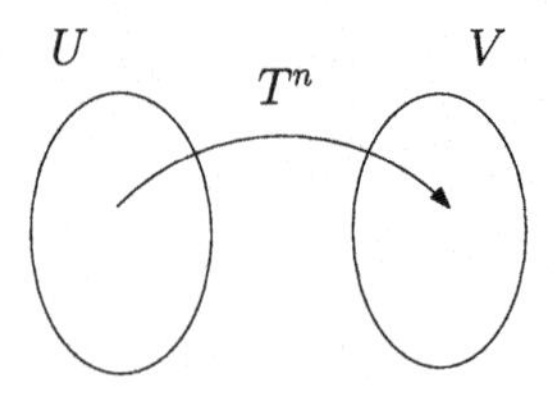

Fig. 1.3 Topological transitivity

Example 1.12. (a) The tent map is topologically transitive. To see this, note that T^n is the piecewise linear map with $T^n(\frac{2k}{2^n}) = 0$, $k = 0, 1, \ldots, 2^{n-1}$, and $T^n(\frac{2k-1}{2^n}) = 1$, $k = 1, \ldots, 2^{n-1}$; see Figure 1.4. Thus, let $U \subset [0,1]$ be nonempty and open. Then U contains some interval $J := [\frac{m}{2^n}, \frac{m+1}{2^n}]$. But since $[0,1] = T^n(J) \subset T^n(U)$, $T^n(U)$ in fact meets every nonempty set V.

(b) The doubling map on the circle, $T : \mathbb{T} \to \mathbb{T}$, $z \to z^2$, is also topologically transitive. In fact, every nonempty open set $U \subset \mathbb{T}$ contains a closed arc of angle $\frac{2\pi}{2^n}$, for some $n \geq 1$. Since the map T doubles angles, we have that $T^n(U)$ contains a closed arc of angle 2π, hence $T^n(U) = \mathbb{T}$ meets every nonempty set V.

(c) No rational rotation $T : \mathbb{T} \to \mathbb{T}$, $z \to e^{i\alpha}z$, is topologically transitive. For example, if $\alpha = \frac{2\pi}{n}$ then the iterates of an open arc γ of angle $\frac{\pi}{n}$ never meet the arc $e^{i\frac{\pi}{n}}\gamma$. In contrast, every irrational rotation is topologically transitive; see Exercise 1.2.9.

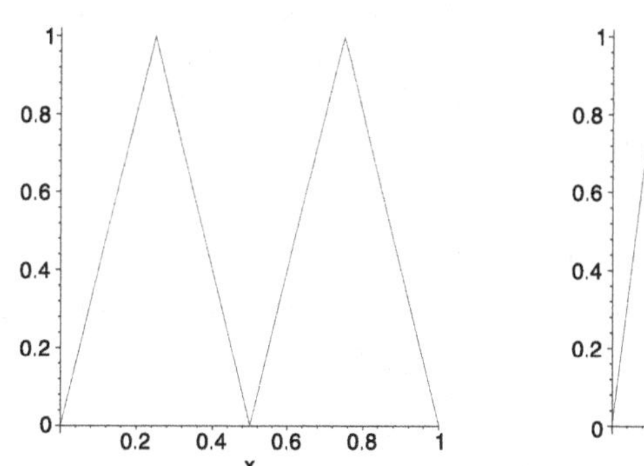

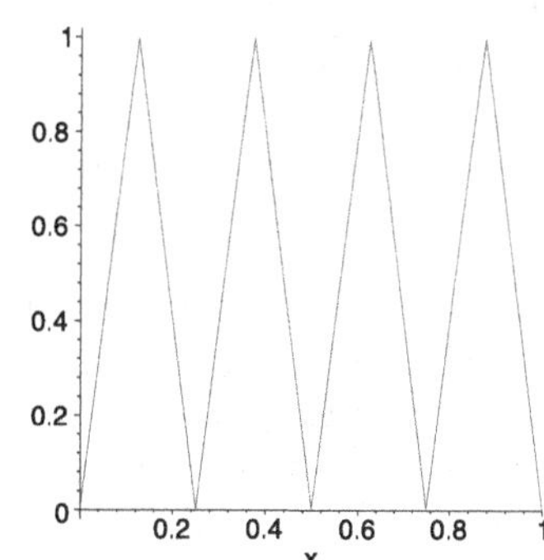

Fig. 1.4 The iterates T^2 and T^3 of the tent map

Proposition 1.13. *Topological transitivity is preserved under quasiconjugacy.*

Proof. Let $T : X \to X$ be quasiconjugate to $S : Y \to Y$ via $\phi : Y \to X$, and let U and V be nonempty open subsets of X. Since ϕ is continuous and of dense range, $\phi^{-1}(U)$ and $\phi^{-1}(V)$ are nonempty and open. Thus there are $y \in \phi^{-1}(U)$ and $n \geq 0$ with $S^n y \in \phi^{-1}(V)$, which implies that $\phi(y) \in U$ and $T^n\phi(y) = \phi(S^n y) \in V$. □

The equivalence of conditions (iv) and (v) in Proposition 1.10 implies the following.

Proposition 1.14. *Let $T : X \to X$ be a dynamical system with continuous inverse T^{-1}. Then T is topologically transitive if and only if T^{-1} is.*

Topological transitivity can be interpreted as saying that T connects all nontrivial parts of X. This is automatically the case whenever there is a point $x \in X$ with dense orbit under T.

Proposition 1.15. *Let T be a continuous map on a metric space X without isolated points.*

(a) *If $x \in X$ has dense orbit under T then so does each $T^n x$, $n \geq 1$.*

(b) *If T has a dense orbit then it is topologically transitive.*

Proof. (a) This follows easily from the fact that $\mathrm{orb}(x,T)\backslash\{x,Tx,\ldots,T^{n-1}x\}$ is contained in $\mathrm{orb}(T^n x,T)$ and that, in every metric space without isolated points, a dense set remains dense even after removing finitely many points.

(b) Suppose that $x \in X$ has dense orbit under T. Let U and V be nonempty open sets in X. Then there is some $n \geq 0$ such that $T^n x \in U$. By (a), also $T^n x$ has dense orbit, so that there is some $m \geq n$ such that $T^m x \in V$. This implies that $T^{m-n}(U) \cap V \neq \varnothing$. □

What is less obvious is that, in separable complete metric spaces, the converse of this result is also true: topologically transitive maps must have a dense orbit. The importance of this result for the theory of dynamical systems can hardly be overemphasized. It was first obtained in 1920 by G. D. Birkhoff in the context of maps on compact subsets of $\mathbb{R}^N$.

Theorem 1.16 (Birkhoff transitivity theorem). *Let T be a continuous map on a separable complete metric space X without isolated points. Then the following assertions are equivalent:*

(i) *T is topologically transitive;*

(ii) *there exists some $x \in X$ such that* $\mathrm{orb}(x,T)$ *is dense in X.*

If one of these conditions holds then the set of points in X with dense orbit is a dense G_δ-set.

Proof. By the previous proposition, (ii) implies (i). For the converse, let T be topologically transitive, and let $\mathcal{D}(T)$ denote the set of points in X that have dense orbit under T. Since X has a countable dense set $\{y_j\ ;\ j \geq 1\}$, the open balls of radius $\frac{1}{m}$ around the y_j, $m,j \geq 1$, form a countable base $(U_k)_{k\geq 1}$ of the topology of X. Hence, x belongs to $\mathcal{D}(T)$ if and only if, for every $k \geq 1$, there is some $n \geq 0$ such that $T^n x \in U_k$. In other words,

$$\mathcal{D}(T) = \bigcap_{k=1}^{\infty} \bigcup_{n=0}^{\infty} T^{-n}(U_k).$$

By continuity of T and Proposition 1.10, each set $\bigcup_{n=0}^{\infty} T^{-n}(U_k)$, $k \geq 0$, is open and dense. The Baire category theorem then implies that $\mathcal{D}(T)$ is a dense G_δ-set, and hence nonempty. □

We note that the absence of isolated points was not needed for the proof that (i) implies (ii).

Example 1.17. It follows from Example 1.12 that the tent map and the doubling map have dense orbits. Irrational rotations (that is, with $\alpha \notin \pi\mathbb{Q}$) have the stronger property that each of their orbits is dense; see Exercise 1.2.9.

Let us briefly reflect on the usefulness of the transitivity theorem. Suppose that we are interested in the existence of a dense orbit under a given map. Sometimes such a point presents itself with little effort, as is the case for irrational rotations. But what if this is not the case? Without further information we will never be likely to stumble on a point with dense orbit. In contrast, topological transitivity seems much easier to prove, as we have seen, for example, in the case of the tent map: we need only connect any two nonempty open sets by suitable iterates.

We stress, however, that in more general spaces topological transitivity and the existence of a dense orbit need not coincide.

Example 1.18. Let X be the set of all points on the unit circle that are 2^nth roots of unity, for some $n \geq 1$. By Example 1.12(b) the doubling map, restricted to X, is topologically transitive, but clearly has no dense orbits.

This example shows that the completeness assumption in the Birkhoff transitivity theorem cannot be dropped.

Proposition 1.19. *The property of having a dense orbit is preserved under quasiconjugacy.*

Proof. Let $T : X \to X$ be quasiconjugate to $S : Y \to Y$ via $\phi : Y \to X$, and let $y \in Y$ have dense orbit under S. If U is a nonempty open subset of X then $\phi^{-1}(U)$ is nonempty and open, so that some $S^n y$, $n \geq 0$, belongs to $\phi^{-1}(U)$. But then $T^n \phi(y) = \phi(S^n y)$ belongs to U. □

Example 1.20. By Examples 1.6(c) and 1.17, the logistic map L_4 on $[0, 1]$ has a dense orbit.

1.3 Chaos

What is chaos? Even when we restrict the meaning of this word to deterministic chaos, that is, chaotic behaviour of a dynamical system, mathematicians have come up with different answers to this question. We will follow here the definition that was suggested by Devaney in 1986. It has three ingredients, which we discuss in turn.

The first ingredient tries to capture the idea of the so-called butterfly effect: small changes in the initial state may lead, after some time, to large discrepancies in the orbit. In order to be able to perturb points we consider only spaces without isolated points.

Definition 1.21. Let (X, d) be a metric space without isolated points. Then a dynamical system $T : X \to X$ is said to have *sensitive dependence on initial conditions* if there exists some $\delta > 0$ such that, for every $x \in X$ and $\varepsilon > 0$, there exists some $y \in X$ with $d(x, y) < \varepsilon$ such that, for some $n \geq 0$, $d(T^n x, T^n y) > \delta$. The number δ is called a *sensitivity constant* for T.

We stress that the definition involves the metric of the space. In the following examples we will always work with the usual metric.

Example 1.22. (a) Using our knowledge of the iterates of the tent map (see Example 1.12), we can easily show that it has sensitive dependence on initial conditions with sensitivity constant 1/4, say. Indeed, if $x \in [0,1]$ and $\varepsilon > 0$ then there is some $n \geq 0$ such that the open ball of radius ε around x contains points y_1 and y_2 with $T^n y_1 = 0$ and $T^n y_2 = 1$; thus $|T^n x - T^n y_j| \geq 1/2$ for some $j \in \{1,2\}$.

(b) A similar argument, based on the fact that the doubling map doubles angles, shows that it has sensitive dependence on initial conditions.

(c) No circle rotation has sensitive dependence on initial conditions because we clearly have that $|T^n z_1 - T^n z_2| = |z_1 - z_2|$ for any $z_1, z_2 \in \mathbb{T}$.

The second ingredient of chaos demands that the system is irreducible in the sense that the map T connects any nontrivial parts of the space. We saw in Section 1.2 that this idea is well captured by the notion of topological transitivity of the system.

The third ingredient demands that the system has many orbits with a regular behaviour; more precisely, there should be a dense set of points with periodic orbit.

Definition 1.23. Let $T : X \to X$ be a dynamical system.

(a) A point $x \in X$ is called a *fixed point* of T if $Tx = x$.

(b) A point $x \in X$ is called a *periodic point* of T if there is some $n \geq 1$ such that $T^n x = x$. The least such number n is called the *period* of x. The set of periodic points is denoted by $\mathrm{Per}(T)$.

A point is periodic if and only if it is a fixed point of some iterate T^n, $n \geq 1$. Thus, for real functions T, one easily detects them by searching for the points where the graphs of T^n and the identity function meet.

Example 1.24. (a) Considering the iterates of the tent map (see Example 1.12), we find that in every interval $[\frac{m}{2^n}, \frac{m+1}{2^n}]$ there is a periodic point of period n. Thus, the tent map has a dense set of periodic points.

(b) The periodic points of the doubling map on the circle are exactly the $(2^n - 1)$st roots of unity, $n \geq 1$, so that also the doubling map has a dense set of periodic points.

(c) For any rational rotation T there is some $N \geq 1$ such that $T^N = I$, so that every point is periodic. In contrast, irrational rotations have no periodic points at all.

Proposition 1.25. *The property of having a dense set of periodic points is preserved under quasiconjugacy.*

Proof. Let $T : X \to X$ be quasiconjugate to $S : Y \to Y$ via $\phi : Y \to X$, and let $U \subset X$ be a nonempty open set. Then $\phi^{-1}(U)$, being also open and nonempty, contains a point y with $S^n y = y$ for some $n \geq 1$. Hence $\phi(y) \in U$ and $T^n \phi(y) = \phi(S^n y) = \phi(y)$. □

Summarizing, we are led to Devaney's definition of chaos.

Definition 1.26 (Devaney chaos – preliminary version). Let (X, d) be a metric space without isolated points. Then a dynamical system $T : X \to X$ is said to be *chaotic (in the sense of Devaney)* if it satisfies the following conditions:

(i) T has sensitive dependence on initial conditions;
(ii) T is topologically transitive;
(iii) T has a dense set of periodic points.

Example 1.27. By Examples 1.12, 1.22 and 1.24, the tent map and the doubling map are chaotic, but no circle rotation is chaotic.

The definition of chaos has a serious blemish: sensitive dependence on initial conditions is not preserved under conjugacy, or, which is the same, it depends on the metric on the underlying space. We illustrate this by an example.

Example 1.28. Let $T :]1, \infty[\to]1, \infty[$ be given by $Tx = 2x$. Since $|T^n x - T^n y| = 2^n |x - y| \to \infty$ whenever $x \neq y$, we have that T has sensitive dependence on initial conditions with respect to the usual metric on $]1, \infty[$. But if we define $d(x, y) = |\log x - \log y|$ then d is an equivalent metric for which $d(T^n x, T^n y) = d(x, y)$ for all $x, y \in]1, \infty[$, which shows that T does not have sensitive dependence on initial conditions with respect to d. On the other hand, the two versions of T are conjugate when we take the identity map as the linking homeomorphism.

Fortunately, one can drop sensitive dependence from Devaney's definition because it is implied by the other two conditions.

Theorem 1.29 (Banks–Brooks–Cairns–Davis–Stacey). *Let X be a metric space without isolated points. If a dynamical system $T : X \to X$ is topologically transitive and has a dense set of periodic points then T has sensitive dependence on initial conditions with respect to any metric defining the topology of X.*

Proof. We fix a metric d defining the topology of X. We first show that there exists some constant $\eta > 0$ such that, for any point $x \in X$ there is a periodic point p such that

$$d(x, T^n p) \geq \eta \quad \text{for all } n \in \mathbb{N}_0.$$

Indeed, since X has no isolated points it is an infinite set, so that we can find two periodic points p_1, p_2 whose orbits are disjoint. Hence,

$$\eta := \inf_{m,n \in \mathbb{N}_0} d(T^m p_1, T^n p_2)/2 > 0.$$

It then follows from the triangle inequality that, for any $x \in X$, either for $j = 1$ or for $j = 2$ we have that $d(x, T^n p_j) \geq \eta$ for all $n \in \mathbb{N}_0$.

We now claim that T has sensitive dependence on initial conditions with sensitivity constant $\delta := \eta/4 > 0$. To this end, let $x \in X$ and $\varepsilon > 0$. By assumption there is a periodic point q such that

$$d(x, q) < \min(\varepsilon, \delta). \tag{1.1}$$

Let q have period N. As we have seen above there is also a periodic point p such that

$$d(x, T^n p) \geq \eta = 4\delta \quad \text{for } n \in \mathbb{N}_0. \tag{1.2}$$

Since T is continuous there is some neighbourhood V of p such that

$$d(T^n p, T^n y) < \delta \quad \text{for } n = 0, 1, \ldots, N \text{ and } y \in V. \tag{1.3}$$

Finally, by topological transitivity of T we can find a point z and some $k \in \mathbb{N}_0$ such that $d(x, z) < \varepsilon$ and $T^k z \in V$. Let $j \in \mathbb{N}_0$ be such that $k \leq jN < k + N$. The triangle inequality, together with (1.2), (1.3) and (1.1), then yields that

$$\begin{aligned} d\left(T^{jN} q, T^{jN} z\right) &= d\left(T^{jN} q, T^{jN-k} T^k z\right) = d\left(q, T^{jN-k} T^k z\right) \\ &\geq d\left(x, T^{jN-k} p\right) - d\left(T^{jN-k} p, T^{jN-k} T^k z\right) - d(x, q) \\ &> 4\delta - \delta - \delta = 2\delta. \end{aligned}$$

This implies that either $d(T^{jN} x, T^{jN} q) > \delta$ or $d(T^{jN} x, T^{jN} z) > \delta$. Since both z and q have a distance less than ε from x, the claim follows. □

This allows us to drop sensitive dependence from Devaney's definition of chaos; for simplicity we also extend it to all metric spaces.

Definition 1.30 (Devaney chaos). A dynamical system $T : X \to X$ is said to be *chaotic (in the sense of Devaney)* if it satisfies the following conditions:
(i) T is topologically transitive;
(ii) T has a dense set of periodic points.

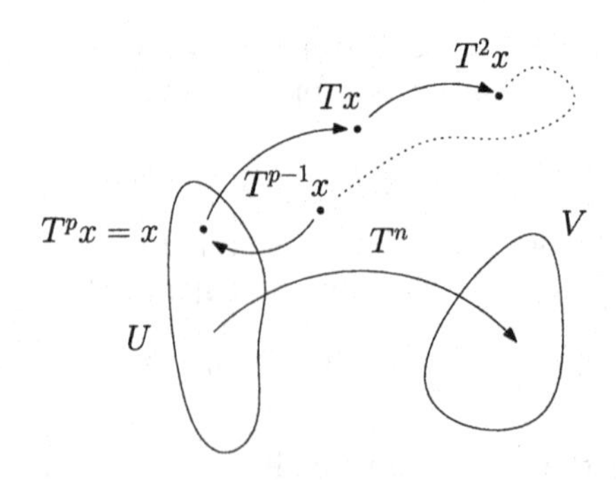

Fig. 1.5 Devaney chaos

By Theorem 1.29, this is consistent with Definition 1.26. Proposition 1.13 and 1.25 immediately show the following.

Proposition 1.31. *Devaney chaos is preserved under quasiconjugacy.*

As an application of this fact we obtain the following from Examples 1.6 and 1.27.

Example 1.32. The logistic map L_4 is chaotic on $[0,1]$.

For the tent map and the doubling map we have shown explicitly that they satisfy the defining conditions of chaos. But their dynamics remain mysterious. To get a better feeling for the origin of chaos in a dynamical system we now want to discuss briefly a special chaotic dynamical system whose dynamical behaviour is quite transparent. On the space

$$\Sigma_2 = \{(x_n)_{n \in \mathbb{N}_0} \ ; \ x_n \in \{0,1\}\}$$

of all 0-1-sequences we study the map

$$\sigma : \Sigma_2 \to \Sigma_2, \quad \sigma(x_0, x_1, x_2, \ldots) = (x_1, x_2, x_3, \ldots).$$

As usual, sequences $(x_n)_n, (y_n)_n, \ldots$ will be denoted by $x, y, \ldots$. We define a topology on Σ_2 by the metric

$$d(x,y) = \sum_{n=0}^{\infty} \frac{|x_n - y_n|}{2^n}.$$

Under this metric, σ is clearly a continuous map.

Definition 1.33. The dynamical system (Σ_2, σ) is called the *shift on two symbols.*

For topological considerations in Σ_2 the following easy result will be of constant use.

Lemma 1.34. *Let $x, y \in \Sigma_2$.*
(a) *If $x_j = y_j$ for $j = 0, 1, \ldots, n$ then $d(x,y) \leq \frac{1}{2^n}$.*
(b) *If $d(x,y) < \frac{1}{2^n}$ then $x_j = y_j$ for $j = 0, 1, \ldots, n$.*

In particular, a sequence of points in Σ_2 converges if and only if each coordinate converges. It also follows easily that Σ_2 is a compact metric space without isolated points; in fact it is homeomorphic to the Cantor set.

As promised, the dynamics of this system are completely transparent.

Proposition 1.35. (a) *A point $x \in \Sigma_2$ is periodic under σ if and only if the sequence $x = (x_n)_n$ is periodic.*

(b) *A point $x \in \Sigma_2$ has dense orbit under σ if and only if every finite 0-1-sequence appears as a block in x.*

Proof. (a) is obvious from the definition of the map σ.

For (b), suppose that x has dense orbit. Let $(y_0, \ldots, y_m)$ be a finite 0-1-sequence and let $y = (y_0, \ldots, y_m, 0, 0, \ldots)$. Since there is some $n \geq 0$ such that $d(y, \sigma^n x) < \frac{1}{2^m}$, Lemma 1.34(b) implies that $(x_n, \ldots, x_{n+m}) = (y_0, \ldots, y_m)$. The converse follows similarly with Lemma 1.34(a). □

Theorem 1.36. *The shift on two symbols is chaotic.*

Proof. Since the set of all finite 0-1-sequences is countable, one can construct a 0-1-sequence that contains each finite 0-1-sequence as a block. Hence, by Proposition 1.35(b), σ has a dense orbit, which shows that it is topologically transitive because Σ_2 has no isolated points.

Now, let $x \in \Sigma_2$ and $m \in \mathbb{N}$. Then the sequence y defined by $y_{j(m+1)+k} = x_k$, $0 \leq k \leq m$, $j \geq 0$ is periodic, hence a periodic point for σ. Moreover, by Lemma 1.34, we have that $d(x, y) \leq \frac{1}{2^m}$. This shows that the periodic points are dense in Σ_2.

Altogether, σ is chaotic. □

It will be a recurrent theme in this book that shifts create chaos. And in many cases maps are chaotic precisely because there is an underlying shift. We illustrate this by the doubling map.

Example 1.37. The doubling map is given by $Tz = z^2$, $z \in \mathbb{T}$. If we write $z = \exp(2\pi i\alpha)$ with $0 \leq \alpha < 1$ then we may represent α in binary form as

$$\alpha = \sum_{n=0}^{\infty} \frac{x_n}{2^{n+1}}, \quad x_n \in \{0, 1\}.$$

In this representation we have that T maps z into

$$z^2 = \exp(2\pi i 2\alpha) = \exp\Big(2\pi i x_0 + 2\pi i \sum_{n=1}^{\infty} \frac{x_n}{2^n}\Big) = \exp\Big(2\pi i \sum_{n=0}^{\infty} \frac{x_{n+1}}{2^{n+1}}\Big).$$

In other words, the doubling map acts as a shift on the binary representation of the argument of z. In this form the dynamics of the doubling map become much clearer. To put it formally, the map

$$\phi : \Sigma_2 \to \mathbb{T}, \quad (x_n)_n \to \exp\Big(2\pi i \sum_{n=0}^{\infty} \frac{x_n}{2^{n+1}}\Big)$$

provides a quasiconjugacy from the shift on two symbols to the doubling map. This proves once more that the doubling map is chaotic.

1.4 Mixing maps

We return to the discussion of topologically transitive maps. As we saw in Example 1.12, the tent map and the doubling map have a strong form of transitivity: $T^n(U)$ intersects V not only for some n but for all sufficiently large $n \in \mathbb{N}_0$. This property carries a special name.

Definition 1.38. A dynamical system $T : X \to X$ is called *mixing* if, for any pair U, V of nonempty open subsets of X, there exists some $N \geq 0$ such that

$$T^n(U) \cap V \neq \varnothing \quad \text{for all } n \geq N.$$

Example 1.39. As noted above, both the tent map and the doubling map are mixing. An example of a topologically transitive system that is not mixing will be given in Example 1.43.

As in the case of topological transitivity one obtains the following.

Proposition 1.40. *The mixing property is preserved under quasiconjugacy.*

Mixing maps have a remarkable permanence property. In order to describe this we need to define products of maps.

Let $S : X \to X$ and $T : Y \to Y$ be dynamical systems. The Cartesian product $X \times Y$ is endowed with the product topology, which is induced by the metric $d((x_1, y_1), (x_2, y_2)) = d_X(x_1, x_2) + d_Y(y_1, y_2)$, where d_X and d_Y denote the metrics in X and Y, respectively. A base for the topology is formed by the products $U \times V$ of nonempty open sets $U \subset X$ and $V \subset Y$.

Definition 1.41. Let $S : X \to X$ and $T : Y \to Y$ be dynamical systems. Then the map $S \times T$ is defined by

$$S \times T : X \times Y \to X \times Y, \quad (S \times T)(x, y) = (Sx, Ty).$$

Then $S \times T$ is clearly continuous, and for the iterates we have that

$$(S \times T)^n = S^n \times T^n.$$

Products of more than two spaces or maps are defined similarly.

Proposition 1.42. *Let $S : X \to X$ and $T : Y \to Y$ be dynamical systems. Then we have the following:*

(i) *if $S \times T$ has a dense orbit then so do S and T;*
(ii) *if $S \times T$ is topologically transitive then so are S and T;*
(iii) *if $S \times T$ is chaotic then so are S and T;*
(iv) *if S and T are topologically transitive and at least one of them is mixing then $S \times T$ is topologically transitive;*
(v) *$S \times T$ is mixing if and only if both S and T are.*

Proof. Obviously, S and T are quasiconjugate to $S \times T$ under the maps $(x, y) \to x$ and $(x, y) \to y$, respectively. Thus the preservation results obtained so far imply (i), (ii) and (iii).

The assertions (iv) and (v) follow directly from the identity

$$(S \times T)^n(U_1 \times U_2) \cap (V_1 \times V_2) = (S^n(U_1) \cap V_1) \times (T^n(U_2) \cap V_2),$$

the fact that topological transitivity and the mixing property can be tested on a base of the topology, and, for (iv), the observation of Exercise 1.2.4. □

In general, however, the product of two topologically transitive maps need not be topologically transitive, even if $S = T$, as the following example shows; hence some extra condition, as in the previous proposition, is needed.

Example 1.43. Let $T : \mathbb{T} \to \mathbb{T}$, $z \to e^{i\alpha}z$ be a circle rotation. Then $(T \times T)^n(z_1, z_2) = (e^{in\alpha}z_1, e^{in\alpha}z_2)$, and we observe that the quotient of the two coordinates is z_1/z_2, independent of n. This is easily seen to imply that $T \times T$ cannot be topologically transitive. On the other hand we know that any irrational rotation is topologically transitive; see Exercise 1.2.9. This also shows that no irrational rotation is mixing.

1.5 Weakly mixing maps

Having looked at the question of when a product of two topologically transitive systems is topologically transitive, one may wonder when the product of a topologically transitive map with itself is again topologically transitive. We saw in Example 1.43 that this is not always the case. On the other hand, for any mixing map T the product $T \times T$ is topologically transitive. This leads us to the following notion.

Definition 1.44. A dynamical system $T : X \to X$ is called *weakly mixing* if $T \times T$ is topologically transitive.

Since the products $U \times V$ of nonempty open sets $U, V \subset X$ form a base of the topology of $X \times X$, T is weakly mixing if and only if, for any 4-tuple $U_1, U_2, V_1, V_2 \subset X$ of nonempty open sets, there exists some $n \geq 0$ such that

$$T^n(U_1) \cap V_1 \neq \varnothing \quad \text{and} \quad T^n(U_2) \cap V_2 \neq \varnothing.$$

Observation 1.45. *For any dynamical system,*

$$\textit{mixing} \implies \textit{weak mixing} \implies \textit{topological transitivity}.$$

Remark 1.46. By Example 1.43, any irrational circle rotation is topologically transitive but not weakly mixing. On the other hand, it is not an easy matter to construct examples of weakly mixing maps that are not mixing. However, in the context of linear operators that will be discussed in the following chapters, such examples will appear abundantly as consequences of our investigations there; see, for example, Remark 4.10.

Proposition 1.47. *The weak mixing property is preserved under quasiconjugacy.*

Proof. If $\phi : Y \to X$ defines a quasiconjugacy from $S : Y \to Y$ to $T : X \to X$, then $\phi \times \phi$ defines a quasiconjugacy from $S \times S$ to $T \times T$. Now the result follows from Proposition 1.13. □

As a consequence, we have the following as in Proposition 1.42.

Proposition 1.48. *Let $S : X \to X$ and $T : Y \to Y$ be dynamical systems. If $S \times T$ is weakly mixing then so are S and T.*

In order to formulate the arguments involving weakly mixing maps more succinctly we introduce the following useful concept.

Definition 1.49. Let $T : X \to X$ be a dynamical system. Then, for any sets $A, B \subset X$, the *return set from A to B* is defined as

$$N_T(A, B) = N(A, B) = \{n \in \mathbb{N}_0 \ ; \ T^n(A) \cap B \neq \varnothing\}.$$

We usually drop the index T when this causes no ambiguity. In this notation, T is topologically transitive (or mixing) if and only if, for any pair U, V of nonempty open subsets of X, the return set

$$N(U, V) \text{ is nonempty (or cofinite, respectively);}$$

and T is weakly mixing if and only if, for any 4-tuple $U_1, U_2, V_1, V_2 \subset X$ of nonempty open sets,

$$N(U_1, V_1) \cap N(U_2, V_2) \neq \varnothing.$$

Note also that if T is topologically transitive then the return sets $N(U, V)$ are even infinite for any nonempty open sets U, V; see Exercise 1.2.4.

Incidentally, we observe that the larger the sets A and B are, the larger also is the return set $N(A, B)$.

It is our aim to give several characterizations of the weak mixing property. To do this we will provide a useful lemma that, due to its form, we will call the *4-set trick*. We note that the return sets $N(A, B)$ refer to T.

Lemma 1.50 (4-set trick). *Let $T : X \to X$ be a dynamical system, and let $U_1, V_1, U_2, V_2 \subset X$ be nonempty open sets.*

(a) *If there is a continuous map $S : X \to X$ commuting with T such that*

$$S(U_1) \cap U_2 \neq \varnothing \quad \textit{and} \quad S(V_1) \cap V_2 \neq \varnothing,$$

then there exist nonempty open sets $U_1' \subset U_1$, $V_1' \subset V_1$ *such that*

$$N(U_1', V_1') \subset N(U_2, V_2) \quad \text{and} \quad N(V_1', U_1') \subset N(V_2, U_2).$$

If, moreover, T *is topologically transitive then* $N(U_1, V_1) \cap N(U_2, V_2) \neq \varnothing$.

(b) *If* T *is topologically transitive then*

$$N(U_1, U_2) \cap N(V_1, V_2) \neq \varnothing \Longrightarrow N(U_1, V_1) \cap N(U_2, V_2) \neq \varnothing.$$

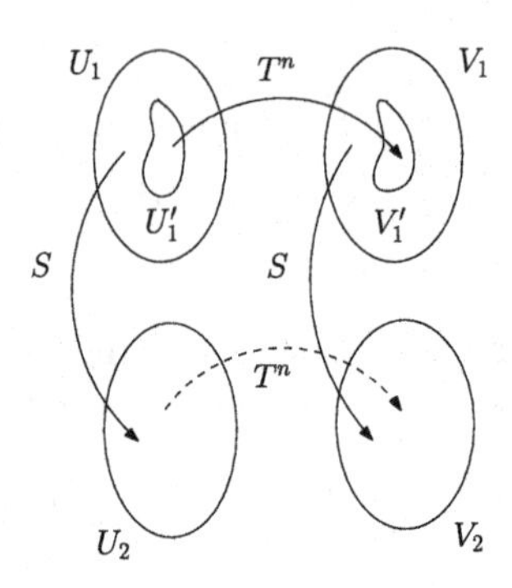

Fig. 1.6 The 4-set trick

Proof. (a) Since S is continuous, it follows from the hypothesis that we can find nonempty open sets $U_1' \subset U_1$ and $V_1' \subset V_1$ such that $S(U_1') \subset U_2$ and $S(V_1') \subset V_2$. If $n \in N(U_1', V_1')$, then there exists some $x \in U_1'$ with $T^n x \in V_1'$. Therefore $T^n Sx = ST^n x \in V_2$ and $Sx \in U_2$, which yields that $n \in N(U_2, V_2)$. By symmetry we also obtain that $N(V_1', U_1') \subset N(V_2, U_2)$. If, moreover, T is topologically transitive then

$$\varnothing \neq N(U_1', V_1') \subset N(U_1, V_1) \cap N(U_2, V_2).$$

(b) If T is topologically transitive and $n \in N(U_1, U_2) \cap N(V_1, V_2)$, then $N(U_1, V_1) \cap N(U_2, V_2) \neq \varnothing$ follows if (a) is applied to $S := T^n$. $\square$

The 4-set trick, simple as it is, already implies an important result that is at first sight quite surprising: as soon as the product $T \times T$ is topologically transitive, every higher product $T \times \cdots \times T$ also is.

Theorem 1.51 (Furstenberg). *Let* $T : X \to X$ *be a weakly mixing dynamical system. Then the n-fold product* $T \times \cdots \times T$ *is weakly mixing for each* $n \geq 2$.

Proof. Since the n-fold product being weakly mixing amounts to the $2n$-fold product being topologically transitive, it suffices to show that every n-fold product $T \times \cdots \times T$ is topologically transitive for $n \geq 2$.

We proceed by induction, the case of $n = 2$ being trivial by definition. Thus, suppose that the n-fold product $T\times \ldots \times T$ is topologically transitive. To prove topological transitivity of the corresponding $(n+1)$-fold product we need to show that, given nonempty open sets $U_k, V_k \subset X$, $k = 1, \ldots, n+1$, we have that

$$\bigcap_{k=1}^{n+1} N(U_k, V_k) \neq \varnothing. \tag{1.4}$$

Indeed, since T is weakly mixing there is some $m \in \mathbb{N}_0$ such that $T^m(U_n) \cap U_{n+1} \neq \varnothing$ and $T^m(V_n) \cap V_{n+1} \neq \varnothing$. The 4-set trick then yields the existence of nonempty open sets $U_n' \subset U_n$, $V_n' \subset V_n$ with $N(U_n', V_n') \subset N(U_n, V_n) \cap N(U_{n+1}, V_{n+1})$. On the other hand, the induction hypothesis implies that

$$\bigcap_{k=1}^{n-1} N(U_k, V_k) \cap N(U_n', V_n') \neq \varnothing,$$

which implies (1.4). □

The next results show that in the definition of weak mixing one may reduce the four open sets to three, and then to two open sets.

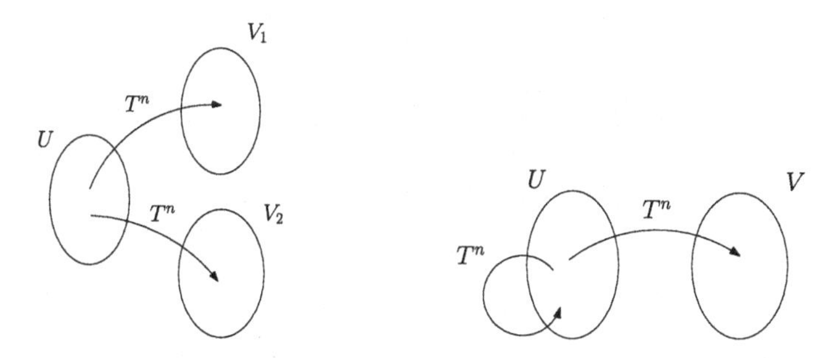

Fig. 1.7 Weak mixing (Propositions 1.52 and 1.53)

Proposition 1.52. *A dynamical system $T : X \to X$ is weakly mixing if and only if, for any nonempty open sets $U, V_1, V_2 \subset X$, we have that*

$$N(U, V_1) \cap N(U, V_2) \neq \varnothing.$$

Proof. We need only show sufficiency of the condition. Thus, let $U_1, U_2, V_1, V_2 \subset X$ be arbitrary nonempty open sets. By hypothesis there is some $n \in N(U_1, U_2) \cap N(U_1, V_2)$. In particular $U := U_1 \cap T^{-n}(U_2)$ and $T^{-n}(V_2)$ are nonempty open sets. By another application of the hypothesis we find some $m \in N(U, V_1) \cap N(U, T^{-n}(V_2))$. In particular, there exists some $x \in U$ with $T^m x \in T^{-n}(V_2)$. We then conclude that $T^m T^n x = T^n T^m x \in V_2$ and $T^n x \in U_2$, which yields that $m \in N(U_1, V_1) \cap N(U_2, V_2)$. □

Proposition 1.53. *A dynamical system $T : X \to X$ is weakly mixing if and only if, for any nonempty open sets $U, V \subset X$, we have that*

$$N(U,U) \cap N(U,V) \neq \varnothing.$$

Proof. It suffices to show that the stated condition implies the condition of Proposition 1.52. Thus, let $U, V_1, V_2 \subset X$ be arbitrary nonempty open sets. By hypothesis there exists some $n \in \mathbb{N}_0$ such that $U_1 := U \cap T^{-n}(V_1)$ is a nonempty open set. Since topologically transitive maps have dense range, the hypothesis also implies that $T^{-n}(V_2)$ is nonempty and open, so that there exists some $m \in N(U_1, U_1) \cap N(U_1, T^{-n}(V_2))$. Therefore there are $x, y \in U_1$ with $T^m x \in U_1$ and $T^n T^m y \in V_2$. We then have that $T^n T^m x \in V_1$, which implies that $n + m \in N(U, V_1) \cap N(U, V_2)$, as desired. □

For more characterizations of weak mixing in the same spirit we refer to Exercise 1.5.1.

We finally characterize the weak mixing property in terms of the size of the return sets $N(U, V)$, or equivalently, in terms of the topological transitivity of certain subsequences $(T^{n_k})_k$; for the notion of topological transitivity for a sequence of maps we refer to the next section.

A strictly increasing sequence $(n_k)_k$ of positive integers is called *syndetic* if

$$\sup_{k \geq 1}(n_{k+1} - n_k) < \infty.$$

Likewise, a subset A of $\mathbb{N}_0$ is called *syndetic* if the increasing sequence of positive integers forming A is syndetic, or equivalently, if its complement does not contain arbitrarily long intervals.

Theorem 1.54. *Let $T : X \to X$ be a dynamical system. Then the following assertions are equivalent:*

(i) *T is weakly mixing;*

(ii) *for any pair $U, V \subset X$ of nonempty open sets, $N(U, V)$ contains arbitrarily long intervals;*

(iii) *for any syndetic sequence $(n_k)_k$, the sequence $(T^{n_k})_k$ is topologically transitive.*

Proof. (i)$\Longrightarrow$(ii). Let $U, V \subset X$ be nonempty open sets, and let $m \in \mathbb{N}$. As in the proof of Proposition 1.53, each set $T^{-k}(V)$, $k = 1, \ldots, m$, is nonempty and open. Since, by Furstenberg's theorem, the m-fold product map $T \times \cdots \times T$ is topologically transitive, there is some $n \in \mathbb{N}$ such that

$$T^n(U) \cap T^{-k}(V) \neq \varnothing \quad \text{for } k = 1, \ldots, m.$$

This implies that $T^{n+k}(U) \cap V \neq \varnothing$ for $k = 1, \ldots, m$.

(ii)$\Longrightarrow$(i). By Proposition 1.52 it suffices to show that, given any nonempty open subsets $U, V_1, V_2 \subset X$, $N(U, V_1) \cap N(U, V_2) \neq \varnothing$. First, by (ii) there is some $m \in N(V_1, V_2)$ and therefore a nonempty open set $V_3 \subset V_1$ such that

$T^m(V_3) \subset V_2$. Also by (ii) there is some $k \in \mathbb{N}_0$ such that $k+j \in N(U,V_3)$ for $j = 0, 1, \ldots, m$. In particular we have that $k+m \in N(U,V_1)$ and

$$T^{k+m}(U) \cap V_2 \supset T^{k+m}(U) \cap T^m(V_3) \supset T^m(T^k(U) \cap V_3) \neq \varnothing.$$

We conclude that $k+m \in N(U,V_1) \cap N(U,V_2)$.

(ii)$\Longleftrightarrow$(iii). This follows immediately from the definitions and the fact that a subset of $\mathbb{N}_0$ contains arbitrarily long intervals if and only if it meets every syndetic sequence. $\square$

Condition (ii) in this result shows nicely how weak mixing sits between topological transitivity and mixing.

1.6 Universality

The basic concepts introduced so far in this chapter allow a far-reaching generalization. The orbit of a point x under a map T is obtained by applying the iterates T^n, $n = 0, 1, 2, \ldots$, of T to x. Instead, one could think of applying arbitrary maps T_n, $n = 0, 1, 2, \ldots$, to x; in this case we need not even have that the T_n are self-maps.

Definition 1.55. Let X and Y be metric spaces, and let $T_n : X \to Y$, $n \in \mathbb{N}_0$, be continuous maps. Then the *orbit* of x under $(T_n)_n$ is defined as

$$\mathrm{orb}(x, (T_n)) = \{T_n x \; ; \; n \in \mathbb{N}_0\}.$$

An element $x \in X$ is called *universal for* $(T_n)_n$ if it has dense orbit under $(T_n)_n$.

An interesting and nontrivial example is provided by *universal Taylor series*: it can be shown that there exists an infinitely differentiable function $f : \mathbb{R} \to \mathbb{R}$ with $f(0) = 0$ such that, for any continuous function $g : \mathbb{R} \to \mathbb{R}$ with $g(0) = 0$, there exists an increasing sequence $(n_k)_k$ of positive integers such that

$$\sum_{\nu=0}^{n_k} \frac{f^{(\nu)}(0)}{\nu!} x^\nu \to g(x) \quad \text{uniformly on any compact subset of } \mathbb{R}.$$

In this case, T_n is the map that associates to f its Taylor polynomial of degree n at 0.

The theory of universality will not be developed in any depth in this book. We note that there is a difference in philosophy between universality and topological dynamics: in the former one is interested in the universal elements and their properties while in the latter the focus is rather on the map and its properties.

However, occasionally the study of the dynamics of a single map requires looking at orbits under general sequences of maps; Theorem 1.54 has already provided such an example. For this reason we consider here briefly how the concepts and results of this chapter can be generalized to universality.

Definition 1.56. Let $T_n : X \to Y$, $n \in \mathbb{N}_0$, be continuous maps between metric spaces X and Y. Then $(T_n)_n$ is called *topologically transitive* if, for any pair $U \subset X$ and $V \subset Y$ of nonempty open sets, there is some $n \geq 0$ such that

$$T_n(U) \cap V \neq \varnothing;$$

it is *mixing* if the same holds for all sufficiently large n, and it is *weakly mixing* if $(T_n \times T_n)_n$ is topologically transitive on $X \times X$.

Now, many of the results in this chapter extend, at least under suitable assumptions, to general sequences. We will content ourselves here with some examples.

Theorem 1.57 (Universality Criterion). *Let X be a complete metric space, Y a separable metric space and $T_n : X \to Y$, $n \in \mathbb{N}_0$, continuous maps. Then the following assertions are equivalent:*

(i) *$(T_n)_n$ is topologically transitive;*

(ii) *there exists a dense set of points $x \in X$ such that* $\operatorname{orb}(x, (T_n))$ *is dense in Y.*

If one of these conditions holds then the set of points in X with dense orbit is a dense G_δ-set.

Proof. Suppose that (ii) holds. If U and V are nonempty open sets of X and Y, respectively, then there exists some $x \in U$ with dense orbit under $(T_n)_n$, so that there exists some $n \geq 0$ with $T_n x \in V$. This implies (i).

The converse implication and the fact that the set of points with dense orbit is a dense G_δ-set can be proved exactly as in the proof of the Birkhoff transitivity theorem. □

Typically, results on iterates of maps have a good chance of extending to sequences $(T_n)_n$ if they consist of commuting self-maps $T_n : X \to X$ of dense range. For example, for such sequences the Birkhoff transitivity theorem has a perfect analogue; see Exercise 1.6.2.

Remark 1.58. If we define the return sets

$$N(A, B) = \{n \in \mathbb{N}_0 \ ; \ T_n(A) \cap B \neq \varnothing\}$$

then part (a) of the 4-set trick remains valid for sequences $(T_n)_n$ of self-maps if the map S commutes with all T_n, $n \geq 0$, as does part (b) for commuting self-maps. As a consequence, Furstenberg's theorem also holds for commuting sequences $(T_n)_n$, as does Proposition 1.52.

Exercises

Exercise 1.1.1. Show that for $\mu = 2$ the iterates of the logistic map L_2 are given by

$$L_2^n x = \tfrac{1}{2}\left(1 - (1-2x)^{2^n}\right), \quad n \geq 0.$$

Deduce from this the long-term behaviour of the orbits $\operatorname{orb}(x, L_2)$ for $x \in \mathbb{R}$.

Exercise 1.1.2. Consider the dynamical system $T :]0,\infty[\to]0,\infty[$, $Tx = \frac{1}{2}(x + \frac{2}{x})$. Show that there is some $q \in]0,1[$ such that $|Tx - Ty| \leq q|x-y|$ for $x, y \geq 1$. Deduce that $|T^n x - \sqrt{2}| \leq q^n|x - \sqrt{2}|$ and hence that $T^n x \to \sqrt{2}$ for all $x \geq 1$.

Exercise 1.1.3. Prove the statements in Example 1.6(a)–(c).

Exercise 1.1.4. Show that the logistic map L_4, when restricted to the interval $[0,1]$, is quasiconjugate to the doubling map on the interval via $\phi(x) = \sin^2(\pi x)$. Also show that the maps are not conjugate.

Exercise 1.2.1. Show that, in general, the implication (i)$\Longrightarrow$(ii) does not hold in Proposition 1.10.

Exercise 1.2.2. Show that the following assertions on a dynamical system $T : X \to X$ are equivalent:

(i) T is topologically transitive;
(ii) for any open set $U \subset X$ with $T^{-1}(U) \subset U$, either $U = \varnothing$ or U is dense in X;
(iii) for any closed set $E \subset X$ with $T(E) \subset E$, either $E = X$ or E is nowhere dense.

Exercise 1.2.3. Suppose that X has at least one isolated point. Prove that, if there is any topologically transitive map $T : X \to X$, then X is finite and $X = \operatorname{orb}(x, T)$ for any $x \in X$.

Exercise 1.2.4. Show that, if $T : X \to X$ is topologically transitive, then for any pair U, V of nonempty open subsets of X, the return set $N(U,V)$ is infinite; see Definition 1.49. (*Hint*: For the trivial case in which X has isolated points apply Exercise 1.2.3. If X has no isolated points, then given $m \in N(U,V)$ and $W := U \cap T^{-m}(V)$, observe that $N(W,W) \cap \mathbb{N} \neq \varnothing$ and $m + N(W,W) \subset N(U,V)$.)

Exercise 1.2.5. Prove that a dynamical system $T : X \to X$ on a metric space X is topologically transitive if and only if, for any $\varepsilon > 0$ and any pair of points $x, y \in X$, we can find $z \in X$ and $n, m \in \mathbb{N}_0$ satisfying $d(T^n z, x) < \varepsilon$ and $d(T^m z, y) < \varepsilon$. (*Hint*: First observe that the above condition is equivalent to the fact that for any pair U, V of nonempty open subsets of X one can find $n, m \in \mathbb{N}_0$ with $T^{-n}(U) \cap T^{-m}(V) \neq \varnothing$. This condition is obviously implied by topological transitivity. For the converse, given nonempty open sets $U, V \subset X$, either find $k \in N(U,V)$ (in that case you are done) or, if $k \in N(V,U)$, set $W = V \cap T^{-k}(U)$, note that $N(W,W)$ is infinite by Exercise 1.2.4, and then find some $j \in N(U,V)$.)

Exercise 1.2.6. Let T be a topologically transitive dynamical system on a separable complete metric space X without isolated points. Prove constructively, not using the Baire category theorem, that T has a dense set of points with dense orbit. (*Hint*: Let $(y_n)_n$ be a dense sequence in X. Start with $x_0 \in X$. Then find x_1 close to x_0 and a positive integer m_1 so that $T^{m_1} x_1$ is close to y_1. Then find x_2 close to x_1 and a positive integer m_2 so that $T^{m_1} x_2$ is close to $T^{m_1} x_1$ and $T^{m_1+m_2} x_2$ is close to y_2. Continue.)

Exercise 1.2.7. Let T be a dynamical system on a metric space X without isolated points. A *backward orbit* of a vector x is a sequence $(x_n)_{n\geq 0}$ in X (if it exists!) such that $x_0 = x$ and $Tx_n = x_{n-1}$, $n \geq 1$. Show the following:

(i) if T is topologically transitive and X is separable and complete then there exists a dense set of points with dense backward orbits;

(ii) if T has a dense backward orbit then T is topologically transitive.

(*Hint*: See the previous exercise.)

Exercise 1.2.8. Let $T : X \to X$ be a dynamical system. For $x \in X$ the *J-set* $J_T(x) = J(x)$ is defined as the set of all points $y \in X$ for which there is a strictly increasing sequence $(n_k)_k$ of positive integers and a sequence $(x_k)_k$ in X such that $x_k \to x$ and $T^{n_k}x_k \to y$ as $k \to \infty$.

(a) Show that $J(x)$ is a closed T-invariant set.

(b) Suppose that X has no isolated points. Show that $J(x) = X$ if and only if, for any pair U, V of nonempty open subsets of X with $x \in U$, there exists some $n \geq 0$ such that $T^n(U) \cap V \neq \varnothing$.

(c) Suppose that X has no isolated points. Show that the following assertions are equivalent:

(i) T is topologically transitive;

(ii) for any $x \in X$, $J(x) = X$;

(iii) there is a dense set of points $x \in X$ such that $J(x) = X$.

Exercise 1.2.9. Show that every orbit under an irrational rotation is dense. (*Hint*: Use the pigeonhole principle to show that, for any $\varepsilon > 0$, some arc of angle ε must contain two iterates of 1, $T^m 1$ and $T^n 1$, $m > n$. Then look at the iterates of T^{m-n}.)

Exercise 1.2.10. A dynamical system $T : X \to X$ is called *minimal* if every orbit under T is dense. Find a characterization of minimality in the spirit of Exercise 1.2.2.

Exercise 1.2.11. Consider the dynamical system $T : [-1, 1] \to [-1, 1]$ given by

$$Tx = \begin{cases} 2 + 2x, & \text{if } -1 \leq x < -1/2, \\ -2x, & \text{if } -1/2 \leq x < 1/2, \\ -2 + 2x, & \text{if } 1/2 \leq x \leq 1. \end{cases}$$

(a) Show that T has a dense orbit but that T^2 does not.

(b) Show that there are two points $x, y \in [-1, 1]$ such that $\text{orb}(x, T^2) \cup \text{orb}(y, T^2)$ is dense in $[-1, 1]$ but neither of them has a dense orbit under T^2.

(c) Show that there is a point $x \in [-1, 1]$ such that $\overline{\text{orb}(x, T^2)}$ contains a nonempty open set but x does not have a dense orbit under T^2.

(*Remark*: We will prove in Chapter 6 that none of these properties can hold in a linear setting.)

Exercise 1.3.1. Show that none of the three conditions in Definition 1.26 alone implies chaos.

Exercise 1.3.2. Let X be a finite set, endowed with the discrete metric. Describe all maps on X that are chaotic. Do the same for countably infinite sets under the discrete metric.

Exercise 1.3.3. Suppose that (X, d) is a metric space without isolated points and $T : X \to X$ is a contracting map, that is $d(Tx, Ty) \leq d(x, y)$ for all $x, y \in X$. Show that if T has one dense orbit then T is minimal (see Exercise 1.2.10); in particular, it cannot be chaotic.

Exercise 1.3.4. Show that $T : X \to X$ is chaotic if and only if every finite family of nonempty open sets shares a periodic orbit, in the following sense: for each finite family $U_j \subset X$, $j = 1, \dots, n$, of nonempty open sets there is a periodic point $x \in U_1$ such that $T^{k_j}x \in U_j$ for some $k_j \geq 0$, $j = 2, \dots, n$; see Figure 1.8. (*Hint:* One implication is trivial; for the other one use continuity of T and an induction process.)

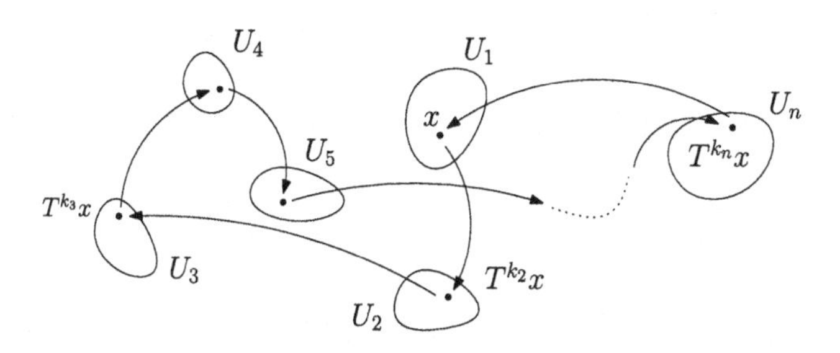

Fig. 1.8 Exercise 1.3.4

Exercise 1.3.5. Show that the space Σ_2 is a complete metric space without isolated points. Show also that the sequences with only finitely many nonzero entries form a dense set.

Exercise 1.3.6. Why is the quasiconjugacy of Example 1.37 between the shift on two symbols and the doubling map not a conjugacy? Use this quasiconjugacy to find a representation of the periodic points and the points with dense orbit for the doubling map.

Exercise 1.4.1. Show that the shift on two symbols is mixing.

Exercise 1.4.2. Prove that a dynamical system T is mixing if and only if, for any strictly increasing sequence $(n_k)_k$ of positive integers, the sequence $(T^{n_k})_k$ is topologically transitive; see Definition 1.56 for the notion of topological transitivity for sequences of maps.

Exercise 1.4.3. Let X be a complete metric space. Prove that a dynamical system $T : X \to X$ is mixing if and only if, for every sequence $(x_n)_n$ in X and for every strictly increasing sequence $(n_k)_k$ of positive integers for which $\{x_{n_k}\ ;\ k \in \mathbb{N}\}$ is relatively compact there exists a dense G_δ-set of points $y \in X$ such that $\liminf_{k\to\infty} d(x_{n_k}, T^{n_k}y) = 0$. (*Hint*: Use the previous exercise; a subset A of a metric space X is relatively compact if and only if every sequence in A has a subsequence that converges in X.)

Exercise 1.4.4. Let $T : X \to X$ be a dynamical system. For $x \in X$, the set $J_T^{\text{mix}}(x) = J^{\text{mix}}(x)$ is defined as the set of all points $y \in X$ for which there is a sequence $(x_n)_n$ in X such that $x_n \to x$ and $T^n x_n \to y$ as $n \to \infty$; see also Exercise 1.2.8.

(a) Show that $J^{\text{mix}}(x)$ is a closed T-invariant set.

(b) Show that $J^{\text{mix}}(x) = X$ if and only if, for any pair U, V of nonempty open subsets of X with $x \in U$, there exists some $N \geq 0$ such that $T^n(U) \cap V \neq \varnothing$ for all $n \geq N$.

(c) Show that the following assertions are equivalent:

(i) T is mixing;
(ii) for any $x \in X$, $J^{\text{mix}}(x) = X$;
(iii) there is a dense set of points $x \in X$ such that $J^{\text{mix}}(x) = X$.

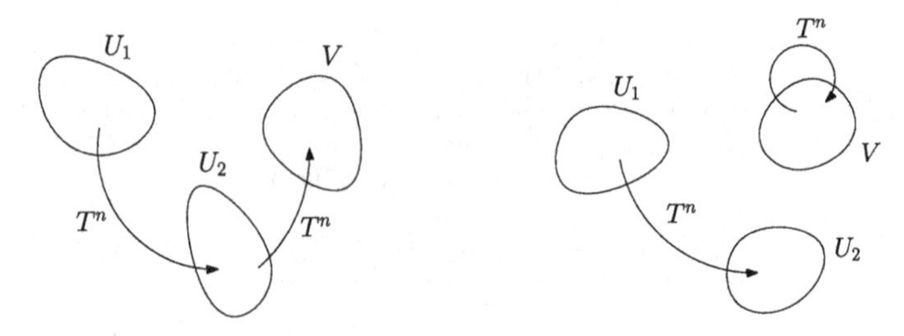

Fig. 1.9 Weak mixing (Exercise 1.5.1(iii) and (iv))

Exercise 1.4.5. Show that minimality and mixing are independent properties of a dynamical system.

Exercise 1.5.1. Let $T : X \to X$ be a dynamical system. In the following, U, U_1, U_2, V will denote arbitrary nonempty open subsets of X. Prove that any of the following conditions is equivalent to T being weakly mixing:

(i) for any $U, V \subset X$ we have $N(U,V) \cap N(V,V) \neq \varnothing$;
(ii) for any $U_1, U_2, V \subset X$ we have $N(U_1,V) \cap N(U_2,V) \neq \varnothing$;
(iii) for any $U_1, U_2, V \subset X$ we have $N(U_1,U_2) \cap N(U_2,V) \neq \varnothing$;
(iv) for any $U_1, U_2, V \subset X$ we have $N(U_1,U_2) \cap N(V,V) \neq \varnothing$;

see Figure 1.9. (*Hint*: To prove sufficiency of (i) use the 4-set trick and Proposition 1.53; this will then imply the sufficiency of the other conditions.)

Exercise 1.5.2. A dynamical system $T : X \to X$ is called *totally transitive* if every power T^p, $p \in \mathbb{N}$, is topologically transitive. Show that any weakly mixing map is totally transitive.

Exercise 1.5.3. Prove that every chaotic and totally transitive dynamical system $T : X \to X$ is weakly mixing. (*Hint*: Verify the hypothesis of Proposition 1.53 by finding a periodic point in U, of period k say, and then by using topological transitivity of T^k.)

Exercise 1.5.4. A dynamical system $T : X \to X$ is called *flip transitive* if, for any pair $U, V \subset X$ of nonempty open sets, $N(U,V) \cap N(V,U) \neq \varnothing$. Show that the map T of Exercise 1.2.11 is flip transitive but not weakly mixing.

Exercise 1.5.5. Show that T is weakly mixing if and only if it is flip transitive and T^2 is topologically transitive. (*Hint*: To prove the weak mixing property use condition (i) in Exercise 1.5.1. To do this, given nonempty open sets $U, V \subset X$, find $k \in \mathbb{N}_0$ with $U' := U \cap T^{-2k}(V) \neq \varnothing$ and then some $m \in N(U', T^{-k}(V)) \cap N(T^{-k}(V), U')$. Consider $m + k$.)

Exercise 1.5.6. A dynamical system $T : X \to X$ is called *topologically ergodic* if, for any pair $U, V \subset X$ of nonempty open sets, $N(U,V)$ is syndetic. Prove the following:

(i) any irrational rotation is topologically ergodic but not weakly mixing;
(ii) every mixing and every chaotic dynamical system is topologically ergodic;
(iii) if $T : X \to X$ is topologically ergodic and $S : Y \to Y$ is weakly mixing, then $T \times S$ is topologically transitive.

Exercise 1.5.7. Let $T : X \to X$ be a dynamical system. Show that any of the following conditions is equivalent to T being weakly mixing:

(i) for any nonempty open sets $U, V \subset X$ and any $m \in \mathbb{N}$ there is some k with $k, k+m \in N(U,V)$;

(ii) for any $m \in \mathbb{N}$ and for any increasing sequence $(n_k)_k$ with $n_{k+1} - n_k \in \{m, 2m\}$, $k \in \mathbb{N}$, we have that $(T^{n_k})_k$ is topologically transitive.

(*Hint*: See the proof of Theorem 1.54.)

Exercise 1.5.8. Let $T : X \to X$ be a dynamical system. Establish the equivalence of the following assertions:

(i) T is weakly mixing;

(ii) for any pair $U, V \subset X$ of nonempty open sets, $N(U, V)$ contains two consecutive integers.

(*Hint*: Proceeding by induction, show that $N(U, V)$ contains arbitrarily long intervals: if $k, k+1 \in N(U, U)$, set $U_1 := U \cap T^{-k}(U)$, $U_2 := U \cap T^{-k-1}(U)$, apply the inductive hypothesis to the pair (U_1, U_2) to find an interval $[j, j+m]$ contained in $N(U_1, U_2)$. By the selection of U_1, obtain that $[j, j+m+1] \subset N(U, U)$.)

Exercise 1.5.9. Given a metric space X, the corresponding *hyperspace* is defined as $\mathcal{K}(X) = \{K \subset X \; ; \; K \text{ is compact}\}$. The space $\mathcal{K}(X)$ is endowed with the (metrizable) *Vietoris topology*, for which a base of open sets is given by the family of sets

$$\mathcal{V}(U_1, \ldots, U_k) := \Big\{K \in \mathcal{K}(X) \; ; \; K \subset \bigcup_{j=1}^{k} U_j \text{ and } K \cap U_j \neq \varnothing, \; j = 1, \ldots, k\Big\},$$

where $U_1, \ldots, U_k$, $k \in \mathbb{N}$, are nonempty open sets in X. If $T : X \to X$ is continuous, then it naturally induces a continuous *hyperextension* $\overline{T} : \mathcal{K}(X) \to \mathcal{K}(X)$ defined by $\overline{T}(K) = T(K) = \{Tx \; ; \; x \in K\}$.

We say that a dynamical system $T : X \to X$ is *hypertransitive* if its hyperextension $\overline{T}$ is topologically transitive. Prove that T is hypertransitive if and only if it is weakly mixing. (*Hint*: For the sufficiency of weak mixing use Furstenberg's theorem; for the necessity use Proposition 1.52.)

Exercise 1.6.1. Let $(x_n)_n$ be a dense sequence in $\mathbb{R}^2$, and let $y_n \in \mathbb{R}^2$, $n \geq 1$, be vectors of length n that are orthogonal to x_n. Consider the maps $T_n : \mathbb{R}^2 \to \mathbb{R}^2$ with $T_n(\alpha, \beta) = \alpha x_n + \beta y_n$. Determine all points in $\mathbb{R}^2$ with dense orbit under $(T_n)_n$. Deduce that $(T_n)_n$ has a dense orbit but is not topologically transitive.

Exercise 1.6.2. Prove the *Birkhoff transitivity theorem* for commuting continuous maps $T_n : X \to X$, $n \in \mathbb{N}_0$, of dense range on a separable complete metric space X, that is, that the following assertions are equivalent:

(i) $(T_n)_n$ is topologically transitive;

(ii) there exists some $x \in X$ such that $\mathrm{orb}(x, (T_n))$ is dense in X.

If one of these conditions holds then the set of points in X with dense orbit is a dense G_δ-set.

Exercise 1.6.3. Let S be a mixing map on a separable complete metric space X and T a map on a metric space Y without isolated points that admits a dense orbit $\mathrm{orb}(y, T)$, $y \in Y$. Show that there exists some $x \in X$ such that (x, y) has a dense orbit under the map $S \times T$. (*Hint*: Use the previous exercise.)

Exercise 1.6.4. A sequence $(T_n)_n$ of continuous maps on a metric space X is called *hereditarily transitive* with respect to an increasing sequence $(n_k)_k$ of positive integers if $(T_{m_k})_k$ is topologically transitive for every subsequence $(m_k)_k$ of $(n_k)_k$. The sequence $(T_n)_n$ is called *hereditarily transitive* if it is so with respect to some sequence $(n_k)_k$. Prove that, if X is separable, then a commuting sequence $(T_n)_n$ is hereditarily transitive if and only if $(T_n)_n$ is weakly mixing. (*Hint*: Note that X has a countable base; use Furstenberg's theorem for sequences of maps.)

Exercise 1.6.5. Show that Theorem 1.54 does not hold for sequences $(T_n)_n$ even if the maps $T_n : X \to X$ commute and have dense range. (*Hint*: A weakly mixing sequence $(T_n)_n$ remains weakly mixing when adding arbitrary maps.)

Sources and comments

Section 1.1. A standard reference for the theory of dynamical systems is Devaney [132]. For more recent textbooks we refer to Brin and Stuck [97] and Robinson [267], while Gulick [188] provides an elementary introduction.

Section 1.2. Kolyada and Snoha [217] give an excellent survey on topological transitivity, with many additional equivalent conditions. The original version of the Birkhoff transitivity theorem can be found in [74, § 62].

Section 1.3. Chaos in the sense of Devaney was introduced in [132]. While there are many other definitions of chaos (see for example Kolyada [216] or Forti [154]), Devaney's definition has become very popular. The theorem of Banks et al. was obtained in [31]; see also Silverman [294] and Glasner and Weiss [163].

Sections 1.4 and 1.5. For Furstenberg's theorem see [157], which also contains the 4-set trick implicitly. Theorem 1.54 is due, independently, to Akin [4], Glasner [162], and Peris and Saldivia [257]; in the context of linear operators on Banach spaces it was also obtained by Grivaux [172]. The remaining characterizations of weak mixing in Section 1.5, including Exercise 1.5.1, are due to Banks [29] and Akin [4]; more precisely, Proposition 1.53 is called the "Furstenberg Intersection Lemma" in Akin's book.

Section 1.6. The universal Taylor series mentioned in Section 1.6 is essentially due to Fekete; see the discussion in [179, Section 3a]. The Universality Criterion was obtained by Grosse-Erdmann [177].

Exercises. Exercises 1.2.8 and 1.4.4 are taken from Costakis and Manoussos [118, 119]. For Exercise 1.3.4 we refer to Touhey [299], for Exercise 1.4.3 to Moothathu [244], for Exercise 1.5.3 to Bauer and Sigmund [32] and to Banks (attributed to Stacey) [28], for Exercise 1.5.5 to Banks [29], and for Exercise 1.5.6 to Moothathu [245]. The two parts of Exercise 1.5.7 are taken from Grivaux [172] and Peris and Saldivia [257], respectively; Exercise 1.5.8 is from Grosse-Erdmann and Peris [187] (see also Grivaux [172] and Bayart and Matheron [45]). The assertion of Exercise 1.5.9 is due to Bauer and Sigmund [32] (one direction), and, independently, to Banks [30] and Peris [256] (the other direction). Exercise 1.6.1 is from Godefroy and Shapiro [165], Exercise 1.6.4 from Bès and Peris [71].

Extensions. Let us add a word on the setting chosen in this chapter. Since the overwhelming majority of linear dynamical systems studied in the literature acts on metric spaces we have restricted our attention to such spaces. In general, however, a *dynamical system* is given by a continuous map $T : X \to X$ on a topological space X. The definitions of *topologically transitive*, *(weakly) mixing* and *chaotic maps* extend verbatim to such systems. The same applies to sequences $(T_n)_n$ of continuous maps $T_n : X \to Y$ between arbitrary topological spaces X and Y.

Then, as the proofs show, all the results in this chapter on general dynamical systems remain true in the setting of arbitrary topological spaces. To be more specific, this concerns all results apart from Proposition 1.15 and the Theorems 1.16, 1.29 and 1.57.

Chapter 2
Hypercyclic and chaotic operators

In this chapter we begin our investigation of linear dynamical systems, that is, dynamical systems that are defined by linear maps. As a simple example one may think of the differentiation operator

$$D : f \to f'.$$

In the language of dynamical systems, the exponential function, for example, is a fixed point of D, while the sine function is periodic with period 4. We will see in this chapter that D is in fact a chaotic operator.

The linearity of a map only makes sense if the underlying space carries, besides its topological structure, a linear structure also. Familiar examples of such spaces are Hilbert spaces and Banach spaces. But some of our main examples demand that we go beyond Banach spaces and allow so-called Fréchet spaces. These spaces will be introduced in the first section of this chapter.

In the subsequent sections we revisit the various topics discussed in Chapter 1 under the influence of linearity, and we contrast linear with nonlinear dynamics.

2.1 Linear dynamical systems

The dynamical systems studied in Chapter 1 were defined by continuous maps on metric spaces. For linear dynamical systems, the underlying space must in addition have a linear structure, as is the case for Hilbert spaces and Banach spaces. In this book we will assume a certain familiarity with such spaces; some of their basic properties are collected in Appendix A.

However, some interesting examples of linear dynamical systems are defined on spaces of a more general type, the so-called Fréchet spaces. In this section we introduce these spaces and describe their operators. The main purpose will be to familiarize the reader with this new concept. Some more

K.-G. Grosse-Erdmann, A. Peris Manguillot, *Linear Chaos*, Universitext,
DOI 10.1007/978-1-4471-2170-1_2,

advanced results that are only used occasionally in this book will be covered in Appendix A.

The following two examples will motivate the concept of a Fréchet space.

Example 2.1. The process of taking derivatives, that is, the operator $D : f \to f'$, provides an interesting linear dynamical system. In order to have the powerful tools of complex analysis at our disposal we regard D as acting on the space of entire functions,

$$H(\mathbb{C}) = \{f : \mathbb{C} \to \mathbb{C}\ ;\ f \text{ holomorphic}\}.$$

The natural concept of convergence for entire functions is that of local uniform convergence, that is, the uniform convergence on all compact sets. In contrast to Banach spaces, convergence is described here by a countably infinite collection of conditions. More precisely, we have that $f_k \to f$ in $H(\mathbb{C})$ if and only if, for all $n \in \mathbb{N}$, $p_n(f_k - f) \to 0$ as $k \to \infty$, where

$$p_n(f) := \sup_{|z| \leq n} |f(z)|.$$

Here, $(p_n)_n$ is an increasing sequence of norms.

Example 2.2. Many natural spaces of sequences are Banach spaces. But the space of *all* (real or complex) sequences,

$$\omega := \mathbb{K}^{\mathbb{N}} = \{(x_n)_n\ ;\ x_n \in \mathbb{K},\ n \in \mathbb{N}\},$$

lies outside this framework, where $\mathbb{K} = \mathbb{R}$ or $\mathbb{C}$. The natural concept of convergence is that of coordinatewise convergence; that is, we have that $x^{(\nu)} \to x$ in ω if and only if, for all $n \in \mathbb{N}$, $p_n(x^{(\nu)} - x) \to 0$ as $\nu \to \infty$, where

$$p_n(x) := \sup_{1 \leq k \leq n} |x_k|, \quad x = (x_k)_k.$$

Here, $(p_n)_n$ is an increasing sequence of seminorms.

We recall the notion of a seminorm.

Definition 2.3. A functional $p : X \to \mathbb{R}_+$ on a vector space X over $\mathbb{K} = \mathbb{R}$ or $\mathbb{C}$ is called a *seminorm* if it satisfies, for all $x, y \in X$ and $\lambda \in \mathbb{K}$,

(i) $p(x + y) \leq p(x) + p(y)$,
(ii) $p(\lambda x) = |\lambda| p(x)$.

A *norm* is a seminorm p for which $p(x) = 0$ implies that $x = 0$. A *Banach space* is a vector space X endowed with a norm, usually denoted by $\|\cdot\|$, whose topology is defined via the metric

$$d(x, y) := \|x - y\|, \quad x, y \in X,$$

and which is complete in that metric. If, moreover, the norm derives from an inner product $\langle \cdot , \cdot \rangle$ via

$$\|x\| := \sqrt{\langle x, x\rangle}, \quad x \in X,$$

then X is called a *Hilbert space*.

We recall here some classical Banach and Hilbert spaces. Further spaces will be introduced as the need arises.

Example 2.4. (a) Let $1 \le p < \infty$. Then the space

$$\ell^p := \Big\{ x = (x_n)_n \in \mathbb{K}^{\mathbb{N}} \ ; \ \sum_{n=1}^{\infty} |x_n|^p < \infty \Big\}$$

of p-summable sequences, endowed with the norm $\|x\| := (\sum_{n=1}^{\infty} |x_n|^p)^{1/p}$, is a Banach space. In particular, ℓ^2 is a Hilbert space with inner product defined by $\langle x, y\rangle = \sum_{n=1}^{\infty} x_n \overline{y_n}$. Occasionally we let the index start with 0. The *finite sequences*, that is, sequences of the form $(x_1, \dots, x_n, 0, 0, \dots)$, $n \ge 1$, constitute a dense subset. Considering only the finite sequences with entries from $\mathbb{Q}$ or $\mathbb{Q} + i\mathbb{Q}$ we see that any ℓ^p, $1 \le p < \infty$, is separable. The space $\ell^p(\mathbb{Z})$ of p-summable sequences, indexed over $\mathbb{Z}$, is defined analogously.

(b) The space $\ell^\infty := \{x = (x_n)_n \in \mathbb{K}^{\mathbb{N}} \ ; \ \sup_{n\in\mathbb{N}} |x_n| < \infty\}$ of bounded sequences, endowed with the sup-norm $\|x\| := \sup_{n\in\mathbb{N}} |x_n|$, is a Banach space. Since it is not separable it will be of less interest to us. Instead, its closed subspace

$$c_0 := \{x = (x_n)_n \in \mathbb{K}^{\mathbb{N}} \ ; \ \lim_{n\to\infty} x_n = 0\}$$

of null sequences is a separable Banach space under the induced norm.

(c) Let $a < b$ and $1 \le p < \infty$. Then the space

$$L^p[a,b] := \Big\{ f : [a,b] \to \mathbb{K} \ ; f \text{ measurable and } \int_a^b |f(t)|^p \, dt < \infty \Big\}$$

of p-integrable functions, endowed with the norm $\|f\| := (\int_a^b |f(t)|^p \, dt)^{1/p}$, is a Banach space; as usual, we identify functions that are equal almost everywhere. In particular, $L^2[a,b]$ is a Hilbert space with inner product defined by $\langle f, g\rangle = \int_a^b f(t)\overline{g(t)}\, dt$. We will occasionally need the fact that the functions $t \to \frac{1}{\sqrt{2\pi}} e^{int}$, $n \in \mathbb{Z}$, form an orthonormal basis in $L^2[0, 2\pi]$.

(d) Let $a < b$. Then the space

$$C[a,b] := \Big\{ f : [a,b] \to \mathbb{K} \ ; f \text{ continuous} \Big\}$$

of continuous functions, endowed with the sup-norm $\|f\| := \sup_{t\in[a,b]} |f(t)|$, is a Banach space.

The concept of a Fréchet space generalizes that of a Banach space by defining the topology via a sequence $(p_n)_n$ of seminorms, which we can always assume to be increasing (by considering $\max_{k\leq n} p_k$, if necessary). Moreover, the sequence is supposed to be *separating*, that is, $p_n(x) = 0$ for all $n \geq 1$ implies that $x = 0$. Then it is easy to see that

$$d(x,y) := \sum_{n=1}^{\infty} \frac{1}{2^n} \min(1, p_n(x-y)), \quad x, y \in X \tag{2.1}$$

defines a metric on X; see Exercise 2.1.1. An important feature of this metric is that it is *translation-invariant*, that is,

$$d(x,y) = d(x+z, y+z) \text{ for all } x, y, z \in X.$$

Definition 2.5. A *Fréchet space* is a vector space X, endowed with a separating increasing sequence $(p_n)_n$ of seminorms, which is complete in the metric given by (2.1).

The following result will be of constant use. We leave its proof as a useful exercise to the reader; see Exercise 2.1.2.

Lemma 2.6. *Let X be a Fréchet space with a defining increasing sequence $(p_n)_n$ of seminorms. Let $x_k, x \in X$, $k \geq 1$, and $U \subset X$. Then:*

(i) *$x_k \to x$ if and only if $p_n(x_k - x) \to 0$ as $k \to \infty$, for all $n \geq 1$;*

(ii) *$(x_k)_k$ is a Cauchy sequence if and only if $p_n(x_k - x_l) \to 0$ as $k, l \to \infty$, for all $n \geq 1$;*

(iii) *U is a neighbourhood of x if and only if there are $n \geq 1$ and $\varepsilon > 0$ such that $\{y \in X \ ; \ p_n(y-x) < \varepsilon\} \subset U$.*

Example 2.7. (a) Let X be a Banach space with norm $\|\cdot\|$. Setting $p_n = \|\cdot\|$, $n \geq 1$, it follows from (i) and (ii) that X is also a Fréchet space according to Definition 2.5.

(b) With the seminorms defined in Example 2.1, the space $H(\mathbb{C})$ of entire functions is a Fréchet space; in view of Lemma 2.6(ii), completeness follows in the usual way. Since the Taylor series expansion of an entire function converges on every compact set, the polynomials form a dense subset of $H(\mathbb{C})$. Considering polynomials with coefficients from $\mathbb{Q} + i\mathbb{Q}$ we see that $H(\mathbb{C})$ is separable.

(c) The space $\omega = \mathbb{K}^{\mathbb{N}}$ of all real or complex sequences, endowed with the seminorms given in Example 2.2, is a Fréchet space. Since the finite sequences with entries from $\mathbb{Q}$ or $\mathbb{Q} + i\mathbb{Q}$ form a dense subset, ω is separable. More generally, if X is a separable Fréchet space then, in a canonical way, also $X^{\mathbb{N}}$ is a separable Fréchet space; see Exercise 2.1.3.

Further Fréchet spaces will be introduced in the course of the book.

Looking at the way the metric is defined in a Banach space it is tempting to introduce, also in a Fréchet space X, a norm-like functional by setting

$$\|x\| := \sum_{n=1}^{\infty} \frac{1}{2^n} \min(1, p_n(x)), \quad x \in X, \tag{2.2}$$

so that $d(x, y) = \|x - y\|$. We summarize its characteristic properties; see Exercise 2.1.4.

Proposition 2.8. *The functional* $\|\cdot\| : X \to \mathbb{R}_+$ *given by* (2.2) *satisfies, for all* $x, y \in X$ *and* $\lambda \in \mathbb{K}$,

(i) $\|x + y\| \leq \|x\| + \|y\|$;
(ii) $\|\lambda x\| \leq \|x\|$ *if* $|\lambda| \leq 1$;
(iii) $\lim_{\lambda \to 0} \|\lambda x\| = 0$;
(iv) $\|x\| = 0$ *implies that* $x = 0$.

Definition 2.9. A functional $\|\cdot\| : X \to \mathbb{R}_+$ on a vector space X that satisfies conditions (i)–(iv) of Proposition 2.8 is called an *F-norm*.

The notion of an F-norm has the advantage that one can largely argue as if one was working in a Banach space. One need only be aware of the fact that the positive homogeneity of a norm is no longer available. In fact, in many cases, this property is not needed at all or it can be replaced by the following weaker property that follows directly from conditions (i) and (ii): for all $x \in X$ and $\lambda \in \mathbb{K}$,

$$\|\lambda x\| \leq (|\lambda| + 1)\,\|x\|. \tag{2.3}$$

Having discussed Fréchet spaces and their topology we now turn to the concept of operators on them.

Definition 2.10. Let X and Y be Fréchet spaces. Then a continuous linear map $T : X \to Y$ is called an *operator*. The space of all such operators is denoted by $L(X, Y)$. If $Y = X$ we say that T is an *operator on* X, with $L(X) = L(X, X)$.

The following extends a familiar result from Banach spaces to Fréchet spaces; see also Exercise 2.1.7.

Proposition 2.11. *Let* X *and* Y *be Fréchet spaces with defining increasing sequences of seminorms* $(p_n)_n$ *and* $(q_n)_n$, *respectively. Then a linear map* $T : X \to Y$ *is an operator if and only if, for any* $m \geq 1$, *there are* $n \geq 1$ *and* $M > 0$ *such that*

$$q_m(Tx) \leq M p_n(x), \quad x \in X.$$

Proof. The condition is obviously sufficient because, by Lemma 2.6, it implies that when $x_k \to x$ in X then $Tx_k \to Tx$ in Y.

Conversely, let $m \geq 1$. By Lemma 2.6, the set $W := \{y \in Y \,;\, q_m(y) < 1\}$ is a 0-neighbourhood in Y. By continuity there is a 0-neighbourhood W' in X such that $T(W') \subset W$. Hence there are $n \geq 1$ and $\varepsilon > 0$ such that $p_n(x) < \varepsilon$

implies that $x \in W'$, and therefore $q_m(Tx) < 1$. Now let $x \in X$. Then, for any $\delta > 0$, we have that

$$p_n\Big(\frac{\varepsilon}{p_n(x)+\delta}x\Big) < \varepsilon$$

and hence

$$q_m(Tx) < \frac{p_n(x)+\delta}{\varepsilon}.$$

Since $\delta > 0$ is arbitrary we obtain the result with $M = 1/\varepsilon$. □

In contrast to Banach space operators one cannot associate a norm with a Fréchet space operator.

Example 2.12. (a) The map

$$D : f \to f'$$

is an operator on $H(\mathbb{C})$. This follows from the Cauchy estimates by which, for any $n \geq 1$, $\sup_{|z|\leq n} |f'(z)| \leq \sup_{|z|\leq n+1} |f(z)|$.

(b) Let $X = H(\mathbb{C})$. Then the translation map T_a is defined by

$$T_a f(z) = f(z+a), \quad a \in \mathbb{C}.$$

This is clearly an operator on X.

(c) Let $X = \ell^p$, $1 \leq p < \infty$, or c_0. Then the *backward shift* $B : X \to X$, defined by

$$B(x_1, x_2, \dots) = (x_2, x_3, \dots)$$

is an operator on X, of norm $\|B\| = 1$, and it is also an operator on ω.

In Chapter 1 we associated to any two dynamical systems $S : X \to X$ and $T : Y \to Y$ a new dynamical system $S \times T$ on $X \times Y$. In a linear setting one usually employs a different, additive notation. More specifically, let X and Y be Fréchet spaces with defining increasing sequences of seminorms $(p_n)_n$ and $(q_n)_n$, respectively. Then the space

$$X \oplus Y := \{(x, y) \; ; \; x \in X, \; y \in Y\}$$

will be endowed with the seminorms $(x, y) \to p_n(x) + q_n(y)$, $n \geq 1$, which induce the product topology on $X \oplus Y$. This space then becomes a Fréchet space, which is separable if X and Y are.

Definition 2.13. Let $S : X \to X$ and $T : Y \to Y$ be operators on Fréchet spaces X and Y. Then the operator $S \oplus T$ is defined by

$$S \oplus T : X \oplus Y \to X \oplus Y, \quad (S \oplus T)(x, y) = (Sx, Ty).$$

With this we end our introduction to Fréchet spaces and their operators. We will show in Chapter 12 that several important results in linear dynamics hold true in the wider context of operators on so-called topological vector spaces. However, other results require the existence of a complete metric, and most operators in linear dynamics are naturally defined on Fréchet spaces. In addition, since we are primarily interested in operators with a dense orbit, we will assume that the space is separable. Thus, with Chapter 12 the only exception, we adopt the following point of view throughout this book.

Definition 2.14. A *linear dynamical system* is a pair (X, T) consisting of a separable Fréchet space X and an operator $T : X \to X$.

Usually we simply call T or $T : X \to X$ a linear dynamical system. *From now on, all operators will be defined on separable Fréchet spaces, if nothing else is said.*

2.2 Hypercyclic operators

We begin our study of the dynamics of linear operators by considering dynamical systems with a dense orbit. In the presence of linearity, such systems are given their own name.

Definition 2.15. An operator $T : X \to X$ is called *hypercyclic* if there is some $x \in X$ whose orbit under T is dense in X. In such a case, x is called a *hypercyclic vector* for T. The set of hypercyclic vectors for T is denoted by $HC(T)$.

The origin of this terminology is easily explained. For a long time, operator theorists have been studying so-called cyclic vectors in connection with the invariant subspace problem. Vectors with a more restrictive property were then called supercyclic.

Definition 2.16. Let $T : X \to X$ be an operator. A vector $x \in X$ is called *cyclic* for T if the linear span of its orbit,

$$\operatorname{span}\{T^n x\ ;\ n \geq 0\}$$

is dense in X. A vector $x \in X$ is called *supercyclic* for T if its projective orbit,

$$\{\lambda T^n x\ ;\ \ n \geq 0,\ \lambda \in \mathbb{K}\}$$

is dense in X.

Operators that possess a cyclic (or supercyclic) vector are called *cyclic* (or *supercyclic*, respectively).

This suggested the name of hypercyclicity for the case when the orbit itself is dense. Cyclic and supercyclic vectors will not be studied in detail in this book.

The *invariant subspace problem*, which is open to this day, asks whether every Hilbert space operator possesses an invariant closed subspace other than the trivial ones given by $\{0\}$ and the whole space. Counterexamples do exist for operators on non-reflexive spaces like ℓ^1.

Obviously, the smallest closed T-invariant subspace of X that contains a given point x coincides with the closure of the span of its orbit. Therefore, an operator has no nontrivial invariant closed subspace precisely if every nonzero vector is cyclic. By the same token we have a link between hypercyclicity and the *invariant subset problem*: does every Hilbert space operator possess an invariant closed subset other than the trivial ones given by $\{0\}$ and the whole space?

Observation 2.17. *An operator has no nontrivial invariant closed subsets if and only if every nonzero vector is hypercyclic.*

Having explained the historical interest in hypercyclicity, our first question has to be if hypercyclic operators exist. That is, does the additional requirement of linearity still allow us to find maps with dense orbits? Indeed, a very simple operator on the Hilbert space ℓ^2 turns out to be hypercyclic.

Example 2.18. Let $T : \ell^2 \to \ell^2$ be twice the backward shift, that is,

$$T = 2B : (x_1, x_2, x_3, \dots) \to 2(x_2, x_3, x_4, \dots).$$

The space ℓ^2 has a countable dense set $\{y^{(k)}\ ;\ k \geq 1\}$ consisting of finite sequences; for each $k \geq 1$, let m_k be the greatest index with $y^{(k)}_{m_k} \neq 0$. By S we denote half the forward shift operator,

$$S = \tfrac{1}{2}F : (x_1, x_2, x_3, \dots) \to \tfrac{1}{2}(0, x_1, x_2, \dots).$$

Then, by induction, we can find a sequence $(n_k)_k$ of positive integers such that, for any $k > j \geq 1$,

$$n_k \geq m_j + n_j \quad \text{and} \quad 2^{n_k} \geq 2^{n_j + k}\, \|y^{(k)}\|.$$

We claim that the vector

$$x := \sum_{k=1}^{\infty} S^{n_k} y^{(k)}$$

is hypercyclic for T. First, since $\|S^{n_k} y^{(k)}\| = 2^{-n_k} \|y^{(k)}\| \leq 2^{-k}$ for $k \geq 2$, the series converges and $x \in \ell^2$. Now, let $k \geq 1$. Then

$$\begin{aligned} T^{n_k} x &= \sum_{j=1}^{k-1} 2^{n_k - n_j} B^{n_k - n_j} y^{(j)} + y^{(k)} + \sum_{j=k+1}^{\infty} 2^{n_k - n_j} F^{n_j - n_k} y^{(j)} \\ &= y^{(k)} + \sum_{j=k+1}^{\infty} 2^{n_k - n_j} F^{n_j - n_k} y^{(j)}, \end{aligned}$$

where we have used that $n_k - n_j \geq m_j$ for $j < k$. From

$$\sum_{j=k+1}^{\infty} 2^{n_k - n_j} \|F^{n_j - n_k} y^{(j)}\| = \sum_{j=k+1}^{\infty} 2^{n_k - n_j} \|y^{(j)}\| \leq \sum_{j=k+1}^{\infty} 2^{-j} = 2^{-k}$$

we deduce that $\|T^{n_k}x - y^{(k)}\| \leq 2^{-k}$. Since the $y^{(k)}$ form a dense set, x has a dense orbit under T.

Instead of an explicit construction of a hypercyclic vector one can also apply the Birkhoff transitivity theorem to show that an operator is hypercyclic. This leads to more transparent proofs, in particular when the operator is complicated. For ease of reference we restate Birkhoff's theorem in our new setting; note that Fréchet spaces clearly have no isolated points.

Theorem 2.19 (Birkhoff transitivity theorem). *An operator T is hypercyclic if and only if it is topologically transitive. In that case, the set $HC(T)$ of hypercyclic vectors is a dense G_δ-set.*

Directly or indirectly, the transitivity theorem will be our main tool for proving the hypercyclicity of an operator.

The first examples of hypercyclic operators were found by G.D. Birkhoff in 1929, G.R. MacLane in 1952 and S. Rolewicz in 1969. These operators will accompany us throughout the book as they will serve as a testing ground for any new concept in linear dynamics; indeed, Example 2.18 was already a special Rolewicz operator.

Example 2.20. **(Birkhoff's operators)** On the space $H(\mathbb{C})$ of entire functions we consider the translation operators given by

$$T_a f(z) = f(z+a), \quad a \neq 0.$$

Let $U, V \subset H(\mathbb{C})$ be arbitrary nonempty open sets, and fix $f \in U$, $g \in V$. By the definition of the topology on $H(\mathbb{C})$ there is a closed disk K centred at 0 and an $\varepsilon > 0$ such that an entire function h belongs to U (or to V) whenever $\sup_{z\in K} |f(z)-h(z)| < \varepsilon$ (or $\sup_{z\in K} |g(z)-h(z)| < \varepsilon$, respectively). Let $n \in \mathbb{N}$ be any integer such that K and $K + na$ are disjoint disks. Considering the function that is defined as f on a neighbourhood of K and by $z \to g(z - na)$ on a neighbourhood of $K + na$, Runge's theorem (see Appendix A), tells us that there exists a polynomial p such that

$$\sup_{z\in K} |f(z) - p(z)| < \varepsilon \quad \text{and} \quad \sup_{z\in K+na} |g(z - na) - p(z)| < \varepsilon,$$

and hence also

$$\sup_{z\in K} |g(z) - (T_a^n p)(z)| = \sup_{z\in K} |g(z) - p(z + na)| < \varepsilon.$$

This shows that $p \in U$ and $T_a^n p \in V$, so that T_a is topologically transitive. Since $H(\mathbb{C})$ is a separable Fréchet space, T_a is hypercyclic.

Example 2.21. **(MacLane's operator)** We next consider the differentiation operator

$$D : f \to f'$$

on $H(\mathbb{C})$. Since the polynomials are dense in $H(\mathbb{C})$, given arbitrary nonempty open sets $U, V \subset H(\mathbb{C})$, there are polynomials $p \in U$ and $q \in V$, $p(z) = \sum_{k=0}^{N} a_k z^k$ and $q(z) = \sum_{k=0}^{N} b_k z^k$. Let $n \geq N+1$ be arbitrary. Then the polynomial

$$r(z) = p(z) + \sum_{k=0}^{N} \frac{k!\, b_k}{(k+n)!} z^{k+n}$$

has the property that $D^n r = q$. Moreover, for any $R > 0$ we have that

$$\sup_{|z| \leq R} |r(z) - p(z)| \leq \sum_{k=0}^{N} \frac{k!|b_k|}{(k+n)!} R^{k+n} \to 0$$

as $n \to \infty$. Thus, if n is sufficiently large, then $r \in U$ and $D^n r \in V$. This implies that D is hypercyclic.

Example 2.22. **(Rolewicz's operators)** On the spaces $X := \ell^p$, $1 \leq p < \infty$, or $X := c_0$ we consider the multiple

$$T = \lambda B : X \to X, \quad (x_1, x_2, x_3, \ldots) \to \lambda(x_2, x_3, x_4, \ldots)$$

of the backward shift, where $\lambda \in \mathbb{K}$. First, if $|\lambda| \leq 1$ then $\|T^n x\| = |\lambda|^n \|B^n x\| \leq \|x\|$ for all $x \in X$ and $n \geq 0$. Thus T cannot be hypercyclic in this case.

On the other hand, T is hypercyclic whenever $|\lambda| > 1$. Indeed, if $U, V \subset X$ are nonempty open sets, we can find $x \in U$ and $y \in V$ of the form

$$x = (x_1, x_2, \ldots, x_N, 0, 0, \ldots), \quad y = (y_1, y_2, \ldots, y_N, 0, 0, \ldots),$$

for some $N \in \mathbb{N}$. Let $n \geq N$ be arbitrary. Defining $z \in X$ by $z_k = x_k$ if $1 \leq k \leq N$, $z_k = \lambda^{-n} y_{k-n}$ if $n+1 \leq k \leq n+N$, and $z_k = 0$ otherwise, we obtain a sequence with $T^n z = y$. Moreover, $\|x - z\| = |\lambda|^{-n} \|y\| \to 0$ as $n \to \infty$. Thus, if n is sufficiently large, then $z \in U$ and $T^n z \in V$. This shows that T is topologically transitive; since the underlying spaces are separable Banach spaces, T is hypercyclic.

There are various ways to derive the hypercyclicity of one operator from that of another. The first result of this type follows from Proposition 1.14 by the Birkhoff transitivity theorem. Recall that by the inverse mapping theorem (see Appendix A), any bijective operator has a continuous inverse and is therefore.

Proposition 2.23. *Let T be an invertible operator. Then T is hypercyclic if and only if T^{-1} is.*

Birkhoff's operators provide examples of invertible hypercyclic operators. Next, in our present context, Proposition 1.19 reads as follows.

Proposition 2.24. *Hypercyclicity is preserved under quasiconjugacy.*

We emphasize that the map ϕ defining the quasiconjugacy need not be linear; it may, for instance, be defined between a complex space and a real space; see Exercise 2.2.5.

We also formulate part of Proposition 1.42 in the linear setting.

Proposition 2.25. *Let $S : X \to X$ and $T : Y \to Y$ be operators. If $S \oplus T$ is hypercyclic then so are S and T.*

We will see in Remark 4.17 that the converse fails in general. As an application of Proposition 2.25 we obtain an interesting transference principle that is specific to the linear setting. Let X be a real separable Fréchet space. Then the *complexification* $\widetilde{X}$ of X is defined formally as

$$\widetilde{X} = \{x + iy \ ; \ x, y \in X\},$$

which will be identified with $X \oplus X$. If multiplication by complex scalars is defined by $(a+ib)(x+iy) = (ax-by)+i(ay+bx)$, then $\widetilde{X}$ becomes a complex separable Fréchet space.

Moreover, let $T : X \to X$ be a (real-linear) operator on X. Then its *complexification* $\widetilde{T} : \widetilde{X} \to \widetilde{X}$ is defined by

$$\widetilde{T}(x + iy) = Tx + iTy.$$

An easy computation shows that $\widetilde{T}$ is a (complex-linear) operator on $\widetilde{X}$. For all of these statements see Exercise 2.2.7. Since $\widetilde{T}$ is nothing but the operator $T \oplus T$ on $X \oplus X$, Proposition 2.25 implies the following.

Proposition 2.26. *Let T be an operator on a real separable Fréchet space. If its complexification $\widetilde{T}$ is hypercyclic then so is T.*

In fact, since $\widetilde{T}$ coincides with $T \oplus T$, hypercyclicity of $\widetilde{T}$ is equivalent to weak mixing of T; see also Section 2.5.

Example 2.27. The complexification of the real Rolewicz operator $T = \lambda B$, $\lambda \in \mathbb{R}$, $|\lambda| > 1$, on the spaces $X := \ell^p$, $1 \leq p < \infty$, or $X := c_0$, of real sequences can be identified with the same operator as understood on the corresponding spaces of complex sequences. Therefore, hypercyclicity for the complex Rolewicz operators implies the same for the real operators.

As a second application of Proposition 2.25 we consider restrictions of hypercyclic operators to invariant subspaces. Let M_1 and M_2 be closed subspaces of a (real or complex) Fréchet space X such that, algebraically, $X = M_1 \oplus M_2$, that is, $X = M_1 + M_2$ and $M_1 \cap M_2 = \{0\}$.

Note that, at the end of Section 2.1, we had defined $M_1 \oplus M_2$ as the topological product of the two spaces. But since the map $\phi : (x_1, x_2) \to x_1 + x_2$ defines an algebraic isomorphism between $M_1 \times M_2$ and $M_1 + M_2$, it is also a topological isomorphism by the inverse mapping theorem (see Appendix A), so that the two forms of $M_1 \oplus M_2$ can be identified.

Now suppose that T is an operator on X that leaves M_1 and M_2 invariant. Then we have that

$$Tx = Tx_1 + Tx_2 \quad \text{if } x = x_1 + x_2,\ x_1 \in M_1,\ x_2 \in M_2.$$

In the sense of the isomorphism ϕ we therefore have that

$$T = T|_{M_1} \oplus T|_{M_2};$$

hence Proposition 2.25 implies the following.

Proposition 2.28. *Let $T : X \to X$ be a hypercyclic operator, and let M_1 and M_2 be T-invariant closed subspaces of X such that $X = M_1 \oplus M_2$. Then the restrictions $T|_{M_1}$ and $T|_{M_2}$ are hypercyclic.*

2.3 Linear chaos

As defined in Chapter 1, chaos in the sense of Devaney consists in demanding topological transitivity and the density of the set of periodic points. In view of the Birkhoff transitivity theorem we can rephrase this definition in our present setting.

Definition 2.29 (Linear chaos). An operator T is said to be *chaotic* if it satisfies the following conditions:
 (i) T is hypercyclic;
 (ii) T has a dense set of periodic points.

We recall that sensitive dependence on initial conditions was a consequence of chaos for metric spaces without isolated points; see Theorem 1.29. For operators, hypercyclicity in itself already implies sensitive dependence.

Proposition 2.30. *Let T be a hypercyclic operator. Then T has sensitive dependence on initial conditions (with respect to any translation-invariant metric defining the topology of X).*

Proof. Let d be any translation-invariant metric on X that induces its topology. Let $\delta, \varepsilon > 0$ and $x \in X$ be arbitrary. We then consider the nonempty open sets

$$U = \{z \in X \ ; \ d(0, z) < \varepsilon\}, \quad V = \{z \in X \ ; \ d(0, z) > \delta\}.$$

By the topological transitivity of T, there are $n \in \mathbb{N}_0$ and $z \in U$ such that $T^n z \in V$. For the point $y := x + z$ we then obtain that $d(x, y) = d(0, z) < \varepsilon$ and $d(T^n x, T^n y) = d(0, T^n z) > \delta$, which implies the result. □

Remark 2.31. Devaney's notion of chaos has been generally accepted in linear dynamics. There are, however, also other definitions of chaos. We mention here that a continuous map $T : X \to X$ on a metric space (X, d) is called *chaotic in the sense of Auslander and Yorke* if it is topologically transitive and it has sensitive dependence on initial conditions. By the previous proposition, every hypercyclic operator is Auslander–Yorke chaotic.

In some cases, the periodic points of an operator are easily determined. As a first example we consider the multiples of backward shifts.

Example 2.32. **(Rolewicz's operators)** Let $T = \lambda B$, $|\lambda| > 1$, be Rolewicz's operator on $X = \ell^p$, $1 \leq p < \infty$, or $X = c_0$. One easily verifies that $x \in X$ is periodic if and only if there are $N \in \mathbb{N}$ and $x_k \in \mathbb{K}$, $k = 1, \ldots, N$, such that

$$x = \big(x_1, \ldots, x_N, \lambda^{-N} x_1, \ldots, \lambda^{-N} x_N, \lambda^{-2N} x_1, \ldots, \lambda^{-2N} x_N, \ldots\big).$$

In order to see that the set of periodic points is dense in X it suffices to approximate any finite sequence $y = (y_1, \ldots, y_n, 0, \ldots)$. By choosing a periodic point whose $N \geq n$ first coordinates coincide with those of y we see that $\|x - y\| \leq \sum_{j=1}^{\infty} |\lambda|^{-jN} \|y\| \to 0$ as $N \to \infty$. Therefore Rolewicz's operators are chaotic.

Let us observe that, for linear maps T on arbitrary vector spaces X, the set of periodic points of T is a subspace of X. Indeed, let $x, y \in X$ be periodic points for T. Then we have that $T^n x = x$ and $T^m y = y$ for certain $n, m \in \mathbb{N}$. Thus $T^{nm}(ax + by) = a(T^n)^m x + b(T^m)^n y = ax + by$, for any $a, b \in \mathbb{K}$, so that also $ax + by$ is periodic.

There is, in fact, a nice and very useful description of the space of periodic points in terms of eigenvectors to unimodular eigenvalues, that is, eigenvalues of absolute value 1, provided that we have a complex space. The corresponding result is of a purely algebraic nature.

Proposition 2.33. *Let T be a linear map on a complex vector space X. Then the set of periodic points of T is given by*

$$\mathrm{Per}(T) = \mathrm{span}\{x \in X \ ; \ Tx = e^{\alpha\pi i} x \textit{ for some } \alpha \in \mathbb{Q}\}.$$

Proof. If $Tx = e^{\alpha\pi i}x$ with $\alpha = \frac{k}{n}$, $k \in \mathbb{Z}$, $n \in \mathbb{N}$, then $T^{2n}x = x$, so that x is periodic. This yields one inclusion.

For the other one, suppose that $T^n x = x$, $n \in \mathbb{N}$. We then decompose the polynomial $z^n - 1$ into a product of monomials,

$$z^n - 1 = (z - \lambda_1)(z - \lambda_2)\cdots(z - \lambda_n).$$

Since all the roots λ_k, $k = 1, \ldots, n$, are different, the system $\{p_1, \ldots, p_n\}$ of polynomials with $p_k(z) := \prod_{j \neq k}(z - \lambda_j)$, $1 \leq k \leq n$, is a basis of the space of polynomials of degree strictly less than n. In particular, there are $\alpha_k \in \mathbb{C}$, $k = 1, \ldots, n$, such that

$$1 = \sum_{k=1}^{n} \alpha_k p_k(z), \quad z \in \mathbb{C}.$$

This means that, when we substitute z by T, then

$$I = \sum_{k=1}^{n} \alpha_k p_k(T).$$

We therefore have that $x = \sum_{k=1}^{n} \alpha_k y_k$ with $y_k := p_k(T)x$, $k = 1, \ldots, n$. Since $(T - \lambda_k)y_k = (T^n - I)x = 0$ with $\lambda_k^n = 1$, we see that x belongs to the desired span. □

In many concrete situations, this proposition leads to a simple verification that a given operator has a dense set of periodic points. As an example we consider Birkhoff's and MacLane's operators. We first need the following result, where e_λ denotes the exponential function

$$e_\lambda(z) = e^{\lambda z}, \quad z \in \mathbb{C}.$$

Lemma 2.34. *Let $\Lambda \subset \mathbb{C}$ be a set with an accumulation point. Then the set*

$$\operatorname{span}\{e_\lambda \ ; \ \lambda \in \Lambda\}$$

is dense in $H(\mathbb{C})$.

Proof. By assumption there are $\lambda \in \mathbb{C}$ and $\lambda_n \in \Lambda$ with $\lambda_n \to \lambda$ and $\lambda_n \neq \lambda$ for all $n \geq 1$. Writing

$$e^{\lambda_n z} = e^{\lambda z} e^{(\lambda_n - \lambda)z} = e^{\lambda z} + e^{\lambda z}(\lambda_n - \lambda)z + e^{\lambda z}\frac{(\lambda_n - \lambda)^2 z^2}{2!} + \ldots \tag{2.4}$$

we see that

$$e^{\lambda_n z} \to e^{\lambda z} \quad \text{uniformly on compact sets,}$$

which, incidentally, also follows directly. Therefore, $e_\lambda \in \overline{\operatorname{span}}\{e_{\lambda_n} \ ; \ n \geq 1\}$.

But now (2.4) also shows that

$$\frac{e^{\lambda_n z} - e^{\lambda z}}{\lambda_n - \lambda} = e^{\lambda z} z + e^{\lambda z} \frac{(\lambda_n - \lambda) z^2}{2!} + \ldots,$$

and hence

$$\frac{e^{\lambda_n z} - e^{\lambda z}}{\lambda_n - \lambda} \longrightarrow z e^{\lambda z} \quad \text{uniformly on compact sets,}$$

so that also the function $z \longrightarrow z e^{\lambda z}$ belongs to $\overline{\text{span}}\{e_{\lambda_n} \; ; \; n \geq 1\}$.

Continuing in this way we find that all functions $z \longrightarrow z^k e^{\lambda z}$, $k \geq 0$, belong to $\overline{\text{span}}\{e_{\lambda_n} \; ; \; n \geq 1\}$.

Now let $f \in H(\mathbb{C})$. Then we have that

$$f(z) = e^{\lambda z}(e^{-\lambda z} f(z)) = e^{\lambda z}\Big(\sum_{k=0}^{\infty} a_k z^k\Big) = \sum_{k=0}^{\infty} a_k z^k e^{\lambda z}$$

with suitable coefficients $a_k \in \mathbb{C}$, $k \geq 0$, where convergence takes place in $H(\mathbb{C})$. Thus we also have that $f \in \overline{\text{span}}\{e_{\lambda_n} \; ; \; n \geq 1\}$, which had to be shown. □

The lemma allows us to show that Birkhoff's and MacLane's operators are chaotic on $H(\mathbb{C})$.

Example 2.35. **(Birkhoff's and MacLane's operators)** For the differentiation operator D, any function e_λ is an eigenvector of D to the eigenvalue λ. Thus, since the subspace

$$\text{span}\{e_\lambda \; ; \; \lambda = e^{\alpha \pi i} \text{ for some } \alpha \in \mathbb{Q}\}$$

is dense in $H(\mathbb{C})$ by Lemma 2.34, Proposition 2.33 tells us that $\text{Per}(T)$ is dense. Since we already know that D is hypercyclic, it is also chaotic.

For the translation operators T_a, $a \in \mathbb{C} \setminus \{0\}$, any function e_λ is an eigenvector of T_a to the eigenvalue $e^{a\lambda}$. Thus, since also the subspace

$$\text{span}\{e_\lambda \; ; \; e^{a\lambda} = e^{\alpha \pi i} \text{ for some } \alpha \in \mathbb{Q}\} = \text{span}\{e_\lambda \; ; \lambda = \tfrac{\alpha}{a} \pi i, \alpha \in \mathbb{Q}\}$$

is dense in $H(\mathbb{C})$, we conclude as before that each T_a is chaotic.

The restriction that Proposition 2.33 only holds for complex spaces can sometimes be overcome by using complexifications: an operator on a real space is chaotic if and only if its complexification is, as we will see in Corollary 2.51 below.

2.4 Mixing operators

In this and the following section we study the mixing and weak mixing properties in the light of linearity.

Since we are in the setting of Fréchet spaces, the proofs could be formulated in terms of their seminorms or their metric. However, the arguments become particularly transparent when we use the topological language of open sets and 0-neighbourhoods. For this we only need the following simple result.

As usual, we set $A+B=\{a+b\ ;\ a\in A, b\in B\}$ for subsets A, B of a vector space.

Lemma 2.36. *Let X be a Fréchet space. If $U\subset X$ is a nonempty open set then there is a nonempty open subset $U_1\subset U$ and a 0-neighbourhood W such that $U_1+W\subset U$. If W is a 0-neighbourhood then there is a 0-neighbourhood W_1 such that $W_1+W_1\subset W$.*

Proof. Let $\|\cdot\|$ be an F-norm defining the topology of X. Then there is some $x_0\in U$ and some $\varepsilon>0$ such that $U_\varepsilon(x_0)=\{x\in X\ ;\|x-x_0\|<\varepsilon\}$ is contained in U. One may then take $U_1=U_{\varepsilon/2}(x_0)$ and $W=U_{\varepsilon/2}(0)$. The second claim follows similarly upon taking $x_0=0$. □

We recall that the mixing property consists in demanding the cofiniteness of the return sets $N(U,V)$ for each pair U,V of nonempty open subsets of X. For operators this requirement can be weakened.

Proposition 2.37. *An operator T is mixing if and only if, for any nonempty open set $U\subset X$ and any 0-neighbourhood W, the return sets*

$$N(U,W) \text{ and } N(W,U)$$

are cofinite.

Proof. It suffices to show sufficiency of the condition. Let $U,V\subset X$ be nonempty open sets. By Lemma 2.36 there are nonempty open sets U_1, V_1 and a 0-neighbourhood W such that $U_1+W\subset U$ and $V_1+W\subset V$. By hypothesis, there exists some $N\in\mathbb{N}$ such that, for any $n\geq N$, there are $u\in U_1$ and $w\in W$ so that $T^nu\in W$ and $T^nw\in V_1$. But then $u+w\in U$ and $T^n(u+w)=T^nu+T^nw\in V$, which implies that $N(U,V)$ is cofinite. □

By following the proofs of the hypercyclicity of Rolewicz's, Birkhoff's and MacLane's operators, one immediately obtains that they are even mixing.

Example 2.38. **(Birkhoff's, MacLane's and Rolewicz's operators)** The three classical hypercyclic operators are mixing.

We have also seen that these mixing operators are even chaotic. This is not always the case, as the following example shows.

Example 2.39. We consider the weighted shift $T : \ell^1 \to \ell^1$ given by

$$T(x_1, x_2, \dots) = \left(2x_2, \tfrac{3}{2}x_3, \tfrac{4}{3}x_4, \dots\right).$$

Let U be a nonempty open subset of ℓ^1 and W a 0-neighbourhood. By density of the finite sequences in ℓ^1 there is a sequence of the form $u = (u_1, \dots, u_N, 0, 0, \dots)$ in U.

Since $T^n u = 0$ whenever $n \geq N$, the set $N(U, W)$ is cofinite. On the other hand, we have that

$$T^n x = \left((n+1)x_{n+1}, \left(\tfrac{n+2}{2}\right)x_{n+2}, \left(\tfrac{n+3}{3}\right)x_{n+3}, \dots\right), \quad n \geq 1.$$

Thus, if we define $w \in \ell^1$ by $w_k = \frac{k-n}{k} u_{k-n}$ for $k = n+1, \dots, n+N$, and $w_k = 0$ otherwise, then $T^n w = u$ and

$$\|w\| \leq \frac{N}{n+1}\|u\|,$$

so that also $N(W, U)$ is cofinite. Hence T is a mixing operator.

We now show that T has no nontrivial periodic points and therefore cannot be chaotic. Indeed, let us suppose that $x \neq 0$ is periodic for T, that is, there is some $n \in \mathbb{N}$ with $T^n x = x$, hence also $T^{jn} x = x$ for all $j \in \mathbb{N}$. Using the above formula for $T^n x$ we obtain that $\frac{jn+k}{k} x_{jn+k} = x_k$, for all $k, j \in \mathbb{N}$. Now, since $x \neq 0$ there is some $k \in \mathbb{N}$ with $x_k \neq 0$. Hence

$$\|x\| \geq \sum_{j=1}^{\infty} |x_{jn+k}| = |x_k| \sum_{j=1}^{\infty} \frac{k}{jn+k} = \infty,$$

which is a contradiction.

We reformulate part of a previous result, Proposition 1.42, for operators.

Proposition 2.40. *Let $S : X \to X$ and $T : Y \to Y$ be hypercyclic operators. If at least one of them is mixing then $S \oplus T$ is hypercyclic. Moreover, $S \oplus T$ is mixing if and only if both S and T are.*

For a later application (see Proposition 8.5), we also need to consider direct sums of countably many operators on Banach spaces. Thus, let T_n be operators on separable Banach spaces X_n, $n \geq 1$. For $1 \leq p < \infty$, we define the *direct ℓ^p-sum* of these spaces as

$$\left(\bigoplus_{n=1}^{\infty} X_n\right)_{\ell^p} = \left\{(x_n)_{n\geq 1}\ ;\ \ x_n \in X_n,\ n \geq 1, \text{ and } \sum_{n=1}^{\infty} \|x_n\|^p < \infty\right\};$$

endowed with the norm $\|(x_n)_n\| = (\sum_{n=1}^{\infty} \|x_n\|^p)^{1/p}$ this space turns into a separable Banach space. The *direct c_0-sum* $(\bigoplus_{n=1}^{\infty} X_n)_{c_0}$ is defined similarly.

Now suppose that $\sup_{n\in\mathbb{N}} \|T_n\| < \infty$. Then the direct sum of the operators T_n, defined by

$$\Big(\bigoplus_{n=1}^{\infty} T_n\Big)(x_n)_n = (T_n x_n)_n,$$

is an operator on $(\bigoplus_{n=1}^{\infty} X_n)_{\ell^p}$ and on $(\bigoplus_{n=1}^{\infty} X_n)_{c_0}$.

Proposition 2.41. *Let T_n be operators on separable Banach spaces X_n, $n \geq 1$, with $\sup_{n\in\mathbb{N}} \|T_n\| < \infty$. Let $1 \leq p < \infty$. Then $\bigoplus_{n=1}^{\infty} T_n$ is mixing on $(\bigoplus_{n=1}^{\infty} X_n)_{\ell^p}$ if and only if each operator T_n, $n \geq 1$, is mixing.*
The same result holds for the direct c_0-sum.

Proof. For the necessity part one need only note that each T_n, $n \geq 1$, is quasiconjugate to $\bigoplus_{k=1}^{\infty} T_k$ via the map $\phi : (x_k)_k \to x_n$.

Now suppose that each T_n, $n \geq 1$, is mixing. Let $U, V \subset (\bigoplus_{n=1}^{\infty} X_n)_{\ell^p}$ be nonempty open sets. It follows from the definition of the norm on this space that there are $\varepsilon > 0$, $m \geq 1$, and points $x := (x_1, \ldots, x_m, 0, 0, \ldots) \in U$ and $y := (y_1, \ldots, y_m, 0, 0, \ldots) \in V$ such that the open balls of radius ε around these points belong to U and V, respectively. Since each T_k is mixing there is some $N \geq 1$ such that, for each $1 \leq k \leq m$ and $n \geq N$, there are $x_k^{(n)} \in X_k$ such that $\|x_k^{(n)} - x_k\| < \varepsilon/m^{1/p}$ and $\|T_k^n x_k^{(n)} - y_k\| < \varepsilon/m^{1/p}$. Then, for all $n \geq N$, $x^{(n)} := (x_1^{(n)}, \ldots, x_m^{(n)}, 0, 0, \ldots) \in U$ and $(\bigoplus_{k=1}^{\infty} T_k)^n x^{(n)} \in V$, which implies that $\bigoplus_{k=1}^{\infty} T_k$ is mixing. The proof for direct c_0-sums is similar. □

2.5 Weakly mixing operators

In our present context, an operator $T : X \to X$ is weakly mixing if and only if $T \oplus T$ is hypercyclic, and if and only if, for any nonempty open subsets U_1, U_2, V_1 and V_2 of X, $N(U_1, V_1) \cap N(U_2, V_2) \neq \varnothing$.

Observation 2.42. *For any linear dynamical system,*

$$\textit{mixing} \implies \textit{weak mixing} \implies \textit{hypercyclicity}.$$

The study of specific hypercyclic operators in Chapter 4 will lead to many simple examples of weakly mixing, non-mixing operators; see Remark 4.10.

In contrast, the strictness of the second implication turned out to be much more delicate and was posed as an open problem by D. Herrero in 1992. The problem has only recently been solved.

Theorem 2.43 (De la Rosa–Read). *There are hypercyclic operators on Banach spaces that are not weakly mixing.*

Refining the techniques of De la Rosa and Read, Bayart and Matheron have shown that such operators even exist on any of the spaces $\ell^p, 1 \leq p < \infty$, and c_0, in particular on Hilbert spaces. The proof is, however, beyond the scope of this book.

The answer to Herrero's question raises the problem of finding (weak) conditions on a hypercyclic operator to be weakly mixing.

To this end we first derive a useful property of hypercyclic operators involving open sets and 0-neighbourhoods.

Lemma 2.44. *Let T be a hypercyclic operator. Then, for any nonempty open sets U and V in X and any 0-neighbourhood W, there is a nonempty open set $U_1 \subset U$ and a 0-neighbourhood $W_1 \subset W$ such that*

$$N(U_1, W_1) \subset N(V, W) \quad \textit{and} \quad N(W_1, U_1) \subset N(W, V).$$

Proof. Using topological transitivity and continuity of T one finds $m \in \mathbb{N}_0$, a nonempty open set $U_1 \subset U$ and a 0-neighbourhood $W_1 \subset W$ such that $T^m(U_1) \subset V$ and $T^m(W_1) \subset W$. Now, if $n \in N(U_1, W_1)$, then there exists some $x \in U_1$ with $T^n x \in W_1$. It follows that $T^n T^m x = T^m T^n x \in W$, so that $n \in N(V, W)$. In the same way we also obtain that $N(W_1, U_1) \subset N(W, V)$. □

This proof copies our proof of the 4-set trick (see Lemma 1.50); a direct application of that trick would not have given us that W_1 contains 0.

Theorem 2.45. *Let T be a hypercyclic operator. If, for any nonempty open set $U \subset X$ and any 0-neighbourhood W, there is a continuous map $S : X \to X$ commuting with T such that*

$$S(U) \cap W \neq \varnothing \textit{ and } S(W) \cap U \neq \varnothing, \tag{2.5}$$

then T is weakly mixing.

Proof. First, the 4-set trick and topological transitivity of T yield that, for any nonempty open set $U \subset X$ and for any 0-neighbourhood W,

$$N(U, W) \cap N(W, U) \neq \varnothing.$$

By Proposition 1.53 it suffices to show that, given any pair U, V of nonempty open subsets of X, there is $n \in N(U, U) \cap N(U, V)$. To do this, using Lemma 2.36, we fix nonempty open sets $U_1 \subset U$, $V_1 \subset V$ and a 0-neighbourhood W_1 such that $U_1 + W_1 \subset U$ and $V_1 + W_1 \subset V$. Lemma 2.44 implies the existence of a 0-neighbourhood $W_2 \subset W_1$ and a nonempty open set $U_2 \subset U_1$ such that $N(W_2, U_2) \subset N(W_1, V_1)$. We fix $n \in N(U_2, W_2) \cap N(W_2, U_2)$; then there are $u_2 \in U_2$ with $T^n u_2 \in W_2$, $w_1 \in W_1$ with $T^n w_1 \in V_1$, and $w_2 \in W_2$ with $T^n w_2 \in U_2$. If we set $u_3 = u_2 + w_2 \in U$ and $u_4 = u_2 + w_1 \in U$, then we obtain that $T^n u_3 \in W_2 + U_2 \subset U$ and $T^n u_4 \in W_2 + V_1 \subset V$. That is, $n \in N(U, U) \cap N(U, V)$. □

An operator $T : X \to X$ is called *flip transitive* if, for any pair U, V of nonempty open subsets of X,

$$N(U,V)\cap N(V,U)\neq \varnothing;$$

see also Exercises 1.5.4 and 1.5.5. By the Birkhoff transitivity theorem, such an operator is hypercyclic. It follows from the previous result that even more is true.

Corollary 2.46. *Every flip transitive operator is weakly mixing.*

As another consequence we obtain a useful characterization of the weak mixing property; this result should also be compared with Proposition 2.37.

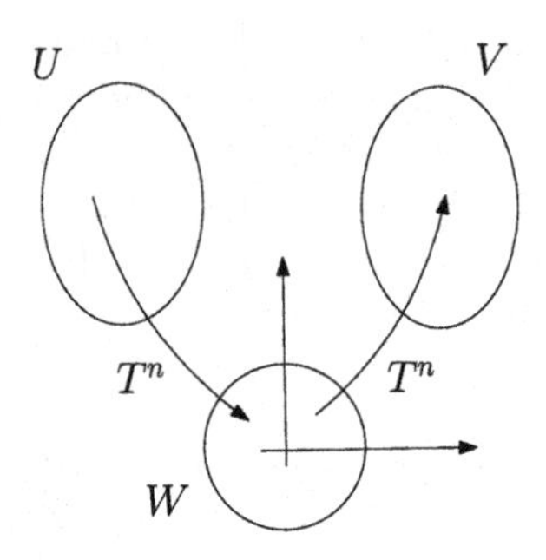

Fig. 2.1 Weak mixing (Theorem 2.47)

Theorem 2.47. *An operator T is weakly mixing if and only if, for any nonempty open sets $U,V\subset X$ and any 0-neighbourhood W,*

$$N(U,W)\cap N(W,V)\neq \varnothing.$$

Proof. It suffices to show sufficiency. As in the proof of Proposition 2.37 the condition implies that T is topologically transitive, hence hypercyclic. Then an application of Theorem 2.45 yields the result. □

Theorem 2.45 provides us with a rather weak condition, in terms of open sets and 0-neighbourhoods, for making a hypercyclic operator weakly mixing. As an application we can show that weak mixing is already implied by the existence of a dense set of points with tame orbits. For the precise formulation we need the notion of bounded sets in Fréchet spaces; see Appendix A.

Theorem 2.48. *Let T be a hypercyclic operator. If there exists a dense subset X_0 of X such that the orbit of each $x\in X_0$ is bounded, then T is weakly mixing.*

Proof. Let U be a nonempty open subset of X and W a 0-neighbourhood. If $(p_n)_n$ is an increasing sequence of seminorms defining the topology of X then there is some $k\in\mathbb{N}$ and some $\varepsilon>0$ such that $p_k(x)<\varepsilon$ implies that

$x \in W$. Now, by assumption, we can find a point $x \in X_0 \cap U$; it follows that $M := \sup_{n\in\mathbb{N}_0} p_k(T^n x) < \infty$. Hence $\frac{\varepsilon}{2M}T^n x \in W$, for all $n \in \mathbb{N}_0$. On the other hand, by topological transitivity of T, there is some $n \in \mathbb{N}_0$ with $\left(\frac{\varepsilon}{2M}T^n(W)\right) \cap U = T^n(\frac{\varepsilon}{2M}W) \cap U \neq \varnothing$. Thus, condition (2.5) is satisfied for $S = \frac{\varepsilon}{2M}T^n$. □

Of course, every periodic point has bounded orbit, as does every point whose orbit converges. The latter holds, a fortiori, for all points from the *generalized kernel*

$$\bigcup_{n=0}^{\infty} \ker T^n$$

of T. Thus we have the following.

Corollary 2.49. *Any of the following operators are weakly mixing:*
(i) *chaotic operators;*
(ii) *hypercyclic operators that have a dense set of points for which the orbits converge;*
(iii) *hypercyclic operators with dense generalized kernel.*

A typical class of operators with dense generalized kernel is the class of unilateral weighted shifts that will be studied in detail in Section 4.1.

For an additional characterization of weakly mixing operators in terms of multiples of the iterates of T we refer to Theorem 12.29.

We end this section with an application to complexifications.

Proposition 2.50. *Let $T : X \to X$ be an operator. Then:*
(i) *$T \oplus T$ is weakly mixing if and only if T is;*
(ii) *$T \oplus T$ is chaotic if and only if T is.*

Proof. Suppose that T is chaotic. Then, by Corollary 2.49, $T \oplus T$ is hypercyclic. Moreover, the set of all points $(x, y) \in X \oplus X$ with periodic points x and y for T provides a dense set of periodic points for $T \oplus T$. Thus $T \oplus T$ is chaotic. The remaining implications are special cases of Propositions 1.42 and 1.48 and Theorem 1.51. □

More generally, an arbitrary direct sum $S \oplus T$ is chaotic if and only if both S and T are; see Exercise 2.5.7. In contrast, assertion (i) cannot be generalized in the same way; see Remark 4.17.

The discussion before Proposition 2.26, together with Proposition 2.40, yields the following.

Corollary 2.51. *An operator T on a real separable Fréchet space is mixing, weakly mixing or chaotic, respectively, if and only if its complexification $\widetilde{T}$ is.*

This result can be applied, for instance, to Rolewicz's operators, just as in Example 2.27; see also Example 3.2.

For extensions of the results of the last two sections to sequences of operators we refer to Section 3.4.

2.6 The set of hypercyclic vectors

A natural question that arises in hypercyclicity is this: which kind of structures can we find in the set of hypercyclic vectors? By the Birkhoff transitivity theorem we already know that the set $HC(T)$ of hypercyclic vectors of a hypercyclic operator T is always a dense G_δ-set. Almost trivially this leads to a somewhat surprising representation result.

Proposition 2.52. *Let T be a hypercyclic operator on X. Then*

$$X = HC(T) + HC(T),$$

that is, every vector $x \in X$ can be written as the sum of two hypercyclic vectors.

Proof. Let $x \in X$. Since both $HC(T)$ and $x - HC(T)$ are dense G_δ-sets, their intersection must be nonempty by the Baire category theorem, which implies that $x \in HC(T) + HC(T)$. □

As a consequence, the set $HC(T)$ of hypercyclic vectors can only then be a linear subspace, except for the zero vector, if any nonzero vector is hypercyclic, in which case the operator has no nontrivial invariant closed subset; see Observation 2.17. Such an operator exists, for example, on ℓ^1, but the construction is highly nontrivial.

Weakening the requirement, it is natural to ask if, for a general hypercyclic operator T, $HC(T)$ contains a large linear subspace, except for 0. In this section we will interpret largeness as being dense. A different sense of largeness will be studied in Chapter 10.

We first need some auxiliary results that are also important in their own right. For the definition of the adjoint of an operator and the notation $\langle x, x^* \rangle$ we refer to Appendix A.

Lemma 2.53. (a) *Let T be a hypercyclic operator. Then its adjoint T^* has no eigenvalues. Equivalently, every operator $T - \lambda I$, $\lambda \in \mathbb{K}$, has dense range.*

(b) *Let T be a hypercyclic operator on a real separable Fréchet space. Then the adjoint $\widetilde{T}^*$ of its complexification $\widetilde{T}$ has no eigenvalues. Equivalently, every operator $\widetilde{T} - \lambda I$, $\lambda \in \mathbb{C}$, has dense range.*

Proof. (a) Let $x \in X$ be a hypercyclic vector for T. Suppose, by way of contradiction, that T^* has an eigenvalue λ, that is,

$$T^* x^* = \lambda x^*$$

for some $x^* \in X^*$, $x^* \neq 0$. Then we have that, for any $n \geq 0$,

$$\langle T^n x, x^* \rangle = \langle x, (T^*)^n x^* \rangle = \lambda^n \langle x, x^* \rangle.$$

Since $x^* \neq 0$, the hypercyclicity of x implies that the left-hand side is dense in $\mathbb{K}$, while the right-hand side clearly is not, which is the desired contradiction.

Moreover, by the Hahn–Banach theorem (see Appendix A), $T - \lambda I$ has dense range precisely when

$$\langle x, T^*x^* - \lambda x^* \rangle = \langle (T - \lambda I)x, x^* \rangle = 0 \text{ for all } x \in X$$

entails that $x^* = 0$, which is equivalent to λ not being an eigenvalue of T^*.

(b) Now let X be a space over the real scalar field, and let $\widetilde{T}$ be the complexification of T. Let $x \in X$ be hypercyclic for T, and suppose that $\widetilde{T}^*$ has an eigenvector $\widetilde{x}^* \in \widetilde{X}^*$, $\widetilde{x}^* \neq 0$, to an eigenvalue λ. Then we have that

$$|\langle T^n x, \widetilde{x}^* \rangle| = |\langle \widetilde{T}^n x, \widetilde{x}^* \rangle| = |\langle x, (\widetilde{T}^*)^n x^* \rangle| = |\lambda|^n |\langle x, \widetilde{x}^* \rangle|, \qquad (2.6)$$

for $n \geq 0$. Since $\langle x_1 + ix_2, \widetilde{x}^* \rangle = \langle x_1, \widetilde{x}^* \rangle + i\langle x_2, \widetilde{x}^* \rangle$ for all $x_1, x_2 \in X$, and since $\widetilde{x}^* \neq 0$, there is some $y \in X$ such that $|\langle y, \widetilde{x}^* \rangle| > 0$. Hence $|\widetilde{x}^*|$ can take every positive value on X. By the hypercyclicity of x, the left-hand side of (2.6) is dense in $\mathbb{R}_+$, while the right-hand side clearly is not, which is a contradiction. The remainder of the proof can be given as in (a). □

As a consequence we obtain one of the cornerstones of the theory of linear dynamical systems. We recall that for any polynomial $p(z) = \sum_{n=0}^{N} a_n z^n$ the operator $p(T)$ is defined as $p(T) = \sum_{n=0}^{N} a_n T^n$.

Theorem 2.54 (Bourdon). *If T is a hypercyclic operator and p is a nonzero polynomial, then the operator $p(T)$ has dense range.*

Proof (complex case). We can assume that $p(z) = \sum_{n=0}^{N} a_n z^n$ with $a_N \neq 0$, $N \geq 1$. For spaces X over the complex field the result follows immediately from Lemma 2.53(a) and the fact that p can be written as a product of linear factors, so that

$$p(T) = a_N (T - \lambda_1 I) \cdots (T - \lambda_N I)$$

with certain $\lambda_k \in \mathbb{C}$, $k = 1, \ldots, N$.

(Real case). If X is a real space we consider the complexification $\widetilde{T}$ of T. With Lemma 2.53(b), it follows as in the complex case that, for any complex polynomial p, $p(\widetilde{T})$ has dense range on $\widetilde{X}$. Now if p has real coefficients, then

$$p(\widetilde{T})(x + iy) = p(T)x + ip(T)y, \quad x, y \in X,$$

which implies that also $p(T) : X \to X$ has dense range. □

We are now ready to deduce an important result on the algebraic structure of the set of hypercyclic vectors.

Theorem 2.55 (Herrero–Bourdon). *If x is a hypercyclic vector for T, then*

$$\{p(T)x \ ; \ p \text{ is a polynomial}\} \setminus \{0\}$$

is a dense set of hypercyclic vectors.

In particular, any hypercyclic operator admits a dense invariant subspace consisting, except for zero, of hypercyclic vectors.

Proof. Let $x \in X$ be a hypercyclic vector for T. Then

$$M = \{p(T)x \; ; \; p \text{ is a polynomial}\} = \operatorname{span} \operatorname{orb}(x, T)$$

is a dense T-invariant subspace of X. Moreover, if $y = p(T)x \in M \setminus \{0\}$ then $p \neq 0$ and

$$T^n y = p(T)(T^n x), \quad n \in \mathbb{N}_0.$$

Since x is hypercyclic and, by Theorem 2.54, $p(T)$ has dense range, also y has dense orbit under T. □

The Herrero–Bourdon theorem allows us to deduce an additional topological structure of the set of hypercyclic vectors: it is always a connected set. This observation comes from the fact that, if $A \subset B \subset \overline{A} \subset X$ and A is connected, then also B is connected. We apply this to $A = M \setminus \{0\}$, and $B = HC(T)$, where M is the dense subspace of the Herrero–Bourdon theorem. Note that M is of dimension greater than 1 because, otherwise, x would be an eigenvector, which is not hypercyclic; hence $A = M \setminus \{0\}$ is connected.

Corollary 2.56. *The set $HC(T)$ of hypercyclic vectors for a hypercyclic operator T is a connected subset of X.*

2.7 Linear vs nonlinear maps, and finite vs infinite dimension

We have seen that chaotic linear operators exist. This is contrary to the common belief that (deterministic) chaos is necessarily connected to the nonlinearity of a system. In this section we want to explore the connection between linear and nonlinear chaos.

The dynamics of linear operators on a finite-dimensional space $X = \mathbb{K}^N$ are easy to describe, thanks to the Jordan decomposition theorem. We assume that $\mathbb{K}^N$ is endowed with the Euclidean norm.

Proposition 2.57. *Let T be a linear operator on $\mathbb{K}^N$, $N \geq 1$. Then, for any $x \in \mathbb{K}^N$, either $T^n x \to 0$ or $\|T^n x\| \to \infty$ or there are $m, M > 0$ such that $m \leq \|T^n x\| \leq M$ for all $n \geq 0$.*

Proof. Since every operator $T : \mathbb{R}^N \to \mathbb{R}^N$ can be regarded as an operator on $\mathbb{C}^N$ it suffices to consider the complex case.

By the Jordan decomposition theorem, $\mathbb{C}^N$ has a basis with respect to which the matrix of T is in Jordan block form. Since all norms on $\mathbb{C}^N$ are equivalent we can assume that this basis is the canonical basis of $\mathbb{C}^N$, and it suffices to show the result for each operator given by a Jordan block

$$T = \begin{pmatrix} \lambda & 1 & 0 & \ldots\ldots \\ 0 & \lambda & 1 & 0 \ \ldots \\ & \ddots & \ddots & \ddots \ \vdots \\ & & 0 & \lambda \ \ 1 \\ & & & 0 \ \ \lambda \end{pmatrix} : \mathbb{C}^N \to \mathbb{C}^N,$$

with $N \geq 1$. For $n \geq N-1$ we have that

$$T^n = \begin{pmatrix} \lambda^n & n\lambda^{n-1} & \binom{n}{2}\lambda^{n-2} & \ldots & \ldots & \binom{n}{N-1}\lambda^{n-N+1} \\ 0 & \lambda^n & n\lambda^{n-1} & \ldots & \ldots & \binom{n}{N-2}\lambda^{n-N+2} \\ & \ddots & \ddots & \ddots & \ldots & \vdots \\ & & 0 & \lambda^n & n\lambda^{n-1} & \binom{n}{2}\lambda^{n-2} \\ & & & 0 & \lambda^n & n\lambda^{n-1} \\ & & & & 0 & \lambda^n \end{pmatrix}.$$

We apply T^n to the vector $x = (x_1, \ldots, x_N) \in \mathbb{C}^N$, $x \neq 0$.

Case 1: $|\lambda| > 1$. Let x_k be the last nonzero entry. Then the kth entry of $T^n x$ is $\lambda^n x_k$, hence $\|T^n x\| \to \infty$.

Case 2: $|\lambda| < 1$. Then all entries of $T^n x$ tend to zero, hence $T^n x \to 0$.

Case 3: $|\lambda| = 1$. If $x = (x_1, 0, \ldots, 0)$ then $\|T^n x\| = \|\lambda^n x\| = \|x\|$ for $n \geq 0$. Otherwise, the first entry of $T^n x$ is

$$\lambda^n x_1 + n\lambda^{n-1} x_2 + \ldots + \binom{n}{N-1}\lambda^{n-N+1} x_N,$$

which tends to infinity in absolute value; hence, again, $\|T^n x\| \to \infty$. □

As an immediate consequence we obtain the following.

Theorem 2.58. *There are no hypercyclic operators on* $\mathbb{K}^N, N \geq 1$.

Of course, this also follows directly from Lemma 2.53 since every operator on $\mathbb{C}^N$ has an eigenvalue. Further proofs of this result are suggested in Exercises 2.7.1 and 2.7.2.

Since every finite-dimensional Fréchet space is isomorphic to some $\mathbb{K}^N, N \geq 1$ (see Appendix A), the theorem extends to such spaces.

Corollary 2.59. *There are no hypercyclic operators on a finite-dimensional Fréchet space.*

The result also implies an interesting property of the orbit of a hypercyclic vector.

Proposition 2.60. *The orbit of any hypercyclic vector forms a linearly independent set.*

Proof. Let x be a hypercyclic vector for T and suppose that there are scalars $\alpha_k \in \mathbb{K}$, $k = 0, \ldots, N$, such that

$$T^{N+1}x = \sum_{k=0}^{N} \alpha_k T^k x.$$

Then $F := \operatorname{span}\{T^k x \; ; \; k = 0, \ldots, N\}$ is a finite-dimensional T-invariant subspace of X. Since x is hypercyclic for T, it is also hypercyclic for $T|_F : F \to F$, which contradicts Corollary 2.59. □

Alternatively, the proof can be based on Bourdon's theorem; see Exercise 2.7.3.

Proposition 2.57 tells us that the dynamics of linear maps on finite-dimensional spaces are quite restrictive. On the other hand, in an infinite-dimensional setting, linear dynamics can be arbitrarily complicated. Indeed, as we now show, every continuous map on a compact metric space is conjugate to the restriction of a linear operator on some invariant set. Even more strikingly, the same operator can be taken for all nonlinear systems, and the operator is even chaotic. In other words: the dynamics of any (compact) nonlinear dynamical system can be described by the dynamics of a single chaotic operator.

Theorem 2.61. *There exists a chaotic operator T on a separable Hilbert space H with the following property.*

For any continuous map f on any compact metric space K there exists a T-invariant subset L of H such that f is conjugate to the restriction $T|_L$ of T to L.

In other words, there is a homeomorphism $\phi : K \to L$ so that the diagram

$$\begin{array}{ccc} K & \xrightarrow{\;f\;} & K \\ \phi\big\downarrow & & \big\downarrow\phi \\ L & \xrightarrow{\;T|_L\;} & L \end{array}$$

commutes.

Proof. Let $H = (\bigoplus_{n=0}^{\infty} \ell^2)_{\ell^2}$ be the space of all sequences $x = (x_n)_{n\geq 0}$ of elements $x_n = (x_{n,k})_{k\geq 0}$ in ℓ^2 such that

$$\|x\| := \Big(\sum_{n=0}^{\infty} \|x_n\|^2\Big)^{1/2} < \infty;$$

see the discussion before Proposition 2.41. Then H is a separable Hilbert space when endowed with the canonical inner product. On H we consider the multiple $T = 2B$ of the backward shift B given by

$$B(x_0, x_1, x_2, \ldots) = (x_1, x_2, x_3, \ldots).$$

The proof that T has the desired properties will be split into three steps.

Step 1. We define a suitable embedding $\phi : K \to H$.

First, we can assume that the metric d on K is bounded by 1, since otherwise we can replace it by the equivalent metric $d'(x, y) = \min(1, d(x, y))$. We then fix a dense sequence $(y_k)_{k \geq 0}$ in K; this is possible because K is a compact metric space. Based on this sequence we define, for any $x \in K$,

$$\phi(x) = \left(\left(\frac{1}{2^{k+n}} d(y_k, f^n(x))\right)_k\right)_n \in H.$$

Then we have for any $x, y \in K$ and $N \geq 0$ that

$$\begin{aligned}\|\phi(x) - \phi(y)\|^2 &= \sum_{k,n=0}^{\infty} \frac{1}{2^{2(k+n)}} |d(y_k, f^n(x)) - d(y_k, f^n(y))|^2 \\ &\leq \sum_{k,n \leq N} \frac{1}{2^{2(k+n)}} |d(y_k, f^n(x)) - d(y_k, f^n(y))|^2 + \sum_{\substack{k>N \text{ or} \\ n>N}} \frac{4}{2^{2(k+n)}}.\end{aligned}$$

This can be made arbitrarily small by first choosing N sufficiently large and then y sufficiently close to x. Thus $\phi : K \to H$ is continuous.

Moreover, ϕ is injective; indeed, $\phi(x) = \phi(y)$ implies that

$$\frac{1}{2^k} d(y_k, x) = \frac{1}{2^k} d(y_k, y) \quad \text{for all } k \geq 0,$$

hence $x = y$ by density of the y_k.

As a continuous injection on a compact space, ϕ is a homeomorphism onto a (compact) subset, L say, of H.

Step 2. We show that $\phi \circ f = T \circ \phi$ and that L is invariant under T.

In fact, for any $x \in K$ we have that

$$\begin{aligned}\phi(f(x)) &= \left(\left(\frac{1}{2^{k+n}} d(y_k, f^{n+1}(x))\right)_k\right)_n \\ &= 2\left(\left(\frac{1}{2^{k+n+1}} d(y_k, f^{n+1}(x))\right)_k\right)_n = T(\phi(x)),\end{aligned}$$

which also implies that

$$T(L) = T(\phi(K)) = \phi(f(K)) \subset \phi(K) = L.$$

Step 3. The operator T is chaotic on H.

The proof is the same as the proof that Rolewicz's operators are chaotic; see Examples 2.22 and 2.32. □

In summary we have found that

- *linear chaos exists;*
- *linear chaos is an infinite-dimensional phenomenon;*
- *linear dynamics can be as complicated as nonlinear dynamics.*

2.8 Hypercyclicity and complex dynamics

In this section we present a connection between the hypercyclicity of weighted backward shifts and the Julia sets of polynomials in one complex variable. In some sense we continue the theme of the previous section by comparing infinite-dimensional dynamics with nonlinear but finite-dimensional dynamics.

Let us first give a brief introduction to complex dynamics. Given a polynomial p in one complex variable, of degree $m \geq 2$, a periodic point z of p is called *repelling* if $|p'(z)| > 1$. The *Julia set* of p can be defined as

$$\mathcal{J}(p) = \overline{\{z \in \mathbb{C}\ ;\ z \text{ is a repelling periodic point of } p\}}.$$

While this is not the common definition of the Julia set, which is more complicated and involves the behaviour of the iterates of p near $\mathcal{J}(p)$, a result by Fatou and Julia tells us that the two definitions are equivalent.

For instance, for the doubling map $p(z) = z^2$ on $\mathbb{C}$ (see Example 1.37), the periodic points are given by $e^{2\pi i\alpha}$ with $\alpha = \frac{k}{2^n-1}$, $n, k \in \mathbb{N}$, and all of them are repelling, so that $\mathcal{J}(p) = \mathbb{T}$. We have also seen that $p|_{\mathbb{T}}$ is chaotic.

This feature is shared by any complex polynomial of degree $m \geq 2$. More precisely one always has that $\mathcal{J}(p)$ is a compact p-invariant set such that $p|_{\mathcal{J}(p)}$ is chaotic.

The dynamical behaviour of p near the Julia set consists in spreading points. Indeed, the following property, which can be regarded as a multi-point approximation by the iterates of p on points near $\mathcal{J}(p)$, characterizes the Julia set: a point $z \in \mathbb{C}$ belongs to $\mathcal{J}(p)$ if and only if

$$\begin{aligned} &\forall \varepsilon > 0,\ \forall z_1, \ldots, z_k \in \mathbb{C},\ \exists z'_1, \ldots, z'_k \in \mathbb{C},\ \exists n \in \mathbb{N} \text{ such that} \\ &\quad |z'_j - z| < \varepsilon \ \text{ and } \ |p^n(z'_j) - z_j| < \varepsilon, \quad j = 1, \ldots, k. \end{aligned} \tag{2.7}$$

Now, one can deduce this property, for certain polynomials and at the point 0, from the hypercyclic behaviour of Rolewicz's operators.

Example 2.62. Let $p(z) = (z+1)^m - 1$, where $m \geq 2$. We consider the following commutative diagram,

$$\begin{array}{ccc} c_0 & \xrightarrow{mB} & c_0 \\ \phi \downarrow & & \downarrow \phi \\ c_0 & \xrightarrow{P} & c_0, \end{array}$$

where mB is the Rolewicz operator with $\lambda = m$ on the complex space c_0, $P(x_1, x_2, \dots) = (p(x_2), p(x_3), \dots)$, and $\phi(x_1, x_2, \dots) = (e^{x_1} - 1, e^{x_2} - 1, \dots)$.

It is easy to see that P and ϕ are continuous maps and that ϕ has dense range. By Proposition 1.19, it follows from the hypercyclicity of Rolewicz's operators that P has a dense orbit.

In order to verify condition (2.7) at $z = 0$, we fix $\varepsilon > 0$ and arbitrary $z_j \in \mathbb{C}$, $j = 1, \dots k$. Since P has a dense orbit we can find $w \in c_0$ and $n \geq 0$ such that

$$\|w\| < \varepsilon \quad \text{and} \quad \|P^n w - (z_1, \dots, z_k, 0, 0, \dots)\| < \varepsilon.$$

By considering the $(j+n)$th coordinates w_{j+n} of w, we deduce that

$$|w_{j+n}| < \varepsilon \quad \text{and} \quad |p^n(w_{j+n}) - z_j| < \varepsilon, \quad j = 1, \dots, k.$$

This shows that condition (2.7) holds for $z = 0$.

Let us mention that the map P in this example is a polynomial on c_0; see Exercise 2.8.1.

More generally, we have the following connection between infinite-dimensional dynamics and the dynamics of arbitrary complex polynomials; its proof is left to the reader: see Exercise 2.8.2.

Proposition 2.63. *Let p be a complex polynomial of degree $m \geq 2$ such that $p(0) = 0$. Let $P : c_0 \to c_0$ be the continuous map given by $P(x_1, x_2, \dots) := (p(x_2), p(x_3), \dots)$. Then P is topologically transitive if and only if 0 belongs to the Julia set $\mathcal{J}(p)$ of p.*

Exercises

Exercise 2.1.1. Let $(p_n)_n$ be a separating increasing sequence of seminorms on a vector space X. Show that the map d defined in (2.1) is a metric on X. Moreover, show that for any $x, y \in X$ and $n \in \mathbb{N}$:

(i) if $p_n(x-y) < \frac{1}{2^n}$ then $d(x,y) < \frac{2}{2^n}$;

(ii) if $d(x,y) < \frac{1}{2^{2n}}$ then $p_k(x-y) < \frac{1}{2^n}$ for all $k \leq n$.

Exercise 2.1.2. Prove Lemma 2.6. (*Hint*: Use the previous exercise.)

Exercise 2.1.3. Show that the spaces $H(\mathbb{C})$ and ω are Fréchet spaces. Show that for every separable Fréchet space X, $X^{\mathbb{N}}$ is also a separable Fréchet space.

Exercise 2.1.4. Prove Proposition 2.8.

Exercise 2.1.5. Let $X = C^\infty(\mathbb{R})$ be the space of infinitely differentiable (real or complex) functions $f : \mathbb{R} \to \mathbb{K}$, endowed with the seminorms

$$p_n(f) = \max_{0\le k\le n} \sup_{|x|\le n} \left| f^{(k)}(x) \right|.$$

Show that X is a separable Fréchet space. (*Hint*: Use the Weierstrass approximation theorem to approximate $f^{(n)}$ on $[-n, n]$ by a polynomial.)

Exercise 2.1.6. Let $1 \le p < \infty$. Let $v : \mathbb{R}_+ \to \mathbb{R}$ be a strictly positive continuous function. We define the space of weighted p-integrable functions by

$$X = L^p_v(\mathbb{R}_+) = \{f : \mathbb{R}_+ \to \mathbb{K} \ ; \ f \text{ is measurable and } \|f\| < \infty\},$$

where $\|f\| := (\int_0^\infty |f(x)|^p v(x)dx)^{1/p}$. Then X is a separable Banach space. Show that the translation operator $T : X \to X$, $(Tf)(x) = f(x + 1)$, is a well-defined operator if and only if

$$\sup_{x\in\mathbb{R}_+} \frac{v(x)}{v(x+1)} < \infty.$$

Exercise 2.1.7. Let X be a Fréchet space with defining increasing sequence of seminorms $(p_n)_n$. Then a seminorm $p : X \to \mathbb{R}$ is continuous if and only if there are $n \ge 1$ and $M > 0$ such that

$$p(x) \le M p_n(x), \quad x \in X.$$

Show that this immediately implies the nontrivial part of Proposition 2.11.

Exercise 2.2.1. Let $X = C_0(\mathbb{R}_+) = \{f : \mathbb{R}_+ \to \mathbb{R} \ ; \ f \text{ is continuous and } \lim_{x\to\infty} f(x) = 0\}$, endowed with the sup-norm $\|f\| = \sup_{x\in\mathbb{R}_+} |f(x)|$. Given $a > 0$ and $\lambda > 1$, consider the operator $T : X \to X$, $(Tf)(x) = \lambda f(x + a)$, $x \in \mathbb{R}_+$. Show that T is hypercyclic. (*Hint*: Use the fact that the continuous functions with compact support form a dense subset of X.)

Exercise 2.2.2. Let X be a Banach space. Show that there are no operators T on X for which λT is hypercyclic for all $\lambda \neq 0$. (*Hint*: Consider $|\lambda| \le 1/\|T\|$.)

Exercise 2.2.3. Given any sequence of nonzero scalars $(w_n)_n$, we define the operator $B_w : \omega \to \omega$, $(x_1, x_2, x_3, \dots) \to (w_2x_2, w_3x_3, w_4x_4, \dots)$. Prove that B_w is hypercyclic.

Exercise 2.2.4. Let T be the translation operator on the space $L^p_v(\mathbb{R}_+)$ defined in Exercise 2.1.6. We assume that there are constants $M \ge 1$ and $w \in \mathbb{R}$ such that

$$v(x) \le M e^{w(y-x)} v(y) \ \text{ whenever } y \ge x \ge 0;$$

in this case, v is called an *admissible weight function*. Show that T is hypercyclic if and only if $\liminf_{x\to\infty} v(x) = 0$. (*Hint*: If the condition does not hold then define $g(x) = v(x)^{-1/p}$ on $[0, 1]$, and 0 otherwise, and show that, if $\|f\|$ is small enough, then $\|T^n f - g\| \ge \frac{1}{2}$ for $n \ge 1$. For the sufficiency, use the density of the continuous functions of compact support.)

Exercise 2.2.5. Let $X = C^\infty(\mathbb{R})$ be the space of infinitely differentiable real functions $f : \mathbb{R} \to \mathbb{R}$; see Exercise 2.1.5. Show that the (real) differentiation operator $D : X \to X$, $f \to f'$ is hypercyclic by defining a suitable quasiconjugacy $\phi : H(\mathbb{C}) \to C^\infty(\mathbb{R})$.

Exercise 2.2.6. Let $T : X \to X$ be an operator. Suppose that $Y \subset X$ is a T-invariant dense subspace of X. Furthermore, suppose that Y carries a Fréchet space topology such that the embedding $Y \to X$ is continuous and such that $T|_Y : Y \to Y$ is continuous. Show that T is quasiconjugate to $T|_Y$. In particular, if $T|_Y$ is hypercyclic, then so is T, and T has a hypercyclic vector belonging to Y. (*Remark*: this result is known as the *hypercyclic comparison principle*; it shows the interest of hypercyclicity on small spaces.)

Exercise 2.2.7. Show that the complexification $\widetilde{X}$ of a real separable Fréchet space X is a complex separable Fréchet space and that the complexification of a (real-linear) operator on X is a (complex-linear) operator on $\widetilde{X}$.

Exercise 2.2.8. Let $T : X \to X$ be an operator, and let M_1 and M_2 be T-invariant closed subspaces of X such that $X = M_1 \oplus M_2$; see Proposition 2.28. Let P_{M_1} be the projection $X \to M_1$, $x = x_1 + x_2 \to x_1$, where $x_1 \in M_1$, $x_2 \in M_2$, and similarly for M_2. Show that, for $j = 1, 2$, $T|_{M_j}$ is quasiconjugate to T via P_{M_j}.

Exercise 2.2.9. Let T be an operator on a Banach space X and $x \in X$. Given $d > 0$, let us a call the orbit of x under T *d-dense* if for each $y \in X$ we can find $n \in \mathbb{N}_0$ such that $\|T^n x - y\| < d$. Show that if T admits a d-dense orbit then it is hypercyclic. (*Hint*: First observe that X is separable. Then, given a vector x whose orbit is d-dense, prove that, for each $\varepsilon > 0$, the vector $\frac{\varepsilon}{d}x$ has an ε-dense orbit and conclude the result by Exercise 1.2.5 and the Birkhoff transitivity theorem).

Exercise 2.2.10. Let $\varepsilon > 0$. An operator T on a Banach space X is called *ε-hypercyclic* if it admits a vector $x \in X$ such that, for any nonzero vector $y \in X$, we can find $n \in \mathbb{N}_0$ satisfying $\|T^n x - y\| \leq \varepsilon \|y\|$; the vector x is then also called *ε-hypercyclic*. Show that if an operator is ε-hypercyclic for all $\varepsilon > 0$ then it is hypercyclic. (*Hint*: First observe that X is separable. Then conclude the result by Exercise 1.2.5 and the Birkhoff transitivity theorem).

Exercise 2.2.11. An operator T on a Fréchet space X is called a *J-class operator* if there is a vector $x \neq 0$ in X with $J(x) = X$, where $J(x)$ is the J-set of x. By Exercise 1.2.8, every hypercyclic operator is J-class.

(a) Let B be the backward shift on ℓ^2. Show that the operator $T = 2I \oplus 2B$ on $\mathbb{K} \oplus \ell^2$ is a J-class operator that is not hypercyclic. Deduce also that being J-class is not preserved under quasiconjugacy.

(b) Let T be a hypercyclic operator on X. Show that, for $\lambda \in \mathbb{K}$, the operator $\lambda I \oplus T$ is J-class on $\mathbb{K} \oplus X$ if and only if $|\lambda| > 1$. Deduce that there is an invertible J-class operator T whose inverse T^{-1} is not J-class.

(c) Show that the multiple $T = 2B$ of the backward shift is J-class on ℓ^∞. Therefore there exist J-class operators on non-separable Banach spaces. (*Hint*: Consider $(1, 0, 0, \ldots)$.)

Exercise 2.3.1. Let T be an operator on a Banach space X. Show that the following assertions are equivalent:

(i) T has sensitive dependence on initial conditions with respect to the usual metric;
(ii) $\sup_{n \geq 0} \|T^n\| = \infty$;
(iii) T has an unbounded orbit.

(*Hint*: Use the Banach–Steinhaus theorem; see Appendix A.)

Exercise 2.3.2. Let T be the translation operator on $L^p_v(\mathbb{R}_+)$ with an admissible weight function v; see Exercise 2.2.4. Show that T is chaotic if and only if $\int_0^\infty v(x)\,dx < \infty$.

Exercise 2.3.3. Let $T : \mathbb{K}^N \to \mathbb{K}^N$, $N \geq 1$, be an operator.

(a) Show that T has a dense set of periodic points if and only if $T^n = I$ for some $n \geq 1$. (*Hint*: Show that $\mathbb{K}^N$ has a basis consisting of periodic points.)

(b) Deduce from (a) that no operator on $\mathbb{K}^N$ can be chaotic.

Exercise 2.3.4. Let $H(\Omega)$ be the Fréchet space of all holomorphic functions on a domain Ω in $\mathbb{C}$; see Section 4.3. Then $D : f \to f'$ is an operator on $H(\Omega)$. Show that the following assertions are equivalent:

(i) D is chaotic on $H(\Omega)$;
(ii) D is hypercyclic on $H(\Omega)$;
(iii) Ω is simply connected.

(*Hint*: By Runge's theorem, the polynomials are dense in $H(\Omega)$ if Ω is simply connected. On the other hand, if Ω is not simply connected, approximate a suitable function $\frac{1}{z-a}$ by derivatives, and integrate both over closed curves in Ω.)

Exercise 2.4.1. Modify Example 2.39 to obtain a non-chaotic mixing operator on any space ℓ^p, $1 < p < \infty$.

Exercise 2.4.2. Let T be the translation operator on $L^p_v(\mathbb{R}_+)$ with an admissible weight function v; see Exercise 2.2.4. Show that T is mixing if and only if $\lim_{x\to\infty} v(x) = 0$.

Exercise 2.4.3. Using Proposition 2.37, show that λD is mixing on $H(\mathbb{C})$ for any $\lambda \neq 0$. (This should be contrasted with Exercise 2.2.2.)

Exercise 2.4.4. Let T be an operator on a separable Banach space. It can be shown that $T \oplus T^*$ is never hypercyclic on $X \oplus X^*$; see Remark 4.17. Deduce that if T is mixing then T^* cannot be hypercyclic. (See Exercise 5.1.1 for a better result.)

Exercise 2.5.1. Let T be a weakly mixing (or mixing) operator and $\lambda \in \mathbb{K}$, $|\lambda| = 1$. Show that λT is also weakly mixing (or mixing, respectively). (*Hint*: Use Proposition 2.37 and Theorem 2.47.)

Exercise 2.5.2. Establish Theorem 2.45 without the use of Proposition 1.53, that is, show the weak mixing property directly. (*Hint*: See Figure 2.2.)

Exercise 2.5.3. Let T be an operator such that, for any nonempty open sets $U, V \subset X$ and any 0-neighbourhood W, $N(U,W)$ is nonempty and $N(W,V)$ is syndetic. Then prove that T is weakly mixing. Do likewise if the sets $N(U,W)$ are syndetic and the sets $N(W,V)$ are nonempty. (*Hint*: Apply Theorem 2.47.)

Exercise 2.5.4. A subset A of $\mathbb{N}_0$ is called *thickly syndetic* if, for any $n \in \mathbb{N}$, there is a syndetic sequence $(n_k)_k$ of positive integers such that $\{n_k + j\ ;\ k \in \mathbb{N}, j = 0, \ldots, n\} \subset A$. Show the following:

(i) the intersection of two thickly syndetic sets is thickly syndetic;
(ii) let T be an operator and $U \subset X$ a nonempty open set; if, for any 0-neighbourhood W, $N(U,W)$ is syndetic then these sets are all thickly syndetic; and similarly for $N(W,U)$.

Deduce that, for any topologically ergodic operator T, $N(U,V)$ is thickly syndetic for any nonempty open sets $U, V \subset X$; see Exercise 1.5.6 for the definition of topological ergodicity.

Exercise 2.5.5. If $S : X \to X$ and $T : Y \to Y$ are topologically ergodic operators, show that $S \oplus T$ is topologically ergodic on $X \oplus Y$. In particular, topologically ergodic operators are weakly mixing. (*Hint*: Use the previous exercise.)

Exercise 2.5.6. An operator T is called *hereditarily ergodic* if, for any pair U, V of nonempty open subsets of X and any syndetic sequence $(n_k)_k$, there exists a subsequence $(n_{k_j})_j$ of $(n_k)_k$ that is also syndetic such that $T^{n_{k_j}}(U) \cap V \neq \varnothing$ for all $j \geq 1$. Show that an operator is hereditarily ergodic if and only if it is topologically ergodic. Deduce that every power T^p, $p \geq 1$, of a topologically ergodic operator is topologically ergodic. (*Hint*: Use Exercise 2.5.4.)

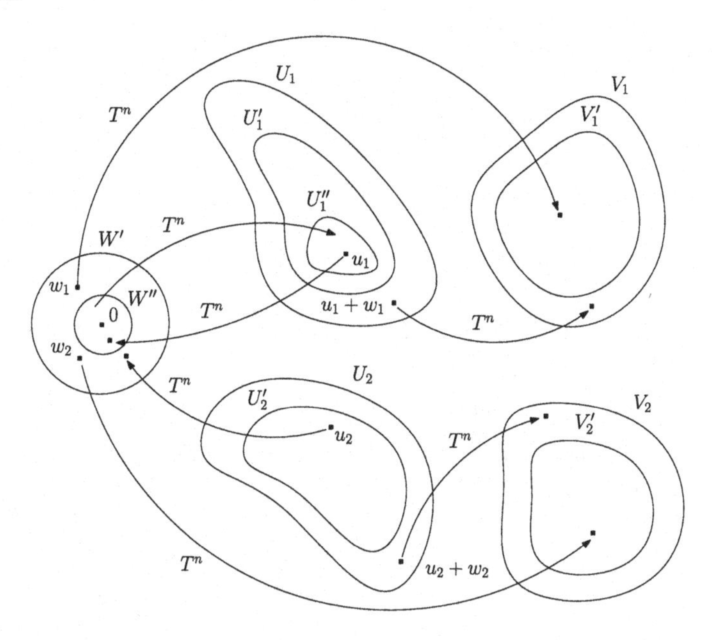

Fig. 2.2 Exercise 2.5.2

Exercise 2.5.7. Prove that the direct sum $S \oplus T$ of two chaotic operators is chaotic. (*Hint*: Use Exercises 1.5.6 and 2.5.5.)

Exercise 2.5.8. Show the following variant of Theorem 2.45: let T be an operator such that, for any nonempty open sets $U, V \subset X$ and any 0-neighbourhood W, there is a continuous map $S : X \to X$ commuting with T such that $S(U) \cap V \neq \varnothing$, $S(W) \cap W \neq \varnothing$ and

$$N(U, W) \cap N(W, U) \neq \varnothing.$$

Then T is weakly mixing. (*Hint*: Use the 4-set trick.)

Exercise 2.5.9. Let T be an operator such that $\lambda_1 T \oplus \lambda_2 T$ is hypercyclic (or mixing), where $|\lambda_1| \leq |\lambda_2|$. Show that then λT is weakly mixing (or mixing, respectively) whenever $|\lambda_1| \leq |\lambda| \leq |\lambda_2|$. (*Hint*: Use Proposition 2.37 and Theorem 2.47.)

Exercise 2.5.10. Let $\mathcal{P}$ denote the set of nonzero polynomials. If T is a hypercyclic operator such that

$$\bigcup_{p \in \mathcal{P}} \ker p(T)$$

is dense in X, then prove that T is weakly mixing. (*Hint*: Given U and W, pick $u \in U$ and $p \in \mathcal{P}$ such that $p(T)u = 0$. Since $p(T)$ has a dense range by Theorem 2.54, there is some $r > 0$ with $p(T)(rW) \cap U \neq \varnothing$. Then define $S = rp(T)$ and apply Theorem 2.45.)

Exercise 2.5.11. An operator T on X is called *upper triangular* if it admits an increasing sequence $(E_n)_n$ of invariant subspaces such that $\dim E_n = n$ for every $n \geq 1$ and X is the closed linear span of the finite-dimensional spaces E_n, $n \geq 1$. If X is a Hilbert space, T is upper triangular if and only if T has an upper triangular matrix with respect to some orthonormal basis of X. Show that every hypercyclic upper triangular operator is weakly mixing. (*Hint*: Use the previous exercise and a well-known result from linear algebra.)

Exercise 2.6.1. Show constructively, as in Example 2.18, that every vector $x \in \ell^2$ is the sum of two vectors that are hypercyclic for $T = 2B$.

Exercise 2.6.2. Let T be a hypercyclic operator on a complex space. Show that its adjoint T^* has no finite-dimensional invariant subspace.

Exercise 2.6.3. Let S and T be commuting operators on X such that T has dense range. Show that $HC(S)$ is T-invariant.

Exercise 2.6.4. (a) Let us call an operator T (on a real or complex space) 2-*hypercyclic* if there are vectors $x, y \in X$ such that

$$\{T^n x + T^m y \ ; \ n, m \geq 0\}$$

is dense in X. Show that the adjoint T^* of a 2-hypercyclic operator has no eigenvalues. (*Hint*: Proceed as in the proof of Lemma 2.53(a).)

(b) Now let T be a hypercyclic operator on a real space. Show that its complexification $\widetilde{T}$ is 2-hypercyclic, and deduce Lemma 2.53(b).

(c) Show that there are 2-hypercyclic operators that are not hypercyclic. (*Hint*: Consider $T = T_1 \oplus T_2$, where T_1 and T_2 are hypercyclic but T is not; see Remark 4.17.)

Exercise 2.6.5. Here is an alternative proof of Bourdon's theorem in the real case. Supply the details.

(a) By the Jordan decomposition theorem, any operator on $\mathbb{R}^2$ has a matrix representation in one of the forms $\left(\begin{smallmatrix} a & 0 \\ 0 & b \end{smallmatrix}\right)$, $\left(\begin{smallmatrix} a & 1 \\ 0 & a \end{smallmatrix}\right)$ or $c\left(\begin{smallmatrix} \cos\varphi & -\sin\varphi \\ \sin\varphi & \cos\varphi \end{smallmatrix}\right)$ with $a, b, c, \varphi \in \mathbb{R}$. Therefore no operator on $\mathbb{R}^2$ can be hypercyclic.

(b) Let T be hypercyclic on X. In view of Lemma 2.53, it remains to prove that every operator $p(T) = T^2 + bT + cI$ has dense range. If this is not the case there is a nonzero $x^* \in X^*$ with $((T^*)^2 + bT^* + cI)x^* = 0$. By Lemma 2.53, x^* and T^*x^* are linearly independent. Then $\phi : X \to \mathbb{R}^2$, $\phi(x) = (\langle x, x^*\rangle, \langle x, T^*x^*\rangle)$ defines a quasiconjugacy from T to an operator on $\mathbb{R}^2$, which is impossible.

Exercise 2.6.6. Let T be a hypercyclic operator on X. Show that $T \oplus I$ is supercyclic on $X \oplus \mathbb{K}$. Show that there is no subspace of $X \oplus \mathbb{K}$ of dimension 2 in which every nonzero vector is supercyclic; in particular, the Herrero–Bourdon theorem fails for supercyclicity.

Exercise 2.7.1. Let T be an operator on $\mathbb{K}^N, N \geq 1$, and let $y \in \mathbb{K}^N$ be a hypercyclic vector for T.

(a) Show that the vectors $y, Ty, T^2y, \ldots, T^{N-1}y$ form a basis for $\mathbb{K}^N$.

(b) Choose $(n_k)_k$ and $(m_k)_k$ such that $T^{n_k}y \to 0$ and $T^{m_k}y \to y$. Show that $T^{n_k}x \to 0$ and $T^{m_k}x \to x$ for all $x \in \mathbb{K}^N$. Deduce that $(\det T)^{n_k} = \det(T^{n_k}) \to 0$ and $(\det T)^{m_k} \to 1$.

(c) Based on (b), give a new proof of Theorem 2.58.

Exercise 2.7.2. Use the Cayley–Hamilton theorem to give a new proof of Theorem 2.58.

Exercise 2.7.3. Use Bourdon's theorem to give a new proof of Proposition 2.60. (*Hint*: In the notation of the stated proof, consider $T^{N+1} - \sum_{k=0}^{N} \alpha_k T^k$.)

Exercise 2.7.4. Let T be a hypercyclic operator. Show that its adjoint T^* has no finite-dimensional invariant subspace. (This complements Exercise 2.6.2.) (*Hint*: Let $M = \operatorname{span}\{x_1^*, \ldots, x_N^*\}$ be T^*-invariant; construct $K = \operatorname{span}\{x_1, \ldots, x_N\}$ such that $x_k^*(x_j) = \delta_{j,k}$; show that $\phi(x) = \sum_{k=1}^{N} x_k^*(x)x_k$ provides a quasiconjugacy between T and $\phi \circ T|_K$ on K.)

Exercise 2.7.5. Show that there are supercyclic operators on $\mathbb{R}$ and on $\mathbb{R}^2$. (*Hint*: For $\mathbb{R}^2$, use a rotation.)

Exercise 2.8.1. Let X be a Fréchet space. A map $Q : X \to X$ is called an *m-homogeneous polynomial*, $m \geq 0$, if there is a continuous multilinear map $A : X^m \to X$ such that $Q(x) = A(x, \ldots, x)$, $x \in X$ (where, for $m = 0$, Q is understood to be a constant map). A map $P : X \to X$ is called a *polynomial* if it can be written as $P = \sum_{m=0}^{N} Q_m$ with m-homogeneous polynomials Q_m. Show that the map P in Example 2.62 is a polynomial on c_0.

Exercise 2.8.2. Prove Proposition 2.63.

Sources and comments

Section 2.1. For introductory texts on functional analysis that also cover Fréchet spaces we refer to Rudin [271] and Meise and Vogt [237]. The notion of an F-norm can be found in Kalton, Peck and Roberts [212].

Section 2.2. The term "hypercyclic" for vectors with a dense orbit was apparently first used around 1986 (Beauzamy [46], [47], [48]) and then extended around 1988 to operators with a dense orbit (Bourdon, Godefroy, Shapiro [94], [165]). Supercyclic vectors were introduced by Hilden and Wallen [202] in 1973.

Beauzamy's work was motivated by the invariant subspace problem. The negative solution for non-Hilbert spaces is due to Enflo [142] and Read [265]. Subsequently, Read [266] even constructed a counterexample to the invariant subset problem.

Theorem 2.64 (Read). *There exists an operator on ℓ^1 all of whose nonzero vectors are hypercyclic.*

Both problems remain open for Hilbert spaces.

The three classical hypercyclic operators were found by Birkhoff [75] in 1929, MacLane [225] in 1952 and Rolewicz [268] in 1969. Example 2.18 reproduces Rolewicz's original proof; Birkhoff and MacLane used very similar constructions.

The first systematic studies of hypercyclicity are due to Kitai [215] in 1982 and Godefroy and Shapiro [165] in 1991. Though never published, Kitai's thesis was widely circulated. Between them, Kitai, Godefroy and Shapiro laid the foundations for what was to become the theory of linear dynamical systems: they clarified the basic concepts, provided a wealth of examples and introduced important techniques like criteria for hypercyclicity that will be discussed in the next chapter. It is difficult to overemphasize the importance of their work for the further development of linear dynamics.

Quasiconjugacies were introduced in hypercyclicity by Herrero [195], Martínez and Peris [229]; the hypercyclic comparison principle (see Exercise 2.2.6), was formulated by Shapiro [279]. Complexifications were first studied in hypercyclicity by Bès and Peris [71].

Section 2.3. Godefroy and Shapiro [165] suggested acceptance of Devaney's definition of chaos for linear operators, and they showed that the three classical operators are chaotic. Chaos in the sense of Auslander and Yorke was introduced for continuous maps on metric spaces in [18]. Proposition 2.30 is due to Godefroy and Shapiro [165]. Proposition 2.33 seems to be folklore, but see Herrero [195] and Bonet, Martínez and Peris [83].

We have taken the proof of Lemma 2.34 from Aron and Markose [15].

Sections 2.4. Proposition 2.37 appears in Grosse-Erdmann and Peris [187].

It does not seem to be easy to find a chaotic operator that is not mixing. The first such operator was constructed by Badea and Grivaux [19].

Sections 2.5. In 1992, Herrero [195] posed the problem of whether every hypercyclic operator (on a Hilbert space) is weakly mixing. De la Rosa and Read [126] constructed a counterexample in a suitable Banach space, while Bayart and Matheron [43] showed that counterexamples exist on many classical Banach spaces like ℓ^p, $1 \leq p < \infty$, c_0, $C[0,1]$ and $L^1[0,1]$, as well as on the Fréchet space $H(\mathbb{C})$.

The results of the section appear in Grosse-Erdmann and Peris [187]; see also Bayart and Matheron [44], [45] and Moothathu [245]. Theorem 2.47 is due to Bernal and Grosse-Erdmann [62] and León [219]; the characterizing condition first appeared in Godefroy and Shapiro [165]. Unlike the proofs in [62] and [219], our argument avoids the Baire category theorem, as does another proof by Yousefi and Rezaei [303].

Theorem 2.48 is due to Grivaux [172]. The fact that every chaotic operator is weakly mixing, Corollary 2.49, is due Bès and Peris [71]; it also follows directly from a result that is due to Bauer and Sigmund [32] and to Stacey (see [28]) using Ansari's theorem (see Section 6.1).

Section 2.6. Proposition 2.52 was observed, for example, by Grosse-Erdmann [177], Godefroy (see [94]) and Kahane (see [249]). Lemma 2.53(a) is due to Kitai [215], while part (b) can be found in Bonet and Peris [85]. Herzog and Lemmert [199] have characterized the hypercyclic operators on $\omega = \mathbb{C}^{\mathbb{N}}$; Conejero [108, p. 123] noted that their condition can be rephrased as $\sigma_p(T^*) = \varnothing$.

Theorem 2.65 (Herzog–Lemmert). *An operator T on $\omega = \mathbb{C}^{\mathbb{N}}$ is hypercyclic if and only if T^* has no eigenvalues.*

In the complex case, Theorem 2.54 is due to Bourdon [90]. The real case was added by Bès [66]; our proof is due to Martínez [227]. Theorem 2.55 is due to Herrero [194] and Bourdon [90].

A considerable improvement of Corollary 2.56 is due to Fathi [146] and Godefroy (see [44, p. 16]). They showed that for any hypercyclic operator T on a Fréchet space X, the set $HC(T)$ of hypercyclic vectors for T is homeomorphic to X.

Section 2.7. Kitai [215] and Rolewicz [268] observed that there are no hypercyclic operators on finite-dimensional spaces. This prompted Rolewicz to pose the problem of whether every infinite-dimensional separable Banach space admits a hypercyclic operator, to which we will turn in Chapter 8. Theorem 2.61 is due to Feldman [147].

The title of this section was inspired by Protopopescu [262], where one finds discussions on the relationship between linear and nonlinear chaos; see also Protopopescu and Azmy [263].

Section 2.8. In this section we follow Peris [255]. For an introduction to the dynamics of complex polynomials we refer to Devaney [132] and Carleson and Gamelin [98].

Exercises. Exercise 2.2.1 can be found in Aron, Seoane and Weber [16]. The result of Exercise 2.2.4 is essentially due to Desch, Schappacher and Webb [131]. Exercise 2.2.9 is taken from Feldman [148], Exercise 2.2.10 from Badea, Grivaux and Müller [21], and Exercise 2.2.11 from Costakis and Manoussos [119]. Exercise 2.3.1 can be found in Feldman [147], Exercise 2.3.2 in deLaubenfels and Emamirad [128], Exercise 2.3.4 in Shapiro [280], and Exercise 2.4.2 in Bermúdez, Bonilla, Conejero, and Peris [49]. Exercises 2.5.1, 2.5.3 and 2.5.8 are extracted from Grosse-Erdmann and Peris [187]. The fact that topologically ergodic operators are weakly mixing (part of Exercise 2.5.5) is implicit in Grosse-Erdmann and Peris [185]. The main part of Exercise 2.5.5 is due to

Desch and Schappacher [129], who introduce the concept of operators satisfying the Recurrent Hypercyclicity Criterion, which is equivalent to topological ergodicity. Hereditarily ergodic operators (see Exercise 2.5.6), were introduced as hereditarily syndetic operators by Badea and Grivaux [19]. Exercise 2.5.9 is taken from Badea, Grivaux and Müller [20], Exercises 2.5.10 and 2.5.11 from Grivaux [172], Exercises 2.6.5 and 2.7.1 from Bès [66], and Exercise 2.6.6 from Bourdon [90].

Extensions. Let us add again some remarks on the setting chosen for this chapter (and for most of the book). Typically, the basic results in linear dynamics either use a Baire category argument, in which case they are often valid in all F-spaces (see below), or they hold in all topological vector spaces (see Chapter 12). Only in more specialized results do structural properties such as being locally convex, being normable or having an inner product play a role. Since our aim has not been to always provide the best possible result but to offer a widely accessible introduction to the main ideas of linear dynamics we have chosen to restrict ourselves to the setting of Fréchet spaces.

The larger class of F-spaces consists of all vector spaces that are endowed with an F-norm and that are complete under the induced metric. For example, the spaces ℓ^p with $0 < p < 1$ are F-spaces. One can show that a vector space is an F-space if and only if it carries a complete translation-invariant metric; see [212].

We finish with a citation that is representative of the common, and as we now know erroneous, belief that chaos and nonlinearity go hand in hand.

> Chaotic systems not only exhibit sensitive dependence, but two other properties as well: they are ***deterministic***, and they are ***nonlinear***.

(L.A. Smith, Chaos: A very short introduction, [295, p. 1]; emphasis in the original.)

Chapter 3
The Hypercyclicity Criterion

The Birkhoff transitivity theorem reduces hypercyclicity to the (formally) simpler condition of topological transitivity. Nonetheless, in many concrete situations it is not obvious how to verify this condition for a given operator.

The main purpose of this chapter is to derive several easily applicable criteria under which an operator is chaotic, mixing or weakly mixing (and, in particular, hypercyclic). These criteria have their origin in specific applications to operators on spaces of sequences or functions. Some of them may appear technical, while others are easier to understand, but all of them are very useful for particular examples. We will proceed from the easiest to the most sophisticated criterion.

Throughout this chapter, X will denote a separable Fréchet space and T an operator on X.

3.1 Criteria for chaos and mixing

The message of our first criterion is quite simple: a large supply of (appropriate) eigenvectors is conducive to chaos. This extends the earlier observation that linked periodic points with certain eigenvectors of modulus 1; see Proposition 2.33.

The criterion, like others in this chapter, is named after its authors.

Theorem 3.1 (Godefroy–Shapiro criterion). *Let T be an operator. Suppose that the subspaces*

$$X_0 := \operatorname{span}\{x \in X \ ; \ Tx = \lambda x \text{ for some } \lambda \in \mathbb{K} \text{ with } |\lambda| < 1\},$$

$$Y_0 := \operatorname{span}\{x \in X \ ; \ Tx = \lambda x \text{ for some } \lambda \in \mathbb{K} \text{ with } |\lambda| > 1\}$$

are dense in X. Then T is mixing, and in particular hypercyclic.

If, moreover, X is a complex space and also the subspace

K.-G. Grosse-Erdmann, A. Peris Manguillot, *Linear Chaos*, Universitext,
DOI 10.1007/978-1-4471-2170-1_3, © Springer-Verlag London Limited 2011

$$Z_0 := \text{span}\{x \in X \ ; \ Tx = e^{\alpha\pi i}x \ \textit{for some} \ \alpha \in \mathbb{Q}\}$$

is dense in X, then T is chaotic.

Proof. Let U, V be a pair of nonempty open subsets of X. By hypothesis we can find $x \in X_0 \cap U$ and $y \in Y_0 \cap V$. Hence these vectors can be expressed in the form

$$x = \sum_{k=1}^{m} a_k x_k \quad \text{and} \quad y = \sum_{k=1}^{m} b_k y_k,$$

where $Tx_k = \lambda_k x_k$, $Ty_k = \mu_k y_k$, for certain scalars $a_k, b_k, \lambda_k, \mu_k \in \mathbb{K}$ with $|\lambda_k| < 1$, $|\mu_k| > 1$, $k = 1, \dots, m$. Since

$$T^n x = \sum_{k=1}^{m} a_k \lambda_k^n x_k \to 0 \quad \text{and} \quad u_n := \sum_{k=1}^{m} b_k \frac{1}{\mu_k^n} y_k \to 0$$

as $n \to \infty$ and $T^n u_n = y$ for all $n \geq 0$, there is some $N \in \mathbb{N}$ so that, for all $n \geq N$,

$$x + u_n \in U \quad \text{and} \quad T^n(x + u_n) = T^n x + y \in V.$$

This shows that T is mixing and therefore hypercyclic.

According to Proposition 2.33, in the complex case, Z_0 is precisely the set of periodic points of T. Thus, if also Z_0 is dense then T is even chaotic. □

The Godefroy–Shapiro criterion provides a new proof for the chaotic behaviour of the three classical operators.

Example 3.2. **(Rolewicz's operators)** Let $T = \mu B$, $|\mu| > 1$, be the multiple of the backward shift on any of the spaces $X = \ell^p$, $1 \leq p < \infty$, or $X = c_0$. By Corollary 2.51 (see also Example 2.27), it suffices to consider the complex case.

One easily determines the eigenvectors of B as being the nonzero multiples of the sequences

$$e_\lambda := (\lambda, \lambda^2, \lambda^3, \dots), \quad |\lambda| < 1,$$

with corresponding eigenvalue λ; the condition $|\lambda| < 1$ ensures that $e_\lambda \in X$. Therefore, e_λ is an eigenvector of $T = \mu B$ to the eigenvalue $\mu\lambda$.

We claim that, for any subset Λ of the unit disk $\mathbb{D} = \{\lambda \in \mathbb{C} \ ; \ |\lambda| < 1\}$ that has an accumulation point inside the disk, the set

$$\text{span}\{e_\lambda \ ; \ \lambda \in \Lambda\}$$

is dense in X. By the Hahn–Banach theorem it suffices to show that any continuous linear functional x^* on X that vanishes on each e_λ, $\lambda \in \Lambda$, vanishes on X. Now, since $x^* \in X^*$ is given, via the canonical representation, by a sequence $y = (y_n)_n \in \ell^q$ for a certain q with $1 \leq q \leq \infty$, we have that

$$x^*(e_\lambda) = \langle e_\lambda, x^* \rangle = \sum_{n=1}^{\infty} y_n \lambda^n \quad \text{if } |\lambda| < 1.$$

But since $(y_n)_n$ is necessarily bounded, this defines a holomorphic function in λ, and it vanishes, by assumption, on the set Λ with an accumulation point. The identity theorem for holomorphic functions implies that each y_n is zero and therefore $x^* = 0$.

In particular, the subspace

$$\begin{aligned} X_0 &= \operatorname{span}\{x \in X \; ; \; Tx = \eta x \text{ for some } \eta \in \mathbb{K} \text{ with } |\eta| < 1\} \\ &= \operatorname{span}\{e_\lambda \; ; \; |\lambda| < 1/|\mu|\} \end{aligned}$$

is dense in X, as are the subspaces Y_0 and Z_0 of the Godefroy–Shapiro criterion; note that $1/|\mu| < 1$. This implies that Rolewicz's operators are mixing and chaotic.

Example 3.3. **(Birkhoff's and MacLane's operators)** Let us first consider the translation operators T_a, $a \neq 0$, on the space $H(\mathbb{C})$ of entire functions. Here all the work has already been done in Section 2.3. Denoting by e_λ the exponential functions $e_\lambda(z) = e^{\lambda z}$, every e_λ, $\lambda \in \mathbb{C}$, is an eigenvector of T_a to the eigenvalue $e^{a\lambda}$. Therefore the subspace X_0 of the Godefroy–Shapiro criterion contains the subspace

$$\operatorname{span}\left\{e_\lambda \; ; \; |e^{a\lambda}| < 1\right\},$$

which is dense in $H(\mathbb{C})$ by Lemma 2.34. The density of Y_0 and Z_0 follows similarly (the latter has, in fact, already been shown in Example 2.35). Hence Birkhoff's operators are mixing and chaotic.

In the same way one can give a new proof of the differentiation operator D being mixing and chaotic.

The essential point in the mixing part of the proof of Theorem 3.1 was the fact that $T^n x \to 0$, for each $x \in X_0$, and that for each $y \in Y_0$ we could find a sequence $(u_n)_n$ in X such that $u_n \to 0$ and $T^n u_n = y$ for $n \geq 0$. The second condition can be achieved, for example, via a map $S : Y_0 \to Y_0$ such that $S^n y \to 0$ and $TSy = y$ for any $y \in Y_0$; one need only define $u_n = S^n y$. This leads us to another important criterion in linear dynamics.

Theorem 3.4 (Kitai's criterion). *Let T be an operator. If there are dense subsets $X_0, Y_0 \subset X$ and a map $S : Y_0 \to Y_0$ such that, for any $x \in X_0$, $y \in Y_0$,*

(i) $T^n x \to 0$,
(ii) $S^n y \to 0$,
(iii) $TSy = y$,

then T is mixing.

Remark 3.5. We emphasize that the map S need not have any properties other than satisfying (ii) and (iii). In particular, it need not be a linear or continuous map. The same remark will apply to the forthcoming criteria.

Our proofs of the hypercyclicity of MacLane's and Rolewicz's operators in Examples 2.21 and 2.22 were based on Kitai's criterion in disguise. Using that criterion leads to very transparent proofs, which, in addition, differ from the proofs based on the Godefroy–Shapiro criterion.

Example 3.6. **(Rolewicz's operators)** The operator $T = \lambda B$, $|\lambda| > 1$, is mixing on any of the spaces $X = \ell^p$, $1 \leq p < \infty$, or c_0. In fact, taking for $X_0 = Y_0$ the set of finite sequences, which is dense in X, and for $S : Y_0 \to Y_0$ the map $S = \frac{1}{\lambda}F$, where F is the forward shift $F : (x_1, x_2, x_3, \ldots) \to (0, x_1, x_2, \ldots)$, the conditions of Kitai's criterion are clearly satisfied.

Example 3.7. **(MacLane's operator)** The differentiation operator D on $H(\mathbb{C})$ is mixing. In this case we take for $X_0 = Y_0$ the set of polynomials, which is dense in $H(\mathbb{C})$, and for S we consider the integral operator $Sf(z) = \int_0^z f(\zeta)\, d\zeta$. While conditions (i) and (iii) are obvious, we note for condition (ii) that it suffices to verify it for the monomials. But then $S^n(z^k) = \frac{k!}{(k+n)!} z^{k+n} \to 0$ as $n \to \infty$, uniformly on compact sets, as required.

In contrast, the proof of hypercyclicity for Birkhoff's operators given in Example 2.20 does not hide a Kitai-type argument. Indeed, Kitai's criterion seems to be less well adapted to this operator. Still, the following provides us with a third argument for the hypercyclicity of the translation operators.

Example 3.8. **(Birkhoff's operators)** For simplicity we only show that the translation operator $T_1 f(z) = f(z+1)$ on $H(\mathbb{C})$ is mixing. For $X_0 = Y_0$ we choose the set of all functions of the form $f_{p,\alpha,\nu}(z) = p(z)e^{-\alpha(z-\nu)^2}$, where p is a polynomial and $\alpha > 0$, $\nu \in \mathbb{N}_0$. Since $f_{p,\alpha,\nu} \to p$ in $H(\mathbb{C})$ as $\alpha \to 0$, this set is dense in $H(\mathbb{C})$. Moreover, for S we consider the translation operator $Sf(z) = f(z-1)$. Now, if $z = x + iy$ with $|y| \leq \frac{1}{2}|x|$ then we have that $|e^{-\alpha z^2}| = e^{-\alpha(x^2-y^2)} \leq e^{-\frac{3}{4}\alpha x^2}$. This implies that, for any p, α and ν, $f_{p,\alpha,\nu}(z \pm n) \to 0$ uniformly on compact sets as $n \to \infty$, which shows that conditions (i) and (ii) of Kitai's criterion hold, while condition (iii) is trivial.

We end this section with an example showing that Kitai's criterion is a stronger result than the Godefroy–Shapiro criterion.

Example 3.9. Consider the bilateral backward shift $T = B$, given by

$$B(x_n)_{n\in\mathbb{Z}} = (x_{n+1})_{n\in\mathbb{Z}},$$

on the weighted space $\ell^1(\mathbb{Z}, v) = \{(x_n)_{n\in\mathbb{Z}}\ ;\ \|x\| := \sum_{n\in\mathbb{Z}} |x_n| v_n < \infty\}$, where $v_n = \frac{1}{|n|+1}$, $n \in \mathbb{Z}$. The equality $Tx = \lambda x$, $x \neq 0$, implies that

$$x = \left(\ldots, \tfrac{1}{\lambda^2}x_0, \tfrac{1}{\lambda}x_0, x_0, \lambda x_0, \lambda^2 x_0, \ldots\right), \quad \lambda \neq 0, \quad x_0 \neq 0.$$

But then we have that

$$\|x\| = |x_0| \left(1 + \sum_{n\in\mathbb{N}} \frac{|\lambda|^n}{n+1} + \sum_{n\in\mathbb{N}} \frac{1}{|\lambda|^n(n+1)}\right) = \infty,$$

whatever the value of λ. Therefore T has no eigenvalues and, in particular, it does not satisfy the Godefroy–Shapiro criterion.

On the other hand it satisfies Kitai's criterion. Indeed, if we choose for $X_0 = Y_0$ the space of bilateral finite sequences $(\ldots, 0, 0, x_{-m}, \ldots, x_n, 0, 0, \ldots)$, $m, n \geq 0$, and for $S : Y_0 \to Y_0$ the forward shift, then one easily verifies the conditions of Kitai's criterion for T.

This example shows, in particular, that hypercyclic operators need not have eigenvectors.

3.2 Weak mixing and the Hypercyclicity Criterion

The next step in the sophistication of criteria for hypercyclicity is to replace the full sequence (n) by an increasing sequence $(n_k)_k$ of positive integers for the iterates of T and S in Kitai's criterion. But in doing so we lose the mixing property.

Theorem 3.10 (Gethner–Shapiro criterion). *Let T be an operator. If there are dense subsets $X_0, Y_0 \subset X$, an increasing sequence $(n_k)_k$ of positive integers, and a map $S : Y_0 \to Y_0$ such that, for any $x \in X_0$, $y \in Y_0$,*
(i) $T^{n_k}x \to 0$,
(ii) $S^{n_k}y \to 0$,
(iii) $TSy = y$,
then T is weakly mixing.

Proof. Let U_1, U_2, V_1 and V_2 be nonempty open sets. By assumption we can find vectors $x_j \in U_j \cap X_0$ and $y_j \in V_j \cap Y_0$, $j = 1, 2$. Then, by (iii),

$$T^{n_k}(x_j + S^{n_k}y_j) = T^{n_k}x_j + y_j, \quad j = 1, 2.$$

It follows from (i) and (ii) that, for sufficiently large k, $x_j + S^{n_k}y_j \in U_j$ and $T^{n_k}x_j + y_j \in V_j$ for $j = 1, 2$. This shows that $N(U_1, V_1) \cap N(U_2, V_2) \neq \varnothing$. □

The requirements of this criterion are clearly weaker than those of Kitai's criterion. In fact we will provide an operator satisfying the Gethner–Shapiro criterion but not Kitai's criterion. This is also the first example that we present of a weakly mixing operator that is not mixing; see also Remark 4.10.

Example 3.11. We consider the weighted backward shift $T = B_w : c_0 \to c_0$ given by

$$B_w(x_1, x_2, \ldots) = (w_2x_2, w_3x_3, w_4x_4, \ldots),$$

with weight sequence

$$w = (w_1, w_2, \dots) = \left(1, 2, 2^{-1}, 2, 2, 2^{-1}, 2^{-1}, 2, 2, 2, 2^{-1}, 2^{-1}, 2^{-1}, \dots\right)$$

(see also Section 4.1); note that the value of w_1 is irrelevant. Since w is bounded, T is continuous. Let $(m_k)_k$ be the increasing sequence of all the integers with $w_{m_k} = 2^{-1}$ and $w_{m_k+1} = 2$, $k \geq 1$. Since

$$T^n x = \left(\left(\prod_{\nu=2}^{n+1} w_\nu\right) x_{n+1}, \dots\right), \quad n \geq 1,$$

we have that $T^{m_k - 1} x = (x_{m_k}, \dots)$ for each $k \geq 1$. In particular, if we define $U = \{x \in c_0 \; ; \; \|x\| < 1\}$ and $V = \{x \in c_0 \; ; \; |x_1| > 1\}$, which are nonempty open sets, then we get that $T^{m_k - 1}(U) \cap V = \varnothing$, for each $k \geq 1$, which shows that T is not mixing. Therefore it cannot satisfy Kitai's criterion.

On the other hand, let us take for $X_0 = Y_0$ the space of finite sequences and for S the weighted forward shift $S : Y_0 \to Y_0$, $S(x_1, x_2, \dots) = (0, w_2^{-1} x_1, w_3^{-1} x_2, \dots)$. It is clear that $TSy = y$, $y \in Y_0$, and that $T^n x \to 0$, $x \in X_0$. It remains to find a suitable increasing sequence $(n_k)_k$ of positive integers so that $S^{n_k} y \to 0$ for each $y \in Y_0$ in order to satisfy all the conditions of the Gethner–Shapiro criterion. Indeed, let $n_k = m_k + k - 1$, $k \in \mathbb{N}$, and denote by e_n the vector in c_0 that has 1 in the nth position, and 0 otherwise. Then we have that

$$S^{n_k} e_1 = \left(0, \dots, 0, \prod_{\nu=2}^{m_k+k} w_\nu^{-1}, 0, \dots\right) = (0, \dots, 0, 2^{-k}, 0, \dots),$$

so that $S^{n_k} e_1 \to 0$. On the other hand, since the weights w_n are bounded away from 0, S is continuous too. Therefore, for any $j \geq 1$,

$$S^{n_k} e_j = S^{n_k}\left(\left(\prod_{\nu=2}^{j} w_\nu\right) S^{j-1} e_1\right) = \left(\prod_{\nu=2}^{j} w_\nu\right) S^{j-1}(S^{n_k} e_1) \to 0.$$

From this we conclude that $S^{n_k} y \to 0$ for each $y \in Y_0$, which had to be shown.

The final step in our search for a very general, but reasonable, criterion for hypercyclicity will be to weaken the requirement of the existence of a right inverse S for T on the dense subset Y_0. The proof of Theorem 3.10 shows that all we need is a sequence of maps S_{n_k} with $S_{n_k} y \to 0$ and $T^{n_k} S_{n_k} y \to y$ for all $y \in Y_0$. These maps S_{n_k} need not even be self-maps of Y_0. We thus have obtained the following.

Theorem 3.12 (Hypercyclicity Criterion). *Let T be an operator. If there are dense subsets $X_0, Y_0 \subset X$, an increasing sequence $(n_k)_k$ of positive integers, and maps $S_{n_k} : Y_0 \to X$, $k \geq 1$, such that, for any $x \in X_0$, $y \in Y_0$,*

(i) $T^{n_k}x \to 0$,
(ii) $S_{n_k}y \to 0$,
(iii) $T^{n_k}S_{n_k}y \to y$,
then T is weakly mixing, and in particular hypercyclic.

Remark 3.13. (a) If the Gethner–Shapiro criterion or the Hypercyclicity Criterion is satisfied for the full sequence $(n_k)_k = (n)_n$, then the proofs show that the operator T is even mixing; see also Exercise 3.1.1.

(b) By Furstenberg's theorem, the Hypercyclicity Criterion even implies that every n-fold direct sum $T \oplus \ldots \oplus T$ is hypercyclic. But this follows also directly because $T \oplus \ldots \oplus T$ satisfies the Hypercyclicity Criterion as well.

Originally, Kitai had shown that the conditions in Theorem 3.4 imply hypercyclicity. She did this by constructing, through a recursive procedure, a hypercyclic vector. While the proofs we have presented here are shorter and give better results, Kitai's construction is useful in many other situations where more abstract methods fail. We therefore give here a second proof of the Hypercyclicity Criterion using Kitai's approach.

Alternative proof of Theorem 3.12. Let $\|\cdot\|$ denote an F-norm defining the topology of the Fréchet space X. Since X is separable we can choose a sequence $(y_j)_j$ from Y_0 that is dense in X. We first show that there are x_j in X and positive integers k_j such that

$$x = x_1 + S_{n_{k_1}}y_1 + x_2 + S_{n_{k_2}}y_2 + x_3 + \ldots$$

exists and is hypercyclic. To this end we construct the x_j and k_j recursively such that we have, for $j \geq 1$,

$$\|x_j\| < \frac{1}{2^j} \quad \text{and} \quad \|T^{n_{k_l}}x_j\| < \frac{1}{2^j} \quad (l = 1, \ldots, j-1), \tag{3.1}$$

$$\|S_{n_{k_j}}y_j\| < \frac{1}{2^j} \quad \text{and} \quad \|T^{n_{k_l}}S_{n_{k_j}}y_j\| < \frac{1}{2^j} \quad (l = 1, \ldots, j-1), \tag{3.2}$$

$$\|T^{n_{k_j}}S_{n_{k_j}}y_j - y_j\| < \frac{1}{2^j} \quad \text{and} \quad \Big\|T^{n_{k_j}}\Big(\sum_{l=1}^{j-1}(x_l + S_{n_{k_l}}y_l) + x_j\Big)\Big\| < \frac{1}{2^j}. \tag{3.3}$$

Indeed, for $j = 1$ these inequalities are satisfied for $x_1 = 0$ and for a suitable k_1, using conditions (ii) and (iii). Now let $j \geq 2$ and assume that $x_1, \ldots, x_{j-1}$ and $k_1, \ldots, k_{j-1}$ have been constructed. Then, by density of X_0, there is some $x_j \in X$ such that (3.1) holds and

$$\sum_{l=1}^{j-1}(x_l + S_{n_{k_l}}y_l) + x_j \in X_0.$$

Choosing k_j sufficiently large, conditions (i), (ii) and (iii) imply that one can also achieve (3.2) and (3.3).

It then follows from the first inequalities in (3.1) and (3.2) that the series defining x converges. The remaining inequalities imply that, for $j \geq 1$,

$$\begin{aligned}\|T^{n_{k_j}}x - y_j\| &= \Big\| T^{n_{k_j}}\Big(\sum_{l=1}^{j-1}(x_l + S_{n_{k_l}}y_l) + x_j\Big) + T^{n_{k_j}}S_{n_{k_j}}y_j - y_j \\ &\qquad + \sum_{l=j+1}^{\infty} T^{n_{k_j}}x_l + \sum_{l=j+1}^{\infty} T^{n_{k_j}}S_{n_{k_l}}y_l\Big\| \\ &\leq \frac{1}{2^j} + \frac{1}{2^j} + \sum_{l=j+1}^{\infty}\Big(\frac{1}{2^l}+\frac{1}{2^l}\Big) = \frac{4}{2^j}.\end{aligned}$$

Thus, since the y_j are dense in X, x is hypercyclic for T.

To see that T is even weakly mixing we need only observe that also $T\oplus T$ satisfies the conditions (i)–(iii), with the dense subsets $X_0 \oplus X_0$, $Y_0 \oplus Y_0$ of $X \oplus X$ and the maps $S_{n_k} \oplus S_{n_k}$. Thus, as we have just seen, $T \oplus T$ is hypercyclic and T is therefore weakly mixing. □

What is a remarkable, and rather unexpected, feature of the Hypercyclicity Criterion is that it actually characterizes when an operator is weakly mixing. The proof of this result will be achieved via the intermediate notion of a hereditarily hypercyclic operator, which, within our present framework, is equivalent to the notion of hereditary transitivity; see Exercises 1.6.2 and 1.6.4.

Definition 3.14. Let $(n_k)_k$ be an increasing sequence of positive integers. Then an operator T is called *hereditarily hypercyclic with respect to* $(n_k)_k$ if, for each subsequence $(m_k)_k$ of $(n_k)_k$, there is some $x \in X$ such that $\{T^{m_k}x\ ;\ k \geq 1\}$ is dense in X.

An operator T is called *hereditarily hypercyclic* if it is so with respect to some sequence $(n_k)_k$.

For dynamical properties of sequences $(T^{n_k})_k$ we use the terminology and the results of Section 1.6; see also Section 3.4 below.

Theorem 3.15 (Bès–Peris). *Let T be an operator. Then the following assertions are equivalent:*

(i) *T satisfies the Hypercyclicity Criterion;*
(ii) *T is weakly mixing;*
(iii) *T is hereditarily hypercyclic.*

Proof. (i)$\Longrightarrow$(ii). This is the result of Theorem 3.12.

(ii)$\Longrightarrow$(iii). By separability, X has a countable base $(O_n)_n$ of open sets. Let $(U_j, V_j)_j$ be an enumeration of all pairs (O_m, O_n), $m, n \geq 1$. Since T is weakly mixing, Furstenberg's theorem implies that, for any $k \geq 1$, there is a positive integer n_k such that

$$T^{n_k}(U_j) \cap V_j \neq \varnothing \quad \text{for } j = 1, \ldots, k,$$

and we can choose $(n_k)_k$ to be increasing. Now let $(m_k)_k$ be a subsequence of $(n_k)_k$. Let U, V be nonempty open sets. Then, by construction, there is some $l \geq 1$ such that $U_l \subset U$ and $V_l \subset V$ and hence

$$T^{m_k}(U) \cap V \supset T^{m_k}(U_l) \cap V_l \neq \varnothing$$

whenever k is sufficiently large. It follows from the Universality Criterion, Theorem 1.57, that $(T^{m_k})_k$ admits a dense orbit.

(iii)$\Longrightarrow$(i). Let T be hereditarily hypercyclic with respect to a certain increasing sequence $(m_k)_k$ of positive integers. In particular, we can find some $x \in X$ such that $\{T^{m_k}x \ ; \ k \geq 1\}$ is dense in X. Let $(q_k)_k$ be a subsequence of $(m_k)_k$ such that $T^{q_k}x \to 0$. Then we can find some $y \in X$ such that $\{T^{q_k}y \ ; \ k \geq 1\}$ is dense in X. In particular, if U_k is the open ball of radius $1/k$ around x then we can find a subsequence $(n_k)_k$ of $(q_k)_k$ such that

$$T^{n_k}y \in kU_k, \quad k \geq 1.$$

Setting $x_k = y/k$, $k \geq 1$, we have that $x_k \to 0$ and $T^{n_k}x_k \to x$. Let $X_0 = Y_0 = \operatorname{orb}(x, T)$, which is dense in X. Since $(n_k)_k$ is a subsequence of $(q_k)_k$, we have by continuity of T that

$$T^{n_k}(T^n x) = T^n(T^{n_k}x) \to T^n 0 = 0, \quad n \geq 0,$$

which gives condition (i) of the Hypercyclicity Criterion. On the other hand, let $y \in Y_0$. Then there is some $n \geq 0$ such that $y = T^n x$, and we define $S_{n_k}y = T^n x_k$, $k \geq 1$. Then we have that

$$S_{n_k}y = T^n x_k \to 0, \quad T^{n_k}S_{n_k}y = T^n(T^{n_k}x_k) \to T^n x = y,$$

yielding also conditions (ii) and (iii) of the Hypercyclicity Criterion. □

The completeness of the space X is essential in the previous theorem, as the following example shows.

Example 3.16. Let T be a chaotic operator on a Fréchet space X (as, for example, Birkhoff's operator on $H(\mathbb{C})$). We consider the T-invariant subspace $Y = \operatorname{Per}(T)$ of periodic points of T. Since, by Corollary 2.49, T is weakly mixing and Y is dense in X, we also have that $T|_Y : Y \to Y$ is weakly mixing. On the other hand, $T|_Y$ cannot satisfy the Hypercyclicity Criterion since the only point $y \in Y$ such that $T^{n_k}y \to 0$ for some $(n_k)_k$ is $y = 0$.

A careful look at the proof of Theorem 3.15 yields a more precise connection between hereditary hypercyclicity and the Hypercyclicity Criterion.

Proposition 3.17. *An operator T is hereditarily hypercyclic with respect to an increasing sequence $(n_k)_k$ of positive integers if and only if every subsequence $(m_k)_k$ of $(n_k)_k$ admits a subsequence $(q_k)_k$ such that T satisfies the Hypercyclicity Criterion with respect to $(q_k)_k$.*

Proof. If T is hereditarily hypercyclic with respect to $(n_k)_k$, then, given a subsequence $(m_k)_k$ of $(n_k)_k$, the argument of Theorem 3.15 yields the existence of a subsequence $(q_k)_k$ of $(m_k)_k$ such that T satisfies the Hypercyclicity Criterion with respect to $(q_k)_k$.

Conversely, let $(n_k)_k$ be an increasing sequence of positive integers such that, for every subsequence $(m_k)_k$ of $(n_k)_k$, T satisfies the Hypercyclicity Criterion with respect to some subsequence $(q_k)_k$ of $(m_k)_k$. But then it follows as in the proof of the Hypercyclicity Criterion (see also Theorem 3.24), that $(T^{q_k})_k$ admits a dense orbit, and hence so does $(T^{m_k})_k$. Therefore T is hereditarily hypercyclic with respect to $(n_k)_k$. □

In view of the equivalence of the weak mixing property with the Hypercyclicity Criterion, Theorem 2.48 and Corollary 2.49 provide us with a rich source of operators that satisfy the Hypercyclicity Criterion.

Theorem 3.18. *Every hypercyclic operator with a dense set of points with bounded orbits satisfies the Hypercyclicity Criterion.*

In particular, any of the following operators satisfy the Hypercyclicity Criterion:

(i) *chaotic operators;*
(ii) *hypercyclic operators that have a dense set of points for which the orbits converge;*
(iii) *hypercyclic operators with dense generalized kernel.*

Since the complexification $\widetilde{T}$ of an operator T on a real space can be identified with $T \oplus T$, Corollary 2.51 and Theorem 3.15 imply the following.

Proposition 3.19. *Let T be an operator on a real separable Fréchet space and $\widetilde{T}$ its complexification. Then the following assertions are equivalent:*

(i) *T satisfies the Hypercyclicity Criterion;*
(ii) *$\widetilde{T}$ is hypercyclic;*
(iii) *$\widetilde{T}$ satisfies the Hypercyclicity Criterion.*

In particular, any hypercyclic operator on a complex space that can be regarded as a complexification of a suitable (real) operator satisfies the Hypercyclicity Criterion.

In spite of all these positive results, the theorem of De la Rosa and Read, Theorem 2.43, tells us that not every hypercyclic operator satisfies the Hypercyclicity Criterion. According to Bayart and Matheron, such operators exist even on any of the spaces ℓ^p, $1 \le p < \infty$, and c_0.

3.3 Equivalent formulations of the Hypercyclicity Criterion

In the last two sections we have discussed some popular criteria for hypercyclicity. But many other criteria have been suggested, some formally stronger than the Hypercyclicity Criterion, some formally weaker. So the problem arises of determining whether they are indeed stronger or weaker than the Hypercyclicity Criterion. We address this problem here for two particular criteria, one of which we have already encountered. Additional results along this line will be discussed in the exercises.

Our first result says that, in the Hypercyclicity Criterion, one may, without loss of generality, assume that the sets X_0 and Y_0 coincide, are linear, and that each S_{n_k} is linear.

Proposition 3.20. *An operator T satisfies the Hypercyclicity Criterion if and only if it satisfies the following criterion.*

There is a dense subspace $X_0 \subset X$, an increasing sequence $(n_k)_k$ of positive integers, and linear maps $S_{n_k} : X_0 \to X$, $k \geq 1$, such that, for any $x \in X_0$,

(i) $T^{n_k} x \to 0$,
(ii) $S_{n_k} x \to 0$,
(iii) $T^{n_k} S_{n_k} x \to x$.

Proof. Suppose that T satisfies the Hypercyclicity Criterion. Then, by the proof of Theorem 3.15, there exists a hypercyclic vector $x \in X$ such that T satisfies the Hypercyclicity Criterion for the dense set $\operatorname{orb}(x,T)$ (taken twice) and certain maps S_{n_k} on $\operatorname{orb}(x,T)$. But since the vectors from $\operatorname{orb}(x,T)$ form a linearly independent set (see Proposition 2.60), one can extend each S_{n_k} linearly to $\operatorname{span}\operatorname{orb}(x,T)$, and the Hypercyclicity Criterion will hold with $X_0 = Y_0 := \operatorname{span}\operatorname{orb}(x,T)$. $\square$

A more delicate question is whether the criteria of Theorems 3.10 and 3.12 are equivalent, that is, if every operator T satisfying the Hypercyclicity Criterion also satisfies the Gethner–Shapiro criterion. It is far from evident whether every operator that satisfies the Hypercyclicity Criterion has a right inverse S of the type demanded. The main purpose of this section is to show that both criteria are indeed equivalent. For this we will need an abstract version of the Mittag-Leffler Theorem.

Theorem 3.21 (Mittag-Leffler). *Let $(X_n)_n$ be a sequence of complete metric spaces and let $f_n : X_{n+1} \to X_n$, $n \in \mathbb{N}$, be continuous maps with dense range. Then, for every nonempty open subset $U \subset X_1$, there exists a sequence $(x_n)_n$ with $x_n \in X_n$, $n \in \mathbb{N}$, such that $x_1 \in U$ and $f_n(x_{n+1}) = x_n$, $n \in \mathbb{N}$.*

Proof. Let d_n be the metric of X_n, $n \in \mathbb{N}$. Given a nonempty open set $U \subset X_1$, we fix $\varepsilon > 0$ and $x \in U$ such that

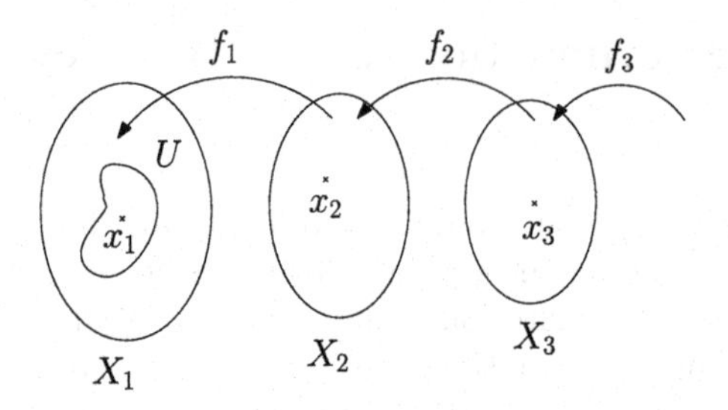

Fig. 3.1 The Mittag-Leffler theorem

$$\{y \in X_1 \ ; \ d_1(x, y) < \varepsilon\} \subset U.$$

Set $x_{1,1} = x$. Since f_1 has dense range, there are $x_{1,2} \in X_1$ and $x_{2,2} \in X_2$ such that $f_1(x_{2,2}) = x_{1,2}$ and $d_1(x_{1,1}, x_{1,2}) < \varepsilon/2$. Proceeding by induction we will find a countable family $\mathcal{A} = \{x_{j,k} \in X_j \ ; \ k \in \mathbb{N}, \ 1 \leq j \leq k\}$ satisfying

$$f_j(x_{j+1,k+1}) = x_{j,k+1} \ \text{ and } \ d_j(x_{j,k}, x_{j,k+1}) < \frac{\varepsilon}{2^k}, \quad k \in \mathbb{N}, \ 1 \leq j \leq k.$$

Indeed, suppose that we already have constructed the finite family $\{x_{j,k} \in X_j \ ; \ 1 \leq k \leq n, \ 1 \leq j \leq k\}$ with the above property. Since f_n has dense range and each f_j, $j < n$, is continuous, we can find $x_{n,n+1} \in f_n(X_{n+1})$ close enough to $x_{n,n}$ so that, for $x_{j,n+1} := f_j(x_{j+1,n+1})$, $j < n$, we have

$$d_j(x_{j,n}, x_{j,n+1}) < \frac{\varepsilon}{2^n}, \quad 1 \leq j \leq n.$$

The induction process is completed if we pick $x_{n+1,n+1} \in X_{n+1}$ with $f_n(x_{n+1,n+1}) = x_{n,n+1}$.

The defining property of the family $\mathcal{A}$ yields that for each $j \in \mathbb{N}$ the sequence $(x_{j,k})_{k \geq j}$ is a Cauchy sequence in X_j. We define $x_j = \lim_{k\to\infty} x_{j,k} \in X_j$, $j \in \mathbb{N}$, to conclude that

$$d_1(x, x_1) = \lim_{k\to\infty} d_1(x_{1,1}, x_{1,k}) \leq \lim_{k\to\infty} \sum_{l=1}^{k-1} d_1(x_{1,l}, x_{1,l+1}) < \lim_{k\to\infty} \sum_{l=1}^{k-1} \frac{\varepsilon}{2^l} = \varepsilon,$$

so that $x_1 \in U$, and

$$f_j(x_{j+1}) = f_j(\lim_{k\to\infty} x_{j+1,k+1}) = \lim_{k\to\infty} x_{j,k+1} = x_j, \quad j \in \mathbb{N},$$

which had to be shown. $\square$

This result implies the aforementioned equivalence of two important criteria.

Theorem 3.22. *An operator satisfies the Hypercyclicity Criterion if and only if it satisfies the Gethner–Shapiro criterion.*

Proof. Let T be an operator satisfying the Hypercyclicity Criterion.

By setting $X_n = X$, $f_n = T$, $n \in \mathbb{N}$, in Mittag-Leffler's theorem, we obtain that

$$Y := \{x \in X \; ; \; \exists (x_n)_n \in X^{\mathbb{N}} \text{ with } x_1 = x, \quad Tx_{n+1} = x_n, \; n \in \mathbb{N}\}$$

is a dense subspace of X. We consider $X^{\mathbb{N}}$ endowed with the product topology. Then

$$\mathcal{X} = \{(x_n)_n \in X^{\mathbb{N}} \; ; \; Tx_{n+1} = x_n, \quad n \in \mathbb{N}\}$$

is a closed subspace of $X^{\mathbb{N}}$ and therefore a separable Fréchet space under the induced topology; see Exercise 2.1.3.

The operator T induces an operator $\mathcal{T} : \mathcal{X} \to \mathcal{X}$,

$$\mathcal{T}(x_1, x_2, x_3, \dots) := (Tx_1, Tx_2, Tx_3, \dots) = (Tx_1, x_1, x_2, \dots).$$

Then $\mathcal{T}$ is an invertible operator whose inverse is the backward shift $\mathcal{B} : \mathcal{X} \to \mathcal{X}$, $\mathcal{B}(x_1, x_2, \dots) = (x_2, x_3, \dots)$.

By Theorem 3.15 we know that T is hereditarily hypercyclic with respect to some increasing sequence $(m_k)_k$ of positive integers. We divide the remainder of the proof into three steps.

Step 1. $(\mathcal{T}^{m_k})_k$ is a topologically transitive sequence of operators; see Section 1.6 (and Section 3.4 below).

To see this, let $\mathcal{U}, \mathcal{V} \subset \mathcal{X}$ be nonempty open sets. Since it suffices to consider sets from a base of the topology of $\mathcal{X}$ we can assume that

$$\mathcal{U} = \{(x_n)_n \in \mathcal{X} \; ; \; x_n \in U_n, \quad n = 1, \dots, N\},$$

$$\mathcal{V} = \{(x_n)_n \in \mathcal{X} \; ; \; x_n \in V_n, \quad n = 1, \dots, N\}$$

with $N \in \mathbb{N}$ and nonempty open sets $U_n, V_n \subset X$, $n = 1, \dots, N$.

Fix $x = (x_n)_n \in \mathcal{U}$ and $y = (y_n)_n \in \mathcal{V}$. By continuity of T we can find open neighbourhoods $U'_N \subset U_N$, $V'_N \subset V_N$ of x_N and y_N, respectively, such that $T^j(U'_N) \subset U_{N-j}$ and $T^j(V'_N) \subset V_{N-j}$, $j = 1, \dots, N-1$.

Since $(T^{m_k})_k$ is topologically transitive (see Exercise 1.6.2), there exists $k \geq 1$ and a nonempty open set $U''_N \subset U'_N$ such that $T^{m_k}(U''_N) \subset V'_N$. By the density of Y there exists some $u_N \in Y \cap U''_N$. Hence there are $u_n \in X$, $n > N$, such that $Tu_{n+1} = u_n$, $n \geq N$; when we now set $u_n = T^{N-n}u_N \in U_n$, $n = 1, \dots, N-1$, then $u := (u_n)_n \in \mathcal{U}$; see Figure 3.2.

On the other hand, $\mathcal{T}^{m_k}u = (T^{m_k}u_1, \dots, T^{m_k}u_N, \dots)$. Since $T^{m_k}u_N \in V'_N$, we get for $n = 1, \dots, N-1$ that

$$T^{m_k}u_n = T^{m_k}(T^{N-n}u_N) = T^{N-n}(T^{m_k}u_N) \in T^{N-n}(V'_N) \subset V_n,$$

and therefore $\mathcal{T}^{m_k} u \in \mathcal{V}$.

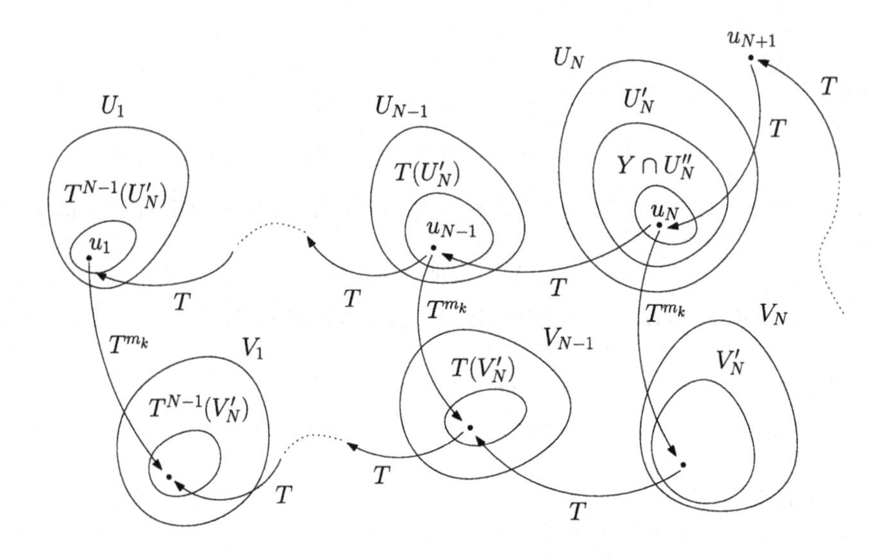

Fig. 3.2 Proof of Theorem 3.22, Step 1

Step 2. There exists a dense subset $Y_0 \subset X$, a map $S : Y_0 \to Y_0$ and a subsequence $(q_k)_k$ of $(m_k)_k$ such that $S^{q_k} y \to 0$ and $TSy = y$, for each $y \in Y_0$.

Indeed, given that the backward shift $\mathcal{B}$ is the inverse of the operator $\mathcal{T}$, it follows from Step 1 that, for each pair of nonempty open sets $\mathcal{U}, \mathcal{V} \subset \mathcal{X}$, there exists $k \in \mathbb{N}$ with

$$\mathcal{T}^{m_k}(\mathcal{U}) \cap \mathcal{V} \neq \varnothing,$$

and therefore

$$\mathcal{B}^{m_k}(\mathcal{V}) \cap \mathcal{U} \neq \varnothing.$$

By the Universality Criterion, Theorem 1.57, both sequences $(\mathcal{T}^{m_k})_k$ and $(\mathcal{B}^{m_k})_k$ have dense G_δ-sets of vectors with dense orbits. Hence there is a vector $(y_n)_n \in \mathcal{X}$ such that

$$\overline{\{(y_{m_k+1}, y_{m_k+2}, \dots) \ ; \ k \in \mathbb{N}\}} = \mathcal{X} = \overline{\{(T^{m_k} y_1, T^{m_k} y_2, \dots) \ ; \ k \in \mathbb{N}\}}.$$

We now set $Y_0 = \{y_n \ ; \ n \in \mathbb{N}\}$; since $\{(y_{m_k+1}, y_{m_k+2}, \dots) \ ; \ k \in \mathbb{N}\}$ is dense in $\mathcal{X}$, projecting onto the first coordinate yields that Y_0 is a dense subset of X. Moreover we define $S : Y_0 \to Y_0$ by $Sy_n = y_{n+1}$, $n \in \mathbb{N}$. This is well defined since $y_m \neq y_n$ if $m \neq n$. Otherwise, we would have $y_m = y_n$ for some $m > n$, hence $T^{m-n} y_n = y_n$, which implies that $T^{m-n} y_1 = y_1$, so that y_1 would be a periodic point for T; this would contradict the density of $\{(T^{m_k} y_1, \dots) \ ; \ k \in \mathbb{N}\}$ in $\mathcal{X}$, again by projecting onto the first coordinate.

Since $(y_n)_n \in \mathcal{X}$ we also have that $TSy = y$ for all $y \in Y_0$. Finally, let $(q_k)_k$ be a subsequence of $(m_k)_k$ such that $\mathcal{B}^{q_k} y = (y_{1+q_k}, y_{2+q_k}, \dots) \to 0$ in $\mathcal{X}$. Then

$$S^{q_k} y_n = y_{n+q_k} \to 0, \quad n \in \mathbb{N}.$$

Step 3. The operator T satisfies the Gethner–Shapiro criterion.

In fact, since T is hereditarily hypercyclic with respect to $(m_k)_k$, there is a hypercyclic vector $w \in X$ and a subsequence $(n_k)_k$ of $(q_k)_k$ satisfying $T^{n_k} w \to 0$. If we set $X_0 = \mathrm{orb}(w, T)$, we obtain that, for any $x \in X_0$, $y \in Y_0$,

(i) $T^{n_k} x \to 0$,
(ii) $S^{n_k} y \to 0$,
(iii) $TSy = y$.

This concludes the proof. □

3.4 Hypercyclic sequences of operators

We resume here the discussion of Section 1.6 in the light of linearity. As we saw in the previous sections, the study of the dynamics of iterates of an operator T sometimes leads to questions concerning subsequences $(T^{n_k})_k$, which forces us to consider the dynamics of general sequences of operators.

Throughout this section, X and Y will denote separable Fréchet spaces and $T_n : X \to Y$, $n \geq 1$, operators between these spaces.

We recall that in this setting the orbit of a vector $x \in X$ is defined as $\mathrm{orb}(x, (T_n)) = \{T_n x \ ; \ n \in \mathbb{N}_0\}$; see Definition 1.55.

Definition 3.23. A sequence $(T_n)_n$ of operators is called *hypercyclic* if there is some $x \in X$ whose orbit under $(T_n)_n$ is dense in Y. In such a case, x is called a *hypercyclic* or *universal vector* for $(T_n)_n$.

The sequence $(T_n)_n$ is called *hereditarily hypercyclic* if there is an increasing sequence $(n_k)_k$ of positive integers such that $(T_{m_k})_k$ is hypercyclic for every subsequence $(m_k)_k$ of $(n_k)_k$.

Topological transitivity, mixing and weak mixing for $(T_n)_n$ are defined in Definition 1.56. A notion of chaos can be defined, but seems less natural in this context.

Now, the Birkhoff transitivity theorem extends to commuting sequences of operators $T_n : X \to X$ with dense range; see Exercise 1.6.2. For general sequences of operators one should instead turn to the Universality Criterion; see Theorem 1.57.

The results of Sections 2.4 and 2.5 carry over to commuting sequences of operators $T_n : X \to X$ (if, in Theorem 2.45, the map $S : X \to X$ commutes with each T_n).

The Hypercyclicity Criterion extends, with the previous proof, to arbitrary sequences of operators.

Theorem 3.24 (Hypercyclicity Criterion for sequences). *Let $(T_n)_n$ be a sequence of operators. If there are dense subsets $X_0 \subset X$ and $Y_0 \subset Y$, an increasing sequence $(n_k)_k$ of positive integers, and maps $S_{n_k} : Y_0 \to X$, $k \geq 1$, such that, for any $x \in X_0$, $y \in Y_0$,*
(i) $T_{n_k} x \to 0$,
(ii) $S_{n_k} y \to 0$,
(iii) $T_{n_k} S_{n_k} y \to y$,
then $(T_n)_n$ is weakly mixing, and in particular hypercyclic.

The (simple) example of Exercise 1.6.1 shows that not every hypercyclic sequence of operators satisfies the Hypercyclicity Criterion.

The Bès–Peris theorem extends likewise to commuting sequences. It is also of interest to add the notion of hereditary transitivity; see Exercise 1.6.4.

Theorem 3.25 (Bès–Peris). *Let $(T_n)_n$ be a commuting sequence of operators $T_n : X \to X$. Then the following assertions are equivalent:*
(i) *$(T_n)_n$ satisfies the Hypercyclicity Criterion;*
(ii) *$(T_n)_n$ is weakly mixing;*
(iii) *$(T_n)_n$ is hereditarily transitive;*
(iv) *$(T_n)_n$ is hereditarily hypercyclic.*

The result breaks down for noncommuting operators; see Exercise 3.4.4.

Exercises

Exercise 3.1.1. Prove the following version of Kitai's criterion. If there are dense subsets $X_0, Y_0 \subset X$ such that
(i) $T^n x \to 0$ for any $x \in X_0$,
(ii) for any $y \in Y_0$ there is a sequence $(u_n)_n$ in X such that $u_n \to 0$ and $T^n u_n \to y$,
then T is a mixing operator.

Exercise 3.1.2. Let T be an operator that satisfies Kitai's criterion with $X_0 = Y_0$. Construct an increasing sequence $(n_k)_k$ of positive integers and a dense sequence $(y_k)_k$ in Y_0 such that $x = \sum_{k=1}^{\infty} S^{n_k} y_k$ is a hypercyclic vector for T. (*Hint*: See Example 2.18 and work with an F-norm on X.)

Exercise 3.1.3. Show that an operator T satisfies Kitai's criterion if and only if there is a syndetic increasing sequence $(n_k)_k$ of positive integers such that T satisfies the Gethner-Shapiro criterion with respect to $(n_k)_k$.

Exercise 3.1.4. Let $C : L^p[0,1] \to L^p[0,1]$, $1 < p < \infty$, be the *Cesàro operator* defined as $Cf(t) = \frac{1}{t}\int_0^t f(s)\,ds$. Show that C is mixing and chaotic. (*Hint*: Use the Hahn–Banach theorem to show the density of span$\{t^\alpha\ ;\ \alpha \in A\}$ in $L^p[0,1]$, for any set $A \subset H := \{z \in \mathbb{C}\ ;\ \operatorname{Re} z > -1/p\}$ with accumulation points in H. Then apply the Godefroy–Shapiro criterion.)

Exercise 3.2.1. Give a new proof of Theorem 3.10 based on Theorem 2.47.

Exercise 3.2.2. Consider the weighted backward shift $B_w : c_0 \to c_0$, $B_w(x_1, x_2, \dots) = (w_2 x_2, w_3 x_3, \dots)$ with weight sequence

$$w = (w_1, w_2, \dots) = (1, 2, 2^{-1}, 2, 2, 2^{-2}, 2, 2, 2, 2^{-3}, \dots).$$

Show that B_w is weakly mixing but not mixing. (*Hint*: Proceed as in Example 3.11; note that this time S is not continuous, so calculate $S^{n_k} e_j$ directly.)

Exercise 3.2.3. Let T be an operator. Consider the assertions:
(i) T satisfies the Hypercyclicity Criterion with respect to the full sequence (n);
(ii) T is mixing;
(iii) T is hereditarily hypercyclic with respect to the full sequence (n).
Show that (i) $\Longrightarrow$ (ii) $\Longleftrightarrow$ (iii). (*Hint*: The proof is simple.)

Exercise 3.2.4. Show that the following "blow-up/collapse" criterion is a reformulation of the Hypercyclicity Criterion: there are dense subsets $X_0, Y_0 \subset X$ and an increasing sequence $(n_k)_k$ of positive integers such that
(i) $T^{n_k} x \to 0$ for any $x \in X_0$, and
(ii) for any $y \in Y_0$ there is a sequence $(u_k)_k$ in X such that $u_k \to 0$ and $T^{n_k} u_k \to y$.

Exercise 3.2.5. Show that every weakly mixing operator T satisfies the Hypercyclicity Criterion, without using the hereditary transitivity property of T. To do this, fix a hypercyclic vector (x, y) for $T \oplus T$ and proceed as follows:
(i) for every $k \geq 1$, the vector $(x, T^k y)$ is also hypercyclic for $T \oplus T$;
(ii) if U_k is the open ball of radius $1/k$ around 0, $k \geq 1$, then there is an increasing sequence $(n_k)_k$ of positive integers and a sequence of vectors $x_k \in U_k \cap \operatorname{orb}(y, T)$ such that $T^{n_k} x \in U_k$ and $T^{n_k} x_k \in x + U_k$, $k \geq 1$;
(iii) the dense set $X_0 = Y_0 := \operatorname{orb}(x, T)$, the sequences $(n_k)_k$ and the maps $S_{n_k} : Y_0 \to X$, $S_{n_k}(T^n x) := T^n x_k$, $k, n \geq 1$, satisfy the hypotheses of the Hypercyclicity Criterion.

Exercise 3.2.6. Show that every hypercyclic operator on the space $\omega = \mathbb{C}^{\mathbb{N}}$ of all complex sequences satisfies the Hypercyclicity Criterion. (*Hint*: Use the Herzog–Lemmert theorem, Theorem 2.65.)

Exercise 3.3.1. Prove that an operator T satisfies the Hypercyclicity Criterion if and only if there exists an increasing sequence $(n_k)_k$ of positive integers such that the following two conditions are satisfied:
(i) there exists a dense subset $X_0 \subset X$ such that $T^{n_k} x \to 0$ for any $x \in X_0$;
(ii) for any 0-neighbourhood W of X, $\bigcup_{k=1}^{\infty} T^{n_k}(W)$ is dense in X.
In the particular case when X is a Banach space, the second condition can be simplified to $\bigcup_{k=1}^{\infty} T^{n_k}(B_X)$ being dense in X, where B_X denotes the unit ball of X.

Exercise 3.3.2. A subset A of a metric space is *precompact* if for every $\varepsilon > 0$ the set A can be covered by a finite union of open balls of radius ε in the space. If the metric space is complete, then any sequence in a precompact set admits a convergent subsequence.

Let T be an operator satisfying the following criterion: there are dense subsets $X_0, Y_0 \subset X$, an increasing sequence $(n_k)_k$ of positive integers, and maps $S_{n_k} : Y_0 \to X$, $k \in \mathbb{N}$, such that, for any $x \in X_0$, $y \in Y_0$,
(i) the set $\{T^{n_k} x \; ; \; k \in \mathbb{N}\}$ is precompact,
(ii) $S_{n_k} y \to 0$,
(iii) $T^{n_k} S_{n_k} y \to y$.
Show that T satisfies the Hypercyclicity Criterion. (*Hint*: Prove that T is weakly mixing.)

Exercise 3.3.3. Consider the bilateral backward shift $T = B$ on the weighted space $\ell^1(\mathbb{Z}, v) = \{(x_n)_{n\in\mathbb{Z}} \; ; \; \|x\| := \sum_{n\in\mathbb{Z}} |x_n| v_n < \infty\}$, where $v_n = \frac{1}{n}$, $n \geq 1$, and $v_n = 1$, $n \leq 0$. Show that T is not hypercyclic. Deduce that in condition (i) of Exercise 3.3.2 one cannot replace precompactness by boundedness.

Exercise 3.3.4. Repeat Exercise 3.3.2 with conditions (i) and (ii) replaced by
(i) $T^{n_k} x \to 0$ for any $x \in X_0$,
(ii) the set $\{S_{n_k} y \; ; \; k \in \mathbb{N}\}$ is precompact for any $y \in Y_0$.

Exercise 3.3.5. Let T be a hypercyclic operator with a dense subset $Y_0 \subset X$ such that each $y \in Y_0$ admits a precompact backward orbit; see Exercise 1.2.7. Show that T satisfies the Hypercyclicity Criterion. (*Hint*: Define $X_0 = \text{orb}(x, T)$ with a hypercyclic vector x and apply Exercise 3.3.4.)

Exercise 3.3.6. An invertible operator T satisfies the Hypercyclicity Criterion if and only if T^{-1} does. Give two proofs of this result, one using Theorem 3.15, another one using Theorem 3.22. (*Hint*: Consider $\bigcup_{n=0}^{\infty} T^n(X_0)$.)

Exercise 3.3.7. Prove the following strengthened version of the Mittag-Leffler theorem. Let $(X_n)_n$ be a sequence of complete metric spaces, let $U_n \subset X_n$, $n \in \mathbb{N}$, be dense open sets and let $f_n : X_{n+1} \to X_n$, $n \in \mathbb{N}$, be continuous maps with dense range. Then, for every nonempty open subset $U \subset X_1$, there exists a sequence $(x_n)_n$ with $x_n \in U_n$, $n \in \mathbb{N}$, such that $x_1 \in U$ and $f_n(x_{n+1}) = x_n$, $n \in \mathbb{N}$.

Deduce from this the Baire category theorem.

Exercise 3.4.1. Two operators T_1, T_2 on a separable Fréchet space X are called *disjoint hypercyclic* if there is some $x \in X$ such that $\{(T_1^n x, T_2^n x) \; ; \; n \geq 0\}$ is dense in $X \oplus X$.

(a) Confirm the following obvious facts. No operator T is disjoint hypercyclic with itself. The operators T_1, T_2 are disjoint hypercyclic if and only if some $(x, x) \in X \oplus X$ is hypercyclic for $T_1 \oplus T_2$, and if and only if the sequence $(\mathcal{T}_n)_n$ of operators $\mathcal{T}_n : X \to X \oplus X$, $\mathcal{T}_n x = (T_1^n x, T_2^n x)$ is hypercyclic.

(b) Derive a corresponding transitivity theorem. The operators T_1, T_2 are disjoint hypercyclic if, for any nonempty open subsets U, V_1, V_2 of X, there exists some $n \geq 0$ such that $T_1^n(U) \cap V_1 \neq \varnothing$ and $T_2^n(U) \cap V_2 \neq \varnothing$.

(c) Derive a corresponding *Disjoint Hypercyclicity Criterion*. If there are dense subsets $X_0, Y_1, Y_2 \subset X$, an increasing sequence $(n_k)_k$ of positive integers, and maps $S_{j,n_k} : Y_j \to X$, $k \geq 1$, $j = 1, 2$, such that, for any $x \in X_0$, $y_1 \in Y_1$, $y_2 \in Y_2$,
(i) $T_j^{n_k} x \to 0$, $j = 1, 2$,
(ii) $S_{j,n_k} y_j \to 0$, $j = 1, 2$,
(iii) $T_j^{n_k} S_{j,n_k} y_j \to y_j$, $j = 1, 2$,
(iv) $T_l^{n_k} S_{j,n_k} y_j \to 0$, $j = 1, l = 2$ or $j = 2, l = 1$,
then T_1, T_2 are disjoint hypercyclic.

Exercise 3.4.2. Show that the following operators are disjoint hypercyclic:
(i) λB, μB^2, where $1 < |\lambda| < |\mu|$ and B is the backward shift on any of the spaces ℓ^p, $1 \leq p < \infty$, or c_0;
(ii) T_a, T_b, where $a, b \in \mathbb{C}$, $a \neq 0$, $b \neq 0$, $a \neq b$, and T_a is the translation operator on $H(\mathbb{C})$ given by $T_a f(z) = f(z + a)$.

Exercise 3.4.3. Formulate and prove an analogue of the Gethner–Shapiro criterion for sequences of operators $T_n : X \to X$. Show that the operators $T_n : \ell^2 \to \ell^2$, $T_n x = 2^n(x_{n+1}, \ldots, x_{2n}, 0, 0, \ldots)$, $n \geq 1$, satisfy the Hypercyclicity Criterion but not the Gethner–Shapiro criterion, so that the two criteria are no longer equivalent.

Exercise 3.4.4. Let B be the backward shift operator on ℓ^2. Define $T_n : \ell^2 \to \ell^2$ by $T_n x = 2^n B^{n+1} x + 3^n (x_1, 0, 0, \ldots)$, $n \geq 1$. Characterize the hypercyclic vectors for $(T_n)_n$. Show that $(T_n)_n$ is hereditarily hypercyclic but not topologically transitive, and therefore not hereditarily transitive.

Exercise 3.4.5. Let $(T_n)_n$ be a sequence of operators. Consider the assertions:
(i) $(T_n)_n$ satisfies the Hypercyclicity Criterion with respect to the full sequence (n);
(ii) $(T_n)_n$ is mixing;
(iii) $(T_n)_n$ is hereditarily transitive with respect to the full sequence (n).
Show that (i) $\Longrightarrow$ (ii) $\Longleftrightarrow$ (iii).

Exercise 3.4.6. Let B be the backward shift operator on ℓ^2 and consider the operators $T_n : \ell^2 \to \ell^2$, $T_n x = 2^n B^n x - x$, $n \geq 1$. Show that $(T_n)_n$ is mixing and that the only sequences x for which $T_n x \to 0$ are multiples of the sequence $(\frac{1}{2^n})_n$. Deduce that the implication (ii) $\Longrightarrow$ (i) does not hold in the previous exercise.

Exercise 3.4.7. Give an example of a hypercyclic noncommuting sequence of operators for which Lemma 2.44 does not hold.

Exercise 3.4.8. Give an example of a noncommuting sequence of operators that satisfies the hypothesis of Theorem 2.47, but is not weakly mixing.

Exercise 3.4.9. Let $(T_{j,n})_n$, $j \geq 1$, be a countable family of hereditarily transitive sequences of operators between separable Fréchet spaces X and Y. Show that there exists a dense subset X_0 of X and increasing sequences $(n_{j,k})_k$ of positive integers such that, for any $x \in X_0$, $T_{j,n_{j,k}} x \to 0$ as $k \to \infty$, $j \geq 1$. (*Hint*: Use Theorem 1.57 to produce common hypercyclic vectors for suitable sequences of operators. Note that any hypercyclic vector has a suborbit that tends to 0.)

Exercise 3.4.10. Let $(T_{j,n})_n$, $j \geq 1$, be a countable family of hereditarily transitive sequences of operators between separable Fréchet spaces X and Y. Show that there exists a dense subspace M of X such that every nonzero vector from M is hypercyclic for each sequence $(T_{j,n})_n$, $j \geq 1$.

Deduce that for every countable family of weakly mixing operators T_j, $j \geq 1$, there exists a dense subspace M of X such that every nonzero vector from M is hypercyclic for each operator T_j, $j \geq 1$.

(*Hint*: The proof is similar to that of the previous exercise; take the span of a suitably constructed dense sequence of vectors.)

Sources and comments

Sections 3.1 and 3.2. The earliest forms of the Hypercyclicity Criterion were found independently by Kitai [215] (who in addition demanded that $X_0 = Y_0$) and by Gethner and Shapiro [161, Remark 2.3(b)]. In its general form it is due to Bès and Peris [71]. The Godefroy–Shapiro criterion is contained implicitly in their paper [165] and was isolated by Bernal [55].

Intended originally as simple tools for obtaining the hypercyclicity of an operator, these criteria have become tremendously important for the understanding of linear dynamics. While the Hypercyclicity Criterion, very unexpectedly, turned out to be equivalent to the weak mixing property and therefore to be close to hypercyclicity, the Godefroy–Shapiro criterion was further developed by Bayart and Grivaux in their study of frequent hypercyclicity; see Chapter 9.

Grivaux [172] showed that there is a mixing operator T for which only $x = 0$ satisfies that $T^n x \to 0$ as $n \to \infty$. Thus, not every mixing operator satisfies the Hypercyclicity Criterion for the full sequence (n), nor Kitai's criterion; see Exercise 3.2.3.

We remark that the main idea of Example 3.8, that of considering the functions $z \to p(z)e^{-\alpha z^2}$, was taken from Birkhoff's original proof in [75].

Theorem 3.15 is due Bès and Peris [71], as is Proposition 3.19. The notion of hereditary hypercyclicity with respect to the full sequence was introduced by Shapiro [279] (who called it strong hypercyclicity) and by Ansari [9].

Section 3.3. Theorem 3.22 is due to Peris [253]; for the abstract Mittag-Leffler theorem we refer to Arens [11, Theorem 2.4]; see also Esterle [145] for an interesting discussion.

It seems to be an open problem whether the strengthened conditions of the Gethner–Shapiro criterion and of Proposition 3.20 can be combined, that is, if every operator satisfying the Hypercyclicity Criterion admits a dense subspace $X_0 \subset X$, an increasing sequence $(n_k)_k$ of positive integers, and a linear map $S : X_0 \to X_0$ such that, for any $x \in X_0$, $T^{n_k} x \to 0$, $S^{n_k} x \to 0$, and $TSx = x$.

Section 3.4. Much closer studies of the Hypercyclicity Criterion for sequences of operators were undertaken by Bernal and Grosse-Erdmann [62], Bermúdez, Bonilla, and Peris [51] and León and Müller [223].

Exercises. For Exercise 3.1.3 we refer to Bermúdez, Bonilla, Conejero, and Peris [49]. Exercise 3.1.4 is taken from León, Piqueras and Seoane [224]; the continuity of the Cesàro operators is due to Hardy's inequality [193]. We will return to these operators in Example 12.20. Exercise 3.2.5 follows the original argument in Bès and Peris [71]. Exercise 3.3.1 is taken from León and Müller [223]; see also Müller [247]. Exercises 3.3.2, 3.3.4 and 3.3.5 are due to Bermúdez, Bonilla, and Peris [51]. The notion of disjoint hypercyclic operators is due to Bernal [58] and Bès and Peris [72], where one can also find the results of Exercises 3.4.1 and 3.4.2. The example in Exercise 3.4.4 is a special case of a class of operators studied by Bernal [56]. For Exercise 3.4.9 we refer to Aron, Bès, León and Peris [14], for Exercise 3.4.10 to Bernal and Calderón [61]. Concerning the latter exercise, it is interesting to note that by a remarkable result by Grivaux [169], any countable family of hypercyclic operators on a Banach space has a common dense subspace of hypercyclic vectors, except for 0.

Extensions. Repeating a remark we made in Chapter 2 we note that the results in this chapter remain valid in separable F-spaces. For further extensions we refer to Chapter 12.

Chapter 4
Classes of hypercyclic and chaotic operators

In this chapter we study in detail some important classes of hypercyclic and chaotic operators. Each of them has its origin in the three classical hypercyclic operators. Rolewicz's multiples of backward shifts lead naturally to the study of arbitrary weighted shifts. MacLane's differentiation operator and Birkhoff's translation operators are both special cases of differential operators, while the translation operators can also be regarded as composition operators. Finally, Rolewicz's operators reappear as adjoint multipliers.

4.1 Weighted shifts

The basic model of all shifts is the *backward shift*

$$B(x_1, x_2, x_3, \dots) = (x_2, x_3, x_4, \dots).$$

In order to distinguish this shift from the bilateral shift that we will discuss later one also speaks of the *unilateral backward shift.*

Rolewicz has shown that, for any λ with $|\lambda| > 1$, the multiples of B, $\lambda B(x_n)_n = (\lambda x_{n+1})_n$, are hypercyclic on the sequence space ℓ^2. It is then a small step to let the weights vary from coordinate to coordinate, which leads to the *(unilateral) weighted shift*

$$B_w(x_1, x_2, x_3, \dots) = (w_2 x_2, w_3 x_3, w_4 x_4, \dots),$$

where

$$w = (w_n)_n$$

is a sequence of nonzero scalars, called a *weight sequence*. Note that the value of w_1 is irrelevant.

We may also generalize these operators in a different direction. Rolewicz had already replaced ℓ^2 by any of the spaces ℓ^p, $1 \leq p < \infty$, and c_0. More

K.-G. Grosse-Erdmann, A. Peris Manguillot, *Linear Chaos*, Universitext,
DOI 10.1007/978-1-4471-2170-1_4, © Springer-Verlag London Limited 2011

generally, one may take as the underlying space an arbitrary *sequence space* X, that is, a linear space of sequences or, in other words, a subspace of $\omega = \mathbb{K}^{\mathbb{N}}$. In addition, X should carry a topology that is compatible with the sequence space structure of X. We interpret this as demanding that the embedding $X \to \omega$ is continuous, that is, convergence in X should imply coordinatewise convergence. A Banach (Fréchet, ...) space of this kind is called a *Banach (Fréchet, ...) sequence space.* The terms of a sequence $x, y, z, \ldots$ will be denoted by $x_n, y_n, z_n, \ldots, n \geq 1$.

By e_n, $n \in \mathbb{N}$,

$$e_n = (\delta_{n,k})_{k\in\mathbb{N}} = (0, \ldots, 0, \underset{n}{1}, 0, \ldots)$$

we denote the canonical unit sequences. If the e_n are contained in X and span a dense subspace then an alternative way of describing weighted shifts is by saying that

$$B_w e_n = w_n e_{n-1},\ n \geq 1, \quad \text{with } e_0 := 0.$$

The continuity of the embedding $X \to \omega$ amounts to requiring the continuity of each coordinate functional

$$X \to \mathbb{K},\ x \to x_n,\ n \geq 1,$$

which implies that each weighted shift has closed graph. From the closed graph theorem (see Appendix A) we thus obtain that a weighted shift defines an operator on a Fréchet sequence space X as soon as it maps X into itself.

Proposition 4.1. *Let X be a Fréchet sequence space. Then every weighted shift $B_w : X \to X$ is continuous.*

We start by studying the (unweighted) backward shift B. Our results will then extend immediately to all weighted shifts via a simple conjugacy.

The following technical result will help us simplify the condition characterizing hypercyclicity of B.

Lemma 4.2. *Let X be a metric space, $v_n \in X$, $n \geq 1$, and $v \in X$. Suppose that there is a strictly increasing sequence $(n_k)_k$ of positive integers such that*

$$v_{n_k - j} \to v \quad \textit{for every } j \in \mathbb{N}.$$

Then there exists a strictly increasing sequence $(m_k)_k$ of positive integers such that

$$v_{m_k + j} \to v \quad \textit{for every } j \in \mathbb{N}.$$

Proof. Let d denote the metric in X. It follows from the assumption that, for any $k \geq 1$, there is some $N_k \geq k+2$ such that

$$d(v_{N_k - j}, v) < \frac{1}{k}, \quad j = 1, \ldots, k.$$

Setting $m_k = N_k - k - 1$, $k \geq 1$, we see that $d(v_{m_k+k+1-j}, v) < \frac{1}{k}$ for $j = 1, \ldots, k$, hence

$$d(v_{m_k+j}, v) < \frac{1}{k}, \quad j = 1, \ldots, k;$$

this implies the assertion when we pass to a strictly increasing subsequence of $(m_k)_k$, if necessary. □

We recall that the sequence $(e_n)_n$ is a basis in the space X if each e_n, $n \in \mathbb{N}$, belongs to X and, for any $x \in X$,

$$x = \lim_{N\to\infty} (x_1, x_2, \ldots, x_N, 0, 0, \ldots) = \sum_{n=1}^{\infty} x_n e_n.$$

Clearly, $(e_n)_n$ is a basis in any of the sequence spaces ℓ^p, $1 \leq p < \infty$, c_0 and ω.

Theorem 4.3. *Let X be a Fréchet sequence space in which $(e_n)_n$ is a basis. Suppose that the backward shift B is an operator on X. Then the following assertions are equivalent:*

(i) *B is hypercyclic;*

(ii) *B is weakly mixing;*

(iii) *there is an increasing sequence $(n_k)_k$ of positive integers such that $e_{n_k} \to 0$ in X as $k \to \infty$.*

Proof. Let $\|\cdot\|$ stand for an F-norm that induces the topology of X; see Section 2.1.

(i)$\Longrightarrow$(iii). Suppose that B is hypercyclic. Let $N \in \mathbb{N}$ and $\varepsilon > 0$. We show that there exists some $n \geq N$ with $\|e_n\| < \varepsilon$.

It follows from the basis assumption that, for any $x \in X$, the sequence $(x_n e_n)_n$ converges to 0 in X. By the Banach–Steinhaus theorem (see Appendix A), applied to the operators $x \to x_n e_n$, $n \geq 1$, there is some $\delta > 0$ such that, for all $x \in X$,

$$\|x\| < \delta \quad \Longrightarrow \quad \|x_n e_n\| < \tfrac{\varepsilon}{2} \quad \text{for all } n \geq 1. \tag{4.1}$$

Moreover, since convergence in X implies coordinatewise convergence, there is some $\eta > 0$ such that, for all $x \in X$,

$$\|x\| < \eta \quad \Longrightarrow \quad |x_1| \leq \tfrac{1}{2}. \tag{4.2}$$

Now, since B is hypercyclic and therefore topologically transitive, there are $x \in X$ and $n \geq N$ such that

$$\|x\| < \delta \quad \text{and} \quad \|B^{n-1}x - e_1\| < \eta.$$

Hence, by (4.1) and (4.2),

$$\|x_n e_n\| < \tfrac{\varepsilon}{2} \quad \text{and} \quad |x_n - 1| \le \tfrac{1}{2}; \tag{4.3}$$

the latter implies that x_n is closer to 1 than to 0 and hence that

$$|x_n^{-1} - 1| = \left|\frac{1 - x_n}{x_n}\right| \le 1.$$

From this and (4.3) we deduce, using the properties of an F-norm, that

$$\|e_n\| = \big\|(x_n^{-1} - 1)x_n e_n + x_n e_n\big\| \le \big\|(x_n^{-1} - 1)x_n e_n\| + \|x_n e_n\big\| < \varepsilon, \tag{4.4}$$

which had to be shown.

(iii)$\Longrightarrow$(ii). We apply the Hypercyclicity Criterion. For $X_0 = Y_0$ we take the set of finite sequences, which by the basis assumption is dense in X. For S_n we take the nth iterate of the forward shift

$$F : (x_1, x_2, x_3, \ldots) \to (0, x_1, x_2, \ldots),$$

that is, $S_n = F^n : Y_0 \to X$, $n \ge 1$. With this, conditions (i) and (iii) of the Hypercyclicity Criterion hold even for the full sequence (n).

As for condition (ii) note that, by continuity of B,

$$e_{n_k - j} = B^j e_{n_k} \to 0 \quad \text{as } k \to \infty,$$

for all $j \ge 1$. Since $(n_k)_k$ must be strictly increasing, it follows from Lemma 4.2 that there is an increasing sequence $(m_k)_k$ of positive integers such that

$$e_{m_k + j} \to 0 \quad \text{as } k \to \infty,$$

for all $j \ge 1$. But since $S_{m_k} e_j = e_{m_k + j}$, we have by linearity that

$$S_{m_k} y \to 0$$

for any $y \in Y_0$. This shows that conditions (i)–(iii) of the Hypercyclicity Criterion hold for the sequence $(m_k)_k$, so that B is weakly mixing.

(ii)$\Longrightarrow$(i) holds for all operators on X. □

We point out that B being an operator on X is part of the hypothesis. By Proposition 4.1 this can be restated simply as saying that $(x_{n+1})_n \in X$ whenever $(x_n)_n \in X$, which is usually easily verified for concrete spaces.

Example 4.4. (a) Let

$$\ell^p(v) = \Big\{(x_n)_{n\ge 1} \,;\, \sum_{n=1}^{\infty} |x_n|^p v_n < \infty\Big\}, \quad 1 \le p < \infty,$$

be a weighted ℓ^p-space, where $v = (v_n)_n$ is a positive weight sequence. Then B is an operator on $\ell^p(v)$ if and only if there is an $M > 0$ such that, for all

$x \in \ell^p(v)$,

$$\Big(\sum_{n=1}^{\infty} |x_{n+1}|^p v_n\Big)^{1/p} \leq M\Big(\sum_{n=1}^{\infty} |x_n|^p v_n\Big)^{1/p},$$

which is equivalent to $\sup_{n\in\mathbb{N}} \frac{v_n}{v_{n+1}} < \infty$. Theorem 4.3 tells us that, under this condition, the hypercyclicity of B is characterized by

$$\inf_{n\in\mathbb{N}} v_n = 0.$$

The same conditions also characterize the continuity and hypercyclicity of the backward shift B on the weighted c_0-space

$$c_0(v) = \big\{(x_n)_{n\geq 1}\ ;\ \lim_{n\to\infty} |x_n| v_n = 0\big\}.$$

(b) Spaces of holomorphic functions constitute a rich and interesting source of sequence spaces via the identification of a holomorphic function with its sequence of Taylor coefficients. As a first example we consider here the *Bergman space* A^2 of all holomorphic functions f on the unit disk $\mathbb{D} = \{z \in \mathbb{C}\ ;\ |z| < 1\}$ such that

$$\|f\|^2 := \frac{1}{\pi}\int_{\mathbb{D}} |f(z)|^2\, d\lambda(z) < \infty,$$

where λ denotes two-dimensional Lebesgue measure. Using polar coordinates and writing $f(z) = \sum_{n=0}^{\infty} a_n z^n$ we obtain that

$$\begin{aligned}\frac{1}{\pi}\int_{\mathbb{D}} |f(z)|^2\, d\lambda(z) &= \frac{1}{\pi}\int_0^1\int_0^{2\pi} \Big|\sum_{n=0}^{\infty} a_n (re^{it})^n\Big|^2 dt\, r\, dr\\ &= 2\int_0^1 \Big(\int_0^{2\pi} \Big|\sum_{n=0}^{\infty} a_n r^n \frac{1}{\sqrt{2\pi}}\, e^{int}\Big|^2 dt\Big) r\, dr\\ &= 2\int_0^1 \sum_{n=0}^{\infty} |a_n|^2 r^{2n} r\, dr = \sum_{n=0}^{\infty} |a_n|^2 \frac{1}{n+1},\end{aligned}$$

where we have applied Parseval's identity in $L^2[0, 2\pi]$ for the orthonormal basis $(\frac{1}{\sqrt{2\pi}}\, e^{int})_{n\in\mathbb{Z}}$. As a consequence, A^2 is isometrically isomorphic to the weighted space $\ell^2(\frac{1}{n+1})$ (with indices running from 0). By (a), the backward shift is therefore hypercyclic on A^2. When acting on functions, B is the operator

$$Bf(z) = \sum_{n=0}^{\infty} a_{n+1} z^n = \frac{1}{z}(f(z) - f(0)) \quad \text{with } Bf(0) = f'(0).$$

Further Banach and Hilbert spaces of holomorphic functions will be studied in Section 4.4.

(c) As in (b) we can consider the space $H(\mathbb{C})$ of entire functions (see Example 2.1) as a sequence space by identifying the entire function $f(z) = \sum_{n=0}^{\infty} a_n z^n$ with the sequence $(a_n)_{n\geq 0}$. By the formula for the radius of convergence of Taylor series, this sequence space is given by

$$\Big\{(a_n)_{n\geq 0}\ ;\ \lim_{n\to\infty} |a_n|^{1/n} = 0\Big\} = \Big\{(a_n)_{n\geq 0}\ ;\ \sum_{n=0}^{\infty} |a_n| m^n < \infty,\ m \geq 1\Big\}.$$

Since $|a_{n+1}|^{1/n} = (|a_{n+1}|^{1/(n+1)})^{(n+1)/n} \to 0$ if $|a_n|^{1/n} \to 0$, we have that the backward shift B is an operator on $H(\mathbb{C})$; see Proposition 4.1. Moreover, the unit sequences e_n correspond to the monomials $z \to z^n$, $n \geq 0$. It then follows from Theorem 4.3 that B is not hypercyclic on $H(\mathbb{C})$.

Using the same arguments as in the proof of Theorem 4.3, but employing Kitai's criterion instead of the Hypercyclicity Criterion, we obtain a characterization of the mixing property for B.

Theorem 4.5. *Let X be a Fréchet sequence space in which $(e_n)_n$ is a basis. Suppose that the backward shift B is an operator on X. Then the following assertions are equivalent:*
(i) *B is mixing;*
(ii) *$e_n \to 0$ in X as $n \to \infty$.*

For chaos we have a curious phenomenon. Proceeding as before, but with somewhat stronger assumptions on the space X, we easily obtain a condition that characterizes chaos for B. But it turns out that this condition is already implied by the existence of a single *nontrivial periodic point*, that is, a periodic point other than 0. Hence, this fact alone implies chaos.

For this result we will require that $(e_n)_n$ is an unconditional basis, that is, it is a basis in X such that, for any $(x_n)_n \in X$ and any 0-1-sequence $(\varepsilon_n)_n$, the series

$$\sum_{n=1}^{\infty} \varepsilon_n x_n e_n$$

converges in X; see Appendix A.

Theorem 4.6. *Let X be a Fréchet sequence space in which $(e_n)_n$ is an unconditional basis. Suppose that the backward shift B is an operator on X. Then the following assertions are equivalent:*
(i) *B is chaotic;*
(ii) *$\sum_{n=1}^{\infty} e_n$ converges in X;*
(iii) *the constant sequences belong to X;*
(iv) *B has a nontrivial periodic point.*

Proof. (i)$\Longrightarrow$(iv) is trivial.

(iv)$\Longrightarrow$(iii). Let $x = (x_1, x_2, x_3, \dots) \neq 0$ be periodic for B, that is, a periodic sequence. Let N be its period. Then there is some $j \leq N$ such that

$x_j \neq 0$, and we have $x_{j+\nu N} = x_j$ for $\nu \geq 0$. Setting all coordinates with indices other than $j + \nu N$ to zero and dividing the result by x_j we obtain, by unconditionality of the basis, that

$$\sum_{\nu=0}^{\infty} e_{j+\nu N} \in X.$$

Applying the backward shift $N - 1$ times and adding the results we obtain (iii).

(iii)$\Longrightarrow$(ii) follows from our assumptions.

(ii)$\Longrightarrow$(i). First, by Theorem 4.3, condition (ii) implies that B is hypercyclic.

Next, since $(1, 1, 1, \ldots) \in X$, the unconditionality of the basis implies that all the periodic 0-1-sequences belongs to X, and hence also all the periodic sequences, which are exactly the periodic points for B. It remains to show that these form a dense set in X.

To see this, let $x = (x_n)_n \in X$ and $\varepsilon > 0$. Since $(e_n)_n$ is a basis, there is some $N \geq 1$ such that

$$\widetilde{x} := \sum_{n=1}^{N} x_n e_n$$

has distance less than $\varepsilon/2$ from x. The associated periodic sequence

$$\sum_{\nu=0}^{\infty} \sum_{n=1}^{N} x_n e_{n+\nu N}$$

belongs to X. The unconditionality of the basis implies that there is some $m \geq 1$ such that

$$\Big\| \sum_{\nu=m}^{\infty} \sum_{n=1}^{N} x_n \varepsilon_{n+\nu N} e_{n+\nu N} \Big\| < \frac{\varepsilon}{2}$$

for any 0-1-sequence $(\varepsilon_n)_n$; see Theorem A.16. In particular we have that

$$\Big\| \sum_{\mu=1}^{\infty} \sum_{n=1}^{N} x_n e_{n+\mu m N} \Big\| < \frac{\varepsilon}{2}.$$

This shows that the periodic point

$$\sum_{\mu=0}^{\infty} \sum_{n=1}^{N} x_n e_{n+\mu m N}$$

has distance less than $\varepsilon/2$ from $\widetilde{x}$, hence less than ε from x. □

Example 4.7. (a) We consider again the space $\ell^p(v)$ of Example 4.4(a). Under the assumption that B is an operator on $\ell^p(v)$ we have that B is mixing if and only if

$$\lim_{n\to\infty} v_n = 0,$$

and it is chaotic if and only if

$$\sum_{n=1}^{\infty} v_n < \infty.$$

In this example, mixing is implied by chaos. In particular, the backward shift on the Bergman space A^2 is mixing but not chaotic.

(b) It is not difficult to give an example that shows that Theorem 4.6 does not remain valid if one drops the unconditionality assumption on the basis $(e_n)_n$; see Exercise 4.1.3.

It is now an easy matter to transfer our results so far to arbitrary weighted shifts by means of a suitable conjugacy. Let B_w be a weighted shift on some sequence space X. We define new weights v_n by

$$v_n = \Big(\prod_{\nu=1}^{n} w_\nu\Big)^{-1}, \quad n \geq 1,$$

and consider the sequence space

$$X_v = \{(x_n)_n \; ; \; (x_n v_n)_n \in X\}.$$

The map $\phi_v : X_v \to X$, $(x_n)_n \to (x_n v_n)_n$ is a vector space isomorphism.

We may use ϕ_v to transfer a topology from X to X_v: a set U is open in X_v if and only if $\phi_v(U)$ is open in X. If X is a Banach (Fréchet, ...) sequence space then so is X_v. And if $(e_n)_n$ is a basis in X then it is also a basis in X_v.

Finally, a simple calculation shows that $B_w \circ \phi_v = \phi_v \circ B$, that is, the following diagram commutes:

$$\begin{array}{ccc} X_v & \xrightarrow{\;B\;} & X_v \\ {\scriptstyle \phi_v}\downarrow & & \downarrow{\scriptstyle \phi_v} \\ X & \xrightarrow{\;B_w\;} & X. \end{array}$$

Thus $B_w : X \to X$ and $B : X_v \to X_v$ are conjugate operators.

Since conjugacies preserve hypercyclicity, (weak) mixing and chaos, our previous results immediately yield the following.

Theorem 4.8. *Let X be a Fréchet sequence space in which $(e_n)_n$ is a basis. Suppose that the weighted shift B_w is an operator on X.*

(a) *The following assertions are equivalent:*

  (i) *B_w is hypercyclic;*

  (ii) *B_w is weakly mixing;*

(iii) *there is an increasing sequence* $(n_k)_k$ *of positive integers such that*

$$\Big(\prod_{\nu=1}^{n_k} w_\nu\Big)^{-1} e_{n_k} \to 0$$

in X *as* $k \to \infty$.

(b) *The following assertions are equivalent:*

(i) B_w *is mixing;*

(ii) *we have that*

$$\Big(\prod_{\nu=1}^{n} w_\nu\Big)^{-1} e_{n} \to 0$$

in X *as* $n \to \infty$.

(c) *Suppose that the basis* $(e_n)_n$ *is unconditional. Then the following assertions are equivalent:*

(i) B_w *is chaotic;*

(ii) *the series*

$$\sum_{n=1}^{\infty}\Big(\prod_{\nu=1}^{n} w_\nu\Big)^{-1} e_{n}$$

converges in X;

(iii) *the sequence*

$$\Big(\Big(\prod_{\nu=1}^{n} w_\nu\Big)^{-1}\Big)_n$$

belongs to X;

(iv) B_w *has a nontrivial periodic point.*

Example 4.9. (a) A weighted shift B_w is an operator on a sequence space ℓ^p, $1 \le p < \infty$, or c_0 if and only if the weights w_n, $n \ge 1$, are bounded. The respective characterizing conditions for B_w to be hypercyclic, mixing or chaotic on ℓ^p, $1 \le p < \infty$, are

$$\sup_{n\ge 1}\prod_{\nu=1}^{n}|w_\nu| = \infty, \quad \lim_{n\to\infty}\prod_{\nu=1}^{n}|w_\nu| = \infty, \quad \sum_{n=1}^{\infty}\frac{1}{\prod_{\nu=1}^{n}|w_\nu|^p} < \infty.$$

We remark that only the third condition depends on the parameter p. The first condition also characterizes when B_w is hypercyclic on c_0, and the second when it is mixing or, equivalently, chaotic on c_0.

In particular, for Rolewicz's operator $T = \lambda B$, $|\lambda| > 1$, we have that $\prod_{\nu=1}^{n}|w_\nu| = \lambda^n$, which implies once more that this operator is chaotic.

As another specific example we consider, for $\alpha > 0$, the weights $w_n = (\frac{n+1}{n})^\alpha$, $n \ge 1$. Then $\prod_{\nu=1}^{n}|w_\nu| = (n+1)^\alpha$, and the corresponding weighted shift is mixing; it is even chaotic on c_0, and it is chaotic on ℓ^p exactly when $\alpha > 1/p$. We note that, for $w_n = (\frac{n+1}{n})^{1/2}$, $n \ge 1$, the weighted shift B_w

on ℓ^2 is conjugate to the backward shift on the Bergman space; see Example 4.4(b).

(b) We consider the Fréchet space $H(\mathbb{C})$ of all entire functions as a sequence space; see Example 4.4(c). It is easy to see that a weighted shift B_w defines an operator on $H(\mathbb{C})$ if and only if $\sup_{n\geq 1} |w_n|^{1/n} < \infty$; moreover, we have that $a_n e_n \to 0$ in $H(\mathbb{C})$ if and only if $|a_n|^{1/n} \to 0$; see Exercise 4.1.1. Theorem 4.8 then shows that a weighted shift B_w on $H(\mathbb{C})$ is mixing if and only if it is chaotic, and that the characterizing conditions for hypercyclicity and mixing/chaos are, respectively,

$$\sup_{n\geq 1} \Big(\prod_{\nu=1}^{n} |w_\nu|\Big)^{1/n} = \infty, \quad \lim_{n\to\infty} \Big(\prod_{\nu=1}^{n} |w_\nu|\Big)^{1/n} = \infty.$$

In particular, for the differentiation operator we have that $D(\sum_{n=0}^{\infty} a_n z^n) = \sum_{n=0}^{\infty}(n+1)a_{n+1}z^n$, so that D is a weighted shift with weight sequence $w_n = n$, $n \geq 1$. Since $(n!)^{1/n} \to \infty$ we obtain MacLane's theorem that D is hypercyclic; in fact, it is even a mixing and chaotic operator.

But in order to prove chaos for D it suffices, as we have seen, to come up with a nontrivial periodic point; the easiest such example is $f(z) = e^z$. Thus, one might be tempted to say that *the exponential function makes D chaotic.*

(c) In the space $\omega = \mathbb{K}^{\mathbb{N}}$, every series $\sum_{n=1}^{\infty} a_n e_n$ converges. Thus, every weighted shift B_w defines an operator on ω and is, indeed, mixing and chaotic on ω.

Remark 4.10. By Example 4.9(a), any bounded weight sequence $(w_n)_n$ with

$$\liminf_{n\to\infty} \prod_{\nu=1}^{n} |w_\nu| < \limsup_{n\to\infty} \prod_{\nu=1}^{n} |w_\nu| = \infty$$

defines a weighted shift B_w on ℓ^p, $1 \leq p < \infty$, or c_0 that is weakly mixing but not mixing. This provides us with a large supply of operators of this kind; see also Example 3.11.

Remark 4.11. One might wonder why we have studied backward shifts and not forward shifts. The simple truth is that a forward shift is never hypercyclic. More precisely, a *(unilateral) weighted forward shift* is given by

$$F_w(x_1, x_2, x_3, \ldots) = (0, w_1 x_1, w_2 x_2, \ldots)$$

with a weight sequence $w = (w_n)_n$. The first coordinate of every point in the orbit of x is either x_1 or 0. By the assumption that convergence in the space implies coordinatewise convergence no orbit can be dense.

Some new and interesting phenomena arise when we study shifts on sequence spaces indexed over $\mathbb{Z}$. The *bilateral backward shift* is given by

$$B(x_n)_{n\in\mathbb{Z}} = (x_{n+1})_{n\in\mathbb{Z}},$$

and the *bilateral weighted backward shifts* are given by

$$B_w(x_n)_{n\in\mathbb{Z}} = (w_{n+1}x_{n+1})_{n\in\mathbb{Z}},$$

where $w = (w_n)_{n\in\mathbb{Z}}$ is a *weight sequence*, that is, a sequence of nonzero scalars. The underlying space is then supposed to be a *Banach (Fréchet, ...) sequence space over* $\mathbb{Z}$, that is, a subspace of $\omega(\mathbb{Z}) := \mathbb{K}^{\mathbb{Z}}$ that carries a Banach (Fréchet, ...) space topology under which the embedding $X \to \omega(\mathbb{Z})$ is continuous.

In this new setting, we say that the unit sequences

$$e_n = (\delta_{n,k})_{k\in\mathbb{Z}}, \quad n \in \mathbb{Z},$$

form a basis in X if they are contained in X and if every sequence $x = (x_n)_{n\in\mathbb{Z}} \in X$ satisfies

$$x = \lim_{M,N\to\infty} (\ldots, 0, 0, x_{-M}, x_{-M+1}, \ldots, x_{N-1}, x_N, 0, 0, \ldots).$$

The *finite sequences* are the sequences in $\operatorname{span}\{e_n\ ;\ n \in \mathbb{Z}\}$.

Theorem 4.12. *Let X be a Fréchet sequence space over $\mathbb{Z}$ in which $(e_n)_{n\in\mathbb{Z}}$ is a basis. Suppose that the bilateral shift B is an operator on X.*

(a) *The following assertions are equivalent:*
  (i) *B is hypercyclic;*
  (ii) *B is weakly mixing;*
  (iii) *there is an increasing sequence $(n_k)_k$ of positive integers such that, for any $j \in \mathbb{Z}$, $e_{j-n_k} \to 0$ and $e_{j+n_k} \to 0$ in X as $k \to \infty$.*

(b) *The following assertions are equivalent:*
  (i) *B is mixing;*
  (ii) *$e_{-n} \to 0$ and $e_n \to 0$ in X as $n \to \infty$.*

(c) *Suppose that the basis $(e_n)_n$ is unconditional. Then the following assertions are equivalent:*
  (i) *B is chaotic;*
  (ii) *$\sum_{n=-\infty}^{\infty} e_n$ converges in X;*
  (iii) *the constant sequences belong to X;*
  (iv) *B has a nontrivial periodic point.*

Proof. (a), (i)$\Longrightarrow$(iii). Let $\|\cdot\|$ be an F-norm that induces the topology of X. We will derive the following equivalent formulation of (iii): *for any $\varepsilon > 0$ and any $N \in \mathbb{N}$ there exists some $n \geq N$ such that if $|j| \leq N$, then*

$$\|e_{j-n}\| < \varepsilon \quad \textit{and} \quad \|e_{j+n}\| < \varepsilon.$$

To this end, we fix $\varepsilon > 0$ and $N \in \mathbb{N}$. As in the unilateral case we can find some $\delta > 0$ such that, for all $x \in X$,

$$\|x\| < \delta \implies \|x_n e_n\| < \tfrac{\varepsilon}{2} \quad (n \in \mathbb{Z}) \quad \text{and} \quad |x_j| \leq \tfrac{1}{2} \quad (|j| \leq N). \quad (4.5)$$

Now, by the topological transitivity of B, we can find some $x \in X$ and $n > 2N$ such that

$$\Big\| x - \sum_{|j| \le N} e_j \Big\| < \delta \quad \text{and} \quad \Big\| B^n x - \sum_{|j| \le N} e_j \Big\| < \delta. \tag{4.6}$$

From (4.5) and (4.6) we obtain that

$$\|x_n e_n\| < \tfrac{\varepsilon}{2} \quad (|n| > N) \quad \text{and} \quad |x_{n+j} - 1| \le \tfrac{1}{2} \quad (|j| \le N),$$

hence

$$\|x_{j+n} e_{j+n}\| < \tfrac{\varepsilon}{2} \quad (|j| \le N) \quad \text{and} \quad |(x_{n+j})^{-1} - 1| \le 1 \quad (|j| \le N);$$

here we have used that $n > 2N$. As in (4.4) this implies that

$$\|e_{j+n}\| < \varepsilon \quad \text{for } |j| \le N.$$

On the other hand, (4.5) and (4.6) yield that

$$|x_j - 1| \le \tfrac{1}{2} \quad (|j| \le N) \quad \text{and} \quad \|x_{n+k} e_k\| < \tfrac{\varepsilon}{2} \quad (|k| > N),$$

hence

$$\big|(2x_j)^{-1}\big| \le 1 \quad (|j| \le N) \quad \text{and} \quad \|x_j e_{j-n}\| < \tfrac{\varepsilon}{2} \quad (|j| \le N),$$

whence

$$\|e_{j-n}\| = \big\|(2x_j)^{-1} 2x_j \, e_{j-n}\big\| < \varepsilon \quad \text{for } |j| \le N.$$

(iii)$\Longrightarrow$(ii). One need only observe that for the forward shift

$$F(x_n)_{n\in\mathbb{Z}} = (x_{n-1})_{n\in\mathbb{Z}}$$

we have that $BFx = x$ for any finite sequence x, and for any $j \in \mathbb{Z}$

$$B^{n_k} e_j = e_{j-n_k} \to 0, \quad F^{n_k} e_j = e_{j+n_k} \to 0,$$

so that the Hypercyclicity Criterion gives the required implication.

The implication (ii)$\Longrightarrow$(i) holds for all operators on X.

(b) The proof here is the same as that for hypercyclicity; for the sufficiency of condition (ii) one applies Kitai's criterion instead of the Hypercyclicity Criterion, while the proof of the necessity of this condition simplifies as we have only to consider the case of $j = 0$.

(c) This proof is much the same as that in the unilateral case. □

Using a suitable conjugacy this result can again be generalized immediately to weighted shifts. The conjugacy here is given by

$$\begin{array}{ccc} X_v & \xrightarrow{B} & X_v \\ \phi_v \downarrow & & \downarrow \phi_v \\ X & \xrightarrow{B_w} & X, \end{array}$$

where

$$X_v = \{(x_n)_{n\in\mathbb{Z}}\ ;\ (x_n v_n)_n \in X\}$$

and $\phi_v : X_v \to X$, $(x_n)_{n\in\mathbb{Z}} \to (x_n v_n)_{n\in\mathbb{Z}}$ with

$$v_n = \Big(\prod_{\nu=1}^{n} w_\nu\Big)^{-1} \quad \text{for } n \geq 1, \quad v_n = \prod_{\nu=n+1}^{0} w_\nu \quad \text{for } n \leq -1, \quad v_0 = 1.$$

Theorem 4.13. *Let X be a Fréchet sequence space over $\mathbb{Z}$ in which $(e_n)_{n\in\mathbb{Z}}$ is a basis. Suppose that the weighted shift B_w is an operator on X.*

(a) *The following assertions are equivalent:*
   (i) *B_w is hypercyclic;*
   (ii) *B_w is weakly mixing;*
   (iii) *there is an increasing sequence $(n_k)_k$ of positive integers such that, for any $j \in \mathbb{Z}$,*

$$\Big(\prod_{\nu=j-n_k+1}^{j} w_\nu\Big) e_{j-n_k} \to 0 \quad \textit{and} \quad \Big(\prod_{\nu=j+1}^{j+n_k} w_\nu\Big)^{-1} e_{j+n_k} \to 0$$

   in X as $k \to \infty$.

(b) *The following assertions are equivalent:*
   (i) *B_w is mixing;*
   (ii) *we have that*

$$\Big(\prod_{\nu=-n+1}^{0} w_\nu\Big) e_{-n} \to 0 \quad \textit{and} \quad \Big(\prod_{\nu=1}^{n} w_\nu\Big)^{-1} e_n \to 0$$

   in X as $n \to \infty$.

(c) *Suppose that the basis $(e_n)_{n\in\mathbb{Z}}$ is unconditional. Then the following assertions are equivalent:*
   (i) *B_w is chaotic;*
   (ii) *the series*

$$\sum_{n=-\infty}^{0} \Big(\prod_{\nu=n+1}^{0} w_\nu\Big) e_n + \sum_{n=1}^{\infty} \Big(\prod_{\nu=1}^{n} w_\nu\Big)^{-1} e_n$$

   converges in X;
   (iii) *the sequence $(x_n)_{n\in\mathbb{Z}}$ with*

$$x_n = \prod_{\nu=n+1}^{0} w_\nu \ (n \le 0), \quad x_n = \Big(\prod_{\nu=1}^{n} w_\nu\Big)^{-1} \ (n \ge 1)$$

belongs to X;

(iv) *B_w has a nontrivial periodic point.*

We see that the absence of an analogue of Lemma 4.2 leads to a more complicated characterization of hypercyclic bilateral shifts. However, for invertible bilateral shifts a simplified characterization is available; see Exercises 4.1.4 and 4.1.5.

Remark 4.14. In the bilateral case, forward shifts can be hypercyclic. A *bilateral weighted forward shift* is given by an operator

$$F_w : X \to X, \quad (x_n)_{n\in\mathbb{Z}} \to (w_{n-1}x_{n-1})_{n\in\mathbb{Z}},$$

where $w = (w_n)_{n\in\mathbb{Z}}$ is a weight sequence. It is easily seen to be conjugate to a suitable backward shift. As a result one obtains, under the same assumptions as in Theorem 4.13, that F_w is hypercyclic if and only if there is an increasing sequence $(n_k)_k$ of positive integers such that, for any $j \in \mathbb{Z}$,

$$\Big(\prod_{\nu=j-n_k}^{j-1} w_\nu\Big)^{-1} e_{j-n_k} \to 0 \quad \text{and} \quad \Big(\prod_{\nu=j}^{j+n_k-1} w_\nu\Big) e_{j+n_k} \to 0$$

in X as $k \to \infty$. The corresponding characterizations hold for the mixing property and chaos.

Example 4.15. A weighted backward shift B_w is an operator on a sequence space $\ell^p(\mathbb{Z})$, $1 \le p < \infty$ if and only if the weights w_n, $n \in \mathbb{Z}$, are bounded. Such an operator is then hypercyclic, mixing or chaotic if and only if the following conditions, respectively, are satisfied:

$$\exists (n_k)_k\ \forall j \in \mathbb{Z} : \lim_{k\to\infty} \prod_{\nu=j-n_k+1}^{j} w_\nu = 0 \quad \text{and} \quad \lim_{k\to\infty} \prod_{\nu=j+1}^{j+n_k} |w_\nu| = \infty;$$

$$\lim_{n\to\infty} \prod_{\nu=-n+1}^{0} w_\nu = 0 \quad \text{and} \quad \lim_{n\to\infty} \prod_{\nu=1}^{n} |w_\nu| = \infty;$$

$$\sum_{n=0}^{\infty} \prod_{\nu=-n+1}^{0} |w_\nu|^p < \infty \quad \text{and} \quad \sum_{n=1}^{\infty} \frac{1}{\prod_{\nu=1}^{n} |w_\nu|^p} < \infty.$$

In particular, a symmetric weight (that is, one with $w_{-n} = w_n$ for all $n \ge 0$) never defines a hypercyclic weighted shift B_w on these spaces.

As a concrete example, the weight

$$w = (\ldots, \tfrac{1}{2}, \tfrac{1}{2}, \tfrac{1}{2}, 2, 2, 2, \ldots)$$

induces a chaotic weighted backward shift on each $\ell^p(\mathbb{Z})$.

One reason for studying shifts is that they provide a rich source of examples. As a first illustration we construct a hypercyclic operator whose adjoint is also hypercyclic.

Proposition 4.16. *There exists an operator T on $\ell^2(\mathbb{Z})$ such that T and its adjoint T^* are weakly mixing, and hence hypercyclic.*

Proof. As usual, we identify the dual of $\ell^2(\mathbb{Z})$ with itself; indeed, every continuous linear functional x^* on $\ell^2(\mathbb{Z})$ is of the form

$$x^*(x) = \langle x, x^* \rangle = \sum_{n\in\mathbb{Z}} x_n y_n, \quad (x_n)_{n\in\mathbb{Z}} \in \ell^2(\mathbb{Z})$$

for a suitable sequence $y = (y_n)_{n\in\mathbb{Z}} \in \ell^2(\mathbb{Z})$.

Now let $T = B_w$ be a bilateral shift. It defines an operator on $\ell^2(\mathbb{Z})$ if and only if the w_n, $n \in \mathbb{Z}$, are bounded. Since

$$\langle B_w x, y \rangle = \sum_{n\in\mathbb{Z}} w_{n+1} x_{n+1} y_n = \sum_{n\in\mathbb{Z}} x_n w_n y_{n-1} = \langle x, F_{(w_{n+1})} y \rangle,$$

we see that the adjoint $T^* = B_w^*$ of B_w is the forward shift $F_{(w_{n+1})}$.

When we define

$$v_n = \Big(\prod_{\nu=1}^{n} w_\nu\Big)^{-1} \ (n \geq 1), \quad v_n = \prod_{\nu=n+1}^{0} w_\nu \ (n \leq -1), \quad v_0 = 1,$$

then Theorem 4.13 and Remark 4.14 tell us that B_w and $F_{(w_{n+1})}$ are weakly mixing if and only if there are increasing sequences $(n_k)_k$ and $(m_k)_k$ of positive integers such that, for any $j \in \mathbb{Z}$,

$$\begin{aligned} v_{j-n_k} &\to 0, \quad v_{j+n_k} \to 0, \\ v_{j-m_k} &\to \infty, \quad v_{j+m_k} \to \infty, \end{aligned}$$

and the continuity of B_w requires that v_n / v_{n+1}, $n \in \mathbb{Z}$, is bounded. But such a sequence is easy to find: we choose the symmetric sequence $(v_n)_{n\in\mathbb{Z}}$ with

$$(v_n)_{n\geq 0} = \left(1, 1, 2, 1, \tfrac{1}{2}, 1, 2, 4, 2, 1, \tfrac{1}{2}, \tfrac{1}{4}, \tfrac{1}{2}, 1, 2, 4, 8, 4, 2, 1, \tfrac{1}{2}, \tfrac{1}{4}, \tfrac{1}{8}, \tfrac{1}{4}, \ldots\right),$$

and the n_k are the indices of the local minima, the m_k the indices of the local maxima of this sequence. Note that B_w is even invertible. □

Remark 4.17. This proposition provides us with an example of two weakly mixing, hence hypercyclic operators $S, T : X \to X$ whose direct sum $S \oplus T$ is not hypercyclic.

We show more generally that for any operator T on a Banach space X the operator $T \oplus T^*$ cannot be hypercyclic on $X \oplus X^*$. Indeed, suppose that (x, x^*) is a hypercyclic vector for $T \oplus T^*$. If we consider $-x \in X$ as a continuous linear functional on X^* then we have for $n \geq 0$ that

$$\langle (T^n x, (T^*)^n x^*), (x^*, -x) \rangle = \langle T^n x, x^* \rangle - \langle x, (T^*)^n x^* \rangle = 0,$$

which is impossible since the left-hand side must be dense in $\mathbb{K}$; note that $(x^*, -x)$ cannot be the zero vector.

It was this observation that motivated Herrero's problem if $T \oplus T$ is hypercyclic whenever T is; see Section 2.5.

4.2 Differential operators

As the last section demonstrates, Rolewicz's result on the hypercyclicity of multiples of the backward shift has seen far-reaching generalizations. Let us turn, in the same spirit, to Birkhoff's theorem and MacLane's theorem. At first glance, the operators

$$Df(z) = f'(z) \quad \text{and} \quad T_a f(z) = f(z+a),\ a \in \mathbb{C},$$

on the space $H(\mathbb{C})$ of entire functions have little in common. But there is a surprisingly simple connection. Since

$$f(z+a) = \sum_{n=0}^{\infty} \frac{f^{(n)}(z)}{n!} a^n = \sum_{n=0}^{\infty} \frac{a^n D^n f}{n!}(z)$$

we have, at least formally, that

$$T_a = e^{aD}.$$

In fact, this representation can be justified rigorously. We will need the following notion from complex analysis: an entire function φ is said to be of *exponential type* if there are constants $M, A > 0$ such that

$$|\varphi(z)| \leq M e^{A|z|} \quad \text{for all } z \in \mathbb{C}. \tag{4.7}$$

Lemma 4.18. *An entire function $\varphi(z) = \sum_{n=0}^{\infty} a_n z^n$ is of exponential type if and only if there are $M, R > 0$ such that, for $n \geq 0$,*

$$|a_n| \leq M \frac{R^n}{n!}. \tag{4.8}$$

Proof. On the one hand, if (4.7) holds, then by the Cauchy estimates we have for any $\rho > 0$ that

$$|a_n| = \left|\frac{\varphi^{(n)}(0)}{n!}\right| \leq \frac{1}{\rho^n} \max_{|z|\leq\rho} |\varphi(z)| \leq \frac{M}{\rho^n} e^{A\rho}.$$

Setting $\rho = n/A$ and using Stirling's formula we get, with some $C > 0$,

$$|a_n| \leq \frac{MA^n}{n^n} e^n \leq CM \frac{\sqrt{n}A^n}{n!} \leq CM \frac{(2A)^n}{n!}.$$

Conversely, if (4.8) holds then

$$|\varphi(z)| \leq \sum_{n=0}^{\infty} |a_n z^n| \leq M \sum_{n=0}^{\infty} \frac{(R|z|)^n}{n!} = Me^{R|z|},$$

so that φ is of exponential type. □

Proposition 4.19. *Let*

$$\varphi(z) = \sum_{n=0}^{\infty} a_n z^n$$

be an entire function of exponential type. Then

$$\varphi(D)f = \sum_{n=0}^{\infty} a_n D^n f$$

converges in $H(\mathbb{C})$ for every entire function f and defines an operator on $H(\mathbb{C})$.

Proof. Let $f \in H(\mathbb{C})$ and $|z| \leq m$. By the Cauchy estimates and Lemma 4.18 there are $M, R > 0$ such that

$$\left|a_n f^{(n)}(z)\right| \leq |a_n| \frac{n!}{m^n} \max_{|\zeta|\leq 2m} |f(\zeta)| \leq M\left(\frac{R}{m}\right)^n \max_{|\zeta|\leq 2m} |f(\zeta)|. \tag{4.9}$$

Therefore, if $m > R$ then $\sum_{n=0}^{\infty} a_n f^{(n)}(z)$ converges uniformly on $|z| \leq m$. Hence

$$\varphi(D)f = \sum_{n=0}^{\infty} a_n D^n f$$

converges in $H(\mathbb{C})$. Moreover, by (4.9), writing $p_m(f) = \max_{|z|\leq m} |f(z)|$, we have for $m > R$ that

$$p_m(\varphi(D)f) \leq M \frac{1}{1 - R/m} p_{2m}(f).$$

This shows that $\varphi(D)$ is an operator on $H(\mathbb{C})$; see Proposition 2.11. □

We will call the operators $\varphi(D)$ simply *differential operators* on $H(\mathbb{C})$. They include all finite-order differential operators

$$T = a_0 I + a_1 D + \ldots + a_m D^m.$$

Proposition 4.19, in particular, justifies our earlier calculation concerning Birkhoff's operators that

$$T_a = \varphi(D) \quad \text{with } \varphi(z) = e^{az}. \tag{4.10}$$

The following result gives a useful description of the differential operators $\varphi(D)$ among the operators on $H(\mathbb{C})$.

Proposition 4.20. *Let T be an operator on $H(\mathbb{C})$. Then the following assertions are equivalent:*

(i) *$T = \varphi(D)$ for some entire function φ of exponential type;*
(ii) *T commutes with D;*
(iii) *T commutes with each $T_a, a \in \mathbb{C}$.*

Proof. (i)$\Longrightarrow$(ii). Let $T = \varphi(D)$. By the continuity of D we have for $f \in H(\mathbb{C})$

$$TDf = \sum_{n=0}^{\infty} a_n D^n(Df) = \sum_{n=0}^{\infty} D(a_n D^n f) = DTf.$$

(ii)$\Longrightarrow$(iii). By the same token, using (4.10), we obtain for $f \in H(\mathbb{C})$ that

$$TT_a f = T \sum_{n=0}^{\infty} \frac{a^n}{n!} D^n f = \sum_{n=0}^{\infty} \frac{a^n}{n!} TD^n f = \sum_{n=0}^{\infty} \frac{a^n}{n!} D^n(Tf) = T_a T f.$$

(iii)$\Longrightarrow$(i). By continuity of $f \to (Tf)(0)$ there is some $M > 0$ and some $R \in \mathbb{N}$ such that

$$|(Tf)(0)| \leq M \max_{|z| \leq R} |f(z)|, \quad f \in H(\mathbb{C}).$$

Denoting by e_n, $n \geq 0$, the monomials $e_n(z) = z^n$ we define

$$a_n = \frac{(Te_n)(0)}{n!}$$

and deduce that

$$|a_n| \leq M \frac{R^n}{n!}.$$

It follows from Lemma 4.18 and Proposition 4.19 that $\varphi(z) = \sum_{n=0}^{\infty} a_n z^n$ defines an entire function φ of exponential type and that $\varphi(D) = \sum_{n=0}^{\infty} a_n D^n$ defines an operator on $H(\mathbb{C})$. Then

$$(\varphi(D) e_n)(0) = a_n n! = (Te_n)(0), \; n \geq 0.$$

Since the monomials span a dense subspace of $H(\mathbb{C})$, we obtain that

$$(\varphi(D)f)(0) = (Tf)(0) \quad \text{for } f \in H(\mathbb{C}).$$

By what we have shown above we also know that $\varphi(D)$ commutes with each T_a. Thus we get with (iii) for any $z \in \mathbb{C}$ and $f \in H(\mathbb{C})$, using the definition of T_z,

$$\begin{aligned}(\varphi(D)f)(z) &= (T_z\varphi(D)f)(0) = (\varphi(D)T_zf)(0)\\ &= (TT_zf)(0) = (T_zTf)(0) = Tf(z),\end{aligned}$$

so that $T = \varphi(D)$. □

With this in hand we can prove a remarkably general common extension of the theorems of Birkhoff and MacLane.

Theorem 4.21 (Godefroy–Shapiro). *Suppose that* $T : H(\mathbb{C}) \to H(\mathbb{C})$, $T \neq \lambda I$, *is an operator that commutes with* D, *that is,*

$$TD = DT.$$

Then T *is mixing and chaotic.*

Proof. By Proposition 4.20 we can write $T = \varphi(D)$ for some entire function

$$\varphi(z) = \sum_{n=0}^{\infty} a_n z^n$$

of exponential type. Our additional assumption implies that φ is nonconstant. It is now easy to verify that T satisfies the conditions of the Godefroy–Shapiro criterion. In fact, considering the exponential functions

$$e_\lambda(z) = e^{\lambda z}, \quad \lambda \in \mathbb{C},$$

we calculate that

$$Te_\lambda = \varphi(D)e_\lambda = \sum_{n=0}^{\infty} a_n \lambda^n e_\lambda = \varphi(\lambda)e_\lambda.$$

Thus each e_λ is an eigenvector of T to the eigenvalue $\varphi(\lambda)$. Consequently,

$$\text{span}\{f \in H(\mathbb{C}) \ ; \ Tf = \mu f \text{ for some } \mu \in \mathbb{C} \text{ with } |\mu| < 1\}$$

contains $\text{span}\{e_\lambda \ ; \ |\varphi(\lambda)| < 1\}$, which is dense in $H(\mathbb{C})$ by Lemma 2.34; indeed, since any nonconstant entire function has dense range (see Appendix A), $\{\lambda \in \mathbb{C} \ ; \ |\varphi(\lambda)| < 1\}$ is a nonempty open set and therefore has an accumulation point. For the same reason, the eigenvectors of T to eigenvalues μ with $|\mu| > 1$ span a dense set in $H(\mathbb{C})$. For the density of

$$\text{span}\{f \in H(\mathbb{C}) \ ; \ Tf = e^{\alpha\pi i}f \text{ for some } \alpha \in \mathbb{Q}\}$$

it suffices to observe that also the set $\{\lambda \in \mathbb{C}\ ;\ \varphi(\lambda) = e^{\alpha\pi i} \text{ for some } \alpha \in \mathbb{Q}\}$ has an accumulation point. Indeed, since $\varphi(\mathbb{C})$ is connected and dense, it must intersect the unit circle. And since nonconstant holomorphic functions are open mappings, infinitely many preimages under φ of roots of unity lie in some bounded subset of $\mathbb{C}$ and therefore have an accumulation point. □

Having established the hypercyclicity of every differential operator $T = \varphi(D) \neq \lambda I$ we now want to focus our attention on properties of the corresponding hypercyclic functions. MacLane had already addressed such a problem: he showed that there exists a D-hypercyclic entire function f of exponential type 1, which means that for every $\varepsilon > 0$ there is some $M > 0$ such that

$$|f(z)| \leq M e^{(1+\varepsilon)r} \quad \text{for all } z \in \mathbb{C}.$$

Here we follow the usual convention of writing $r = |z|$. MacLane's growth condition can be improved, and one can even determine the least possible rate of growth.

Theorem 4.22. (a) *Let $\phi :]0,\infty[\to [1,\infty[$ be a function with $\phi(r) \to \infty$ as $r \to \infty$. Then there exists an entire function f that is hypercyclic for D and that satisfies*

$$|f(z)| \leq M\phi(r)\frac{e^r}{\sqrt{r}} \quad \textit{for } |z| = r > 0$$

with some $M > 0$.

(b) *There is no entire function f that is hypercyclic for D and that satisfies*

$$|f(z)| \leq M\frac{e^r}{\sqrt{r}} \quad \textit{for } |z| = r > 0$$

with some $M > 0$.

Proof. (a) The assertion suggests consideration of the space

$$X = \Big\{ f \in H(\mathbb{C})\ ;\ \|f\| := \sup_{r=|z|>0} \frac{\sqrt{r}\,|f(z)|}{\phi(r)e^r} < \infty \Big\}.$$

Proving our assertion then amounts to showing that the sequence of operators

$$T_n : X \to H(\mathbb{C}), \quad f \to f^{(n)}, \quad n \geq 0$$

admits a dense orbit in the sense of Section 3.4; note that the T_n are indeed operators because the inclusion map $X \to H(\mathbb{C})$ is obviously continuous.

To prove that $(T_n)_n$ is hypercyclic we apply the Hypercyclicity Criterion for sequences of operators, Theorem 3.24.

It is an easy exercise to see that X is a Banach space. We would like to take as X_0 the set of polynomials, but we cannot guarantee that the polynomials are dense in X. Thus we replace X by the closure $\overline{X_0}$ of X_0 in X. Clearly $T_n f \to 0$ for any $f \in X_0$. For Y_0 we take the set of polynomials in $H(\mathbb{C})$, and we define $S_n = S^n : Y_0 \to \overline{X}_0$ using the antiderivative operator $Sf(z) = \int_0^z f(\zeta)d\zeta$. Then $T_nS_nf = f$, $n \in \mathbb{N}_0$, for any $f \in Y_0$.

It remains to show that $S_nf \to 0$ in $\overline{X}_0$ for any polynomial f. By linearity we may assume that f is a monomial $e_n(z) = z^n$, and because $S_ne_k = \frac{k!}{(n+k)!}e_{n+k} = k!S_{n+k}e_0$ it suffices to consider $f = e_0$. For this we find that

$$\|S_ne_0\| = \left\|\frac{e_n}{n!}\right\| = \sup_{r>0}\frac{r^{n+1/2}}{n!\phi(r)e^r}.$$

A simple calculation shows that

$$\sup_{r>0}\frac{r^{n+1/2}}{n!e^r} = \frac{(n+1/2)^{n+1/2}}{n!e^{n+1/2}},$$

and Stirling's formula implies that this is bounded in $n \geq 0$ by some constant C. Fixing $\varepsilon > 0$, and letting $R > 0$ be such that $\phi(r) > 1/\varepsilon$ for $r \geq R$ we obtain that

$$\|S_ne_0\| \leq \frac{R^{n+1/2}}{n!} + \sup_{r\geq R}\frac{r^{n+1/2}\varepsilon}{n!e^r} \leq \frac{R^{n+1/2}}{n!} + C\varepsilon,$$

which implies that $S_ne_0 \to 0$ in X and therefore in $\overline{X}_0$.

(b) Let $f \in H(\mathbb{C})$. Under the assumed growth condition we have by the Cauchy estimates that, for any $n \in \mathbb{N}_0$ and $\rho > 0$,

$$|f^{(n)}(0)| \leq \frac{n!}{\rho^n}\max_{|z|\leq\rho}|f(z)| \leq M\frac{n!}{\rho^n\sqrt{\rho}}e^{\rho}.$$

Choosing $\rho = n$ we get

$$|f^{(n)}(0)| \leq M\frac{n!}{n^{n+1/2}}e^n,$$

which is bounded by Stirling's formula. Thus, f cannot be hypercyclic for D.
□

Exercise 4.2.5 explains how the critical rate of growth $e^r/\sqrt{r}$ is related to the differentiation operator.

In contrast to the result for MacLane's operator, entire functions that are hypercyclic for Birkhoff's operators can grow arbitrarily slowly. The proof requires a different technique and will be provided in Chapter 8; see Exercise 8.1.3.

Theorem 4.23 (Duyos-Ruiz). *Let $a \neq 0$. Let $\phi :]0,\infty[\to [1,\infty[$ be a function so that, for any $N \geq 1$, $\phi(r)/r^N \to \infty$. Then there exists an entire function f that is hypercyclic for T_a and that satisfies*

$$|f(z)| \leq M\phi(r) \quad \textit{for } |z| = r > 0$$

with some $M > 0$.

By the method used in the proof of Theorem 4.21 one can derive certain possible rates of growth for an arbitrary operator $\varphi(D)$; see Exercise 4.2.4.

4.3 Composition operators I

As we have seen, operators may often be interpreted in various ways. MacLane's operator is both a differential operator and a weighted shift. Birkhoff's operators are differential operators as well. Here now we have another interpretation of Birkhoff's operators T_a: they are special composition operators. Writing

$$\tau_a(z) = z + a$$

we see that τ_a is an entire function such that

$$T_a f = f \circ \tau_a.$$

In fact, τ_a is even an automorphism of $\mathbb{C}$, that is, a bijective entire function. These observations serve as the starting point of another major investigation: the hypercyclicity of general composition operators.

Let Ω be an arbitrary domain in $\mathbb{C}$, that is, a nonempty connected open set. An *automorphism* of Ω is a bijective holomorphic function

$$\varphi : \Omega \to \Omega;$$

its inverse is then also holomorphic. The set of all automorphisms of Ω is denoted by $\mathrm{Aut}(\Omega)$. Now, for $\varphi \in \mathrm{Aut}(\Omega)$ the corresponding *composition operator* is defined as

$$C_\varphi f = f \circ \varphi,$$

that is, $(C_\varphi f)(z) = f(\varphi(z))$, $z \in \Omega$.

What about the underlying space? Following Birkhoff we consider the space $H(\Omega)$ of all holomorphic functions on Ω which we endow, as in the case $\Omega = \mathbb{C}$, with the topology of local uniform convergence. To describe this topology by seminorms we need an *exhaustion* of Ω by compact sets, that is, an increasing sequence of compact sets $K_n \subset \Omega$ such that each compact set $K \subset \Omega$ is contained in some K_n.

Lemma 4.24. *Every domain* $\Omega \subset \mathbb{C}$ *has an exhaustion of compact sets.*

Proof. For each $n \in \mathbb{N}$ we consider the grid of all points $x + iy$ in $\mathbb{C}$ so that either x or y is an integer multiple of $\frac{1}{2^n}$; then let K_n be the (finite) union of all closed squares that have their sides lying on the grid and that lie entirely in $\Omega \cap \{z : |z| < n\}$. It is obvious that $(K_n)_n$ is an exhaustion of Ω. □

Now, if $(K_n)_n$ is an exhaustion of Ω then we endow $H(\Omega)$ with the topology induced by the seminorms

$$p_n(f) = \sup_{z \in K_n} |f(z)|, \quad n \in \mathbb{N}.$$

In this way $H(\Omega)$ turns into a Fréchet space; note that the topology is independent of the chosen exhaustion. Moreover, by Runge's theorem, $H(\Omega)$ is separable; see Exercise 4.3.1.

Clearly, for any automorphism φ of Ω the composition operator C_φ is continuous on $H(\Omega)$. Let us first note that conformal maps, that is, holomorphic bijections between two domains, induce conjugacies between the corresponding composition operators; the proof is immediate.

Proposition 4.25. *Let Ω_1 and Ω_2 be domains in $\mathbb{C}$ and $\psi : \Omega_1 \to \Omega_2$ a conformal map. If φ_1 and φ_2 are automorphisms of Ω_1 and Ω_2, respectively, such that $\varphi_2 \circ \psi = \psi \circ \varphi_1$ then C_{φ_2} and C_{φ_1} are conjugate via the map $J : H(\Omega_2) \to H(\Omega_1), f \to f \circ \psi$, that is, the diagram*

$$\begin{array}{ccc} H(\Omega_2) & \xrightarrow{C_{\varphi_2}} & H(\Omega_2) \\ {\scriptstyle J}\downarrow & & \downarrow{\scriptstyle J} \\ H(\Omega_1) & \xrightarrow{C_{\varphi_1}} & H(\Omega_1) \end{array}$$

commutes.

Example 4.26. Any two Birkhoff operators T_a, T_b, $a, b \neq 0$, are conjugate. This follows immediately by taking $\psi(z) = \frac{b}{a}z$, $z \in \mathbb{C}$, since $\tau_b \circ \psi = \psi \circ \tau_a$.

We turn to the problem of determining which composition operators are hypercyclic. The crucial concept will be the notion of a run-away sequence.

Definition 4.27. Let Ω be a domain in $\mathbb{C}$ and $\varphi_n : \Omega \to \Omega$, $n \geq 1$, holomorphic maps. Then the sequence $(\varphi_n)_n$ is called a *run-away sequence* if, for any compact subset $K \subset \Omega$, there is some $n \in \mathbb{N}$ such that

$$\varphi_n(K) \cap K = \varnothing.$$

We will usually apply this definition to the sequence $(\varphi^n)_n$ of iterates of an automorphism φ on Ω. Let us consider two examples. Another important example will be studied below; see Proposition 4.36.

Example 4.28. (a) Let $\Omega = \mathbb{C}$. Then the automorphisms of $\mathbb{C}$ are the functions

$$\varphi(z) = az + b, \quad a \neq 0, b \in \mathbb{C},$$

and $(\varphi^n)_n$ is run-away if and only if $a = 1$, $b \neq 0$.

Indeed, let φ be an automorphism of $\mathbb{C}$. If φ is not a polynomial then, by the Casorati–Weierstrass theorem, $\varphi(\{z \in \mathbb{C}\ ;\ |z| > 1\})$ is dense in $\mathbb{C}$ and therefore intersects the set $\varphi(\mathbb{D})$, which is open by the open mapping theorem. Since this contradicts injectivity, φ must be a polynomial. Again by injectivity, its degree must be one, so that φ is of the stated form. Now, if $a = 1$ then $\varphi^n(z) = z + nb$, so that we have the run-away property if and only if $b \neq 0$; while if $a \neq 1$ then $(1-a)^{-1}b$ is a fixed point of φ so that $(\varphi^n)_n$ cannot be run-away.

(b) Let $\Omega = \mathbb{C}^* = \mathbb{C} \setminus \{0\}$, the punctured plane. An argument as in (a) shows that the automorphisms of $\mathbb{C}^*$ are the functions

$$\varphi(z) = az \quad \text{or} \quad \varphi(z) = \frac{a}{z}, \quad a \neq 0.$$

Then $(\varphi^n)_n$ is run-away if and only if $\varphi(z) = az$ with $|a| \neq 1$.

We first show that the run-away property is a necessary condition for the hypercyclicity of the composition operator.

Proposition 4.29. *Let Ω be a domain in $\mathbb{C}$ and $\varphi \in \mathrm{Aut}(\Omega)$. If C_φ is hypercyclic then $(\varphi^n)_n$ is a run-away sequence.*

Proof. If $(\varphi^n)_n$ is not run-away then there exists a compact set $K \subset \Omega$ and elements $z_n \in K$ such that

$$\varphi^n(z_n) \in K, \quad n \in \mathbb{N}. \tag{4.11}$$

Now suppose that $f \in H(\Omega)$ is a hypercyclic vector for C_φ. Let $M = \sup_{z\in K} |f(z)|$. Then, by (4.11), we have that

$$\inf_{z\in K} |((C_\varphi)^n f)(z)| \leq |((C_\varphi)^n f)(z_n)| = |f(\varphi^n(z_n))| \leq M,$$

so that the functions $(C_\varphi)^n f$ cannot approximate, for example, the constant function $M + 1$ uniformly on K, a contradiction. □

Corollary 4.30. *There is no automorphism of $\mathbb{C}^*$ whose composition operator is hypercyclic.*

Proof. By Proposition 4.29 and Example 4.28(b), C_φ can only be hypercyclic on $H(\mathbb{C}^*)$ if $\varphi(z) = az$ with $|a| \neq 1$. Suppose that $f \in H(\mathbb{C}^*)$ is hypercyclic for such a function φ. If $f(z) = \sum_{n\in\mathbb{Z}} c_n z^n$ then

$$\int_{\mathbb{T}} ((C_\varphi)^n f - \tfrac{1}{z})\, dz = \int_{\mathbb{T}} f(a^n z)\, dz - \int_{\mathbb{T}} \tfrac{1}{z}\, dz = 2\pi i\big(\tfrac{c_{-1}}{a^n} - 1\big),$$

where the unit circle $\mathbb{T}$ is positively oriented. By hypercyclicity, there is a sequence $(n_k)_k$ for which the left-hand side converges to zero, unlike the right-hand side, which is a contradiction. □

Thus, when Ω is $\mathbb{C}^*$ then the run-away property is not a sufficient condition for hypercyclicity. Our goal now is to show that in essentially all other cases, hypercyclicity is characterized by the run-away property. To this end we need to introduce some topological properties of plane sets.

We denote by $\widehat{\mathbb{C}} = \mathbb{C} \cup \{\infty\}$ the one-point compactification of $\mathbb{C}$. A domain Ω is called *simply connected* if $\widehat{\mathbb{C}} \setminus \Omega$ is connected. A domain Ω is called *finitely connected* if $\widehat{\mathbb{C}} \setminus \Omega$ contains at most finitely many connected components, otherwise it is *infinitely connected.* If M is any set in $\mathbb{C}$ then one also speaks of a bounded component of $\widehat{\mathbb{C}} \setminus M$ as a *hole.* In that sense, finitely connected domains have only finitely many holes, a simply connected domain has no hole.

We first deal with finitely, not simply connected domains Ω. One can show that unless such a domain is conformally equivalent to $\mathbb{C}^*$, that is, unless there is a holomorphic bijection between Ω and $\mathbb{C}^*$, Ω does not admit an automorphism φ so that $(\varphi^n)_n$ is run-away; we omit the proof. By Proposition 4.29 and Corollary 4.30 we therefore have the following.

Proposition 4.31. *Let Ω be a finitely connected but not simply connected domain in $\mathbb{C}$. Then C_φ is not hypercyclic for any automorphism of Ω.*

In all other cases we have the following.

Theorem 4.32. *Let Ω be a domain in $\mathbb{C}$ that is either simply connected or infinitely connected. Let $\varphi \in \text{Aut}(\Omega)$. Then C_φ is hypercyclic if and only if $(\varphi^n)_n$ is a run-away sequence.*

In view of Proposition 4.29 we only have to prove sufficiency of the run-away property. For this we need to study the geometry of domains more closely, at least for infinitely connected domains. A compact subset K of a domain Ω will be called *Ω-convex* if every hole of K contains a point of $\mathbb{C} \setminus \Omega$; see Figure 4.1. Of course, in a simply connected domain, Ω-convexity only says that K has no holes, in other words, that its complement is connected.

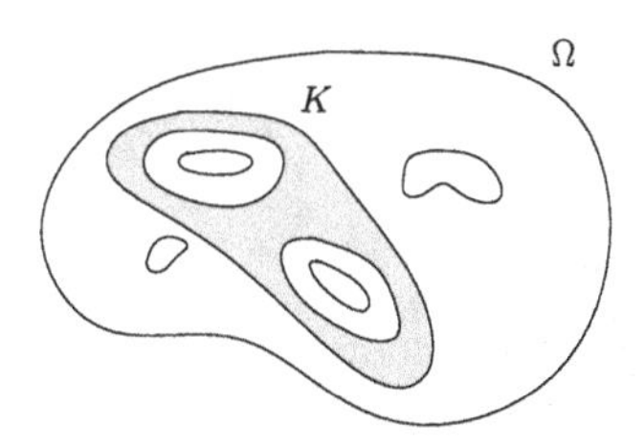

Fig. 4.1 An Ω-convex set K

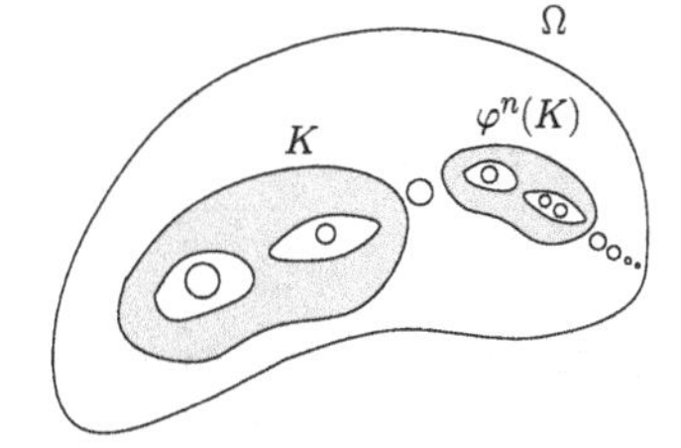

Fig. 4.2 $\varphi^n(K) \cup K$ is Ω-convex

The following auxiliary result will be crucial for the proof of sufficiency in Theorem 4.32. However, since its proof is rather technical we will postpone it to the end of the section. For later use we formulate the lemma for arbitrary sequences $(\varphi_n)_n$ of automorphisms.

Lemma 4.33. *Let Ω be an infinitely connected domain in $\mathbb{C}$ and $(\varphi_n)_n$ a run-away sequence of automorphisms of Ω. Then every compact subset of Ω is contained in some Ω-convex compact subset K of Ω for which there is some $n \in \mathbb{N}$ such that $\varphi_n(K) \cap K = \varnothing$ and $\varphi_n(K) \cup K$ is Ω-convex.*

Proof of Theorem 4.32 *(sufficiency).* Suppose that $(\varphi^n)_n$ is a run-away sequence. We want to show that then C_φ is topologically transitive. Let $f, g \in H(\Omega)$, let L be a compact subset of Ω and $\varepsilon > 0$. Then there is a compact subset K of Ω containing L and an $n \in \mathbb{N}$ such that $\varphi^n(K) \cap K = \varnothing$ and $\varphi^n(K) \cup K$ is Ω-convex (see Figure 4.2); in the simply connected case one can take any Ω-convex compact set K containing L, in the infinitely connected case one applies Lemma 4.33. Then the function $g \circ (\varphi^n)^{-1}$ is holomorphic on some neighbourhood of $\varphi^n(K)$, and f is holomorphic on some neighbourhood of K. It follows from Runge's theorem that there is a function $h \in H(\Omega)$ such that

$$\sup_{z \in K} |f(z) - h(z)| < \varepsilon \quad \text{and} \quad \sup_{z \in \varphi^n(K)} \left| g \circ (\varphi^n)^{-1}(z) - h(z) \right| < \varepsilon,$$

hence

$$\sup_{z \in L} |f(z) - h(z)| < \varepsilon \quad \text{and} \quad \sup_{z \in L} |g(z) - h(\varphi^n(z))| < \varepsilon.$$

As in Example 2.20 this implies that C_φ is topologically transitive. $\square$

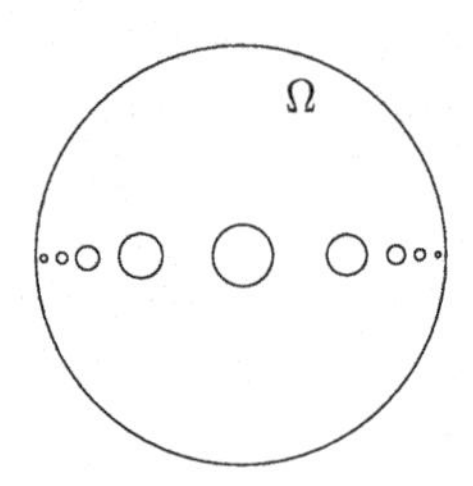

Fig. 4.3 The set Ω (Example 4.34)

Example 4.34. We give an example of a hypercyclic composition operator on an infinitely connected domain. We start with the unit disk $\mathbb{D}$ and the automorphism

$$\varphi(z) = \frac{z - \frac{1}{2}}{1 - \frac{1}{2}z}$$

of $\mathbb{D}$; see also Proposition 4.36. Let $A = \{z : |z| \leq \frac{1}{10}\}$. It is easy to see that the forward and backward iterates $\varphi^n(A)$, $n \in \mathbb{Z}$, of A are pairwise disjoint. Then

$$\Omega := \mathbb{D} \setminus \bigcup_{n\in\mathbb{Z}} \varphi^n(A)$$

is an infinitely connected domain (see Figure 4.3), and the restriction of φ to Ω is an automorphism of Ω. Moreover, a simple calculation shows that

$$\varphi^n(z) = \frac{z - a_n}{1 - a_n z} \quad \text{with} \quad a_n = \frac{3^n - 1}{3^n + 1}, \ n \geq 0,$$

so that $\lim_{n\to\infty} \varphi^n(z) = -1$, uniformly on compact subsets of Ω. Hence, $(\varphi^n)_n$ is a run-away sequence on Ω, which implies that C_φ is hypercyclic on $H(\Omega)$.

Remark 4.35. Any hypercyclic composition operator C_φ on a domain Ω is even weakly mixing. To see this, let $(K_n)_n$ be an exhaustion of Ω by compact sets. Since $(\varphi^n)_n$ is run-away, there is some m_1 such that $\varphi^{m_1}(K_1)\cap K_1 = \varnothing$. If $L = K_2 \cup \bigcup_{k=1}^{m_1} \varphi^k(K_1)$, there is some m_2 such that $\varphi^{m_2}(L)\cap L = \varnothing$. Then, in particular, $\varphi^{m_2}(K_2)\cap K_2 = \varnothing$; moreover, since $\varphi^{m_2}(L)$ contains $\varphi^{m_2}(K_1)$ and L contains $\varphi^k(K_1)$ for $k = 1, \ldots, m_1$, we must have that $m_2 > m_1$. Proceeding inductively we obtain a strictly increasing sequence $(m_n)_n$ such that $\varphi^{m_n}(K_n) \cap K_n = \varnothing$, for any $n \in \mathbb{N}$; as a consequence, $(\varphi^{m_n})_n$ and any of its subsequences is run-away. The proofs in this section then show that every subsequence of $(C_{\varphi^{m_n}})_n$ admits a dense orbit. This tells us that C_φ is hereditarily hypercyclic, and hence weakly mixing by Theorem 3.15.

We want to study the case of simply connected domains in greater detail. If $\Omega = \mathbb{C}$, the automorphisms are given by

$$\varphi(z) = az + b, \quad a \neq 0, b \in \mathbb{C},$$

and C_φ is hypercyclic if and only if $a = 1$, $b \neq 0$; see Example 4.28(a) and Theorem 4.32. Thus the hypercyclic composition operators on $\mathbb{C}$ are precisely Birkhoff's translation operators.

Let us now consider the simply connected domains Ω other than $\mathbb{C}$. By the Riemann mapping theorem, Ω is conformally equivalent to the unit disk, that is, there is a conformal map $\psi : \mathbb{D} \to \Omega$. By Proposition 4.25 it suffices to study the case when $\Omega = \mathbb{D}$.

Proposition 4.36. *The automorphisms of $\mathbb{D}$ are the linear fractional transformations*

$$\varphi(z) = b\frac{a - z}{1 - \overline{a}z}, \quad |a| < 1, \ |b| = 1.$$

Moreover, φ maps $\mathbb{T}$ bijectively onto itself.

Proof. We first consider the maps

$$h_a(z) = \frac{a - z}{1 - \overline{a}z}, \quad |a| < 1.$$

A simple calculation shows that, for $w = h_a(z)$,

$$1 - |w|^2 = \frac{1 - |a|^2}{|1 - \overline{a}z|^2}\,(1 - |z|^2). \tag{4.12}$$

Hence $\mathbb{D}$ and $\mathbb{T}$ are invariant under h_a. Moreover one finds that $h_a \circ h_a = I$ on $\overline{\mathbb{D}}$. This implies that h_a is an automorphism of $\mathbb{D}$ that maps $\mathbb{T}$ bijectively onto itself. The same is then true for bh_a, $|b| = 1$.

Conversely, let φ be an automorphism of $\mathbb{D}$, and let $0 = \varphi(a)$ with $|a| < 1$. Then the map $f := \varphi \circ h_a^{-1}$ is also an automorphism of $\mathbb{D}$ with $f(0) = 0$. The Schwarz lemma then implies that $|f(z)| \leq |z|$ for $z \in \mathbb{D}$. The same argument applied to the inverse of f shows that $|f^{-1}(z)| \leq |z|$, hence $|z| \leq |f(z)|$ for $z \in \mathbb{D}$. Altogether we have that $|f(z)| = |z|$ for $z \in \mathbb{D}$. Again by the Schwarz lemma, f can only be a rotation, that is, there is some b with $|b| = 1$ such that $f(z) = bz$ and therefore $\varphi = bh_a$. □

Now, linear fractional transformations are a very well understood class of holomorphic maps; see Appendix A. Using their properties it is not difficult to determine the dynamical behaviour of the corresponding composition operators; via conjugacy these results can then be carried over to arbitrary simply connected domains.

Theorem 4.37. *Let Ω be a simply connected domain and $\varphi \in \mathrm{Aut}(\Omega)$. Then the following assertions are equivalent:*
(i) *C_φ is hypercyclic;*
(ii) *C_φ is mixing;*
(iii) *C_φ is chaotic;*
(iv) *$(\varphi^n)_n$ is a run-away sequence;*
(v) *φ has no fixed point in Ω;*
(vi) *C_φ is quasiconjugate to a Birkhoff operator.*

Proof. The implications (vi)$\Longrightarrow$(iii) and (vi)$\Longrightarrow$(ii) follow from known properties of the Birkhoff operators, (iii)$\Longrightarrow$(i) and (ii)$\Longrightarrow$(i) hold for all operators on $H(\Omega)$, and (i)$\Longleftrightarrow$(iv) was proved in Theorem 4.32.

(i)$\Longrightarrow$(v). If φ has a fixed point $z_0 \in \Omega$ then, for any $f \in H(\Omega)$ and $n \geq 0$, $((C_\varphi)^n f)(z_0) = f(\varphi^n(z_0)) = f(z_0)$, so that f cannot have a dense orbit.

It remains to prove that (v)$\Longrightarrow$(vi). In the case $\Omega = \mathbb{C}$ the result was shown in Example 4.28(a). In the case $\Omega \neq \mathbb{C}$ we can assume by the discussion leading up to Proposition 4.36 that $\Omega = \mathbb{D}$. The proof then requires certain properties of linear fractional transformations. Since we will have occasion to study them in Section 4.5 we will postpone the proof to the end of that section. □

In particular the final condition in the theorem is of great interest. Any property of the Birkhoff operators that is preserved under quasiconjugacies will transmit to all composition operators on simply connected domains.

It remains to give the proof of Lemma 4.33.

Proof of Lemma 4.33. We consider the exhaustion of Ω by compact sets K_n constructed in the proof of Lemma 4.24. Then each K_n is automatically Ω-convex. But also $\psi(K_n)$ is Ω-convex for every automorphism ψ. Indeed, if some hole of $\psi(K_n)$ contained only points of Ω then one could deform the boundary of that hole continuously in Ω to a point in Ω; applying the map ψ^{-1}, the same would then be true for the corresponding hole of K_n, contradicting the Ω-convexity of K_n.

By the run-away property, we can find a strictly increasing sequence $(m_n)_n$ of positive integers such that $\varphi_{m_n}(K_n) \cap K_n = \varnothing$ for $n \geq 1$; see Remark 4.35.

Now, every compact subset of Ω is contained in some K_N, and since Ω is infinitely connected we can assume that K_N has at least two holes. Then, for all $n \geq N$, $\varphi_{m_n}(K_N) \cap K_N = \varnothing$. Also, each $\varphi_{m_n}(K_N)$ is Ω-convex. To finish the proof it suffices to show that there is some $n \geq N$ such that, in addition, $\varphi_{m_n}(K_N) \cup K_N$ is Ω-convex.

We distinguish three cases. First, if there is some $n \geq N$ such that $\varphi_{m_n}(K_N)$ lies in the unbounded component of the complement of K_N and K_N lies in the unbounded component of the complement of $\varphi_{m_n}(K_N)$ then clearly $\varphi_{m_n}(K_N) \cup K_N$ is Ω-convex.

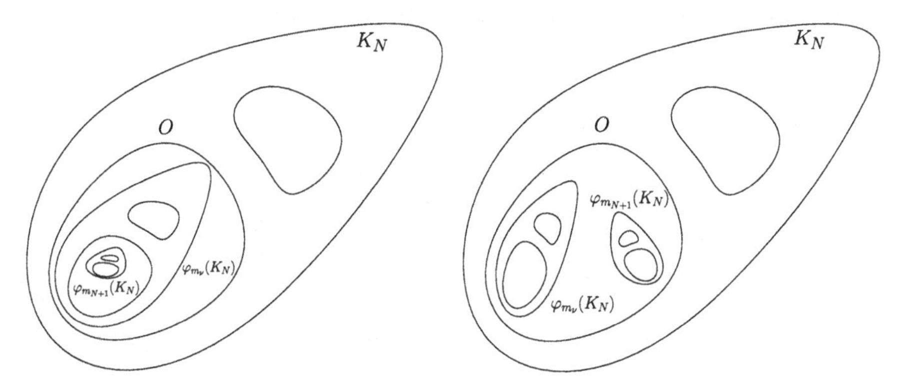

Fig. 4.4 $\varphi_{m_{N+1}}(K_N) \cup K_N$ is Ω-convex

Fig. 4.5 Both $\varphi_{m_{N+1}}(K_N) \cup K_N$ and $\varphi_{m_\nu}(K_N) \cup K_N$ are Ω-convex

Secondly, suppose that infinitely many $\varphi_{m_n}(K_N)$, $n \geq N$, lie in holes of K_N. Since K_N only has a finite number of holes, infinitely many $\varphi_{m_n}(K_N)$, $n \geq N$, must lie in some fixed hole O of K_N; by passing to a subsequence we may assume that this is true for all $n > N$. We then choose some $\nu > N$ such that $\varphi_{m_{N+1}}(K_N) \subset K_\nu$. Since $\varphi_{m_\nu}(K_\nu) \cap K_\nu = \varnothing$ we have that $\varphi_{m_{N+1}}(K_N)$ and $\varphi_{m_\nu}(K_N)$ are disjoint subsets of O. Now one has three possibilities: either $\varphi_{m_{N+1}}(K_N)$ lies in a hole of $\varphi_{m_\nu}(K_N)$ (see Figure 4.4), or $\varphi_{m_\nu}(K_N)$ lies in a hole of $\varphi_{m_{N+1}}(K_N)$, or both sets lie in the unbounded component of the complement of the other set (see Figure 4.5). Since both sets have at least two holes one finds that in each of these cases either $\varphi_{m_{N+1}}(K_N) \cup K_N$ or $\varphi_{m_\nu}(K_N) \cup K_N$ is Ω-convex.

Finally, the remaining case is when for infinitely many $n \geq N$, K_N lies in a hole of $\varphi_{m_n}(K_N)$. Again we can assume that this is true for all $n > N$. We then choose some $\nu > N$ such that $\varphi_{m_{N+1}}(K_N) \subset K_\nu$. As above we find that $\varphi_{m_{N+1}}(K_N)$ and $\varphi_{m_\nu}(K_N)$ are disjoint sets. Since both these sets contain K_N in one of their holes, we have that either $\varphi_{m_{N+1}}(K_N)$ lies in a hole of $\varphi_{m_\nu}(K_N)$, or vice versa. Since both sets have two holes we find that either $\varphi_{m_\nu}(K_N) \cup K_N$ or $\varphi_{m_{N+1}}(K_N) \cup K_N$ is Ω-convex. □

4.4 Adjoint multipliers

In this section we consider an interesting generalization of the backward shift operator. The underlying space will be the Hardy space H^2. Arguably its easiest definition is the following. If $(a_n)_{n\geq 0}$ is a complex sequence such that

$$\sum_{n=0}^{\infty} |a_n|^2 < \infty,$$

then it is, in particular, bounded, and hence

$$f(z) = \sum_{n=0}^{\infty} a_n z^n, \quad z \in \mathbb{C}, |z| < 1,$$

defines a holomorphic function on the complex unit disk $\mathbb{D}$. The Hardy space is then defined as the space of these functions, that is,

$$H^2 = \Big\{ f : \mathbb{D} \to \mathbb{C} \,;\, f(z) = \sum_{n=0}^{\infty} a_n z^n, z \in \mathbb{D}, \text{ with } \sum_{n=0}^{\infty} |a_n|^2 < \infty \Big\}.$$

In other words, the Hardy space is simply the sequence space $\ell^2(\mathbb{N}_0)$, with its elements written as holomorphic functions. It is then clear that H^2 is a Banach space under the norm

$$\|f\| = \Big(\sum_{n=0}^{\infty} |a_n|^2 \Big)^{1/2} \quad \text{when } f(z) = \sum_{n=0}^{\infty} a_n z^n,$$

and it is even a Hilbert space under the inner product

$$\langle f, g \rangle = \sum_{n=0}^{\infty} a_n \overline{b_n} \quad \text{when } f(z) = \sum_{n=0}^{\infty} a_n z^n, \; g(z) = \sum_{n=0}^{\infty} b_n z^n.$$

The polynomials form a dense subspace of H^2.

The following result is an immediate consequence of the definitions.

Proposition 4.38. *For any $\lambda \in \mathbb{D}$ define $k_\lambda : \mathbb{D} \to \mathbb{C}$ by*

$$k_\lambda(z) = \sum_{n=0}^{\infty} \overline{\lambda}^n z^n = \frac{1}{1 - \overline{\lambda} z}.$$

Then $k_\lambda \in H^2$ and, for any $f \in H^2$,

$$f(\lambda) = \langle f, k_\lambda \rangle.$$

This implies that for any $\lambda \in \mathbb{D}$ the point evaluation

$$f \to f(\lambda)$$

is a continuous linear functional on H^2. The functions k_λ, $\lambda \in \mathbb{D}$, are called *reproducing kernels*. They will play the same role here as the exponential functions $e_\lambda \in H(\mathbb{C})$, $\lambda \in \mathbb{C}$, in Section 4.2. In particular we have the following analogue of Lemma 2.34.

Lemma 4.39. *Let $\Lambda \subset \mathbb{D}$ be a set with an accumulation point in $\mathbb{D}$. Then the set*

$$\operatorname{span}\{k_\lambda \ ; \ \lambda \in \Lambda\}$$

is dense in H^2.

Proof. It suffices to show that only the zero function can be orthogonal to $\operatorname{span}\{k_\lambda \ ; \ \lambda \in \Lambda\}$. But that is immediate by the identity theorem for holomorphic functions: if, for $f \in H^2$, $\langle f, k_\lambda \rangle = f(\lambda)$ vanishes for all $\lambda \in \Lambda$, then $f = 0$. $\square$

The operators that we want to study are those that map $f \in H^2$ to φf, where φ is a bounded holomorphic function on $\mathbb{D}$. In order to see that this defines an operator on H^2 we need another representation of the space.

Proposition 4.40. *A holomorphic function $f : \mathbb{D} \to \mathbb{C}$ belongs to H^2 if and only if*

$$\sup_{0 \le r < 1} \int_0^{2\pi} |f(re^{it})|^2 \, dt < \infty.$$

Moreover, for any $f, g \in H^2$,

$$\|f\| = \Big(\sup_{0 \le r < 1} \frac{1}{2\pi} \int_0^{2\pi} |f(re^{it})|^2 \, dt \Big)^{1/2} = \Big(\lim_{r \nearrow 1} \frac{1}{2\pi} \int_0^{2\pi} |f(re^{it})|^2 \, dt \Big)^{1/2}$$

and

$$\langle f, g \rangle = \lim_{r \nearrow 1} \frac{1}{2\pi} \int_0^{2\pi} f(re^{it}) \overline{g(re^{it})} \, dt.$$

Proof. Writing $f(z) = \sum_{n=0}^{\infty} a_n z^n$ we obtain that

$$\frac{1}{2\pi}\int_0^{2\pi}|f(re^{it})|^2\,dt = \frac{1}{2\pi}\int_0^{2\pi}\Big|\sum_{n=0}^{\infty}a_n(re^{it})^n\Big|^2\,dt$$
$$= \int_0^{2\pi}\Big|\sum_{n=0}^{\infty}a_n r^n\frac{1}{\sqrt{2\pi}}e^{int}\Big|^2\,dt = \sum_{n=0}^{\infty}|a_n|^2r^{2n},$$

where we have used Parseval's identity in $L^2[0,2\pi]$ for the orthonormal basis $(\frac{1}{\sqrt{2\pi}}\,e^{int})_{n\in\mathbb{Z}}$; see also Example 4.4(b). Since

$$\sup_{0\le r<1}\sum_{n=0}^{\infty}|a_n|^2r^{2n} = \lim_{r\nearrow 1}\sum_{n=0}^{\infty}|a_n|^2r^{2n} = \sum_{n=0}^{\infty}|a_n|^2 = \|f\|^2,$$

the first part of the assertion follows. In the same way, one obtains the second part by using Parseval's identity for the inner product. □

Now let φ be a bounded holomorphic function on $\mathbb{D}$. Then, for any $f\in H^2$, φf is holomorphic on $\mathbb{D}$, and we have that

$$\sup_{0\le r<1}\frac{1}{2\pi}\int_0^{2\pi}|(\varphi f)(re^{it})|^2\,dt \le \sup_{z\in\mathbb{D}}|\varphi(z)|^2\sup_{0\le r<1}\frac{1}{2\pi}\int_0^{2\pi}|f(re^{it})|^2\,dt,$$

so that also $\varphi f\in H^2$ by the previous proposition. Moreover, we see that

$$M_\varphi f = \varphi f$$

defines an operator on H^2 with $\|M_\varphi\|\le\sup_{z\in\mathbb{D}}|\varphi(z)|$. The function φ is called a *multiplier* of H^2, M_φ is called the corresponding *multiplication operator* or briefly *multiplier*.

Clearly, multiplication operators are never hypercyclic. For if $(M_\varphi)^n f = \varphi^n f$, $n\ge 0$, formed a dense set in H^2 then, by continuity of point evaluations, the same would be true of the sequence $(\varphi(0)^n f(0))_{n\ge 0}$ in $\mathbb{C}$, which is never the case. Instead, we will consider the (Hilbert space) adjoint $M_\varphi^*: H^2\to H^2$ of M_φ, called an *adjoint multiplication operator* or *adjoint multiplier*.

In fact, we already know that some operators M_φ^* are hypercyclic, as we will see now. As usual, B and F denote the backward and forward shifts on $\ell^2(\mathbb{N}_0)$, respectively, which are operators of norm 1. Hence, if $\varphi(z)=\sum_{n=0}^\infty a_nz^n$ is holomorphic on some neighbourhood of $\overline{\mathbb{D}}$ then $\sum_{n=0}^\infty\|a_nB^n\|\le\sum_{n=0}^\infty|a_n|<\infty$, so that

$$\varphi(B)=\sum_{n=0}^{\infty}a_nB^n$$

defines an operator on $\ell^2(\mathbb{N}_0)$, and the same is true for $\varphi(F)=\sum_{n=0}^\infty a_nF^n$; see also Appendix B.

Proposition 4.41. *Let $\varphi(z)=\sum_{n=0}^\infty a_nz^n$ be holomorphic on a neighbourhood of $\overline{\mathbb{D}}$, and set $\varphi^*(z)=\sum_{n=0}^\infty\overline{a_n}z^n$. Then, via the identification of H^2 with $\ell^2(\mathbb{N}_0)$:*

(i) *the multiplier M_φ corresponds to the operator $\varphi(F)$ on $\ell^2(\mathbb{N}_0)$;*
(ii) *the adjoint multiplier M_φ^* corresponds to the operator $\varphi^*(B)$ on $\ell^2(\mathbb{N}_0)$.*

Proof. (i) On the one hand we have that for $f \in H^2$, $f(z) = \sum_{k=0}^\infty b_k z^k$,

$$M_\varphi f(z) = \sum_{n=0}^\infty a_n z^n \sum_{k=0}^\infty b_k z^k = \sum_{n=0}^\infty \Big(\sum_{k=0}^n a_{n-k} b_k\Big) z^n.$$

On the other hand we have for $(b_k)_{k\geq 0} \in \ell^2(\mathbb{N}_0)$,

$$\begin{aligned}\varphi(F)(b_k)_k &= \sum_{n=0}^\infty a_n F^n (b_k)_k = \sum_{n=0}^\infty a_n (0,\ldots,0,b_0,b_1,\ldots)\\ &= (a_0 b_0, a_1 b_0 + a_0 b_1, a_2 b_0 + a_1 b_1 + a_0 b_2, \ldots) = \Big(\sum_{k=0}^n a_{n-k} b_k\Big)_{n\geq 0}.\end{aligned}$$

This implies the result.

(ii) A simple calculation shows that B is the adjoint of F. Hence, by (i), the adjoint M_φ^* corresponds to $\varphi(F)^* = (\sum_{n=0}^\infty a_n F^n)^* = \sum_{n=0}^\infty \overline{a_n} B^n$, where we have used properties of the adjoint; see Proposition A.8. □

In particular, the adjoint multipliers M_φ^* with $\varphi(z) = \lambda z$, $|\lambda| > 1$, correspond to the Rolewicz operators $\overline{\lambda} B$ and are therefore hypercyclic.

The Godefroy–Shapiro criterion allows us to characterize the hypercyclic adjoint multipliers. We can exclude constant multipliers because their adjoint multiplication operators are multiples of the identity.

Theorem 4.42. *Let φ be a nonconstant bounded holomorphic function on $\mathbb{D}$ and let M_φ^* be the corresponding adjoint multiplier on H^2. Then the following assertions are equivalent:*
(i) *M_φ^* is hypercyclic;*
(ii) *M_φ^* is mixing;*
(iii) *M_φ^* is chaotic;*
(iv) *$\varphi(\mathbb{D}) \cap \mathbb{T} \neq \varnothing$.*

Proof. Suppose that condition (iv) holds. Considering the reproducing kernels k_λ, $\lambda \in \mathbb{D}$, we find that, for all $f \in H^2$,

$$\langle f, M_\varphi^* k_\lambda\rangle = \langle \varphi f, k_\lambda\rangle = (\varphi f)(\lambda) = \langle f, \overline{\varphi(\lambda)}\, k_\lambda\rangle,$$

which shows that

$$M_\varphi^*\, k_\lambda = \overline{\varphi(\lambda)}\, k_\lambda.$$

Consequently,

$$\operatorname{span}\{f \in H^2 \;;\; M_\varphi^* f = \mu f \text{ for some } \mu \in \mathbb{C} \text{ with } |\mu| < 1\}$$

contains $\operatorname{span}\{k_\lambda \,;\, |\varphi(\lambda)| < 1\}$, which is dense in H^2 by Lemma 4.39; indeed, since nonconstant holomorphic functions are open mappings, condition (iv) implies that $\{\lambda \in \mathbb{D} \,;\, |\varphi(\lambda)| < 1\}$ is nonempty and open and therefore contains an accumulation point in $\mathbb{D}$. For the same reason the eigenvectors of M_φ^* to eigenvalues of modulus greater than 1 span a dense set in H^2. Finally, the same is true for the eigenvectors of M_φ^* to eigenvalues that are roots of unity. For this it suffices to show that $\{\lambda \in \mathbb{D} \,;\, \varphi(\lambda) \text{ is a root of unity}\}$ has an accumulation point. But since φ is an open mapping, condition (iv) implies that infinitely many preimages of roots of unity lie in some relatively compact subset of $\mathbb{D}$ and therefore have an accumulation point in $\mathbb{D}$. By the Godefroy–Shapiro criterion, therefore, (iv) implies (ii) and (iii).

To finish the proof it suffices to show that (i) implies (iv). Let us suppose that $\varphi(\mathbb{D})$ does not intersect the unit circle. Since $\varphi(\mathbb{D})$ is connected, it must lie entirely inside or entirely outside $\mathbb{D}$. If $\varphi(\mathbb{D}) \subset \mathbb{D}$ then

$$\|M_\varphi^*\| = \|M_\varphi\| \leq \sup_{z\in\mathbb{D}} |\varphi(z)| \leq 1$$

(see Proposition A.8), and hence M_φ^* cannot be hypercyclic. On the other hand, if $\varphi(\mathbb{D}) \subset \mathbb{C} \setminus \overline{\mathbb{D}}$ then $\psi := 1/\varphi$ is a bounded holomorphic function on $\mathbb{D}$ with $\psi(\mathbb{D}) \subset \mathbb{D}$, which implies that M_ψ^* cannot be hypercyclic. But M_φ is the inverse of M_ψ and therefore M_φ^* is the inverse of M_ψ^*; see Proposition A.8. By Proposition 2.23, M_φ^* cannot be hypercyclic. □

This result can easily be extended to more general Hilbert spaces of holomorphic functions, for example the Bergman space (see Exercise 4.4.3); but see also Exercise 4.4.4. An extension to some Banach spaces X of holomorphic functions is also possible, in which case, of course, M_φ^* is the Banach space adjoint on the dual X^*; see Exercise 4.4.5.

We pass on to another, partial generalization of Theorem 4.42 that is motivated by Proposition 4.41. Under the assumptions of that proposition, $\varphi(B) = \sum_{n=0}^\infty a_n B^n$ defines an operator on each of the spaces ℓ^p, $1 \leq p < \infty$, and c_0.

Theorem 4.43. *Let X be one of the complex sequence spaces ℓ^p, $1 \leq p < \infty$, or c_0. Furthermore, let φ be a nonconstant holomorphic function on a neighbourhood of $\overline{\mathbb{D}}$. Then the following assertions are equivalent:*

(i) *$\varphi(B)$ is chaotic;*
(ii) *$\varphi(\mathbb{D}) \cap \mathbb{T} \neq \varnothing$;*
(iii) *$\varphi(B)$ has a nontrivial periodic point.*

Proof. (ii)$\Longrightarrow$(i). We saw in Example 3.2 that any sequence

$$e_\lambda := (\lambda, \lambda^2, \lambda^3, \ldots), \quad |\lambda| < 1$$

is an eigenvector of B to the eigenvalue λ and that, for any set $\Lambda \subset \mathbb{D}$ that has an accumulation point in $\mathbb{D}$, the set

$$\operatorname{span}\{e_\lambda \; ; \; \lambda \in \Lambda\}$$

is dense in X. Now, for any $\lambda \in \mathbb{D}$ we have that

$$\varphi(B)e_\lambda = \sum_{n=0}^{\infty} a_n B^n e_\lambda = \sum_{n=0}^{\infty} a_n \lambda^n e_\lambda = \varphi(\lambda)e_\lambda,$$

so that each e_λ is also an eigenvector of $\varphi(B)$ to the eigenvalue $\varphi(\lambda)$. From here we proceed exactly as in the proof of Theorem 4.42 to show that $\varphi(B)$ is chaotic.

(i)$\Longrightarrow$(iii) is trivial.

(iii)$\Longrightarrow$(ii). For this implication we need to use results from spectral theory; see Appendix B. By condition (iii) there is some point $x \neq 0$ from X and some $N \geq 1$ such that $\varphi^N(B)x = \varphi(B)^N x = x$. Thus, $1 \in \sigma_p(\varphi^N(B))$, the point spectrum of $\varphi^N(B)$. It follows from the point spectral mapping theorem (see Theorem B.7) that $1 = \varphi^N(\lambda)$ for some $\lambda \in \sigma_p(B) = \mathbb{D}$; see Example 3.2. Hence $\varphi(\lambda) \in \mathbb{T}$, which implies (ii). □

Example 4.44. The theorem shows in particular that any operator

$$I + \lambda B \quad \text{and} \quad e^{\lambda B}, \quad \lambda \neq 0$$

is hypercyclic (and even chaotic) on $X = \ell^p$, $1 \leq p < \infty$, or c_0. In Section 8.1 we will see that much more is true: for any weight sequence w for which the backward shift B_w is an operator on X, the operators $I + B_w$ and e^{B_w} are hypercyclic (and even mixing); see Theorems 8.1 and 8.2.

4.5 Composition operators II

In this section we return to the composition operators studied in Section 4.3, but we consider them now on the Hardy space H^2. Thus, let φ be an automorphism of the unit disk $\mathbb{D}$ and let

$$C_\varphi f = f \circ \varphi$$

be the corresponding composition operator, where we now demand that f belongs to H^2. The first problem is that of determining if this defines an operator on H^2.

Proposition 4.45. *For any $\varphi \in \operatorname{Aut}(\mathbb{D})$, C_φ defines an operator on H^2.*

Proof. By Proposition 4.36 there are $a, b \in \mathbb{C}$ with $|a| < 1$ and $|b| = 1$ such that

$$\varphi(z) = b\frac{a - z}{1 - \overline{a}z}.$$

First, let f be a polynomial. Then f and $f\circ\varphi$ are continuous on $\overline{\mathbb{D}}$ so that, by Proposition 4.40,

$$\|f\circ\varphi\|^2 = \frac{1}{2\pi}\int_0^{2\pi} \left|f(\varphi(e^{it}))\right|^2\,dt, \tag{4.13}$$

and similarly for $\|f\|^2$. Also by Proposition 4.36, φ is a bijective self-map on $\mathbb{T}$ so that there is some $u_0\in\mathbb{R}$ and a continuously differentiable function $u:[0,2\pi]\to[u_0,u_0+2\pi]$ such that

$$e^{iu(t)} = \varphi(e^{it}),\quad t\in[0,2\pi].$$

Differentiating with respect to t we obtain that

$$ie^{iu(t)}\frac{du}{dt} = ie^{it}\varphi'(e^{it}),\quad t\in[0,2\pi],$$

so that (4.13) and the substitution $u=u(t)$ yield

$$\|f\circ\varphi\|^2 = \frac{1}{2\pi}\int_{u_0}^{u_0+2\pi} |f(e^{iu})|^2\frac{1}{|\varphi'(e^{it(u)})|}\,du.$$

Since, for $|z|=1$,

$$|\varphi'(z)| = \left|b\frac{a\overline{a}-1}{(1-\overline{a}z)^2}\right| \geq \frac{1-|a|^2}{(1+|a|)^2} = \frac{1-|a|}{1+|a|},$$

we deduce that

$$\|f\circ\varphi\|^2 \leq \frac{1+|a|}{1-|a|}\cdot\frac{1}{2\pi}\int_{u_0}^{u_0+2\pi}|f(e^{iu})|^2\,du = \frac{1+|a|}{1-|a|}\cdot\|f\|^2. \tag{4.14}$$

Now let $f\in H^2$ be arbitrary, and let f_n, $n\geq 0$, be the partial sums of its Taylor series. By Proposition 4.40 and (4.14) we have for $n\geq 0$ and $0\leq r<1$ that

$$\frac{1}{2\pi}\int_0^{2\pi}|f_n(\varphi(re^{it}))|^2\,dt \leq \|f_n\circ\varphi\|^2 \leq \frac{1+|a|}{1-|a|}\cdot\|f_n\|^2.$$

Letting $n\to\infty$ and noting that $f_n\to f$ in H^2 and locally uniformly on $\mathbb{D}$ we deduce that, for $0\leq r<1$,

$$\frac{1}{2\pi}\int_0^{2\pi}|f(\varphi(re^{it}))|^2\,dt \leq \frac{1+|a|}{1-|a|}\cdot\|f\|^2.$$

By Proposition 4.40 this shows that $f\circ\varphi\in H^2$ and that C_φ is continuous on H^2. □

The proof also gives us a norm estimate on the operator C_φ, namely

$$\|C_\varphi\| \le \Big(\frac{1+|a|}{1-|a|}\Big)^{1/2}.$$

Our aim now is to characterize when C_φ is hypercyclic on H^2. To this end we need to study the (nonlinear) dynamical system that is described by the automorphism φ on $\mathbb{D}$. It will be convenient to consider φ as a particular linear fractional transformation; see Appendix A.

Indeed, let

$$\varphi(z) = \frac{az+b}{cz+d}, \quad ad-bc \neq 0,$$

be an arbitrary linear fractional transformation, which we consider as a map on the extended complex plane $\widehat{\mathbb{C}}$. Then φ has either one or two fixed points in $\widehat{\mathbb{C}}$, or it is the identity.

Suppose that φ has a single fixed point z_0, and let σ be a linear fractional transformation that maps z_0 to ∞. Then $\psi := \sigma\circ\varphi\circ\sigma^{-1}$ has ∞ as a unique fixed point, which easily implies that $\psi(z) = z + c$ for some $c \neq 0$.

Now suppose that φ has two distinct fixed points z_0 and z_1, and let σ be a linear fractional transformation that maps z_0 to 0 and z_1 to ∞. Then $\psi := \sigma\circ\varphi\circ\sigma^{-1}$ has fixed points 0 and ∞, which easily implies that $\psi(z) = \lambda z$ for some $\lambda \neq 0$. The constant λ is called the *multiplier* of φ. Replacing σ by $1/\sigma$ one sees that also $1/\lambda$ is a multiplier, which, however, causes no problem in the following.

Definition 4.46. Let φ be a linear fractional transformation that is not the identity.

(a) If φ has a single fixed point then it is called *parabolic*.

(b) Suppose that φ has two fixed points, and let λ be its multiplier. If $|\lambda| = 1$ then φ is called *elliptic*; if $\lambda > 0$ then φ is called *hyperbolic*; in all other cases, φ is called *loxodromic*.

It is now easy to deduce some important dynamical properties of automorphisms φ of $\mathbb{D}$.

Proposition 4.47. *Let $\varphi \in \operatorname{Aut}(\mathbb{D})$, not the identity. Then we have the following:*

(i) *if φ is parabolic then its fixed point z_0 lies in $\mathbb{T}$, and $\varphi^n(z) \to z_0$, $\varphi^{-n}(z) \to z_0$ for all $z \in \widehat{\mathbb{C}}$;*

(ii) *if φ is elliptic then it has a fixed point in $\mathbb{D}$;*

(iii) *if φ is hyperbolic then it has distinct fixed points z_0 and z_1 in $\mathbb{T}$ such that $\varphi^n(z) \to z_0$ for all $z \in \widehat{\mathbb{C}}$, $z \neq z_1$, and $\varphi^{-n}(z) \to z_1$ for all $z \in \widehat{\mathbb{C}}$, $z \neq z_0$;*

(iv) *φ cannot be loxodromic.*

Proof. In the various cases, let σ and ψ be the linear fractional transformations given above. Then σ provides a conjugacy between φ and ψ.

(i) If φ is parabolic then $\psi^n(z) = z + nc \to \infty$ for all $z \in \mathbb{C}$ and hence $\varphi^n(z) \to \sigma^{-1}(\infty) = z_0$ for all $z \in \widehat{\mathbb{C}}$. In the same way, $\varphi^{-n}(z) \to z_0$ for all

$z \in \widehat{\mathbb{C}}$. Since φ maps $\mathbb{T}$ into $\mathbb{T}$ (see Proposition 4.36), we must have that $z_0 \in \mathbb{T}$.

(ii) Let φ be elliptic. Since φ maps $\mathbb{D}$ onto itself, ψ maps $\sigma(\mathbb{D})$ onto itself, which is either a half-plane, or the interior or the exterior of a disk U. Since ψ is a rotation and $\lambda \neq 1$, the first alternative is excluded and U must be centred at 0. Thus, either 0 or ∞ lies in $B = \sigma(\mathbb{D})$, so that either z_0 or z_1 belongs to $\mathbb{D}$.

(iii) Let φ be hyperbolic. Then $\lambda > 0$, and since with λ also $1/\lambda$ is a multiplier we can assume that $\lambda < 1$. Then $\psi^n(z) = \lambda^n z \to 0$ for all $z \in \widehat{\mathbb{C}}$, $z \neq \infty$, and therefore $\varphi^n(z) \to \sigma^{-1}(0) = z_0$ for all $z \in \widehat{\mathbb{C}}$, $z \neq \sigma^{-1}(\infty) = z_1$. It follows as in (i) that $z_0 \in \mathbb{T}$. Moreover, we find that $\psi^{-n}(z) = \lambda^{-n} z \to \infty$ for all $z \in \widehat{\mathbb{C}}$, $z \neq 0$, and hence $\varphi^{-n}(z) \to \sigma^{-1}(\infty) = z_1$ for all $z \in \widehat{\mathbb{C}}$, $z \neq \sigma^{-1}(0) = z_0$. Since $\mathbb{T}$ is also invariant under φ^{-1} we find that also $z_1 \in \mathbb{T}$.

(iv) Let φ be loxodromic. As in (ii), ψ maps $\sigma(\mathbb{D})$ onto itself, which is either a half-plane, or the interior or the exterior of a disk. But this is incompatible with the fact that $|\lambda| \neq 1$ and $\lambda \not> 0$. □

The dynamical properties of φ imply the dynamical properties of C_φ.

Theorem 4.48. *Let $\varphi \in \mathrm{Aut}(\mathbb{D})$ and C_φ be the corresponding composition operator on H^2. Then the following assertions are equivalent:*

(i) *C_φ is hypercyclic;*
(ii) *C_φ is mixing;*
(iii) *φ has no fixed point in $\mathbb{D}$.*

Proof. The implication (ii)$\Longrightarrow$(i) holds for all operators on H^2, and (i)$\Longrightarrow$(iii) follows as in the proof of Theorem 4.37, using the fact that point evaluations are continuous on H^2.

(iii)$\Longrightarrow$(ii). Suppose that φ has no fixed point in $\mathbb{D}$. It suffices to show that C_φ satisfies Kitai's criterion. By Proposition 4.47, φ is either parabolic or hyperbolic, and in both cases there are $z_0, z_1 \in \mathbb{T}$ (possibly with $z_0 = z_1$) such that $\varphi^n(z) \to z_0$ for all $z \in \mathbb{T}\backslash\{z_1\}$ and $\varphi^{-n}(z) \to z_1$ for all $z \in \mathbb{T}\backslash\{z_0\}$.

Now, for X_0 we will take the subspace of H^2 of all functions that are holomorphic on a neighbourhood of $\overline{\mathbb{D}}$ and that vanish at z_0. To see that X_0 is dense in H^2, let $f \in H^2$, $f(z) = \sum_{n=0}^{\infty} a_n z^n$, be orthogonal to any $g \in X_0$. Since, for any $n \geq 0$, the functions $g_n : z \to z_0 z^n - z^{n+1}$ belong to X_0 we have that $\langle f, g_n \rangle = \overline{z_0} a_n - a_{n+1} = 0$ and hence $a_n = a_0 \overline{z_0}^n$, $n \geq 0$. Since $(a_n)_n$ is square summable and $|z_0| = 1$ we must have that $a_0 = 0$, hence $f = 0$. This implies that X_0 is dense in H^2. Moreover, let $f \in X_0$. As in (4.13) we have that

$$\|(C_\varphi)^n f\|^2 = \frac{1}{2\pi} \int_0^{2\pi} \left| f\left(\varphi^n\left(e^{it}\right)\right)\right|^2 dt.$$

Since the integrands are uniformly bounded and convergent to $|f(z_0)|^2 = 0$, for every t with possibly one exception, the dominated convergence theorem implies that $(C_\varphi)^n f \to 0$ for all $f \in X_0$.

Next, for Y_0 we take the subspace of H^2 of all functions that are holomorphic on a neighbourhood of $\overline{\mathbb{D}}$ and that vanish at z_1, and for S we take the map $S = C_{\varphi^{-1}}$. Since z_1 is a fixed point of φ^{-1}, S maps Y_0 into itself, and clearly $TS = I$. It follows as above that Y_0 is dense in H^2 and that $S^n f \to 0$ for all $f \in Y_0$.

Therefore the conditions of Kitai's criterion are satisfied, so that C_φ is mixing. □

Some concrete instances of this result are treated in Exercise 4.5.2.

We end this chapter by returning to Theorem 4.37 of Section 4.3. Proposition 4.47 allows us to give the missing proof of the implication (v)$\Longrightarrow$(vi) for $\Omega = \mathbb{D}$.

Conclusion of the proof of Theorem 4.37. Let φ be an automorphism of $\mathbb{D}$ without fixed points in $\mathbb{D}$. Again, φ is either parabolic or hyperbolic.

First, let φ be parabolic. By the discussion before Definition 4.46 there is a linear fractional transformation σ that provides a conjugacy between φ and $\psi(z) = z + c$, $c \neq 0$. Then ψ is an automorphism of $\sigma(\mathbb{D})$, so that C_φ is conjugate to the operator C_ψ on $H(\sigma(\mathbb{D}))$ by Proposition 4.25. By Runge's theorem, the continuous restriction map $H(\mathbb{C}) \to H(\sigma(\mathbb{D}))$, $f \to f|_{\sigma(\mathbb{D})}$ has dense range. Hence C_φ is quasiconjugate to the Birkhoff operator C_ψ on $H(\mathbb{C})$.

In the hyperbolic case, there is a linear fractional transformation σ such that φ is conjugate to a dilation $\psi(z) = \lambda z$, $\lambda \neq 1$ strictly positive, and ψ is an automorphism of $\sigma(\mathbb{D})$, which therefore must be a half-plane with 0 on its boundary. After conjugation with a suitable rotation we can assume that it is the right half-plane $\mathbb{C}_+$, and ψ remains unchanged. Now, the principal branch log of the logarithm is a conformal map from $\mathbb{C}_+$ to the strip $S = \{z \in \mathbb{C}\ ;\ |\mathrm{Im}(z)| < \frac{\pi}{2}\}$, and conjugating ψ with log gives us the translation $\tau(z) = z + \log\lambda$, $\log\lambda \neq 0$, on S. We conclude as in the parabolic case that C_φ is quasiconjugate to the Birkhoff operator C_τ on $H(\mathbb{C})$. □

Exercises

Exercise 4.1.1. Show that a weighted shift B_w defines an operator on $H(\mathbb{C})$ if and only if $\sup_{n\geq 1} |w_n|^{1/n} < \infty$, and that $a_n e_n \to 0$ in $H(\mathbb{C})$ if and only if $|a_n|^{1/n} \to 0$.

Exercise 4.1.2. Let $T := B_w$ be a weighted shift on ℓ^p, $1 \leq p < \infty$.

(a) Given an increasing sequence $(n_k)_k$ of positive integers, show that the sequence of operators $(T^{n_k})_k$ is hypercyclic if and only if, for each $j \in \mathbb{N}$,

$$\sup_{k\geq 1} \prod_{\nu=1}^{j+n_k} |w_\nu| = \infty.$$

(b) Show that T is hereditarily hypercyclic with respect to $(n_k)_k$ if and only if, for each $j \in \mathbb{N}$,

$$\lim_{k\to\infty} \prod_{\nu=1}^{j+n_k} |w_\nu| = \infty.$$

Exercise 4.1.3. Let X be the Banach space of all sequences $(x_n)_n$ satisfying

$$\|x\| = \sum_{n=1}^{\infty} \left| \frac{x_n}{n} - \frac{x_{n+1}}{n+1} \right| < \infty \quad \text{and} \quad \frac{x_n}{n} \to 0 \text{ as } n \to \infty.$$

Show that the backward shift B is a hypercyclic operator on X and that conditions (ii)–(iv) of Theorem 4.6 are satisfied, but that the only periodic points of B are the constant sequences and that B is therefore not chaotic.

Exercise 4.1.4. Let X be a Fréchet sequence space over $\mathbb{Z}$ in which $(e_n)_{n\in\mathbb{Z}}$ is a basis. Suppose that the bilateral weighted shift B_w is an invertible operator on X. Show that B_w is hypercyclic if and only if there is an increasing sequence $(n_k)_k$ of positive integers such that

$$\Big(\prod_{\nu=-n_k+1}^{0} w_\nu \Big) e_{-n_k} \to 0 \quad \text{and} \quad \Big(\prod_{\nu=1}^{n_k} w_\nu \Big)^{-1} e_{n_k} \to 0$$

in X as $k \to \infty$. (*Hint:* For the sufficiency, look at the proof of (iii)$\Longrightarrow$(ii) in Theorem 4.3.)

Exercise 4.1.5. Find a (necessarily non-invertible) bilateral weighted shift that satisfies the condition stated in the previous exercise but that is not hypercyclic. (*Hint*: See the proof of Proposition 4.16, but choose nonsymmetric v_n.)

Exercise 4.1.6. Prove the results stated in Remark 4.14; instead of using a conjugacy one may also observe that a forward shift on the basis $(e_n)_n$ is a backward shift on the basis $(e_{-n})_n$.

Exercise 4.1.7. Show that the characterizing conditions on a weight w to define a hypercyclic bilateral weighted shift B_w on $\ell^p(\mathbb{Z})$ can also be written as follows: for any $\varepsilon > 0$ and any $M, N \geq 1$ there exists some $n \geq N$ such that whenever $|j| \leq M$ then

$$\prod_{\nu=j-n+1}^{j} |w_\nu| < \varepsilon, \qquad \prod_{\nu=j+1}^{j+n} |w_\nu| > \frac{1}{\varepsilon}.$$

Exercise 4.2.1. An entire function φ is of *exponential type* 0 if for any $\varepsilon > 0$ there is some $M > 0$ such that

$$|\varphi(z)| \leq M e^{\varepsilon |z|} \quad \text{for all } z \in \mathbb{C}.$$

For example, any polynomial but no exponential function $z \to e^{\lambda z}$, $\lambda \neq 0$, is of exponential type 0.

For a domain $\Omega \subset \mathbb{C}$, let $H(\Omega)$ denote the Fréchet space of holomorphic functions on Ω; see Section 4.3. Show the following:

(i) an entire function $\varphi(z) = \sum_{n=0}^{\infty} a_n z^n$ is of exponential type 0 if and only if, for any $\varepsilon > 0$, there is some $M > 0$ such that $|a_n| \leq M \frac{\varepsilon^n}{n!}$;

(ii) for any domain $\Omega \subset \mathbb{C}$ and any entire function $\varphi(z) = \sum_{n=0}^{\infty} a_n z^n$ of exponential type 0, $\varphi(D) = \sum_{n=0}^{\infty} a_n D^n$ defines an operator on $H(\Omega)$;

(iii) for any simply connected domain $\Omega \subset \mathbb{C}$ and any nonconstant entire function φ of exponential type 0, $\varphi(D)$ is chaotic on $H(\Omega)$. (*Hint*: Use the Godefroy–Shapiro theorem and the fact, that, by Runge's theorem, $H(\mathbb{C})$ is dense in $H(\Omega)$.)

Exercise 4.2.2. Let Ω be a domain and P a nonconstant polynomial. Show that the following assertions are equivalent:
(i) $P(D)$ is chaotic on $H(\Omega)$;
(ii) $P(D)$ is hypercyclic on $H(\Omega)$;
(iii) Ω is simply connected.
(*Hint*: If Ω is not simply connected then there is a smooth Jordan curve Γ in Ω surrounding some $a \notin \Omega$. Show that $f \to \int_\Gamma f(\zeta)\,d\zeta$ is an eigenvector of $P(D)^*$, and use Lemma 2.53.)

Exercise 4.2.3. Let $X = C^\infty_{\mathbb{R}}(\mathbb{R})$ be the space of infinitely differentiable real functions $f : \mathbb{R} \to \mathbb{R}$; see Exercise 2.1.5. Show that every (real) differential operator $T : X \to X$, $Tf = \sum_{n=0}^N a_n f^{(n)}$, $T \neq a_0 I$, is chaotic. (*Hint*: See Exercise 2.2.5.)

Exercise 4.2.4. Let φ be a nonconstant entire function of exponential type and $A = \min\{|z| \; ; \; z \in \mathbb{C}, |\varphi(z)| = 1\}$. Show that, for any $\varepsilon > 0$, there is an entire function f that is hypercyclic for $\varphi(D)$ such that

$$|f(z)| \leq Me^{(A+\varepsilon)r} \quad \text{for } |z| = r > 0$$

with some $M > 0$.

For the proof consider the Hilbert spaces

$$E^2_\tau = \Big\{ f \in H(\mathbb{C}) \; ; \; f(z) = \sum_{n=0}^\infty a_n z^n, \; \sum_{n=0}^\infty \Big(\frac{n!}{\tau^n}\Big)^2 |a_n|^2 < \infty \Big\}, \quad \tau > 0.$$

Show that any $f \in E^2_\tau$ satisfies $|f(z)| \leq Me^{\tau r}$; use ideas from Example 3.2 to show that for any $\Lambda \subset \mathbb{D}_\tau$ with an accumulation point, $\text{span}\{e_\lambda \; ; \; \lambda \in \Lambda\}$ is dense in E^2_τ (see Appendix A for the dual of E^2_τ); show that $\varphi(D)$ is an operator on any E^2_τ, and that $\varphi(D)$ is hypercyclic on $E^2_{A+\varepsilon}$ for any $\varepsilon > 0$.

Apply the result to MacLane's and Birkhoff's operators.

Exercise 4.2.5. Let B_w be a chaotic weighted shift on $H(\mathbb{C})$; see Example 4.9(b). Then $\sum_{n=0}^\infty (\prod_{\nu=1}^n w_\nu)^{-1} z^n$ is an entire function, and its *maximum term* is defined by

$$\mu_w(r) = \max_{n \geq 0} \frac{r^n}{\prod_{\nu=1}^n |w_\nu|}, \qquad r \geq 0.$$

(a) Let $\phi :]0, \infty[\to [1, \infty[$ be a function with $\phi(r) \to \infty$ as $r \to \infty$. Show that there exists an entire function f that is hypercyclic for B_w and that satisfies

$$|f(z)| \leq M\phi(r)\mu_w(r) \quad \text{for } |z| = r > 0$$

with some $M > 0$.

(b) Suppose that $|w_n| \to \infty$ monotonically. Show that there is no entire function f that is hypercyclic for B_w and that satisfies

$$|f(z)| \leq M\,\mu_w(r) \quad \text{for } |z| = r > 0$$

with some $M > 0$. (*Hint*: Determine $\mu_w(\rho)$ for $\rho = |w_n|$.)

Deduce Theorem 4.22 from this.

Exercise 4.3.1. Show in detail that $H(\Omega)$ is a separable Fréchet space and that its topology is independent of the exhaustion chosen. (*Hint*: For separability, fix an exhaustion $(K_n)_n$ of Ω by Ω-convex compact sets. In each connected component of the complements of the K_n fix one point outside Ω, possibly ∞; this set is denumerable. Now use Runge's theorem.)

Exercise 4.3.2. Let Ω be a domain and $\varphi : \Omega \to \Omega$ a holomorphic self-map that is not necessarily an automorphism of Ω. Show that $C_\varphi : H(\Omega) \to H(\Omega)$, $C_\varphi f = f \circ \varphi$, defines an operator on $H(\Omega)$, also called a *composition operator*. Moreover, show the following:
(i) if C_φ is hypercyclic then φ is injective;
(ii) let Ω be simply connected; then C_φ is hypercyclic if and only if φ is injective and $(\varphi^n)_n$ is a run-away sequence.

Exercise 4.3.3. Let Ω be a domain, $\varphi : \Omega \to \Omega$ an injective holomorphic self-map and $C_\varphi f = f \circ \varphi$ the corresponding composition operator on $H(\Omega)$; see the previous exercise.
(a) Based on the proof of Theorem 4.32, find a sufficient condition under which C_φ is hypercyclic.
(b) Let $\varphi : \mathbb{D} \to \mathbb{D}$ be given by $\varphi(z) = \frac{z}{4} + \frac{3}{4}$. Let $K = \{z \in \mathbb{C} \; ; \; |z| \leq \frac{1}{2}\}$ and $\Omega = \mathbb{D} \setminus \bigcup_{n=0}^{\infty} \varphi^n(K)$. Show that the restriction of φ to Ω defines a hypercyclic composition operator C_φ on $H(\Omega)$.

Exercise 4.3.4. Let $\Omega = \mathbb{C} \setminus \mathbb{Z}$. Then $\varphi(z) = z+1$ is an automorphism of Ω. Show that the composition operator C_φ is chaotic on $H(\Omega)$. (*Hint*: Show that the linear span of the functions $e^{\lambda z}$, $e^{\lambda N} = 1$ for some $N \geq 1$, and $\lim_{m\to\infty} \sum_{\nu=-m}^{m} \frac{1}{(z-k-\nu N)^\alpha}$, $k \in \mathbb{Z}$, $\alpha > 1$, $N \geq 1$, forms a dense set of periodic points.)

Exercise 4.4.1. Let φ be a holomorphic function on $\mathbb{D}$ such that $\varphi f \in H^2$ for all $f \in H^2$. Use the closed graph theorem to show that the mapping $M_\varphi : f \to \varphi f$ is continuous. Deduce that φ is necessarily bounded, and $\|M_\varphi\| = \sup_{z\in\mathbb{D}} |\varphi(z)|$. (*Hint*: $\varphi^n(z) = \langle (M_\varphi)^n 1, k_z \rangle$.)

Exercise 4.4.2. Let $\Omega \subset \mathbb{C}$ be a domain and $H \neq \{0\}$ a Hilbert space of holomorphic functions on Ω. Suppose that each point evaluation $f \to f(\lambda)$, $\lambda \in \Omega$, is a continuous linear functional on H. Use the closed graph theorem to prove that the canonical embedding $H \hookrightarrow H(\Omega)$ is continuous, so that convergence in H implies locally uniform convergence on Ω.

Exercise 4.4.3. Let $\Omega \subset \mathbb{C}$ be a domain and $H \neq \{0\}$ a Hilbert space of holomorphic functions on Ω such that each point evaluation $f \to f(\lambda)$, $\lambda \in \Omega$, is continuous on H.
(a) By the Riesz representation theorem (see Appendix A), there is a unique function $k_\lambda \in H$, again called a reproducing kernel, such that

$$f(\lambda) = \langle f, k_\lambda \rangle, \quad f \in H.$$

Prove an analogue of Lemma 4.39 and deduce that H is separable.
(b) Now let φ be a nonconstant bounded holomorphic function on Ω for which $M_\varphi f = \varphi f$ defines an operator on H. Let M_φ^* be the corresponding adjoint multiplier on H. Show that M_φ^* is chaotic and mixing as soon as $\varphi(\Omega) \cap \mathbb{T} \neq \varnothing$. Show that, in this case, for some $\lambda \in \mathbb{C}$, λM_φ^* is chaotic. Deduce that M_φ^* is supercyclic.
(c) Finally, suppose that every bounded holomorphic function φ on Ω defines a multiplication operator with $\|M_\varphi\| \leq \sup_{z\in\Omega} |\varphi(z)|$. Show that if φ is a nonconstant bounded holomorphic function on Ω such that M_φ^* is hypercyclic then $\varphi(\Omega) \cap \mathbb{T} \neq \varnothing$.
(d) Deduce that Theorem 4.42 holds also for the Bergman space A^2; see Example 4.4(b).

Exercise 4.4.4. The *Dirichlet space* $\mathcal{D}$ is defined as the space of all holomorphic functions f on $\mathbb{D}$ such that

$$\|f\|^2 := |f(0)|^2 + \frac{1}{\pi} \int_{\mathbb{D}} |f'(z)|^2 \, d\lambda(z) < \infty,$$

where λ denotes two-dimensional Lebesgue measure. Show the following:

(i) if $f(z) = \sum_{n=0}^{\infty} a_n z^n$ then $\|f\|^2 = |a_0|^2 + \sum_{n=1}^{\infty} n|a_n|^2$;
(ii) $\mathcal{D}$ is a Hilbert space with continuous point evaluations;
(iii) $\mathcal{D} \subset H^2 \subset A^2$, where A^2 is the Bergman space (see Example 4.4);
(iv) if φ is a bounded holomorphic function on $\mathbb{D}$ such that φ' is also bounded then M_φ defines an operator on $\mathcal{D}$;
(v) if $\varphi(z) = \sum_{n=0}^{\infty} b_n z^n$ with $\sum_{n=0}^{\infty} |b_n| < \infty$ and $\sum_{n=0}^{\infty} n|b_n|^2 = \infty$ (existence?) then φ is a bounded holomorphic function on $\mathbb{D}$ for which M_φ does not define an operator on $\mathcal{D}$;
(vi) if φ is a nonconstant bounded holomorphic function on $\mathbb{D}$ such that M_φ defines an operator on $\mathcal{D}$ and if $\varphi(\mathbb{D}) \cap \mathbb{T} \neq \varnothing$ then M_φ^* is mixing and chaotic;
(vii) the function $\varphi(z) = z$, $z \in \mathbb{D}$, defines a hypercyclic adjoint multiplier M_φ^* on $\mathcal{D}$, but $\varphi(\mathbb{D}) \cap \mathbb{T} = \varnothing$; is M_φ^* mixing or chaotic? (*Hint*: Identify M_φ^* with a weighted shift on a weighted ℓ^2-space.)

Exercise 4.4.5. Let $X \neq \{0\}$ be a Banach space of holomorphic functions on a domain $\Omega \subset \mathbb{C}$. Suppose that X is reflexive, that is, $X^{**} = X$. Suppose further that the point-evaluations $f \to f(\lambda)$, $\lambda \in \Omega$, are continuous on X and that every bounded holomorphic function φ on Ω defines a multiplication operator M_φ with $\|M_\varphi\| \leq \sup_{z\in\Omega} |\varphi(z)|$. Then show the analogue of Theorem 4.42 for M_φ^*, the (Banach space) adjoint of M_φ. (*Hint*: Use reflexivity and the Hahn–Banach theorem to obtain the analogue of Lemma 4.39.)

Exercise 4.4.6. Let X be one of the complex spaces ℓ^p, $1 \leq p < \infty$, or c_0. Let $a, b \in \mathbb{C}$, $b \neq 0$. Show that the following assertions are equivalent:
(i) $aI + bB$ is chaotic on X;
(ii) $|b| > |1 - |a||$.

Exercise 4.4.7. Generalize part of Theorem 4.43 as follows: let $X = \ell^p(v) = \{(x_n)_n \in \mathbb{C}^{\mathbb{N}};\ \sum_{n=1}^{\infty} |x_n|^p v_n < \infty\}$, $1 \leq p < \infty$, where $v = (v_n)_n$ is a positive weight sequence such that $M := \sup_{n\in\mathbb{N}} \frac{v_n}{v_{n+1}} < \infty$. Let $R := (\limsup_{n\to\infty} v_n^{1/n})^{-1} > 0$, which is finite, and let $\varphi(z) = \sum_{n=0}^{\infty} a_n z^n$ be a nonconstant function that is holomorphic in $\mathbb{D}_r$ for some $r > M$. Then $\varphi(B) = \sum_{n=0}^{\infty} a_n B^n$ defines an operator on X, and if

$$\varphi\big(R^{1/p}\mathbb{D}\big) \cap \mathbb{T} \neq \varnothing \tag{4.15}$$

then $\varphi(B)$ is a chaotic operator on X. (*Hint*: See Appendix A for the dual of X.)

Exercise 4.4.8. In the setting of Exercise 4.4.7, let $v_n = 1/n^2$, $n \geq 1$.
(a) Show that B is chaotic but condition (4.15) does not hold.
(b) Show that the operator $\frac{1}{2}(I+B)$ has a nontrivial periodic point but is not chaotic.

Exercise 4.4.9. Let $w = (w_n)_n$ be a weight sequence, and let f be a nonconstant polynomial. Show that $f(B_w)$ is (well defined and) chaotic on $\omega = \mathbb{K}^{\mathbb{N}}$. (*Hint*: For the density of eigenvectors for B_w show that they are contained and dense in a suitable weighted ℓ^1-space; see Appendix A.)

Exercise 4.5.1. The aim of this exercise is to prove *Littlewood's subordination principle*: if $\varphi : \mathbb{D} \to \mathbb{D}$ is a holomorphic self-map then $C_\varphi : f \to f \circ \varphi$ defines an operator on H^2 with $\|C_\varphi\| \leq (\frac{1+r}{1-r})^{1/2}$, where $r = |\varphi(0)|$.
(a) First prove the result when $\varphi(0) = 0$ by proceeding as follows:
  (i) show that, for any $f \in H^2$, $C_\varphi f = f(0) + M_\varphi C_\varphi B f$, where we write B for M_z^* (see Proposition 4.41(ii));
  (ii) using orthogonality, deduce that, for any polynomial f, $\|C_\varphi f\|^2 \leq |f(0)|^2 + \|C_\varphi B f\|^2$;
  (iii) deduce that $\|C_\varphi f\|^2 \leq \sum_{k=0}^{n} |(B^k f)(0)|^2 + \|C_\varphi B^{n+1} f\|^2$;

(iv) deduce that $\|C_\varphi f\| \le \|f\|$;
(v) conclude that $C_\varphi f \in H^2$ and $\|C_\varphi f\| \le \|f\|$ for all $f \in H^2$.

(b) Prove the result by factorizing $C_\varphi = C_{\varphi_1} C_{\varphi_2}$ with $\varphi_1(0) = 0$ and $\varphi_2 \in \mathrm{Aut}(\mathbb{D})$.

Exercise 4.5.2. For the following linear fractional transformations decide if they are automorphisms of $\mathbb{D}$; in the case of an automorphism, determine if the corresponding composition operator C_φ is hypercyclic on H^2:

(i) $\varphi(z) = \dfrac{2z-1}{2-z}$;

(ii) $\varphi(z) = \dfrac{1+(i-1)z}{(i+1)-z}$;

(iii) $\varphi(z) = \dfrac{4-5z}{5-4z}$;

(iv) $\varphi(z) = \dfrac{z+1}{2}$.

Exercise 4.5.3. Let $\alpha > -1$. Then the *weighted Bergman space* A^2_α is defined as the space of all holomorphic functions f on $\mathbb{D}$ such that

$$\|f\|^2 := \frac{1}{\pi} \int_{\mathbb{D}} |f(z)|^2 \left(1-|z|^2\right)^\alpha \, d\lambda(z) < \infty,$$

where λ denotes two-dimensional Lebesgue measure; see Example 4.4(b).

(a) Let f be a holomorphic function on $\mathbb{D}$ with $f(z) = \sum_{n=0}^\infty a_n z^n$, $z \in \mathbb{D}$. Show that $\|f\|^2 = \sum_{n=0}^\infty |a_n|^2 \frac{\Gamma(\alpha+1)\Gamma(n+1)}{\Gamma(\alpha+n+2)}$ and deduce that $f \in A^2_\alpha$ if and only if $\sum_{n=0}^\infty |a_n|^2 \frac{1}{(n+1)^{\alpha+1}} < \infty$. (*Hint*: Stirling's formula.)

(b) Let $\varphi \in \mathrm{Aut}(\mathbb{D})$. Show that C_φ is an operator on A^2_α with $\|C_\varphi\| \le (\frac{1+r}{1-r})^{1+\alpha/2}$, where $r = |\varphi(0)|$. (*Hint*: Show that if $w = \varphi(z)$ then $d\lambda(z) = |\varphi'(z)|^{-2}\, d\lambda(w)$, and use (4.12).)

Exercise 4.5.4. Let $\alpha > -1$. Then the *weighted Dirichlet space* $\mathcal{D}_\alpha$ is defined as the space of all holomorphic functions f on $\mathbb{D}$ such that

$$\|f\|^2 := |f(0)|^2 + \frac{1}{\pi} \int_{\mathbb{D}} \left|f'(z)\right|^2 \left(1-|z|^2\right)^\alpha \, d\lambda(z) < \infty,$$

where λ denotes two-dimensional Lebesgue measure; see Exercise 4.4.4 and the previous exercise.

(a) Let f be a holomorphic function on $\mathbb{D}$ with $f(z) = \sum_{n=0}^\infty a_n z^n$, $z \in \mathbb{D}$. Show that $f \in \mathcal{D}_\alpha$ if and only if $\sum_{n=0}^\infty |a_n|^2 (n+1)^{1-\alpha} < \infty$.

(b) Let $\varphi \in \mathrm{Aut}(\mathbb{D})$. Show that C_φ is an operator on $\mathcal{D}_\alpha$.

(c) Let $\alpha > 0$. Show that Theorem 4.48 remains true for $\mathcal{D}_\alpha$. (*Hint*: Proceed as in the proof of that theorem; use the change of variables $w = \varphi^n(z)$; note that $1 - |\varphi^{-n}(w)|^2 \to 0$.)

Exercise 4.5.5. Let $\varphi \in \mathrm{Aut}(\mathbb{D})$. By the previous exercise, C_φ is an operator on the Dirichlet space $\mathcal{D}$. Show that, for any $f \in \mathcal{D}$,

$$\|C_\varphi f\|^2 \ge \frac{1}{\pi} \int_{\mathbb{D}} \left|f'(z)\right|^2 \, d\lambda(z).$$

Deduce that C_φ is not hypercyclic on $\mathcal{D}$. (*Hint*: Change of variables $w = \varphi(z)$.)

Exercise 4.5.6. Let $\beta = (\beta_n)_{n\geq 0}$ be a sequence of strictly positive numbers such that $\sum_{n=0}^{\infty} \beta_n^{-2} r^n < \infty$ whenever $0 \leq r < 1$. Then the *weighted Hardy space* $H^2(\beta)$ is defined as the space of all holomorphic functions f on $\mathbb{D}$ such that

$$\|f\|^2 := \sum_{n=0}^{\infty} |a_n|^2 \beta_n^2 < \infty,$$

where $f(z) = \sum_{n=0}^{\infty} a_n z^n$, $z \in \mathbb{D}$. By the assumption on $(\beta_n)_n$, this condition alone implies that $f \in H(\mathbb{D})$.

Let $\varphi \in \text{Aut}(\mathbb{D})$ and suppose that C_φ defines an operator on $H^2(\beta)$. Show the following:

(i) if $\sum_{n=0}^{\infty} \beta_n^{-2} < \infty$ then C_φ is never hypercyclic on $H^2(\beta)$;

(ii) if $\sum_{n=0}^{\infty} \beta_n^{-2} = \infty$ and φ is elliptic then C_φ is not hypercyclic on $H^2(\beta)$.

(*Hint* for (i): Show that all functions in $H^2(\beta)$ have a continuous extension to $\overline{\mathbb{D}}$; and use the fact that φ has a fixed point in $\overline{\mathbb{D}}$.)

Exercise 4.5.7. Let $\nu \in \mathbb{R}$. Then the space $\mathcal{S}_\nu$ is defined as the space of all holomorphic functions f on $\mathbb{D}$ such that

$$\|f\|^2 := \sum_{n=0}^{\infty} |a_n|^2 (n+1)^{2\nu} < \infty,$$

where $f(z) = \sum_{n=0}^{\infty} a_n z^n$, $z \in \mathbb{D}$. In particular, $\mathcal{S}_0$ is the Hardy space H^2, $\mathcal{S}_{-1/2}$ is the Bergman space A^2, and $\mathcal{S}_{1/2}$ is the Dirichlet space $\mathcal{D}$ under an equivalent norm.

(a) Show that $f \in \mathcal{S}_\nu$ if and only if $f' \in \mathcal{S}_{\nu-1}$. If $\nu \in \mathbb{N}$, show that $f \in \mathcal{S}_\nu$ if and only if $f^{(\nu)} \in H^2$.

(b) Show that the multiplier M_z is an operator on each $\mathcal{S}_\nu$, and calculate the norm of M_z^n, $n \geq 0$. More generally, let φ be holomorphic on $\mathbb{D}$, $\varphi(z) = \sum_{n=0}^{\infty} b_n z^n$, such that $\sum_{n=0}^{\infty} |b_n|(n+1)^\nu < \infty$. Show that M_φ is an operator on $\mathcal{S}_\nu$.

(c) Let $\varphi \in \text{Aut}(\mathbb{D})$. Deduce from Exercise 4.5.3 that C_φ defines an operator on $\mathcal{S}_\nu$ for $\nu < 0$. Use parts (a) and (b) to conclude that C_φ defines an operator on $\mathcal{S}_\nu$ for any $\nu \in \mathbb{R}$. (*Hint*: $(C_\varphi f)' = M_{\varphi'} C_\varphi f'$.)

(d) Let $\varphi(z) = \sum_{n=0}^{\infty} b_n z^n$ with $\sum_{n=0}^{\infty} |b_n| \leq 1$ and $\sum_{n=0}^{\infty} |b_n|^2 n = \infty$ (existence?). Show that φ is a holomorphic self-map of $\mathbb{D}$ for which C_φ does not define an operator on the Dirichlet space $\mathcal{D}$.

Exercise 4.5.8. Let $\nu \in \mathbb{R}$ and $\varphi \in \text{Aut}(\mathbb{D})$. By Exercise 4.5.7, C_φ is an operator on S_ν. Deduce the following from the previous exercises:

(i) if $\nu \geq \frac{1}{2}$ then C_φ is never hypercyclic on S_ν;

(ii) if $\nu < \frac{1}{2}$ then C_φ is hypercyclic on S_ν if and only if φ is not the identity and non-elliptic.

Spell out these results for the (weighted) Bergman and Dirichlet spaces.

Sources and comments

Section 4.1. Rolewicz's multiples of the backward shift were the first Banach space operators to be proved hypercyclic [268]. Due to its simple structure, *the class of weighted shifts is a favorite testing ground for operator-theorists* (Salas [274]). Accordingly, whenever a new notion in linear dynamics is introduced it is usually first tested on weighted shifts.

Salas [274] characterized hypercyclic and weakly mixing unilateral and bilateral weighted shifts on ℓ^2 and $\ell^2(\mathbb{Z})$, respectively. The characterizations for more general sequence spaces and of chaos are due to Grosse-Erdmann [180], see also Martínez and Peris [229] in the special case of Köthe sequence spaces, while mixing shifts on ℓ^2 and $\ell^2(\mathbb{Z})$ were characterized by Costakis and Sambarino [124]. The approach chosen here of first studying the unweighted shift and then using suitable conjugacies is due to Martínez and Peris [229].

The first example of a hypercyclic operator whose adjoint is also hypercyclic (see Proposition 4.16) was found by Salas [273]. He later showed [276] that every separable Banach space with separable dual supports such an operator. The observation that $T \oplus T^*$ is never hypercyclic is due to Deddens; see [273].

Section 4.2. The investigation of hypercyclicity for differential operators $\varphi(D)$ is due to Godefroy and Shapiro [165]. Theorem 4.22 on the rate of growth of MacLane's operator was obtained independently by Grosse-Erdmann [178] and Shkarin [283]. The corresponding result for Birkhoff's operators was obtained by Duyos-Ruiz [137]; alternative proofs can be found in Chan and Shapiro [106] and in Exercise 8.1.3.

Translation and differentiation operators have also been studied on spaces of harmonic functions on $\mathbb{R}^N$, $N \geq 2$. Hypercyclicity of these operators and corresponding growth results have been obtained by Dzagnidze [138], Aldred and Armitage [6], [7], [13], and Gómez, Martínez, Peris and Rodenas [166].

Section 4.3. This section draws heavily on the work of Bernal and Montes [64], who also coined the term "run-away sequence", and the work of Shapiro [281]. The material up to Theorem 4.32 can be found in [64], while most of Theorem 4.37 is implicit in [281]. We mention that Seidel and Walsh [278] were the first to study the analogue of Birkhoff's result in the unit disk.

For two different proofs of Proposition 4.31 we refer to Bernal and Montes [64] and to Grosse-Erdmann and Mortini [184]. Example 4.34 is taken from Kim and Krantz [214]; see also Gorkin, León and Mortini [168].

Section 4.4. The study of the dynamical properties of adjoint multipliers was initiated by Godefroy and Shapiro [165], who also obtained Theorem 4.42. Functions of the backward shift on the spaces ℓ^p and c_0 were studied by deLaubenfels and Emamirad [128], who also obtained Theorem 4.43. An interesting related investigation of functions of the backward shift on the Bergman space is due to Bourdon and Shapiro [96].

For a more detailed introduction to Hardy spaces we refer to Duren [136] and Rudin [270].

Section 4.5. In this section we closely follow the book of Shapiro [279]; see also Shapiro [281]. Proposition 4.45 is a special case of the Littlewood subordination principle; see Exercise 4.5.1. Theorem 4.48 is due to Bourdon and Shapiro [94], [95]. This result is only the beginning of a fascinating story on the interplay between operator theory and complex function theory. The extension of Theorem 4.48, first to non-automorphic linear fractional transformations and then to more arbitrary holomorphic self-maps of $\mathbb{D}$, can be found in the cited work of Bourdon and Shapiro. The proofs, however, require a much deeper understanding, for example, of the (nonlinear) dynamics of self-maps of $\mathbb{D}$.

Hosokawa [204] proved that, for any automorphism φ of $\mathbb{D}$, C_φ is chaotic whenever it is hypercyclic; see also Taniguchi [298]. Thus one can add chaos to the equivalent conditions in Theorem 4.48.

We note that Gallardo and Montes [158] have obtained a complete characterization of the cyclic, supercyclic and hypercyclic composition operators C_φ for linear fractional self-maps φ of $\mathbb{D}$ on any of the spaces $\mathcal{S}_\nu$, $\nu \in \mathbb{R}$ (see Exercises 4.5.7 and 4.5.8).

For a more detailed introduction to composition operators on weighted Hardy, Bergman and Dirichlet spaces we refer to Cowen and MacCluer [125].

Exercises. Exercise 4.1.2 is taken from Bès and Peris [71], Exercise 4.1.3 from Grosse-Erdmann [180]. For Exercises 4.1.4 and 4.1.5 we refer to Feldman [150], Exercise 4.1.7 states the condition in the form found originally by Salas [274]. But note that the weighted shifts considered by Feldman and Salas are forward shifts. For Exercise 4.2.1 we refer to Bernal [55] and Shapiro [280], for Exercise 4.2.2 to Shapiro [280], for Exercise 4.2.4 to Chan and Shapiro [106] and to Bernal and Bonilla [60], for Exercise 4.2.5 to Grosse-Erdmann [181]. Exercises 4.3.2 and 4.3.3 follow Montes [241] and Grosse-Erdmann and Mortini [184], while Exercise 4.3.4 is taken from Shapiro [280]. The material for Exercises 4.4.1–4.4.4 can be found in Godefroy and Shapiro [165], with Exercise 4.4.4(vii) being taken from Chan and Seceleanu [104]; for Exercises 4.4.6–4.4.8 we refer to deLaubenfels and Emamirad [128], for Exercise 4.4.9 to Martínez [228]. For Exercise 4.5.1 we have again followed Shapiro [279]; Exercises 4.5.4(c), 4.5.5 and 4.5.8 are taken from Gallardo and Montes [158], Exercise 4.5.6 from Zorboska [304] and Exercise 4.5.7 from Hurst [205].

Chapter 5
Necessary conditions for hypercyclicity and chaos

In Chapter 3 we derived various sufficient conditions for an operator to be hypercyclic, which we then used in Chapter 4 to obtain classes of such operators. In this chapter we are interested in the opposite question: which conditions on an operator rule out its hypercyclicity? In other words, we are searching for necessary conditions for an operator to be hypercyclic. This then leads to classes of operators that do not include any hypercyclic operators.

Many of the known necessary conditions seem to involve, in one way or another, the spectrum of the operator. Hence we begin by studying spectral properties of hypercyclic and chaotic operators.

Spectral considerations show their full strength only for operators on complex Banach spaces. Using the complexification technique one can then often extend the results to the real scalar case. In this chapter we therefore restrict our attention to operators on Banach spaces; in the final section we study, in particular, operators on Hilbert spaces.

We suppose that the reader is familiar with the basics of spectral theory as laid out in Appendix B. The main tool needed in this chapter, the Riesz decomposition theorem, is not usually covered in introductory courses on functional analysis; its proof is contained in the appendix. We have also provided some exercises on spectral theory that the reader might find useful; see Exercises 5.0.1–5.0.9.

5.1 Spectral properties of hypercyclic and chaotic operators

In this section we study the influence of hypercyclicity and chaos on the spectrum of an operator.

We start with two simple observations on the adjoint T^* of a hypercyclic operator, the first of which has already been observed in greater generality in Lemma 2.53.

K.-G. Grosse-Erdmann, A. Peris Manguillot, *Linear Chaos*, Universitext,
DOI 10.1007/978-1-4471-2170-1_5, © Springer-Verlag London Limited 2011

Proposition 5.1. *Let T be a hypercyclic operator on a (real or complex) Banach space X. Then we have:*

(i) *T^* has no eigenvalues, that is, $\sigma_p(T^*) = \varnothing$;*

(ii) *the orbit of every $x^* \neq 0$ in X^* under T^* is unbounded.*

Proof of (ii). Suppose that there exists some $x^* \in X^*$, $x^* \neq 0$, and some $M > 0$ such that $\|(T^*)^n x^*\| \leq M$ for all $n \geq 0$. Let $x \in X$ be a hypercyclic vector for T. Then $\langle T^n x, x^* \rangle$, $n \geq 0$, forms a dense set in $\mathbb{K}$. On the other hand,

$$|\langle T^n x, x^* \rangle| = |\langle x, (T^*)^n x^* \rangle| \leq M\|x\|, \quad n \geq 0,$$

which is impossible. □

For mixing operators T one can even assert that $\|(T^*)^n x^*\| \to \infty$ for any nonzero element $x^* \in X^*$; see Exercise 5.1.1.

Next we study the spectrum $\sigma(T)$ of the operator T, which suggests consideration of complex spaces.

Throughout the remainder of this section, T denotes an operator on a complex Banach space X.

Lemma 5.2. *Let $r > 0$. Then we have:*

(i) *if $\sigma(T) \subset \{z \in \mathbb{C} \; ; \; |z| < r\}$ then there are $\varepsilon > 0$ and $M > 0$ such that $\|T^n x\| \leq M(r-\varepsilon)^n \|x\|$ for all $x \in X$ and $n \in \mathbb{N}_0$;*

(ii) *if $\sigma(T) \subset \{z \in \mathbb{C} \; ; \; |z| > r\}$ then there are $\varepsilon > 0$ and $M > 0$ such that $\|T^n x\| \geq M(r+\varepsilon)^n \|x\|$ for all $x \in X$ and $n \in \mathbb{N}_0$.*

Proof. (i) Since $\sigma(T)$ is a compact set, the assumption and the spectral radius formula imply that $\lim_{n\to\infty} \|T^n\|^{1/n} = r(T) < r$. Hence, if $\varepsilon < r - r(T)$ then there is some $M > 0$ such that $\|T^n\| \leq M(r-\varepsilon)^n$ for all $n \in \mathbb{N}_0$, and the result follows.

(ii) The assumption implies that $0 \notin \sigma(T)$, so that T is invertible. Since $\sigma(T^{-1}) = \sigma(T)^{-1}$ (see Exercise 5.0.7), we obtain that $\sigma(T^{-1}) \subset \{z \in \mathbb{C} \; ; \; |z| < \frac{1}{r}\}$. By (i) there are then some $\eta > 0$ with $\eta < \frac{1}{r}$ and $M > 0$ such that $\|(T^{-1})^n y\| \leq M(\frac{1}{r} - \eta)^n \|y\|$ for all $y \in X$ and $n \in \mathbb{N}_0$. Setting $y = T^n x$ and defining ε by $\frac{1}{r} - \eta = \frac{1}{r+\varepsilon}$ we obtain the result. □

The Riesz decomposition theorem now allows us to deduce a first major necessary condition for an operator to be hypercyclic. The result will be superseded by Kitai's theorem below.

Proposition 5.3. *Let T be a hypercyclic operator. Then $\sigma(T)$ meets the unit circle:*

$$\sigma(T) \cap \mathbb{T} \neq \varnothing.$$

Proof. Suppose, on the contrary, that $\sigma(T)$ does not meet the unit circle. If $\sigma(T) \subset \mathbb{D}$ then, by taking $r = 1$ in Lemma 5.2(i), we see that all orbits of T tend to 0, which is impossible. Similarly, Lemma 5.2(ii) shows that $\sigma(T) \subset \mathbb{C} \setminus \overline{\mathbb{D}}$ is impossible. We therefore have that the sets

$$\sigma_1 := \sigma(T) \cap \mathbb{D} \quad \text{and} \quad \sigma_2 := \sigma(T) \cap (\mathbb{C} \setminus \overline{\mathbb{D}})$$

form a partition of $\sigma(T)$ into nonempty closed sets. By the Riesz decomposition theorem there then exist nontrivial T-invariant closed subspaces M_1 and M_2 such that $X = M_1 \oplus M_2$, $\sigma(T|_{M_1}) = \sigma_1$ and $\sigma(T|_{M_2}) = \sigma_2$. By Proposition 2.28, $T|_{M_1}$ would then be a hypercyclic operator with $\sigma(T|_{M_1}) \subset \mathbb{D}$, which is impossible, as we saw above. □

Thus, the spectrum of a hypercyclic operator must contain at least one point from the unit circle. In Example 8.4 we will see that, indeed, the spectrum need not contain any additional point: there is a hypercyclic, and even a mixing, operator T with $\sigma(T) = \{1\}$.

Example 5.4. **(Volterra operator)** Let X be the space $C[0,1]$ of (complex-valued) continuous functions on the interval $[0,1]$, or one of the spaces $L^p[0,1]$ of (complex-valued) p-integrable functions on $[0,1]$, where $1 \leq p < \infty$; see Example 2.4. For any $f \in X$ we define Vf by

$$Vf(t) = \int_0^t f(s)\,ds, \quad 0 \leq t \leq 1.$$

This obviously defines an operator on $C[0,1]$ with $\|V\| \leq 1$.

On the other hand, let $p > 1$, and let q be the dual exponent to p defined by $\frac{1}{p} + \frac{1}{q} = 1$. It follows from Hölder's inequality that

$$\begin{aligned}\int_0^1 \Big(\int_0^t |f(s)|\,ds\Big)^p dt &\leq \int_0^1 \Big(\int_0^t |f(s)|^p\,ds\Big)^{p/p}\Big(\int_0^t 1^q\,ds\Big)^{p/q} dt\\ &\leq \|f\|_p^p \int_0^1 t^{p/q}\,dt = \frac{1}{\frac{p}{q}+1}\|f\|_p^p.\end{aligned}$$

Thus V is also an operator on $L^p[0,1]$, and since $\frac{p}{q} + 1 = p$ we have that $\|V\| \leq p^{-1/p}$. A similar, and simpler, argument shows that the same result is true for $p = 1$. The operator V is called the *Volterra operator.*

A simple induction shows that the nth iterate of V, $n \geq 1$, is given by

$$V^n f(t) = \frac{1}{(n-1)!}\int_0^t (t-s)^{n-1} f(s)ds, \quad 0 \leq t \leq 1.$$

In the case of $X = C[0,1]$ we therefore have that

$$\|V^n f\| \leq \frac{1}{(n-1)!}\|f\| \max_{0 \leq t \leq 1} \int_0^t (t-s)^{n-1} ds = \frac{1}{n!}\|f\|,$$

so that

$$\|V^n\|^{1/n} \leq \frac{1}{n!^{1/n}} \to 0.$$

The spectral radius formula implies that $r(V) = 0$, so that $\sigma(V) = \{0\}$. The same result holds for any space $X = L^p[0,1]$, $1 \leq p < \infty$; see Exercise 5.1.2.

By Proposition 5.3, V cannot be hypercyclic on any of the spaces considered. Indeed, for the same reason, no multiple λV, $\lambda \in \mathbb{C}$, can be hypercyclic.

We turn to the announced improvement of Proposition 5.3. For its proof we need a topological lemma that we accept without proof.

Lemma 5.5. *Let A and B be compact subsets of a metric space, where B is connected. Then the following assertions are equivalent:*
(i) *every connected component of A meets B;*
(ii) *$A \cup B$ is connected.*

Theorem 5.6 (Kitai). *Let T be a hypercyclic operator. Then every connected component of $\sigma(T)$ meets the unit circle.*

Proof. By the preceding lemma we have to show that the compact set $\sigma(T) \cup \mathbb{T}$ is connected. If this is not the case then $\sigma(T) \cup \mathbb{T}$ can be partitioned into two nonempty closed sets C_1 and C_2. Since $\mathbb{T}$ is connected, it must lie entirely in one of these two sets, C_2 say. Then we define a partition of $\sigma(T)$ into closed sets by setting

$$\sigma_1 := C_1 \cap \sigma(T) \quad \text{and} \quad \sigma_2 := C_2 \cap \sigma(T).$$

Since $\mathbb{T}$ is contained in C_2, we have that $\sigma_1 = C_1$ is nonempty. Moreover, if $C_2 \cap \sigma(T)$ was empty then $\sigma(T)$ would be contained in C_1 and therefore disjoint from $\mathbb{T}$, which is impossible by Proposition 5.3.

Now, by the Riesz decomposition theorem there exist nontrivial T-invariant closed subspaces M_1 and M_2 such that $X = M_1 \oplus M_2$, $\sigma(T|_{M_1}) = \sigma_1$ and $\sigma(T|_{M_2}) = \sigma_2$. By Proposition 2.28, $T|_{M_1}$ is then a hypercyclic operator whose spectrum does not meet $\mathbb{T}$, contradicting Proposition 5.3. □

In particular, the spectrum of a hypercyclic operator cannot have isolated points outside the unit circle. For chaotic operators, the same is true unrestrictedly.

Proposition 5.7. *Let T be a chaotic operator. Then its spectrum has no isolated points and its point spectrum contains infinitely many roots of unity.*

Proof. We first show that $\sigma_p(T)$ contains infinitely many roots of unity. Since T has periodic points, Proposition 2.33 shows that $\sigma_p(T)$ contains at least one root of unity. Now suppose that $\lambda_1, \ldots, \lambda_N$, $N \geq 1$, are the only roots of unity in the point spectrum of T. Consider the polynomial

$$p(z) = (z - \lambda_1) \cdots (z - \lambda_N).$$

Then, for any eigenvector x of T to an eigenvalue λ_k, $k = 1, \ldots, N$, we have that

$$p(T)x = (T - \lambda_1 I) \cdots (T - \lambda_N I)x = 0.$$

Again by Proposition 2.33 and by linearity of $p(T)$ we have that $p(T)x = 0$ for every periodic point of T. Since these points form a dense set in X we obtain that $p(T) = 0$, which contradicts Theorem 2.54.

Now suppose that λ is an isolated point of the spectrum of T. Then by the Riesz decomposition theorem there are nontrivial T-invariant closed subspaces M_1 and M_2 of X such that $X = M_1 \oplus M_2$ and $\sigma(T|_{M_2}) = \{\lambda\}$. It follows as in Proposition 2.28 (see also Exercise 2.2.8), that $T|_{M_2}$ is chaotic, which is impossible since its spectrum is a singleton. □

5.2 Classes of non-hypercyclic operators on Banach spaces

Based on necessary conditions for hypercyclicity we can now show that certain classes of operators cannot contain hypercyclic operators.

Throughout this section, T denotes an operator on a (real or complex) Banach space X.

Power-bounded operators. An operator T is called
a *contraction* if $\|T\| \leq 1$,
quasinilpotent if $\lim_{n\to\infty} \|T^n\|^{1/n} = 0$,
power bounded if $\sup_{n\geq 0} \|T^n\| < \infty$.
Over complex scalars, the spectral radius formula tells us that an operator is quasinilpotent if and only if $\sigma(T) = \{0\}$.

Clearly, every contraction and every quasinilpotent operator is power bounded, and any orbit of a power-bounded operator is bounded and therefore not dense. In fact, for a quasinilpotent operator, all orbits tend to 0.

Proposition 5.8. *No power-bounded operator is hypercyclic. In particular, no contraction and no quasinilpotent operator is hypercyclic.*

By Theorem 4.21, every nonzero multiple λD of the differentiation operator on $H(\mathbb{C})$ is hypercyclic. Such a phenomenon cannot occur in the Banach space setting. More precisely, if T is a nonzero operator on a Banach space and $\lambda \in \mathbb{K}$ with $|\lambda| \leq 1/\|T\|$ then λT is a contraction and therefore not hypercyclic.

We apply this proposition to a specific family of operators.

Example 5.9. **(Volterra integral operator)** Let $k : [0,1] \times [0,1] \to \mathbb{K}$ be a continuous function. Then the operator

$$V_k : C[0,1] \to C[0,1], \quad (Tf)(t) = \int_0^t k(t,s) f(s)\, ds, \quad 0 \leq t \leq 1,$$

is called a *Volterra integral operator*, and k is its *kernel*; note that the values $k(t,s)$, $s > t$, are irrelevant. It is obvious that V_k defines an operator on $C[0,1]$ with $\|V_k\| \leq \max_{t\in[0,1]} \int_0^t |k(t,s)|\,ds < \infty$. In the special case when k is identically 1 we obtain the classical Volterra operator of Example 5.4.

A simple induction shows that every power of a Volterra integral operator is a Volterra integral operator. More precisely, if k_n is the kernel of $(V_k)^n$, $n \geq 1$, then

$$k_{n+1}(t,s) = \int_s^t k(t,u)k_n(u,s)\,du, \ \ s \leq t, \quad n \geq 1;$$

in addition, if $M = \max_{0\leq s\leq t\leq 1} |k(t,s)|$, then

$$|k_n(t,s)| \leq \frac{M^n}{(n-1)!}(t-s)^{n-1}, \ \ s \leq t, \quad n \geq 1,$$

and hence

$$\|(V_k)^n\| \leq \frac{M^n}{n!},$$

which implies that $\lim_{n\to\infty} \|(V_k)^n\|^{1/n} = 0$. We ask the reader to verify these statements; see Exercise 5.2.1. Thus, each Volterra integral operator is quasinilpotent and therefore not hypercyclic.

Finite-rank operators. We already know that there are no hypercyclic operators on finite-dimensional spaces; see Corollary 2.59. This immediately extends to *finite-rank operators*, that is, operators whose range has finite dimension.

Proposition 5.10. *No finite-rank operator is hypercyclic.*

Proof. Let T be an operator with finite-dimensional range $\operatorname{ran} T$, and let x be a hypercyclic vector for T. Since $Tx \in \operatorname{ran} T$ is also hypercyclic for T (see Proposition 1.15), the restriction $T|_{\operatorname{ran} T} : \operatorname{ran} T \to \operatorname{ran} T$ is hypercyclic, in contradiction to Corollary 2.59. □

Compact operators. An operator T is called compact if the image of the closed unit ball under T is relatively compact, that is, if its closure is compact. This is equivalent to saying that, for any sequence $(x_n)_n$ in X with $\|x_n\| \leq 1$, $n \geq 1$, the sequence $(Tx_n)_n$ has a convergent subsequence.

In $\mathbb{K}^N$, $N \geq 1$, and therefore in any finite-dimensional space, every bounded set is relatively compact; hence every finite-rank operator is compact. The following result therefore contains Proposition 5.10.

Theorem 5.11. *No compact operator is hypercyclic.*

For the proof we need several lemmas. The first one is fundamental for the study of compact operators.

Lemma 5.12 (Riesz's lemma). *Let X be a Banach space, M a proper closed subspace of X and $\varepsilon > 0$. Then there is some $x \in X \setminus M$ such that*

$$\|x\| = 1 \quad \text{and} \quad \|x - y\| \geq 1 - \varepsilon \text{ for all } y \in M.$$

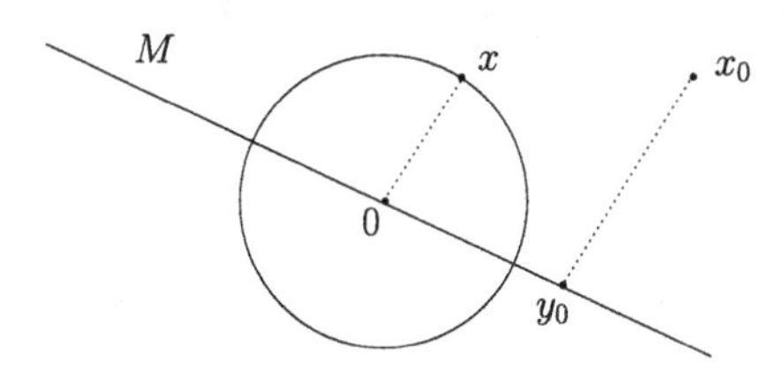

Fig. 5.1 The proof of Riesz's lemma

Proof. Let $0 < \varepsilon < 1$. We choose any $x_0 \in X \setminus M$. Since M is closed we have that

$$d := \inf_{y \in M} \|x_0 - y\| > 0.$$

Thus there is a point $y_0 \in M$ such that

$$\|x_0 - y_0\| \leq \frac{d}{1 - \varepsilon}.$$

We now consider the point

$$x := \frac{x_0 - y_0}{\|x_0 - y_0\|};$$

see Figure 5.1. Then $\|x\| = 1$, and for any $y \in M$ we have that

$$\begin{aligned}
\|x - y\| &= \left\| \frac{x_0 - y_0}{\|x_0 - y_0\|} - y \right\| \\
&= \frac{1}{\|x_0 - y_0\|} \|x_0 - (y_0 + y\|x_0 - y_0\|)\| \\
&\geq \frac{d}{\|x_0 - y_0\|} \geq 1 - \varepsilon,
\end{aligned}$$

where we have used the fact that $y_0 + y\|x_0 - y_0\| \in M$. □

Lemma 5.13. *Let T be a compact operator and $\lambda \in \mathbb{K}$, $\lambda \neq 0$. Then $\lambda I - T$ has closed range.*

Proof. Since $\lambda I - T = \lambda(I - T/\lambda)$ and T/λ is also compact we can assume that $\lambda = 1$.

Thus let $y \in \overline{\operatorname{ran}(I-T)}$. We choose points $y_n \in \operatorname{ran}(I-T)$, $n \geq 1$, so that $y_n \to y$ as $n \to \infty$. We then set

$$d_n = \inf\{\|x\| \ ; \ (I-T)x = y_n\},$$

and we choose points x_n with $(I-T)x_n = y_n$ and $\|x_n\| \leq d_n + 1$, $n \geq 1$.

We first show that $(d_n)_n$ must contain a bounded subsequence. Otherwise we have that $d_n \to \infty$ as $n \to \infty$. We then set

$$z_n = \frac{1}{d_n+1} x_n, \quad n \geq 1.$$

Since $\|x_n\| \leq d_n + 1$, $n \geq 1$, the sequence $(z_n)_n$ belongs to the closed unit ball. By compactness of T, $(Tz_n)_n$ contains a convergent subsequence, which we can assume to be the full sequence. Moreover, since $(y_n)_n$ converges and $d_n \to \infty$ we conclude that

$$z_n - Tz_n = (I-T)z_n = \frac{1}{d_n+1}(I-T)x_n = \frac{1}{d_n+1} y_n \to 0 \tag{5.1}$$

as $n \to \infty$. This implies that the sequence $(z_n)_n$ also converges. We call its limit z. It then follows from (5.1) that $z - Tz = 0$ and hence also that

$$(I-T)(x_n - (d_n+1)z) = (I-T)x_n - (d_n+1)(I-T)z = y_n$$

for $n \geq 1$. Therefore, by the definition of d_n, we have that, for all $n \geq 1$,

$$d_n \leq \|x_n - (d_n+1)z\| = (d_n+1)\|z_n - z\|,$$

which is impossible since $\|z_n - z\| \to 0$ and $d_n \to \infty$ as $n \to \infty$.

Consequently, $(d_n)_n$ contains a bounded subsequence; we can assume that $(d_n)_n$ itself is bounded. Thus there is some $\delta > 0$ such that any δx_n, $n \geq 1$, belongs to the closed unit ball. By compactness of T, the sequence $(T(\delta x_n))_n$ contains a convergent subsequence, which we can assume to be the full sequence; hence also $(Tx_n)_n$ converges. But we also have that

$$x_n - Tx_n = (I-T)x_n = y_n \to y$$

as $n \to \infty$. This implies that the sequence $(x_n)_n$ itself converges. Calling its limit x we obtain that

$$y = \lim_{n\to\infty} (I-T)x_n = (I-T)x,$$

so that $y \in \operatorname{ran}(I-T)$. This shows that $\operatorname{ran}(I-T)$ is closed. □

The next lemma tells us that the eigenvalues of a compact operator can only have 0 as an accumulation point.

Lemma 5.14. *Let T be a compact operator and $\varepsilon > 0$. Then T has only a finite number of eigenvalues λ with $|\lambda| \geq \varepsilon$.*

Proof. We assume that, on the contrary, there is a sequence of pairwise distinct eigenvalues λ_n with $|\lambda_n| \geq \varepsilon$, $n \geq 1$. Let x_n be an eigenvector of T to the eigenvalue λ_n, $n \geq 1$. It is well known from linear algebra that the x_n are linearly independent. Therefore the spaces

$$M_n := \operatorname{span}\{x_1, \dots, x_n\}, \ n \geq 1$$

form a strictly increasing sequence of subspaces of X. Since each M_{n-1} (with $M_0 = \{0\}$) is finite dimensional and hence closed in M_n (see Appendix A), it follows from Riesz's lemma that there are points $y_n \in M_n \setminus M_{n-1}$, $n \geq 1$, such that

$$\|y_n\| = 1 \quad \text{and} \quad \|y_n - y\| \geq \tfrac{1}{2} \text{ for all } y \in M_{n-1}.$$

For $n > m \geq 1$ we have that

$$\|Ty_n - Ty_m\| = \|\lambda_n y_n - (\lambda_n I - T)y_n - Ty_m\|.$$

Now, since y_n is a linear combination of $x_1, \dots, x_n$, and since $(\lambda_n I - T)x_n = 0$, $(\lambda_n I - T)y_n$ belongs to M_{n-1}, as does Ty_m. Hence

$$w := \frac{1}{\lambda_n}((\lambda_n I - T)y_n + Ty_m) \in M_{n-1}.$$

It follows that, for $n > m \geq 1$,

$$\|Ty_n - Ty_m\| = \|\lambda_n y_n - \lambda_n w\| = |\lambda_n| \|y_n - w\| \geq \frac{\varepsilon}{2}.$$

But this is impossible because $(y_n)_n$ belongs to the closed unit ball so that, by compactness of T, $(Ty_n)_n$ must possess a convergent subsequence. □

Our final lemma shows that, in an infinite-dimensional setting, 0 belongs to the spectrum of a compact operator.

Lemma 5.15. *Let T be a compact operator on an infinite-dimensional Banach space. Then $0 \in \sigma(T)$.*

Proof. If 0 is not in the spectrum then T is invertible. Thus the unit ball of X, as the homeomorphic image (under T^{-1}) of a relatively compact set, is itself relatively compact. But this can only happen in finite-dimensional spaces; see Appendix A. □

We are now in a position to show that compact operators are never hypercyclic. Here we only give the proof in the case of complex Banach spaces. In the real case the proof is outlined in Exercise 5.2.5; it uses the same ideas as in the complex case but it is less straightforward.

Proof of Theorem 5.11 *for complex Banach spaces.* Let T be a hypercyclic compact operator on a complex, necessarily infinite-dimensional, Banach space, and let $\lambda \neq 0$ be in the spectrum of T. By Bourdon's theorem and Lemma 5.13, $\lambda I - T$ has closed and dense range, so that it is surjective. Therefore λ must be an eigenvalue of T.

It now follows from Lemmas 5.14 and 5.15 that the spectrum of T consists either of a sequence of points converging to 0, together with the point 0, or of a finite set, including 0. In both cases the spectrum has connected components that do not meet the unit circle. But this contradicts Kitai's theorem. □

Example 5.16. In Example 5.9 we showed that the Volterra integral operators on $C[0,1]$ are quasinilpotent, and therefore not hypercyclic.

We want to show here that they are also compact, implying once more that they are not hypercyclic. To this end let $(f_n)_n$ be a sequence in $C[0,1]$ with $\|f_n\| \leq 1$, $n \geq 1$. We have to construct a subsequence $(g_\nu)_\nu$ of $(f_n)_n$ such that $(Tg_\nu)_\nu$ converges.

We begin the proof by a familiar diagonal process. Let $(t_j)_j$ be a dense sequence in $[0,1]$. Since T is continuous, the sequence $(Tf_n)_n$ is bounded in $C[0,1]$, and hence the sequence $((Tf_n)(t_1))_n$ is bounded in $\mathbb{K}$. We can therefore extract a subsequence $(n_{1,\nu})_\nu$ such that $((Tf_{n_{1,\nu}})(t_1))_\nu$ converges in $\mathbb{K}$. In the same way, since $((Tf_{n_{1,\nu}})(t_2))_\nu$ is bounded there exists a subsequence $(n_{2,\nu})_\nu$ of $(n_{1,\nu})_\nu$ such that $((Tf_{n_{2,\nu}})(t_2))_\nu$ converges. Proceeding inductively, we find subsequences $(n_{j,\nu})_\nu$ of $(n_{j-1,\nu})_\nu$ such that $((Tf_{n_{j,\nu}})(t_j))_\nu$ converges, for $j \geq 2$. Now define $m_\nu = n_{\nu,\nu}$. Since, for any $j \geq 1$, $(m_\nu)_{\nu \geq j}$ is a subsequence of $(n_{j,\nu})_\nu$ we have that $((Tf_{m_\nu})(t_j))_\nu$ converges.

We claim that $g_\nu = f_{m_\nu}$, $\nu \geq 1$, defines the required subsequence of $(f_n)_n$. To show that $(Tg_\nu)_\nu$ converges in $C[0,1]$ it suffices to show that it is a Cauchy sequence. Thus, let $\varepsilon > 0$. Since the kernel k of T is continuous, and hence uniformly continuous, on $[0,1] \times [0,1]$ there is some $\delta > 0$ such that, if $|\tau_1 - \tau_2| < \delta$, $\tau_1, \tau_2 \in [0,1]$, then $\max_{s \in [0,1]} |k(\tau_1, s) - k(\tau_2, s)| < \varepsilon$. Moreover, $M := \max_{t,s \in [0,1]} |k(t,s)| < \infty$. By the density of $(t_j)_j$ there is some $J \in \mathbb{N}$ such that for any $t \in [0,1]$ there is some $j \leq J$ such that $|t - t_j| < \min(\delta, \varepsilon)$. The definition of the sequence $(g_\nu)_\nu$ shows that there is some $N \in \mathbb{N}$ such that, for any $j \leq J$ and $\nu, \mu \geq N$,

$$|(Tg_\nu)(t_j) - (Tg_\mu)(t_j)| \leq \varepsilon.$$

Now let $t \in [0,1]$. Choose $j \leq J$ such that $|t - t_j| < \min(\delta, \varepsilon)$. Then, for any $\nu \geq 1$,

$$\begin{aligned}
|(Tg_\nu)(t) - (Tg_\nu)(t_j)| &\leq \Big| \int_0^{t_j} (k(t,s) - k(t_j,s)) g_\nu(s)\, ds \Big| \\
&\qquad + \Big| \int_{t_j}^{t} k(t,s) g_\nu(s)\, ds \Big| \\
&\leq \varepsilon + M\varepsilon,
\end{aligned}$$

where we have used that $\|g_\nu\| \leq 1$, $\nu \geq 1$. We therefore have that, for any $\nu, \mu \geq N$,

$$\begin{aligned}|(Tg_\nu)(t) - (Tg_\mu)(t)| &\leq |(Tg_\nu)(t) - (Tg_\nu)(t_j)| + |(Tg_\nu)(t_j) - (Tg_\mu)(t_j)| \\ &\qquad + |(Tg_\mu)(t_j) - (Tg_\mu)(t)| \\ &\leq (2M+3)\varepsilon.\end{aligned}$$

This shows that $(Tg_\nu)_\nu$ is a Cauchy sequence in $C[0,1]$, as required.

Power-compact operators. An operator T is called *power compact* if some power T^n, $n \geq 1$, of T is compact.

Proposition 5.17. *No power-compact operator is hypercyclic.*

The proof can be given as in the case of compact operators, taking account of the fact that $\sigma(T^n) = \sigma(T)^n$ by the spectral mapping theorem. Alternatively, the result is also an immediate consequence of the compact case and of Ansari's theorem that we will obtain in the next chapter; see Theorem 6.2. By Ansari, T can only be hypercyclic if T^n is, which is impossible if T^n is compact.

Finite-rank perturbations. We have noted above that finite-rank operators cannot be hypercyclic. The same is true for finite-rank perturbations of multiples of the identity I, that is, operators of the form

$$\lambda I + F, \quad \lambda \in \mathbb{K},$$

where F is a finite-rank operator.

Proposition 5.18. *No finite-rank perturbation of a multiple of the identity is hypercyclic.*

In fact, an even more general result can be established; we leave the proof to the reader (see Exercise 5.2.8).

Compact perturbations. We next consider compact perturbations of multiples of the identity. We first show that hypercyclicity imposes a severe restriction on such an operator.

Lemma 5.19. *Let T be a hypercyclic operator of the form*

$$T = \lambda I + K,$$

where $\lambda \in \mathbb{K}$ and K is compact. Then $|\lambda| = 1$ and K is quasinilpotent.

We restrict our attention again to the complex case and refer to Exercise 5.2.9 for the real case.

Proof (for complex Banach spaces). Let $T = \lambda I + K$ be hypercyclic. For $\mu \in \mathbb{C}$ we have that

$$\mu I - T = (\mu - \lambda)I - K.$$

As in the proof of Theorem 5.11 it then follows from Bourdon's theorem and Lemmas 5.13, 5.14 and 5.15 that $\lambda \in \sigma(T)$ and that every point $\mu \in \sigma(T)$ with $\mu \neq \lambda$ is an isolated point of the spectrum.

Consequently, if $\mu \in \sigma(T)$ with $\mu \neq \lambda$ then $\{\mu\}$ and $\sigma(T) \setminus \{\mu\}$ form a partition of $\sigma(T)$ into nonempty closed subsets. By the Riesz decomposition theorem there are nontrivial T-invariant closed subspaces M_1 and M_2 of X such that $X = M_1 \oplus M_2$ with $\sigma(T|_{M_1}) = \{\mu\}$. It follows from Proposition 2.28 that $T|_{M_1}$ is a hypercyclic operator and hence, by Corollary 2.59, that M_1 is infinite-dimensional.

On the other hand, M_1 is also invariant under $K = T - \lambda I$. Since M_1 is closed, $K|_{M_1}$ is a compact operator, and $\sigma(K|_{M_1}) = \sigma(T|_{M_1} - \lambda I|_{M_1}) = \{\mu - \lambda\}$ does not contain 0. By Lemma 5.15 this is only possible if the underlying space M_1 is finite dimensional, a contradiction.

Therefore the spectrum of T cannot contain points other than λ, and thus $\sigma(T) = \{\lambda\}$. Kitai's theorem then forces $|\lambda|$ to be 1. Moreover, $\sigma(K) = \{0\}$. □

We will show later that there are indeed hypercyclic operators of the form $T = I + K$ with K compact; see Example 8.4. But such an operator cannot be chaotic.

Proposition 5.20. *No compact perturbation of a multiple of the identity is chaotic.*

Proof. Since the complexification of a compact perturbation of a multiple of the identity is again of this type it suffices, by Corollary 2.51, to study the complex case.

Now, by Lemma 5.19, any chaotic operator $T = \lambda I + K$, K compact, must satisfy $\sigma(T) = \{\lambda\}$, which is impossible by Proposition 5.7. □

Lower triangular operators. For operators on Banach spaces with a basis it can be very instructive to study their matrix representations. Specifically, let X be a Banach space with a basis $(e_n)_{n\geq 1}$, and let $(e_n^*)_{n\geq 1}$ be the corresponding coefficient functionals $e_n^*(\sum_{k=1}^\infty x_k e_k) = x_n$, $n \geq 1$; see Appendix A. Then we have for any operator T on X and $x = \sum_{k=1}^\infty x_k e_k \in X$,

$$Tx = \sum_{n=1}^\infty \langle Tx, e_n^*\rangle e_n = \sum_{n=1}^\infty \Big(\sum_{k=1}^\infty \langle Te_k, e_n^*\rangle x_k\Big) e_n = \sum_{n=1}^\infty \Big(\sum_{k=1}^\infty a_{nk} x_k\Big) e_n,$$

where $a_{nk} = \langle Te_k, e_n^*\rangle$, $n, k \geq 1$. Therefore, in terms of the coefficient sequences $(x_n)_n$ of x and $(y_n)_n$ of $y = Tx$, T may be represented by an infinite matrix

$$A = (a_{nk})_{n,k\geq 1}.$$

Now suppose that in the nth row of the matrix A, all off-diagonal elements are 0, that is, $a_{nk} = 0$ for all $k \neq n$. Then we have for $x = \sum_{k=1}^{\infty} x_k e_k \in X$ that

$$\langle x, T^* e_n^* \rangle = \langle Tx, e_n^* \rangle = \sum_{k=1}^{\infty} a_{nk} x_k = a_{nn} x_n = a_{nn} \langle x, e_n^* \rangle.$$

This shows that e_n^* is an eigenvector of T^*. In view of Proposition 5.1, T cannot be hypercyclic. A very particular, but quite common, class of matrices with the stated property consists of the lower triangular matrices, that is, matrices $A = (a_{nk})_{n,k\geq 1}$ with $a_{nk} = 0$ for $k > n$, $n \geq 1$: here, the first row has no nonzero off-diagonal elements. The corresponding operators are called *lower triangular operators*.

Proposition 5.21. *Let T be an operator on a Banach space X with a basis. If the matrix A representing T has a row all of whose off-diagonal elements are 0 then T is not hypercyclic. In particular, no lower triangular operator is hypercyclic.*

As an illustration, on $X = \ell^p$, $1 \leq p < \infty$, and c_0, the weighted forward shifts F_w (never hypercyclic) and the weighted backward shifts B_w (hypercyclic for certain weights) are represented by the matrices

$$\begin{pmatrix} 0 & 0 & 0 & 0 & \dots \\ w_1 & 0 & 0 & 0 & \dots \\ 0 & w_2 & 0 & 0 & \dots \\ 0 & 0 & w_3 & 0 & \dots \\ \vdots & \vdots & \vdots & \vdots & \end{pmatrix} \quad \text{and} \quad \begin{pmatrix} 0 & w_2 & 0 & 0 & \dots \\ 0 & 0 & w_3 & 0 & \dots \\ 0 & 0 & 0 & w_4 & \dots \\ 0 & 0 & 0 & 0 & \dots \\ \vdots & \vdots & \vdots & \vdots & \end{pmatrix},$$

respectively; see Section 4.1. For complex scalars, if $\varphi(z) = \sum_{n=0}^{\infty} a_n z^n$ is holomorphic on a neighbourhood of $\overline{\mathbb{D}}$ then $\varphi(B)$ is represented by the matrix

$$\begin{pmatrix} a_0 & a_1 & a_2 & a_3 & \dots \\ 0 & a_0 & a_1 & a_2 & \dots \\ 0 & 0 & a_0 & a_1 & \dots \\ 0 & 0 & 0 & a_0 & \dots \\ \vdots & \vdots & \vdots & \vdots & \end{pmatrix}$$

(see Section 4.4), which are special Toeplitz matrices; chaos for $\varphi(B)$ is characterized in Theorem 4.43.

Example 5.22. On the sequence spaces ℓ^p, $1 < p < \infty$, or c_0, the *Cesàro operator* C_1 is given by $C_1 x = (\frac{1}{n} \sum_{k=1}^{n} x_k)_n$. It associates to each sequence its sequence of arithmetic means. In the canonical bases, the operator is represented by the lower triangular matrix

$$\begin{pmatrix} 1 & 0 & 0 & 0 & \dots \\ \frac{1}{2} & \frac{1}{2} & 0 & 0 & \dots \\ \frac{1}{3} & \frac{1}{3} & \frac{1}{3} & 0 & \dots \\ \frac{1}{4} & \frac{1}{4} & \frac{1}{4} & \frac{1}{4} & \dots \\ \vdots & \vdots & \vdots & \vdots & \end{pmatrix}.$$

The Cesàro operator is therefore not hypercyclic. Of course, it needs to be verified that C_1 is indeed an operator on the stated sequence spaces. For c_0 this is easily shown, for the spaces ℓ^p this is the content of the so-called Hardy inequality.

5.3 Classes of non-hypercyclic operators on Hilbert spaces

An important feature of Hilbert spaces is that they are self-dual; a more precise statement is contained in the Riesz representation theorem; see Appendix A. As a consequence, the adjoint of an operator can be considered as an operator on the space itself. This leads to some interesting classes of operators that have no analogue for general Banach spaces.

Throughout this section, if T is an operator on a Hilbert space H then T^* denotes the (Hilbert space) adjoint of T; see Appendix A.

Example 5.23. Every operator T on $\mathbb{C}^N$, $N \geq 1$, can be represented by a matrix $A = (a_{nk})_{1\leq n,k\leq N}$ via the usual matrix product:

$$Tx = Ax = \Big(\sum_{k\geq 1} a_{nk}x_k\Big)_n.$$

If $\mathbb{C}^N$ carries its usual inner product then we have for all $x, y \in \mathbb{C}^N$,

$$\langle Tx, y\rangle = \sum_{n\geq 1}\Big(\sum_{k\geq 1} a_{nk}x_k\Big)\overline{y_n} = \sum_{k\geq 1} x_k \overline{\Big(\sum_{n\geq 1} \overline{a_{nk}}y_n\Big)}.$$

This shows that the adjoint T^* is represented by the conjugate transpose $\overline{A}^t$ of the matrix A. In the same way, on the real space $\mathbb{R}^N$, the adjoint corresponds to the transpose of the representing matrix.

In linear algebra, a matrix A is called normal if $\overline{A}^t A = A\overline{A}^t$. In analogy, a general Hilbert space operator T is called *normal* if it commutes with its adjoint,

$$T^*T = TT^*.$$

An important class of normal operators are the *self-adjoint* operators, that is, operators T satisfying

$$T = T^*.$$

In order to explore these notions we need a preliminary result.

Lemma 5.24. *Let T be an operator on a Hilbert space H.*

(a) *Suppose that H is a complex Hilbert space. If*

$$\langle Tx, x\rangle \in \mathbb{R} \quad \textit{for all } x \in H,$$

then T is self-adjoint. If

$$\langle Tx, x\rangle = 0 \quad \textit{for all } x \in H,$$

then $T = 0$.

(b) *Suppose that H is a real Hilbert space. If T is self-adjoint and*

$$\langle Tx, x\rangle = 0 \quad \textit{for all } x \in H,$$

then $T = 0$.

Proof. For any $x, y \in H$ we have that

$$\langle T(x+y), x+y\rangle - \langle Tx, x\rangle - \langle Ty, y\rangle = \langle Tx, y\rangle + \langle Ty, x\rangle. \tag{5.2}$$

In the real case, this already implies (b). In fact, the left-hand side is zero by assumption, and since T is self-adjoint and the inner product is symmetric, we deduce that, for all $x, y \in H$,

$$0 = \langle Tx, y\rangle + \langle Ty, x\rangle = \langle Tx, y\rangle + \langle y, Tx\rangle = \langle Tx, y\rangle + \langle Tx, y\rangle = 2\langle Tx, y\rangle,$$

which can only be true if $T = 0$.

In the complex case, we replace y by iy in (5.2) to obtain that, for any $x, y \in H$,

$$\langle T(x+iy), x+iy\rangle - \langle Tx, x\rangle - \langle T(iy), iy\rangle = i(-\langle Tx, y\rangle + \langle Ty, x\rangle). \tag{5.3}$$

Under the first assumption in (a), the left-hand sides in (5.2) and (5.3) are both real-valued. Hence we deduce that

$$\operatorname{Im}\langle Tx, y\rangle = -\operatorname{Im}\langle Ty, x\rangle \quad \text{and} \quad \operatorname{Re}\langle Tx, y\rangle = \operatorname{Re}\langle Ty, x\rangle$$

and therefore

$$\langle Tx, y\rangle = \overline{\langle Ty, x\rangle} = \langle x, Ty\rangle$$

for all $x, y \in H$. This shows that $T^* = T$.

Under the second assumption in (a) we obtain from (5.2) and (5.3) that

$$\langle Tx, y\rangle + \langle Ty, x\rangle = 0 = -\langle Tx, y\rangle + \langle Ty, x\rangle$$

and hence

$$\langle Tx, y\rangle = 0$$

for all $x, y \in H$. This shows that $T = 0$. □

It follows from Proposition A.8 that $T^*T - TT^*$ is self-adjoint. Hence the lemma tells us that, both in the real and the complex case, $T^*T = TT^*$ if and only if $\langle (T^*T - TT^*)x, x\rangle = 0$ for all $x \in H$. But since

$$\begin{aligned}\langle (T^*T - TT^*)x, x\rangle &= \langle T^*Tx, x\rangle - \langle TT^*x, x\rangle = \langle Tx, Tx\rangle - \langle T^*x, T^*x\rangle \\ &= \|Tx\|^2 - \|T^*x\|^2, \end{aligned} \tag{5.4}$$

we have shown the following.

Proposition 5.25. *An operator T on a Hilbert space H is normal if and only if, for all $x \in H$,*

$$\|Tx\| = \|T^*x\|.$$

An operator T on a Hilbert space H is called *positive* if, for all $x \in H$,

$$\langle Tx, x\rangle \geq 0.$$

One then writes

$$T \geq 0.$$

By Lemma 5.24, every positive operator on a complex Hilbert space is self-adjoint. However, the matrix $\left(\begin{smallmatrix}1 & 2\\ 0 & 1\end{smallmatrix}\right)$ defines a positive operator on the real Hilbert space $\mathbb{R}^2$ that is not self-adjoint; see also Example 5.33.

It seems natural to weaken the normality condition $T^*T - TT^* = 0$ to a mere positivity assumption: an operator T on a Hilbert space H is called *hyponormal* if

$$T^*T - TT^* \geq 0.$$

Remark 5.26. Of course, one may just as well consider operators that satisfy

$$TT^* - T^*T \geq 0.$$

That these are of less interest in the present context will become clear in Exercise 5.3.5.

Clearly, every normal operator is hyponormal. Equation (5.4) implies the following characterization.

Proposition 5.27. *An operator T on a Hilbert space H is hyponormal if and only if, for all $x \in H$,*

$$\|Tx\| \geq \|T^*x\|.$$

Now we derive an interesting property of hyponormal operators. By the Cauchy–Schwarz inequality, any operator T satisfies

$$\|Tx\|^2 = \langle Tx, Tx\rangle = \langle T^*Tx, x\rangle \leq \|T^*Tx\|\|x\|$$

for all $x \in H$. Proposition 5.27, applied to Tx, then shows that for hyponormal operators we have that

$$\|Tx\|^2 \leq \|T^2x\|\|x\|$$

for all $x \in H$. Operators T satisfying this condition are called *paranormal*. Thus every hyponormal operator is paranormal.

In Table 5.1 we collect the various classes of Hilbert space operators that we have introduced, in increasing order of generality, along with their characterizing conditions.

self-adjoint	$T = T^*$	
normal	$T^*T = TT^*$	$\forall\, x \in H,\ \|Tx\| = \|T^*x\|$
hyponormal	$T^*T - TT^* \geq 0$	$\forall\, x \in H,\ \|Tx\| \geq \|T^*x\|$
paranormal		$\forall\, x \in H,\ \|Tx\|^2 \leq \|T^2x\|\|x\|$

Table 5.1 Classes of Hilbert space operators

Remark 5.28. The operator on $\mathbb{K}^2$ that is given by the matrix $A = \left(\begin{smallmatrix} 0 & 1 \\ -1 & 0 \end{smallmatrix}\right)$ is normal, but not self-adjoint. For distinguishing the remaining classes of operators we refer to Exercises 5.3.1 and 5.3.2.

The defining condition for paranormal operators is obviously not specific to the Hilbert space setting, which leads us back to general Banach spaces.

Definition 5.29. An operator T on a Banach space X is called *paranormal* if, for all $x \in X$,

$$\|Tx\|^2 \leq \|T^2x\|\|x\|.$$

The following is the main result of this section.

Theorem 5.30. *No paranormal operator on a Banach space is hypercyclic.*

Proof. We show that the orbits of paranormal operators are too well behaved for the operator to be hypercyclic. Indeed, let T be a paranormal operator on a Banach space X and $x \in X$. Suppose that, for some $n \geq 0$, $\|T^{n+1}x\| > \|T^nx\|$, which implies, in particular, that $T^{n+1}x \neq 0$ and $T^nx \neq 0$. Applying the definition of paranormality to T^nx we obtain that

$$\|T^{n+2}x\| \geq \frac{\|T^{n+1}x\|^2}{\|T^nx\|} > \|T^{n+1}x\|.$$

Therefore, the orbit of any $x \in X$ is either decreasing in norm, or strictly increasing in norm from some index on. As a consequence, no orbit can be dense. $\square$

One may also arrive at the same conclusion from another point of view. Considering $T^n x$ in the definition of paranormality and taking logarithms we see that, if T is a paranormal operator, then, for any $x \in X$ and $n \geq 0$ with $T^{n+2}x \neq 0$,

$$\log \|T^{n+1}x\| \leq \frac{1}{2}(\log \|T^{n+2}x\| + \log \|T^n x\|).$$

Thus, for any $x \in X$ whose orbit does not end up in the origin, the sequence $(\log \|T^n x\|)_n$ is convex. This leads to the behaviour of the sequence $(\|T^n x\|)_n$ found in the proof.

Corollary 5.31. *No self-adjoint operator, no normal operator, no hyponormal operator on a Hilbert space is hypercyclic. No positive operator on a complex Hilbert space is hypercyclic.*

We finish this section, and indeed the chapter, with two examples.

Example 5.32. We consider the bilateral weighted backward shift

$$B_w x = (w_{n+1}x_{n+1})_{n\in\mathbb{Z}}$$

on the Hilbert space $\ell^2(\mathbb{Z})$, where $w = (w_n)_{n\in\mathbb{Z}}$ is a weight sequence; see Section 4.1. An easy computation shows that

$$B_w^* = F_v,$$

the weighted forward shift with weight sequence $v = (\overline{w_{n+1}})_{n\in\mathbb{Z}}$; see Remark 4.14. (Note that this is the Hilbert space adjoint of B_w; its Banach space adjoint was determined in the proof of Proposition 4.16.)

We calculate that, for $x \in \ell^2(\mathbb{Z})$ and $\lambda \geq 0$,

$$\big\langle \big({F_v}^2 B_w^2 - 2\lambda F_v B_w + \lambda^2 I\big)\, x, x\big\rangle = \sum_{n\in\mathbb{Z}} \big(|w_n|^2|w_{n-1}|^2 - 2\lambda|w_n|^2 + \lambda^2\big)\, |x_n|^2.$$

Hence, by Exercise 5.3.4(b), B_w is paranormal if and only if, for all $\lambda \geq 0$ and $n \in \mathbb{Z}$,

$$|w_n|^2|w_{n-1}|^2 - 2\lambda|w_n|^2 + \lambda^2 \geq 0,$$

which by Exercise 5.3.4(a) is equivalent to

$$4|w_n|^4 \leq 4|w_n|^2|w_{n-1}|^2 \quad \text{for all } n \in \mathbb{Z}.$$

Therefore, B_w is paranormal if and only if $(|w_n|)_{n\in\mathbb{Z}}$ is a decreasing sequence. A comparison with Example 4.15 shows that this produces some, but not all by far, non-hypercyclic bilateral weighted shifts.

Example 5.33. We know from Theorem 4.43 that the operator $I + B$, with B the backward shift, is chaotic on ℓ^2. This result holds for real and for complex sequences. In the real case, however, the Cauchy–Schwarz inequality implies that, for all $x \in \ell^2$,

$$\langle (I+B)x, x\rangle = \|x\|^2 + \langle Bx, x\rangle \geq \|x\|^2 - \|Bx\|\|x\| \geq 0,$$

so that $I+B$ is positive. On *real* Hilbert spaces, therefore, positive operators can be hypercyclic.

Exercises

Note: Exercises 5.0.1–5.0.9 concern general spectral theory; see Appendix B.

Exercise 5.0.1. Consider the operator T given by $T(x_1, x_2) = (-x_2, x_1)$ on the *real* Banach space $\mathbb{R}^2$. Show that $\sigma(T) = \varnothing$ and $\sigma(T^2) = \{-1\}$. Thus, some results of Appendix B do not hold in the real case.

Exercise 5.0.2. Let B be the backward shift operator and F the forward shift operator on any of the sequence spaces $X = \ell_p$, $1 \leq p < \infty$, or $X = c_0$. Show that $\sigma_p(B) = \mathbb{D}$ and $\sigma(B) = \overline{\mathbb{D}}$. Determine also $\sigma_p(F)$ and $\sigma(F)$. (*Hint*: Find F^*.)

Exercise 5.0.3. Let B_w be a weighted shift on $X = \ell^p$, $1 \leq p < \infty$, or c_0, given by $B_w(x_n)_n = (w_{n+1}x_{n+1})_n$; see Section 4.1. Suppose that $(w_n)_n$ is a positive, decreasing sequence with $\lim_{n\to\infty} w_n = r \geq 0$. Determine $\sigma(B_w)$.

Exercise 5.0.4. Let K be a nonempty compact subset of $\mathbb{C}$. Show that there is an operator T on the Hilbert space ℓ^2 such that $\sigma(T) = K$. (*Hint*: Choose $T(x_n)_n = (a_n x_n)_n$.)

Exercise 5.0.5. Prove the spectral mapping theorem. (*Hint*: Look at the proof of the point spectral mapping theorem.)

Exercise 5.0.6. Give a direct proof of the spectral mapping theorem for polynomials, without the use of functional calculus. (*Hint*: For $\lambda \in \mathbb{C}$ there are $c, \mu_1, \ldots, \mu_N \in \mathbb{C}$ such that $\lambda - f(z) = c(\mu_1 - z)\cdots(\mu_N - z)$; use the corresponding factorization for $\lambda I - T$.)

Exercise 5.0.7. If T is an invertible operator then $0 \notin \sigma(T)$ and $\sigma(T^{-1}) = \sigma(T)^{-1} := \{\frac{1}{z}\ ;\ z \in \sigma(T)\}$. Give two proofs of the identity, one direct and one using the spectral mapping theorem.

Exercise 5.0.8. Let T be an operator and f a holomorphic function on a neighbourhood of $\sigma(T)$. Show that if $f(T) = 0$ then $f = 0$ on $\sigma(T)$. Use the Volterra operator (see Example 5.4), to show that the converse implication is false.

Exercise 5.0.9. Give a direct proof of the point spectral mapping theorem for polynomials using the ideas of Exercise 5.0.6.

Exercise 5.1.1. Let T be a mixing operator on a (real or complex) Banach space. Then, for every nonzero vector $x^* \in X^*$, $\|(T^*)^n x^*\| \to \infty$ as $n \to \infty$. (*Hint*: Suppose that $\|(T^*)^{n_k} x^*\| \leq M$ and consider $U = \{x\ ;\ \|x\| < 1\}$, $V = \{x \in X\ ;\ |\langle x, x^*\rangle| > M\}$.)

Exercise 5.1.2. (a) Show that the Volterra operator also defines an operator on the space $L^\infty[0,1]$ of essentially bounded measurable functions on $[0,1]$. Moreover, show that $\|V\| = 1$ for V as an operator on L^1 and on L^∞.

(b) Show that, on any space $C[0,1]$ or $L^p[0,1]$, $1 \leq p \leq \infty$, the Volterra operator has no eigenvalues.

(c) Confirm the representation of V^n given in Example 5.4.

(d) For $p = 1$ and $p = \infty$, show that $\|V^n\| \leq \frac{1}{n!}$; deduce that $\sigma(V) = \{0\}$.

(e) For $1 < p < \infty$, show that $\|V^n\| \leq \frac{1}{n!}p^{-1/p}q^{-1/q}(1 - 1/(pn))^{-1/q}$; deduce that $\sigma(V) = \{0\}$.

(f) Show directly that, on $C[0,1]$, $\lambda I - V$ is injective for any $\lambda \in \mathbb{C}$, and that it is surjective precisely for $\lambda \neq 0$. (*Hint*: If $\lambda \neq 0$, show that $\lambda I - V$ maps the function $t \to \frac{1}{\lambda^2}\int_0^t e^{(t-s)/\lambda} g(s)\, ds + \frac{1}{\lambda} g(t)$ to g.)

Exercise 5.1.3. Deduce the following from Kitai's theorem.

(a) Every non-invertible hypercyclic operator has an uncountable spectrum.

(b) If the spectrum of a hypercyclic operator is finite or countable then it is contained in $\mathbb{T}$.

Exercise 5.1.4. A vector $x \in X$ is called an *irregular vector* of an operator T on X if $\liminf_{n\to\infty} \|T^n x\| = 0$ and $\limsup_{n\to\infty} \|T^n x\| = \infty$. For example, any hypercyclic vector is irregular. Show the following.

(a) Let $S : X \to X$ and $T : Y \to Y$ be operators. If $S \oplus T$ has an irregular vector, then so does at least one of the operators S and T.

(b) If T has irregular vectors then $\sigma(T)$ meets the unit circle.

(c) The analogue of Kitai's theorem is not true for operators with irregular vectors. (*Hint*: Consider $T \oplus (\lambda I)$ with T hypercyclic.)

(d) Unilateral weighted forward shifts F_w (see Remark 4.11), can have irregular vectors.

Exercise 5.1.5. Let T be an ε-hypercyclic operator; see Exercise 2.2.10. Show that every connected component of $\sigma(T)$ meets the unit circle.

Exercise 5.1.6. Let T be a J-class operator; see Exercise 2.2.11. Show that $\sigma(T)$ meets the unit circle, but that not necessarily every connected component of $\sigma(T)$ does. (*Hint*: Avoid the problem presented in Exercise 2.2.11(a) by modifying the proof of Proposition 5.3.)

Exercise 5.2.1. Prove the statements on the Volterra integral operator made in Example 5.9.

Exercise 5.2.2. Let X and Y be Banach spaces. An operator $T : X \to Y$ is called a *finite-rank operator* if its range $\operatorname{ran} T$ is finite dimensional. Show that $T : X \to Y$ is a finite-rank operator if and only if there are $y_1, \ldots, y_n \in Y$ and $x_1^*, \ldots, x_n^* \in X^*$, $n \geq 1$, such that $T = \sum_{k=1}^n \langle \cdot\,, x_k^* \rangle y_k$.

Exercise 5.2.3. Let X be a Banach space and $K(X)$ the space of compact operators on X. Show the following:

(i) the identity operator on X is compact if and only if X is finite dimensional; (*Hint*: Use a result from Appendix A.)

(ii) if $K, L \in K(X)$, $\lambda \in \mathbb{K}$ and $T \in L(X)$ then λK, $K + L$, TK and KT all belong to $K(X)$ (one says that $K(X)$ is an *ideal* in $L(X)$).

Exercise 5.2.4. Let X be a Banach space and $K(X)$ the space of all compact operators on X. Show that $K(X)$ is a closed subspace of $L(X)$, the space of all operators on X, endowed with the operator norm topology. Deduce that a limit, in the operator norm, of finite-rank operators is compact. (*Hint*: Look at the argument contained in Example 5.16.)

Exercise 5.2.5. Show that no compact operator on a real Banach space can be hypercyclic by proceeding as follows. Suppose that T is a hypercyclic compact operator on an infinite-dimensional real Banach space X, and let $\widetilde{T} : \widetilde{X} \to \widetilde{X}$ denote its complexification. By Exercise 2.6.4, $\widetilde{T}$ is 2-hypercyclic.

(a) Show that $\widetilde{T}$ is compact.

(b) Show that, for every $\varepsilon > 0$, $\sigma(\widetilde{T}) \cap \{z \in \mathbb{C} \; ; \; |z| \geq \varepsilon\}$ is a finite set. (*Hint*: Use Exercise 2.6.4.)

(c) By Lemma 5.15, $0 \in \sigma(\widetilde{T})$. If $\sigma(\widetilde{T}) = \{0\}$ deduce that all orbits under $\widetilde{T}$ tend to 0, contradicting the 2-hypercyclicity of $\widetilde{T}$.

(d) If $\{0\} \subsetneqq \sigma(\widetilde{T})$, show that there are nontrivial $\widetilde{T}$-invariant closed subspaces M_1 and M_2 with $\widetilde{X} = M_1 \oplus M_2$ such that $\sigma(\widetilde{T}|_{M_1}) \subset \mathbb{D}$. Prove that $\widetilde{T}|_{M_1}$ is 2-hypercyclic and deduce a contradiction as in (c).

Exercise 5.2.6. Prove the *Arzelà–Ascoli theorem*: a subset A of $C[0,1]$ is relatively compact if and only if it is bounded and equicontinuous. Here, A is called equicontinuous if for any $\varepsilon > 0$ there is some $\delta > 0$ such that, for any $f \in A$ and any $s, t \in [0,1]$ with $|s - t| < \delta$, $|f(s) - f(t)| < \varepsilon$. (*Hint*: Look at the argument contained in Example 5.16.)

Exercise 5.2.7. Let $\varphi : [0,1] \to [0,1]$ be a continuous function. Then the *Volterra composition operator* $V_\varphi : C[0,1] \to C[0,1]$ is defined as

$$V_\varphi f(t) = \int_0^{\varphi(t)} f(s)\, ds, \quad 0 \leq t \leq 1.$$

(a) Show that no Volterra composition operator is hypercyclic on $C[0,1]$. (*Hint*: $V_\varphi = C_\varphi \circ V$.)

(b) Now let $\varphi(x) = x^\alpha$, $\alpha > 0$, and consider V_φ as an operator on $C_0[0,1[$, the space of continuous functions f on $[0,1[$ with $f(0) = 0$; it is a Fréchet space under the seminorms $p_n(f) = \sup_{t \in [0,1-1/n]} |f(t)|$, $n \geq 1$. Show the following:

(i) for $\alpha \geq 1$, V_φ is not hypercyclic on $C_0[0,1[$;

(ii) for $\alpha < 1$, V_φ is hypercyclic on $C_0[0,1[$.

(*Hint* for (ii): Use Kitai's criterion with X_0 the set of polynomials in $C_0[0,1[$ and $Y_0 = \operatorname{span}\{x^\nu \; ; \; \nu > \alpha/(1-\alpha)\}$; use the Weierstrass approximation theorem to show that Y_0 is dense.)

Exercise 5.2.8. Show that no hypercyclic operator can commute with a finite-rank operator. Deduce Proposition 5.18 from this. (*Hint*: Use a commutative diagram.)

Exercise 5.2.9. Prove Lemma 5.19 for real Banach spaces. (*Hint*: Use the ideas of Exercise 5.2.5.)

Exercise 5.2.10. Let B_w be a weighted shift on $X = \ell^p$, $1 \leq p < \infty$, or c_0, given by $B_w(x_n)_n = (w_{n+1}x_{n+1})_n$. Show that B_w is compact if and only if $\lim_{n\to\infty} w_n = 0$. Show that B_w is quasinilpotent in that case. (*Hint*: Use Exercise 5.2.4.)

Exercise 5.2.11. The aim of this and the next exercise is to get a better understanding of the spectrum of compact operators. Show here that the adjoint T^* of a compact operator is compact. (*Hint*: Let $\|x_n^*\| \leq 1$. For the unit ball B_X of X, $K := \overline{T(B_X)}$ is a compact metric space. Note that $x_n^*|_K \in C(K)$, the space of continuous functions on K, endowed with the sup-norm. Apply the Arzelà–Ascoli theorem (see Exercise 5.2.6), which remains true for $C(K)$.)

Exercise 5.2.12. Let T be a compact operator on an infinite-dimensional complex Banach space. Show that its spectrum is either finite, including 0, or it consists of a sequence of points converging to 0, again including 0. Also, every nonzero spectral point is an eigenvalue. (*Hint*: Let $|\lambda| \geq \varepsilon$. If λ is not an eigenvalue then $\lambda I - T$ is non-surjective of closed range. By the proof of Lemma 2.53, λ is an eigenvalue of the compact operator T^*; see the previous exercise. By Lemma 5.14 there are only finitely many such λ, so that λ is isolated. Apply the Riesz decomposition theorem to find a restriction of $T|_M$ with spectrum $\{\lambda\}$. By Lemma 5.15, $\dim M < \infty$, so that λ is an eigenvalue of T.)

Exercise 5.3.1. An operator T on a Hilbert space H is called *quasinormal* if $T(T^*T) = (T^*T)T$. Show the following:
(i) any normal operator is quasinormal, and any quasinormal operator is hyponormal;
(ii) the unilateral forward shift on ℓ^2, $F(x_n)_n = (0, x_1, x_2, \ldots)$, is quasinormal but not normal;
(iii) the operator $B + 2F$ on ℓ^2, where B is the backward shift, is hyponormal but not quasinormal, and hence not normal.
(*Hint*: For the nontrivial implication in (i), consider separately $x \in \overline{\operatorname{ran} T}$ and $x \in (\operatorname{ran} T)^\perp$, the orthogonal complement of $\operatorname{ran} T$.)

Exercise 5.3.2. An operator T on a Hilbert space H is called *quasihyponormal* if $T^*(T^*T - TT^*)T \geq 0$. Show the following:
(i) an operator T is quasihyponormal if and only if, for all $x \in H$, $\|TTx\| \geq \|T^*Tx\|$;
(ii) any hyponormal operator is quasihyponormal, and any quasihyponormal operator is paranormal;
(iii) let $H = \ell^2(\mathbb{N}, \mathbb{C}^2)$ be the Hilbert space of all sequences $(x_n)_{n\geq 1}$ with $x_n \in \mathbb{C}^2$, $n \geq 1$, such that $\sum_{n=1}^\infty \|x_n\|^2 < \infty$, where $\|\cdot\|$ is the Euclidean norm in $\mathbb{C}^2$, and define the operator $T : H \to H$ by $T(x_n)_n = (0, Ax_1, Bx_2, Bx_3, Bx_4, \ldots)$, where $A = \left(\begin{smallmatrix} 1 & 0 \\ 0 & 0 \end{smallmatrix}\right)$ and $B = \frac{1}{\sqrt{2}}\left(\begin{smallmatrix} 1 & 1 \\ 1 & 1 \end{smallmatrix}\right)$; show that T is quasihyponormal and therefore also paranormal, but not hyponormal.

Exercise 5.3.3. An operator T on a Banach space X is called *power-regular* if, for any $x \in X$, $(\|T^n x\|^{1/n})_n$ converges. It is called *normaloid* if, for any $n \geq 1$, $\|T^n\| = \|T\|^n$. Show the following:
(i) every paranormal operator is power regular;
(ii) every paranormal operator is normaloid;
(iii) normaloid operators can be hypercyclic.
(*Hint*: In (i), study the behaviour of the sequence $(\|T^{n+1}x\|/\|T^n x\|)_{n\geq 0}$; in (ii), prove inductively that $\|T^n x\| \geq \|Tx\|^n$, for all $n \geq 1$ and all unit vectors x.)

Exercise 5.3.4. (a) Let $p(\lambda) = a\lambda^2 + b\lambda + c$ be a real polynomial with $a \neq 0$ and $b \leq 0$. Show that $p(\lambda) \geq 0$ for all $\lambda \geq 0$ if and only if $a > 0$ and $b^2 \leq 4ac$.

(b) Deduce from (a) that an operator T on a Hilbert space H is paranormal if and only if, for all $\lambda \geq 0$,

$$T^{*2}T^2 - 2\lambda T^*T + \lambda^2 I \geq 0.$$

Exercise 5.3.5. An operator T on a Hilbert space H is called *cohyponormal* if T^* is hyponormal, that is, if $TT^* - T^*T \geq 0$, or, equivalently, $\|T^*x\| \geq \|Tx\|$ for all $x \in H$. Show that a cohyponormal operator can be hypercyclic.

Exercise 5.3.6. Show that for bilateral weighted backward shifts on $\ell^2(\mathbb{Z})$, hyponormality and paranormality are equivalent properties.

Exercise 5.3.7. Show that the Volterra operator on $L^2[0,1]$ is not paranormal.

Sources and comments

Section 5.1. It seems fair to say that all the results in this chapter were either found, or at least initiated, by C. Kitai in her PhD thesis [215]. Some of her results were also found, independently, by Matache [233].

In particular, Theorem 5.6 is due to Kitai [215]. Lemma 5.5 is not as obvious as it may at first seem. While one implication is standard (see for example [250, p. 74, Corollary 2]), the other implication can be derived from [250, p. 83, Corollary 1].

Quite remarkably, Shkarin [287] has recently shown that Kitai's necessary spectral condition actually characterizes spectra of hypercyclic operators on Hilbert spaces.

Theorem 5.34. *Let $K \subset \mathbb{C}$ be a nonempty compact set. There exists a hypercyclic operator T on a complex Hilbert space such that $\sigma(T) = K$ if and only if every connected component of K meets the unit circle.*

Dilworth and Troitsky [135] have shown that Kitai's theorem remains true for *weakly hypercyclic operators*, that is, operators with a weakly dense orbit. Herrero [194] has shown that for any supercyclic operator (on a complex Hilbert space) there is some $r \geq 0$ such that every connected component of its spectrum meets the circle $|z| = r$; see also Feldman, Miller and Miller [151, Theorem 6.2]. Herrero also obtained a spectral characterization of all operators on complex Hilbert spaces that are limits, in the operator norm, of hypercyclic operators.

Concerning the other necessary conditions found in Section 5.1, Proposition 5.1, part (i) is due to Kitai [215] while part (ii) was observed by Bourdon; see [106, p. 1446] and [96, Lemma 3.1]. Proposition 5.7 is implicit in Bonet, Martínez and Peris [82].

Section 5.2. The main result of this section, Theorem 5.11, is once more due to Kitai [215] (for complex Banach spaces); see also Matache [233]. The real case was provided by Bonet and Peris [85]. Kitai's proof is quite short but it draws heavily on the Riesz spectral theory of compact operators by which the spectrum of any compact operator on an infinite-dimensional complex Banach space is either finite, including 0, or it consists of a sequence converging to 0, again including 0; see [116, Chapter VII, § 7]. In the approach chosen here, Bourdon's theorem allows us to avoid some parts of the Riesz theory; but see Exercise 5.2.12.

Proposition 5.17, Proposition 5.18, Lemma 5.19 and Proposition 5.20 are due to Kitai [215], Chan and Shapiro [106, p. 1445, p. 1446] and Martínez and Peris [229].

It is interesting to note that, by Shkarin [289], there are hypercyclic finite-rank perturbations of unitary operators on complex Hilbert spaces; by Grivaux [175], the perturbation may even be of rank 1.

Concerning specific operators, the Volterra operator (on $L^p[0,1]$, $1 \leq p < \infty$) and the Cesàro operator (on ℓ^p, $1 < p < \infty$) are not even supercyclic; see Gallardo and Montes [159] and León, Piqueras and Seoane [224]. As for Example 5.22, Hardy's inequality can be found in [193].

Section 5.3. We refer to Kubrusly [218] for a discussion of various types of normality conditions for Hilbert space operators. It was once more Kitai [215] who showed that no hyponormal operator can be hypercyclic. This was extended in two respects by Bourdon [92, p. 352] who proved that no paranormal operator can even be supercyclic. Concerning Example 5.33, it should be noted that some authors include self-adjointness in the definition of positive operators on real Hilbert spaces. In that case, of course, no positive operator is hypercyclic.

Exercises. We have taken Exercise 5.1.1 from Bonet [79] and Exercise 5.1.3 from Matache [234]. The notion of an irregular vector was introduced by Beauzamy [48, p. 41];

Exercises 5.1.4 and 5.1.5 are due to Prăjitură [261], while Exercise 5.1.6 is taken from Costakis and Manoussos [119]. The result of Exercise 5.2.5 is due to Bonet and Peris [85]. Exercise 5.2.7 is taken from Herzog and Weber [201], Exercise 5.2.8 from Shapiro [281, 2.9]. The result of Exercise 5.2.11 is one half of Schauder's theorem; see [116]. For the background to Exercises 5.3.1, 5.3.2 and 5.3.5 we refer to Kubrusly [218]. As for Exercise 5.3.3, power-regular operators were introduced by Atzmon [17], who also showed that operators with a countable spectrum are power regular. As a consequence, power-regular operators can be hypercyclic; see Example 8.4. That every paranormal operator is power regular and normaloid is due to Bourdon [92, p. 352] and Istrăţescu, Saitô and Yoshino [206], respectively. Incidentally, it was in the latter article that the class of paranormal operators was introduced (as operators of class (N)). The characterization of paranormal operators in Exercise 5.3.4 is due to Ando [8].

Extensions. We have decided to restrict our attention in this chapter to Banach spaces. But the main results in Section 5.2 remain true in arbitrary Fréchet spaces, and even in all locally convex spaces; see Chapter 12. Indeed, no compact (and therefore no power-compact) operator on a locally convex space can be hypercyclic (see Bonet and Peris [85]), and no compact perturbation of a multiple of the identity can be chaotic; see Martínez and Peris [229]. For an idea of the proofs we refer to Exercise 12.2.3. Propositions 5.10 and 5.18 also remain true; their proofs extend immediately.

To finish, we briefly review the hypercyclicity status of further classes of operators that are commonly studied in operator theory. For each class we give a reference to its definition and we state a reason why it contains hypercyclic operators or why it does not.

We begin with classes that do not contain any hypercyclic operators:
absolutely summing operators (are power compact [134, p. 15, p. 50]);
essentially quasinilpotent operators (= Riesz operators [2, p. 302]);
Hilbert–Schmidt operators (are power compact [134, p. 50, p. 84]);
nuclear operators (are compact [134, p. 112, p. 113]);
operators of Schatten–von Neumann class (are compact [134, p. 80]);
Riesz operators (if hypercyclic then quasinilpotent [2, p. 302]);
strictly singular operators (are Riesz operators [259, 1.9.2, 26.6.5]);
strictly cosingular operators (are Riesz operators [259, 1.10.2, 26.6.10]);
trace class operators (= nuclear operators on Hilbert spaces [134, p. 123]).

We next state classes that do contain hypercyclic operators:
completely continuous operators (e.g., operators on ℓ^1 [116, p. 173]);
Dunford–Pettis operators (= completely continuous operators [2, p. 498]);
Fredholm operators (e.g., invertible operators [2, p. 156]);
Toeplitz operators (e.g., Rolewicz's operators [191, p. 136]);
weakly compact operators (e.g., operators on Hilbert spaces [116, p. 183]).

Chapter 6
Connectedness arguments in linear dynamics

This chapter is devoted to some of the most fundamental results in linear dynamics. What is particularly striking is that they hold for all operators, without further technical assumptions.

We have already obtained such a result in Chapter 2 . It says that every hypercyclic operator admits a dense subspace of hypercyclic vectors, except for the zero vector. Note that this property would not make sense in a nonlinear setting.

In this chapter we will consider the following problems, which, a priori, do not involve linearity.

- If T has a dense orbit, does then every power T^p also have a dense orbit?
- Suppose that the union of a finite collection of orbits is dense. Will then at least one of these orbits be actually dense?
- If an orbit is somewhere dense, is it (everywhere) dense?

Each of these questions has a negative answer for arbitrary, nonlinear maps. It is therefore even more surprising that they all have a positive answer for (linear) operators, and that without any restrictions. The proofs depend in a crucial way on connectedness arguments.

In the final section we will consider two more problems.

- Let T be a hypercyclic operator, and let $\lambda \in \mathbb{K}$ with $|\lambda| = 1$. Is then λT also hypercyclic?
- Let $(T_t)_{t\geq 0}$ be a hypercyclic C_0-semigroup on a Banach space. Is then every single operator T_t, $t > 0$, hypercyclic?

Again we will give positive answers to these questions. The proofs can be given within a common framework and use, once more, a connectedness argument, this time via a suitable homotopy.

K.-G. Grosse-Erdmann, A. Peris Manguillot, *Linear Chaos*, Universitext,
DOI 10.1007/978-1-4471-2170-1_6,

6.1 Ansari's theorem

In this section we deal with the question of whether every power T^p, $p \in \mathbb{N}$, of a hypercyclic operator T is again hypercyclic. Since every sequence $(kp)_k$ is syndetic, Theorem 1.54 implies a positive answer if T is even a weakly mixing operator on a separable Fréchet space. We will show here that the answer is positive for all hypercyclic operators.

The following auxiliary result will be crucial.

Lemma 6.1. *The T be a continuous map on a metric space X without isolated points. Then the interiors of the closures of two orbits under T either coincide, or they are disjoint.*

Proof. Suppose that $\operatorname{int}(\overline{\operatorname{orb}(x,T)})\cap\operatorname{int}(\overline{\operatorname{orb}(y,T)}) \neq \varnothing$, $x, y \in X$. Then there is some $n \in \mathbb{N}_0$ such that

$$T^n x \in \overline{\operatorname{orb}(y,T)}.$$

Since $\overline{\operatorname{orb}(y,T)}$ is T-invariant, we have that $T^k x \in \overline{\operatorname{orb}(y,T)}$ for $k \geq n$ and therefore

$$\overline{\{T^k x\ ;\ k \geq n\}} \subset \overline{\operatorname{orb}(y,T)}.$$

Since X has no isolated points, one shows easily that

$$\operatorname{int}(\overline{\operatorname{orb}(x,T)}) \subset \operatorname{int}(\overline{\{T^k x\ ;\ k \geq n\}});$$

see also Exercise 6.2.1. Hence $\operatorname{int}(\overline{\operatorname{orb}(x,T)}) \subset \operatorname{int}(\overline{\operatorname{orb}(y,T)})$. By symmetry, we also have the converse inclusion, so that the two interiors coincide. □

Theorem 6.2 (Ansari). *Let T be an operator on a Fréchet space. Then, for any $p \in \mathbb{N}$, $HC(T) = HC(T^p)$. In particular, if T is hypercyclic then so is every power T^p.*

Proof. Let $p \in \mathbb{N}$. We clearly have that $HC(T^p) \subset HC(T)$.

For the converse inclusion we fix $x \in HC(T)$. From Proposition 1.15 and Corollary 2.56 we know that $D := HC(T)$ is a dense, T-invariant connected subset of X; in particular, it does not have isolated points. For the remainder of the proof we consider the map $T : D \to D$; the topological operations of closure and interior will be understood in D. Since D is dense in X it then suffices to show that $\overline{\operatorname{orb}(x,T^p)} = D$.

To this end we define

$$D_j = \overline{\operatorname{orb}(T^j x, T^p)}, \quad j = 0, \ldots, p-1.$$

We need to show that $D = D_0$. Observe that

$$D = \overline{\operatorname{orb}(x,T)} = \overline{\bigcup_{j=0}^{p-1} \operatorname{orb}(T^j x, T^p)} = \bigcup_{j=0}^{p-1} D_j$$

and

$$T(D_j) \subset D_{j+1(\operatorname{mod} p)}.$$

Let $F \subset \{0, \dots, p-1\}$ be a set of minimal cardinality such that

$$D = \bigcup_{j \in F} D_j.$$

Suppose that F is not a singleton. Let, in addition, $\operatorname{int}(D_j) \cap \operatorname{int}(D_k) \neq \varnothing$ for some $j, k \in F$ with $j \neq k$. By Lemma 6.1, $\operatorname{int}(D_j) = \operatorname{int}(D_k)$. From minimality we deduce that

$$D \setminus \bigcup_{l \in F \setminus \{j\}} D_l$$

is nonempty, and it is an open set contained in D_j and thus in $\operatorname{int}(D_j) \subset D_k$, which is not possible. Therefore, $\operatorname{int}(D_j \cap D_k) = \varnothing$ for any $j, k \in F$ with $j \neq k$.

We now set $F_l = F + l (\operatorname{mod} p)$, $l = 0, \dots, p-1$. We have $D = \overline{T^l(D)} = \overline{\bigcup_{j \in F} T^l(D_j)} = \bigcup_{k \in F_l} D_k$, $l = 0, \dots, p-1$. Since $\operatorname{card}(F_l) = \operatorname{card}(F)$, which is minimal, we also get that $\operatorname{int}(D_j \cap D_k) = \varnothing$ for any $j, k \in F_l$ with $j \neq k$, $l = 0, \dots, p-1$. As a consequence, the set

$$A := \bigcup_{l=0}^{p-1} \bigcup_{\substack{j,k \in F_l \\ j \neq k}} (D_j \cap D_k)$$

is nowhere dense as a finite union of nowhere dense sets; and it is T-invariant. If A were nonempty, with $y \in A$, say, then

$$D = \overline{\operatorname{orb}(y, T)} \subset \overline{A} = A,$$

which is a contradiction. Therefore $A = \varnothing$, which implies that

$$D = \bigcup_{j \in F} D_j$$

is a finite union of pairwise disjoint closed subsets. But this contradicts the connectedness of D.

In conclusion, $F = \{j\}$ is a singleton. Then $D = D_j$, and we obtain that $D = \overline{T^{p-j}(D_j)} = D_0$, which had to be shown. $\square$

The simple example $T : \{-1, 1\} \to \{-1, 1\}$, $Tx = -x$, shows that Ansari's theorem fails for nonlinear dynamical systems; see also Exercise 1.2.11 for an example on a metric space without isolated points. Ansari's theorem does extend to the nonlinear setting if the set of points with dense orbit is connected; see Exercise 6.1.7.

6.2 Somewhere dense orbits

We recall that a set is called *somewhere dense* if its closure contains a nonempty open set.

It was a key point in the proof of Ansari's theorem to write the space D as a finite union of closures of orbits. Then one of these closures must have an interior point, which means that the corresponding orbit is somewhere dense. In the end we concluded that this orbit is, in fact, (everywhere) dense. Do we have a general principle here, that is, is every somewhere dense orbit necessarily dense? We will give a positive answer to this question.

Thus, let T be an operator on a Fréchet space X. For $x \in X$ we write

$$D(x) = \overline{\mathrm{orb}(x,T)} \quad \text{and} \quad U(x) = \mathrm{int}\, D(x).$$

The following properties can be easily deduced from the continuity of T and the fact that X has no isolated points (see Exercise 6.2.1):

(i) if $y \in D(x)$, then $D(y) \subset D(x)$;
(ii) $U(x) = U(T^k x)$ for each $k \in \mathbb{N}$;
(iii) if $R : X \to X$ is a continuous map that commutes with T, then $R(D(x)) \subset D(Rx)$.

We first need a generalization of Theorem 2.54. An easy adaptation of the argument used there gives the result; see Exercise 6.2.2.

Lemma 6.3. *If T admits a somewhere dense orbit and p is a nonzero polynomial, then the operator $p(T)$ has dense range.*

Before proving that a vector whose orbit is somewhere dense is necessarily hypercyclic, we will show that it is cyclic, that is, the linear span of its orbit is dense in X.

Lemma 6.4. *If $\mathrm{orb}(x,T)$ is somewhere dense, then the set $\{p(T)x\ ;\ p \neq 0$ a polynomial$\}$ is connected and dense in X.*

Proof. The set $A := \{p(T)x\ ;\ p \neq 0 \text{ a polynomial}\}$ is path connected. Indeed, let p, q be nonzero polynomials. If q is not a multiple of p then the straight path $t \to tp(T)x + (1-t)q(T)x$, $t \in [0,1]$, is contained in A. Otherwise we select a third nonzero polynomial r that is not a multiple of p, and therefore not of q, and we take the union of the straight paths connecting $p(T)x$ and $q(T)x$ with $r(T)x$.

On the other hand, $\overline{A}$ is a subspace of X that contains $\overline{\mathrm{orb}(x,T)}$. It follows from the hypothesis that there is some $x_0 \in X$ and a 0-neighbourhood W such that $x_0 + W \subset \overline{A}$. Thus, for any $y \in X$, there is a scalar λ with $y \in \lambda W$; hence $y \in \lambda(x_0 + W) - \lambda x_0 \subset \overline{A}$. Consequently, A is dense in X. □

Theorem 6.5 (Bourdon–Feldman). *Let T be an operator on a Fréchet space X and $x \in X$. If $\mathrm{orb}(x,T)$ is somewhere dense in X, then it is dense in X.*

Proof. We have to show that if $U(x) \neq \varnothing$ then $D(x) = X$. The proof will be split into four steps.

Step 1. We have that $T(X \setminus U(x)) \subset X \setminus U(x)$.

We show, equivalently, that $T^{-1}(U(x)) \subset U(x)$. First, since $U(x) \neq \varnothing$ there is some $m \in \mathbb{N}_0$ with $x_m := T^m x \in U(x)$.

Now let $y \in T^{-1}(U(x))$, and let V be an arbitrary neighbourhood of y. Since, by property (ii), x_m also has a somewhere dense orbit, Lemma 6.4 implies that we can find a polynomial p such that $p(T)x_m \in V \cap T^{-1}(U(x))$.

We have, using property (ii), that

$$p(T)x_m \in p(T)(U(x)) = p(T)(U(T^{m+1}x)) \subset p(T)(D(T^{m+1}x)).$$

Moreover, since $Tp(T)x_m \in U(x) \subset D(x)$, properties (iii) and (i) yield that

$$p(T)(D(T^{m+1}x)) \subset D(Tp(T)x_m) \subset D(x).$$

We have therefore shown that $V \cap D(x) \neq \varnothing$. Since V was arbitrary and $D(x)$ is closed, we deduce that $y \in D(x)$ and hence $T^{-1}(U(x)) \subset D(x)$. Continuity of T implies that $T^{-1}(U(x)) \subset U(x)$.

Step 2. For any $z \in X \setminus U(x)$, $D(z) \subset X \setminus U(x)$.

By Step 1, $X \setminus U(x)$ is T-invariant, and it is closed. The claim then follows from the definition of $D(z)$.

Step 3. For any polynomial $p \neq 0$, $p(T)x \in X \setminus \partial D(x)$, where $\partial D(x)$ denotes the boundary of $D(x)$; see Figure 6.1.

Suppose that $p(T)x \in \partial D(x)$ for some polynomial $p \neq 0$. By Lemma 6.3 there is some $y \in X$ such that $p(T)y \in U(x)$. Since $p(T)x \notin U(x)$, property (iii) and Step 2 imply that

$$p(T)(D(x)) \subset D(p(T)x) \subset X \setminus U(x).$$

We therefore have that $y \in X \setminus D(x)$. By Lemma 6.4 there then exists a polynomial q such that $q(T)x$ is close enough to y to satisfy $q(T)x \in X \setminus D(x) \subset X \setminus U(x)$ and $p(T)q(T)x \in U(x)$. Since $p(T)x \in D(x)$, property (iii) and Step 2 imply that

$$p(T)q(T)x = q(T)p(T)x \in q(T)(D(x)) \subset D(q(T)x) \subset X \setminus U(x),$$

which is a contradiction. This proves the claim.

Step 4. We have that $D(x) = X$.

By Step 3,

$$A := \{p(T)x \ ; \ p \neq 0 \text{ a polynomial}\} \subset U(x) \cup (X \setminus D(x)),$$

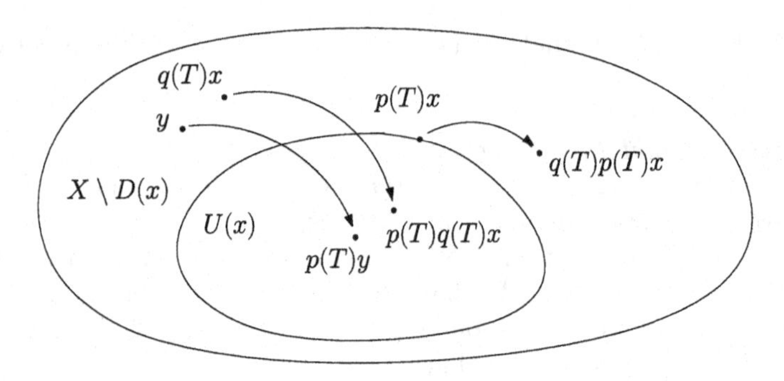

Fig. 6.1 Step 3

which is a disjoint union of open sets. Since, by Lemma 6.4, A is connected and, by density of A, $A \cap U(x) \neq \varnothing$, we must have that $A \cap (X \setminus D(x)) = \varnothing$. Hence, $A \subset D(x)$, which implies that $D(x) = X$. □

6.3 Multi-hypercyclic operators

The Bourdon–Feldman theorem provides us with a very powerful tool for obtaining dense orbits. A particular case occurs when the union of a finite number of orbits under T is dense in X. In this case the operator T is called *multi-hypercyclic.*

Theorem 6.6 (Costakis–Peris). *Let T be an operator on a Fréchet space X and $x_1, \ldots, x_n \in X$. If*

$$\bigcup_{j=1}^{n} \operatorname{orb}(x_j, T)$$

is dense in X, then there is some $j \in \{1, \ldots, n\}$ such that $\operatorname{orb}(x_j, T)$ *is dense in X. In particular, every multi-hypercyclic operator is hypercyclic.*

Proof. The hypothesis says that

$$\bigcup_{j=1}^{n} \overline{\operatorname{orb}(x_j, T)} = \overline{\bigcup_{j=1}^{n} \operatorname{orb}(x_j, T)} = X.$$

Since a finite union of nowhere dense sets is nowhere dense, $\operatorname{orb}(x_j, T)$ must be somewhere dense in X for some $j \in \{1, \ldots, n\}$. By the Bourdon–Feldman theorem, x_j then has a dense orbit. □

Ansari's result can easily be derived from this theorem. Let $x \in HC(T)$ and $p \in \mathbb{N}$. Since

$$\operatorname{orb}(x,T)=\bigcup_{j=0}^{p-1}\operatorname{orb}(T^j x,T^p)$$

is dense in X, Theorem 6.6 implies that there is some $j\in\{0,\ldots,p-1\}$ such that $T^j x$ is hypercyclic for T^p. Since T^{p-j} has dense range and

$$T^{p-j}(\operatorname{orb}(T^j x,T^p))\subset\operatorname{orb}(x,T^p)$$

we obtain that $x\in HC(T^p)$.

These arguments also imply that because Ansari's theorem fails for non-linear dynamical systems the same is true for the theorems of Costakis–Peris and Bourdon–Feldman; see also Exercise 1.2.11.

6.4 Hypercyclic semigroup actions

In this section we will be dealing with two additional important problems in linear dynamics.

The *problem of unimodular multiples* asks whether, given a hypercyclic operator T, is every multiple λT with $\lambda\in\mathbb{K}$, $|\lambda|=1$, also hypercyclic? The operator λT is called a *rotation* of T. In the real setting the answer is positive. Indeed, one only needs to show that if T is hypercyclic then so is $-T$. But this follows from Ansari's theorem because the two operators have a common square, $T^2=(-T)^2$. Thus we will concentrate here on the complex setting.

The *problem of hypercyclic discretizations of semigroups* asks whether, given a hypercyclic C_0-semigroup $(T_t)_{t\geq 0}$ on a Banach space, is every single operator T_t, $t>0$, also hypercyclic? Although C_0-semigroups will only be treated in the next chapter (and we ask the reader to consult the relevant definitions there), there will be no harm in already considering the discretization problem here. The (very basic) proof that a hypercyclic C_0-semigroup satisfies the assumptions imposed in this section will be postponed to Chapter 7.

The main aim of this section is to show that both problems have a positive answer. In analogy with Ansari's theorem, it will even be proved that the corresponding sets of hypercyclic vectors coincide.

Theorem 6.7 (León–Müller). *Let T be an operator on a complex Fréchet space X. If $x\in X$ is such that $\{\lambda T^n x\ ;\ \lambda\in\mathbb{C},\ |\lambda|=1,\ \text{and}\ n\in\mathbb{N}_0\}$ is dense in X then $\operatorname{orb}(x,\lambda T)$ is dense in X for each $\lambda\in\mathbb{C}$ with $|\lambda|=1$.*

In particular, for any $\lambda\in\mathbb{C}$ with $|\lambda|=1$, T and λT have the same hypercyclic vectors, that is,

$$HC(T)=HC(\lambda T).$$

In Exercise 2.5.1 we saw that rotations of mixing (or weakly mixing) operators are mixing (or weakly mixing, respectively). But that result did not say anything about the sets of hypercyclic vectors.

Theorem 6.8 (Conejero–Müller–Peris). *Let $(T_t)_{t\geq 0}$ be a C_0-semigroup on a Banach space X. If $x \in X$ is hypercyclic for $(T_t)_{t\geq 0}$, then it is hypercyclic for each operator T_t, $t > 0$.*

It is particularly gratifying that the two problems can be treated within a common framework, that of semigroup actions. We will also show that a variant of the method leads to a new proof of Ansari's theorem; see Exercise 6.4.5.

Throughout this section we will write

$$G = \mathbb{N}_0 \times \mathbb{R}_+,$$

which is a semigroup under addition. If X is a Fréchet space then a map

$$\Psi : G \to L(X)$$

is called a *(continuous and linear) semigroup action* of G on X if the following properties hold:
(i) $\Psi(0) = I$;
(ii) for any $g_1, g_2 \in G$, $\Psi(g_1 + g_2) = \Psi(g_1)\Psi(g_2)$;
(iii) the map $G\times X \to X$, $(g, x) \to \Psi(g)x$, is continuous, where $G = \mathbb{N}_0 \times \mathbb{R}_+$ and $G \times X$ carry the product topology.

Definition 6.9. A semigroup action Ψ on a Fréchet space X is called *hypercyclic* if there is some $x \in X$ such that $\{\Psi(g)x \ ; \ g \in G\}$ is dense in X. The vector x is then called *hypercyclic* for Ψ, and we write $x \in HC(\Psi)$.

Let us see how our two problems fit into this framework. If T is an operator on a complex Fréchet space X, then we define

$$\Psi(n,t) = e^{2\pi t i}T^n, \quad n \in \mathbb{N}_0, \ t \geq 0.$$

In the second case, if $(T_t)_{t\geq 0}$ is a C_0-semigroup on a Banach space X, then we define

$$\Psi(n,t) = T_t, \quad n \in \mathbb{N}_0, \ t \geq 0.$$

It is easy to see that these are semigroup actions of G on X; we refer to Chapter 7 for the definition of a C_0-semigroup.

Moreover, in both cases, the following properties are satisfied:
(α) either $\Psi(1,0) = I$ or $\Psi(0,1) = I$;
(β) if the semigroup action is hypercyclic then each convex combination of $\Psi(0,s)$ and $\Psi(1,t)$, $s,t \geq 0$, has dense range.

That property (β) is satisfied follows from a simple generalization of Theorem 2.54 and by Theorem 7.16, respectively.

The following theorem therefore immediately implies the Theorems of León–Müller and Conejero–Müller–Peris.

Theorem 6.10. *Let Ψ be a semigroup action on an infinite-dimensional Fréchet space X satisfying properties (α) and (β). If $x \in X$ is hypercyclic for Ψ then it is hypercyclic for each operator $\Psi(1,t)$, $t > 0$.*

Proof. We first note that it suffices to prove the claim for $t = 1$. Indeed, let x be hypercyclic for Ψ, and let $t > 0$ be arbitrary. We distinguish the two subcases of (α). If $\Psi(1,0) = I$ then

$$\widetilde{\Psi}(n,s) := \Psi(n,st)$$

defines a semigroup action that satisfies (α) and (β). Since x is also hypercyclic for $\widetilde{\Psi}$ we can conclude that x is hypercyclic for $\widetilde{\Psi}(1,1) = \Psi(1,t)$. If $\Psi(0,1) = I$ then we define

$$\widetilde{\Psi}(n,s) = \Psi(n,nt+s),$$

and we can conclude as before that x is hypercyclic for $\widetilde{\Psi}(1,1) = \Psi(1,t+1) = \Psi(1,t)$.

As usual, $\mathbb{T} = \{z \in \mathbb{C}\ ;\ |z| = 1\}$ is the unit circle. For ease of notation we introduce the map $\rho : \mathbb{R}_+ \to \mathbb{T}$, given by $\rho(t) := e^{2\pi ti}$. We then define, for every pair $u, v \in X$, the subset $F_{u,v}$ of $\mathbb{T}$ by

$$F_{u,v} := \{\lambda \in \mathbb{T}\ ;\ \exists\,((n_k,t_k))_k \subset G \text{ with } \Psi(n_k,t_k)u \to v \ \text{ and } \rho(t_k) \to \lambda\}.$$

The remainder of the proof will be divided into several steps.

Step 1. If $u \in HC(\Psi)$, then $F_{u,v} \neq \varnothing$ for all $v \in X$.

Since $u \in HC(\Psi)$, we can find sequences $(n_k)_k$ in $\mathbb{N}_0$ and $(t_k)_k$ in $\mathbb{R}_+$ such that $\Psi(n_k,t_k)u \to v$. By passing to a subsequence if necessary, we may assume that $(\rho(t_k))_k$ is convergent. Its limit is an element of $F_{u,v}$.

Step 2. If $\lambda_k \in F_{u,v_k}$, $v_k \to v$ and $\lambda_k \to \lambda$, then $\lambda \in F_{u,v}$. In particular, $F_{u,v}$ is a closed set for each $u, v \in X$.

Let W be a 0-neighbourhood of X and $\varepsilon > 0$. There is a 0-neighbourhood W_1 such that $W_1 + W_1 \subset W$; see Lemma 2.36. By assumption, there is some $k \in \mathbb{N}$ with $v - v_k \in W_1$ and $|\lambda - \lambda_k| < \varepsilon$. Now, by definition, there are $n_k \in \mathbb{N}_0$ and $t_k \in \mathbb{R}_+$ such that $v_k - \Psi(n_k,t_k)u \in W_1$ and $|\lambda_k - \rho(t_k)| < \varepsilon$. We then get that $v - \Psi(n_k,t_k)u \in W_1 + W_1 \subset W$ and $|\lambda - \rho(t_k)| < 2\varepsilon$, so that $\lambda \in F_{u,v}$.

Step 3. If $u, v, w \in X$, $\lambda \in F_{u,v}$, and $\mu \in F_{v,w}$, then $\lambda\mu \in F_{u,w}$.

Given a 0-neighbourhood W, take a 0-neighbourhood W_1 such that $W_1 + W_1 \subset W$. Let $\varepsilon > 0$. Then there are $n_1 \in \mathbb{N}_0$ and $t_1 \in \mathbb{R}_+$ such that $w - \Psi(n_1,t_1)v \in W_1$ and $|\mu - \rho(t_1)| < \varepsilon$. One can then find a 0-neighbourhood

V, $n_2 \in \mathbb{N}_0$ and $t_2 \in \mathbb{R}_+$ satisfying $\Psi(n_1,t_1)(V) \subset W_1$, $v - \Psi(n_2,t_2)u \in V$, and $|\lambda - \rho(t_2)| < \varepsilon$. Consequently we have for $n_3 := n_1 + n_2$ and $t_3 := t_1 + t_2$ that

$$w - \Psi(n_3,t_3)u = w - \Psi(n_1,t_1)v + \Psi(n_1,t_1)(v - \Psi(n_2,t_2)u) \in W_1 + W_1 \subset W,$$

and

$$|\lambda\mu - \rho(t_3)| \leq |\lambda|\,|\mu - \rho(t_1)| + |\rho(t_1)|\,|\lambda - \rho(t_2)| < 2\varepsilon.$$

Hence $\lambda\mu \in F_{u,w}$.

We now fix $x \in HC(\Psi)$. Our aim is to show that $x \in HC(\Psi(1,1))$. By Steps 1, 2 and 3, $F_{x,x}$ is a nonempty closed subsemigroup of the multiplicative group $\mathbb{T}$.

Step 4. If $F_{x,x} = \mathbb{T}$ then x is hypercyclic for $\Psi(1,1)$.

Suppose that $F_{x,x} = \mathbb{T}$. Given any $y \in X$, Steps 1 and 3 imply that $F_{x,y} = \mathbb{T}$. In particular $1 \in F_{x,y}$, which yields the existence of sequences $(n_k)_k$ in $\mathbb{N}_0$ and $(t_k)_k$ in $\mathbb{R}_+$ such that $\Psi(n_k,t_k)x \to y$ and $\rho(t_k) = e^{2\pi t_k i} \to 1$. We can then write $t_k = j_k - 1 + \varepsilon_k$ with $j_k \in \mathbb{N}$ and $\varepsilon_k \in [-1/2, 1/2]$, where $\varepsilon_k \to 0$.

Let W be a 0-neighbourhood, and let W_1 be a 0-neighbourhood such that $W_1 + W_1 \subset W$. By a standard compactness argument, the continuity of the semigroup action implies that there is a 0-neighbourhood V such that $\Psi(0,t)(V) \subset W_1$ if $0 \leq t \leq 2$. Moreover, there is some $k \in \mathbb{N}$ such that $\Psi(n_k,t_k)x - y \in V$ and $\Psi(0,1-\varepsilon_k)y - \Psi(0,1)y \in W_1$. Therefore

$$\begin{aligned}\Psi(n_k,j_k)x - \Psi(0,1)y&\\ &= \Psi(0,1-\varepsilon_k)(\Psi(n_k,t_k)x - y) + (\Psi(0,1-\varepsilon_k) - \Psi(0,1))y\\ &\in \Psi(0,1-\varepsilon_k)(V) + W_1 \subset W_1 + W_1 \subset W.\end{aligned}$$

Observe that, by property (α), $\Psi(n_k,j_k)x \in \operatorname{orb}(x,\Psi(1,1))$. Thus $\Psi(0,1)y \in \overline{\operatorname{orb}(x,\Psi(1,1))}$. Since $\Psi(0,1)$ has dense range by property (β), and $y \in X$ is arbitrary, x is hypercyclic for $\Psi(1,1)$.

For the rest of the proof we can now assume that $F_{x,x} \neq \mathbb{T}$, and we will show that this leads to a contradiction.

Step 5. There exists some $m \in \mathbb{N}$ such that, for each $y \in HC(\Psi)$, there is $\lambda \in \mathbb{T}$ satisfying $F_{x,y} = \{\lambda z\ ;\ z^m = 1\}$.

We first note that $F_{x,x}$ must be of the form $F_{x,x} = \{z \in \mathbb{T}\ ;\ z^m = 1\}$ for some $m \in \mathbb{N}$. Indeed, if $F_{x,x}$ contained points $z = e^{2\pi t i}$ with $t > 0$ arbitrarily small then, being a closed subsemigroup of $\mathbb{T}$, $F_{x,x}$ would be dense, and hence coincide with $\mathbb{T}$, which was excluded. Hence there is a minimal $t_0 \in]0,1]$ such that $z_0 = e^{2\pi t_0 i} \in F_{x,x}$. By the same argument, t_0 cannot be irrational; see

Example 1.17. There is then a minimal $m \in \mathbb{N}$ with $z_0^m = 1$. The minimality of t_0 and m easily imply that $F_{x,x} = \{z \in \mathbb{T} \; ; \; z^m = 1\}$.

Now let $y \in HC(\Psi)$. By Step 1, there exist $\lambda \in F_{x,y}$ and $\mu \in F_{y,x}$. Then, by Step 3, $\lambda F_{x,x} \subset F_{x,y}$ and $\mu F_{x,y} \subset F_{x,x}$, so that $\operatorname{card}(F_{x,y}) = \operatorname{card}(F_{x,x})$. This implies that $F_{x,y} = \lambda F_{x,x}$.

Step 6. There is a continuous function $f : HC(\Psi) \to \mathbb{T}$ such that $f(\Psi(0,t)x) = e^{2\pi mti}$ for every $t \geq 0$.

Let $m \in \mathbb{N}$ be the integer given by Step 5. Then, for any $y \in HC(\Psi)$, we define

$$f(y) = \lambda^m \quad \text{if } \lambda \in F_{x,y}.$$

By Step 5, this is well defined. Moreover, f is continuous. Otherwise there are $y_k \in HC(\Psi)$ and $y \in HC(\Psi)$ such that $y_k \to y$ but $f(y_k) \not\to f(y)$. We choose $\lambda_k \in F_{x,y_k}$. Passing to a subsequence if necessary, we can assume that $f(y_k) \to \mu \neq f(y)$ and $\lambda_k \to \lambda$ for some $\lambda, \mu \in \mathbb{T}$. It follows from Step 2 that $\lambda \in F_{x,y}$ and hence that $f(y_k) = \lambda_k^m \to \lambda^m = f(y)$, which is a contradiction.

Now let $t \geq 0$. By property (β), $\Psi(0,t)$ has dense range and therefore $\Psi(0,t)x \in HC(\Psi)$. Since, by definition, $\rho(t) = e^{2\pi ti} \in F_{x,\Psi(0,t)x}$ we conclude that $f(\Psi(0,t)x) = e^{2\pi mti}$.

Step 7. There is a continuous function $h : \overline{\mathbb{D}} \to \mathbb{T}$, whose restriction to the unit circle is homotopically nontrivial. A contradiction.

This is the decisive, and most difficult part of the proof. We will use here the terminology and some results of homotopy theory; see Appendix A. In order to define the function h we will first define a function $g : \mathbb{T} \to HC(\Psi)$, where we will distinguish the two subcases of (α).

Case 1: $\Psi(0,1) = I$. Here we define $g : \mathbb{T} \to HC(\Psi)$ by

$$g(e^{2\pi ti}) = \Psi(0,t)x, \quad 0 \leq t < 1,$$

which is well defined by property (β), and g is continuous because $\Psi(0,1) = I$. By Step 6, the function $f \circ g : \mathbb{T} \to \mathbb{T}$ satisfies $f(g(e^{2\pi ti})) = e^{2\pi mti}$, $0 \leq t < 1$, so that the index of $f \circ g$ is $m \geq 1$.

We extend the function g to the closed unit disk $\overline{\mathbb{D}}$ by defining $g(z) = (1-r)\Psi(1,0)x + rg(e^{2\pi ti})$ for $z = re^{2\pi ti} \in \overline{\mathbb{D}}$, $r \geq 0$. This extension is clearly continuous on $\overline{\mathbb{D}}$. Since $g(z)$ is a convex combination of $\Psi(1,0)x$ and $\Psi(0,t)x$ for some $t \geq 0$, property (β) implies that $g(z) \in HC(\Psi)$ for every $z \in \overline{\mathbb{D}}$.

To summarize, we have found a continuous function $h := f \circ g : \overline{\mathbb{D}} \to \mathbb{T}$ whose restriction to the unit circle is homotopically nontrivial. In other words, the map $H : \mathbb{T} \times [0,1] \to \mathbb{T}$, $(e^{2\pi ti}, r) \to h(re^{2\pi ti})$ defines a homotopy between the function h on $\mathbb{T}$, which is homotopically nontrivial, and a constant function. This is the desired contradiction.

Case 2: $\Psi(1,0) = I$. Here the construction of g is slightly more delicate. First, since f is continuous and $f(x) = 1$, we can find a 0-neighbourhood W

such that $|f(y) - 1| < 1$ if $y \in HC(\Psi)$ and $y - x \in W$. We can assume that W is balanced, that is, $\mu W \subset W$ whenever $|\mu| \leq 1$; see Lemma 2.6(iii).

Since no 0-neighbourhood in an infinite-dimensional Fréchet space can be relatively compact (see Appendix A), the set $U := W \setminus \{x - \Psi(0,t)x \; ; \; 0 \leq t \leq 1\}$ is open and nonempty. By the hypercyclicity of x there are $n_0 \in \mathbb{N}_0$ and $t_0 \geq 0$ such that $x - \Psi(n_0, t_0)x \in U$. Since $\Psi(1,0) = I$ we also have that $x - \Psi(0, t_0)x \in U$, and therefore $t_0 > 1$ and $x - \Psi(0, t_0)x \in W$. We can now define $g : \mathbb{T} \to HC(\Psi)$ by

$$g(e^{2\pi ti}) = \begin{cases} \Psi(0, 2tt_0)x & \text{if } 0 \leq t < 1/2, \\ (2t-1)x + (2-2t)\Psi(0,t_0)x & \text{if } 1/2 \leq t < 1, \end{cases}$$

which is clearly continuous. The fact that g is well defined is a consequence of property (β); note that $x = \Psi(1,0)x$.

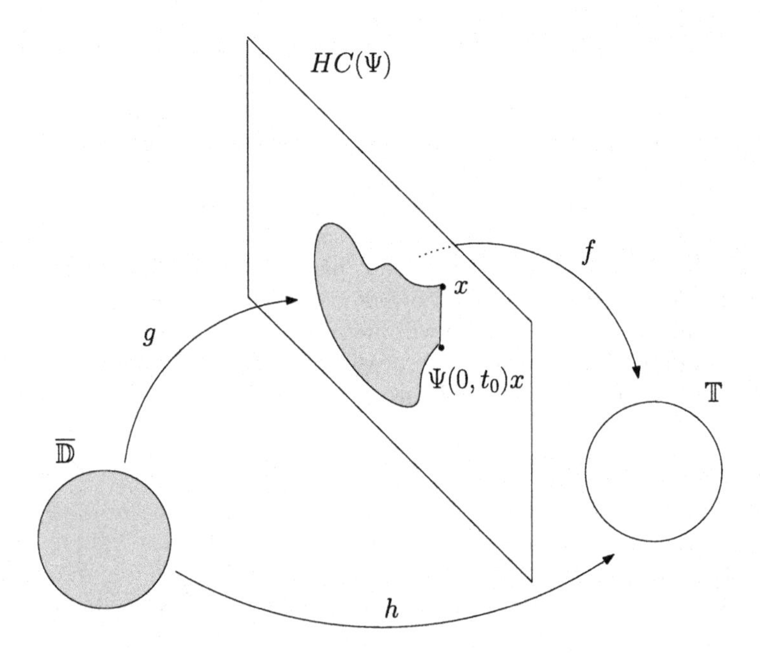

Fig. 6.2 The map h, Case 2

We consider the function $f \circ g : \mathbb{T} \to \mathbb{T}$. Then $f(g(e^{2\pi ti})) = e^{4\pi m t t_0 i}$ for $0 \leq t < 1/2$. Moreover, by the selection of t_0, and since W is balanced, we obtain that $|f(g(e^{2\pi ti})) - 1| < 1$ for $1/2 \leq t < 1$. Thus, as t moves from 0 to $1/2$, $f(g(e^{2\pi ti}))$, starting from 1, moves along the unit circle in a positive direction and covers it $[mt_0]$ times, finishing inside the disk of radius 1 around 1. As t then moves from $1/2$ to 1, $f(g(e^{2\pi ti}))$ stays completely in that disk, returning to 1 for $t = 1$. As a consequence, the path $t \to f(g(e^{2\pi ti}))$ can be deformed homotopically to the path $t \to e^{2\pi n i}$ with either $n = [mt_0] \geq 1$ if

$\mathrm{Im}(e^{2\pi m t_0 i}) \geq 0$ or with $n = [mt_0] + 1 \geq 2$ if $\mathrm{Im}(e^{2\pi m t_0 i}) < 0$. In any case, the index of $f \circ g$ is nonzero.

We extend the function g continuously to the closed unit disk $\overline{\mathbb{D}}$ by defining $g(z) = (1-r)x + rg(e^{2\pi t i})$ for $z = re^{2\pi t i} \in \overline{\mathbb{D}}$, $r \geq 0$. Since $g(z)$ is a convex combination of $x = \Psi(1,0)x$ and $\Psi(0,s)x$ for some $s \geq 0$, property (β) implies that $g(z) \in HC(\Psi)$ for every $z \in \overline{\mathbb{D}}$. We define again the map $h : \overline{\mathbb{D}} \to \mathbb{T}$ by $h = f \circ g$; see Figure 6.2. Since the restriction of h to the unit circle is homotopically nontrivial we obtain a contradiction as in Case 1. □

When we combine Theorem 6.10 with Ansari's theorem, by which $\Psi(n,t) = \Psi(1,t/n)^n$ is hypercyclic whenever $\Psi(1,t/n)$ is, we obtain the following.

Corollary 6.11. *Let Ψ be a semigroup action on a Fréchet space X satisfying properties (α) and (β). If $x \in X$ is hypercyclic for Ψ then it is hypercyclic for each operator $\Psi(n,t)$, $n,t > 0$.*

Exercises

Exercise 6.1.1. In a metric space, show that a finite union of nowhere dense sets is nowhere dense.

Exercise 6.1.2. Let $T : X \to X$ be a (not necessarily linear) weakly mixing dynamical system. Show that any T^p, $p \in \mathbb{N}$, is also weakly mixing. (*Hint*: Theorem 1.54.)

Exercise 6.1.3. Let T be an operator on a separable Fréchet space X that satisfies the Hypercyclicity Criterion. Give two proofs of the fact that any T^p, $p \in \mathbb{N}$, also satisfies the Hypercyclicity Criterion.

Exercise 6.1.4. Let T be a chaotic operator on a Fréchet space X. Without the use of Theorem 6.2, show that any T^p, $p \in \mathbb{N}$, is also chaotic. This is not true for nonlinear maps by the example of Exercise 1.2.11.

Exercise 6.1.5. Let $S : X \to X$, $T : Y \to Y$ be topologically ergodic operators on Fréchet spaces X and Y. Show that any operator $S^p \oplus T^q$, $p, q \in \mathbb{N}$, is topologically ergodic on $X \oplus Y$. (*Hint*: Exercises 2.5.5 and 2.5.6.)

Exercise 6.1.6. Let T be an operator on a Fréchet space X and x a hypercyclic vector for T. Show that there exists an increasing sequence $(n_k)_k$ of positive integers with $\sup_{k\geq 1}(n_{k+1} - n_k) = 2$ such that x does not have dense orbit under $(T^{n_k})_k$. (*Hint*: Show that there is some $y \in X$ and a 0-neighbourhood W such that $z \in y + W$ implies that $Tz \notin y + W$.)

In the following two exercises, let $T : X \to X$ be a (not necessarily linear) dynamical system, that is, a continuous map T on a metric space X. Suppose that X does not have isolated points, and let $D = \{x \in X \; ; \; \mathrm{orb}(x,T)$ is dense in $X\}$.

Exercise 6.1.7. Show the following generalization of Ansari's theorem. If D contains a connected and dense set then T and T^p, $p \in \mathbb{N}$, have the same points of dense orbits. (*Hint*: Follow the proof of Ansari's theorem and note that D itself must be connected; see the proof of Corollary 2.56.)

Exercise 6.1.8. An alternative proof of Exercise 6.1.7 (and thus of Ansari's theorem) is the following. With the notation of Theorem 6.2, let

$$A_k := \bigcup_{0\leq j_1<\cdots<j_k\leq p-1} (D_{j_1} \cap \cdots \cap D_{j_k}),$$

where $k = 1, \ldots, p$. Prove the following assertions:

(i) $A_1 = D$, $A_p = \bigcap_{j=0}^{p-1} D_j$, and $A_{k+1} \subset A_k$, $k = 1, \ldots, p-1$;

(ii) $T(A_k) \subset A_k$, $k = 1, \ldots, p$;

(iii) if $A_k = D$, then $A_{k+1} = D$, $k = 1, \ldots, p-1$.

In particular, $\text{orb}(x, T^p)$ is dense in X for every $x \in D$. (*Hint*: For (iii), observe that if $A_{k+1} \neq D$, then $A_{k+1} = \varnothing$, since it is closed, T-invariant, and $T : D \to D$ is minimal; hence A_k is a finite union of pairwise disjoint closed sets.)

Exercise 6.2.1. Prove assertions (i), (ii) and (iii) before Lemma 6.3. (*Hint*: See the proof of Proposition 1.15.)

Exercise 6.2.2. Prove Lemma 6.3. (*Hint*: Follow the argument of Theorem 2.54.)

Exercise 6.2.3. Let T be a continuous map on a metric space X without isolated points, and let $x \in X$. With the notation of this section, prove that if $U(x) \neq \varnothing$ and $T(X \setminus U(x)) \subset X \setminus U(x)$, then $D(x) = \overline{U(x)}$. (*Hint*: Show that $\text{orb}(x, T) \subset U(x)$.)

Exercise 6.2.4. Let T be an operator on a Fréchet space X and $x \in X$. With the notation of this section, prove directly that if X is a complex (or real) space, then $U(x) = U(\lambda x) = \lambda U(x)$ for $\lambda \neq 0$ (or for $\lambda > 0$, respectively). Deduce that $D(x) = X$ if $0 \in U(x)$. (*Hint*: Use Lemma 6.1.)

Exercise 6.2.5. Let $T : X \to X$ be a (not necessarily linear) topologically transitive dynamical system and $x \in X$. Show that if $\text{orb}(x, T)$ is somewhere dense in X, then it is dense in X.

Exercise 6.2.6. Let T be an operator on a Fréchet space X and x a hypercyclic vector for T. Show that there exists an increasing sequence $(n_k)_k$ of positive integers with $\sup_{k\geq 1}(n_{k+1} - n_k) = 2$ such that the orbit of x under $(T^{n_k})_k$ is somewhere dense but not dense. (*Hint*: See Exercise 6.1.6.)

Exercise 6.3.1. Let T be an invertible operator on a Fréchet space X and $x \in X$ such that $\{T^n x \,;\, n \in \mathbb{Z}\}$ is dense in X. Show that x is either hypercyclic for T or for T^{-1}; in particular, both T and T^{-1} are hypercyclic. For the proof,

(i) either use the Bourdon–Feldman theorem,

(ii) or proceed directly.

(*Hint*: For (i); see the proof of Theorem 6.6. For (ii), suppose that $T^n x \notin U$ for all $n \geq 0$; for any V, find $k \in \mathbb{Z}$ and $U' \subset U$ such that $T^k(U') \subset V$; find $m < -|k|$ such that $T^m x \in U'$; then $T^{m+k} x \in V$, $m + k < 0$.)

Exercise 6.3.2. Let T be an operator on a Fréchet space X admitting a countable set $\{x_1, x_2, \ldots\}$ of vectors such that

$$\bigcup_{j=1}^{\infty} \overline{\text{orb}(x_j, T)} = X.$$

Show that some vector x_j, $j \geq 1$, is hypercyclic for T. Give an example of an operator on a normed space for which this assertion fails.

Exercise 6.3.3. An operator T on a Banach space X is called *countably hypercyclic* if it admits a countable bounded set $\{x_1, x_2, \dots\}$ of vectors with $\inf_{j\neq k} \|x_j - x_k\| > 0$ such that

$$\overline{\bigcup_{j=1}^{\infty} \operatorname{orb}(x_j, T)} = X.$$

Show that the operator $T = 2(I \oplus B)$ on $X = \ell^2 \oplus \ell^2$ is countably hypercyclic but not hypercyclic, where B is the backward shift. (*Hint*: Take $x_j = (0, e_j) + 2^{-n_j}(I \oplus F)^{n_j} y_j$, where F is the forward shift.)

Exercise 6.3.4. Let T be a countably hypercyclic operator on a Banach space X. Show the following.

(a) The spectrum $\sigma(T)$ meets the unit circle.

(b) The orbit of every $x^* \neq 0$ in X^* under T^* is unbounded.

(*Hint*: See Section 5.1.)

Exercise 6.3.5. Let $T = B_w$ be a weighted backward shift on $X = \ell^p$, $1 \leq p < \infty$, or c_0; see Section 4.1. Show that if T is countably hypercyclic then it is hypercyclic. (*Hint*: Apply Exercise 6.3.4(b) to $x^* = e_1$.)

Exercise 6.4.1. Let $T = B_w$ be the weighted bilateral backward shift on $\ell^2(\mathbb{Z})$ with weights $w_n = \frac{n+1}{n}$ if $n \geq 1$ and $w_{-n} = \frac{n+1}{n+2}$ if $n \geq 0$; see Section 4.1. Show that λT, $\lambda \in \mathbb{C}$, is hypercyclic if and only if $|\lambda| = 1$. Discuss this result in the light of Kitai's theorem, showing first that $\sigma(T) \subset \mathbb{T}$. (*Hint*: For the first part note that $\lambda B_w = B_{\lambda w}$; for the second part use the spectral radius formula for T and T^{-1} and Exercise 5.0.7.)

Exercise 6.4.2. Let T_j be operators on complex Fréchet spaces X_j, $j = 1, \dots, n$, such that $T_1 \oplus \dots \oplus T_n$ is hypercyclic. Show that, for any $\lambda_j \in \mathbb{C}$ with $|\lambda_j| = 1$, $j = 1, \dots, n$, the operator $\lambda_1 T_1 \oplus \dots \oplus \lambda_n T_n$ is also hypercyclic and, moreover, that it shares the set of hypercyclic vectors with $T_1 \oplus \dots \oplus T_n$. (*Hint*: Set $\Psi(n,t) = S_1^n \oplus e^{2\pi t i} S_2^n$ for suitable operators S_1, S_2, deduce that $HC(\Psi) = HC(S_1 \oplus S_2)$, and apply this result repeatedly.)

Exercise 6.4.3. Let T_j be operators on complex Fréchet spaces X_j, $j = 1, \dots, n$, and let $x_j \in X_j$, $j = 1, \dots, n$, be such that

$$\left\{(\lambda_1 T_1^k x_1, \dots, \lambda_n T_n^k x_n)\ ;\ k \in \mathbb{N}_0,\ (\lambda_1, \dots, \lambda_n) \in \mathbb{T}^n\right\}$$

is dense in $X_1 \oplus \dots \oplus X_n$. Show that $x := (x_1, \dots, x_n)$ is hypercyclic for $T_1 \oplus \dots \oplus T_n$. (*Hint*: Let $(U_m)_m$ be a countable base of open sets in $X_1 \oplus \dots \oplus X_n$. Show that the sets $\{(\mu_1, \dots, \mu_n) \in \mathbb{T}^n\ ;\ \exists k \in \mathbb{N}_0 \text{ with } (\mu_1^k T_1^k x_1, \dots, \mu_n^k T_n^k x_n) \in U_m\}$ are open and dense in $\mathbb{T}^n$. By a Baire argument, find $(\mu_1, \dots, \mu_n) \in \mathbb{T}^n$ such that x is hypercyclic for $\mu_1 T_1 \oplus \dots \oplus \mu_n T_n$, and conclude by using Exercise 6.4.2.)

Exercise 6.4.4. Let $X = C_0(\mathbb{R}_+)$, the space of continuous functions on $\mathbb{R}_+$ that vanish at ∞, endowed with the sup-norm. Consider the semigroup action $\Psi(n,t)f(x) := 2^{n-t} f(x+t)$, $n \in \mathbb{N}_0$, $t \in \mathbb{R}_+$. Then Ψ is hypercyclic but the operator $\Psi(1,1)$ is not hypercyclic. Which hypothesis of Theorem 6.10 is not satisfied?

Exercise 6.4.5. Give a new proof of Ansari's theorem by proceeding as follows. Let T be a hypercyclic operator on a Fréchet space X, x a hypercyclic vector for T and $p \in \mathbb{N}$. For $u, v \in X$ define the subset $F_{u,v}$ of $\mathbb{T}$ by

$$F_{u,v} = \left\{e^{2\pi j i/p}\ ;\ \exists\, (n_k)_k \subset \mathbb{N}_0 \text{ with } T^{n_k p + j} u \to v, j = 0, \dots, p-1\right\}.$$

Show the following:

(i) if $u \in HC(T)$, then $F_{u,v} \neq \varnothing$ for all $v \in X$;
(ii) if $u, v, w \in X$, $\lambda \in F_{u,v}$, and $\mu \in F_{v,w}$, then $\lambda\mu \in F_{u,w}$;
(iii) there is a divisor $m \geq 1$ of p such that $F_{x,x} = \{e^{2\pi mi/p} \ ; \ j = 0, \ldots, p/m - 1\}$;
(iv) for every $y \in HC(T)$ there is some j, $0 \leq j \leq m-1$, such that $F_{x,y} = e^{2\pi ji/p} F_{x,x}$.

Now let $D_j = \{y \in HC(T) \ ; \ F_{x,y} = e^{2\pi ji/p} F_{x,x}\}$, $j = 0, \ldots, m-1$. Then finish the proof as follows:
(v) show that the D_j form a partition of $HC(T)$ into closed (and open) sets;
(vi) deduce that $m = 1$ and hence that $x \in HC(T^p)$.

In the following two exercises, let $T : X \to X$ be a (not necessarily linear) dynamical system, where X does not have isolated points.

Exercise 6.4.6. Show the following *separation theorem.* If $x \in X$ has dense orbit under T but not under T^p, $p > 1$, then there is a divisor $m > 1$ of p and a partition $D_0, \ldots, D_{m-1}$ of $D = \{x \in X \ ; \ \mathrm{orb}(x, T) \text{ is dense in } X\}$ into closed (and open) subsets with the following properties:
(i) $T(D_j) \subset D_{j+1 (\mathrm{mod}\, m)}$, $j = 0, \ldots, m-1$;
(ii) for $j = 0, \ldots, m-1$, the orbit of $T^j x$ under T^p is contained and dense in D_j.
(*Hint*: Proceed as in the previous exercise.)

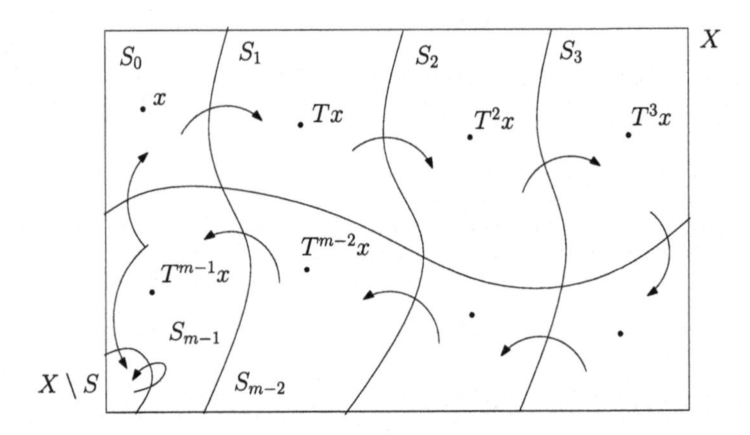

Fig. 6.3 Nonlinear dynamics if x has dense orbit under T but not under T^p, $m|p$

Exercise 6.4.7. Show the following *decomposition theorem.* If $x \in X$ has dense orbit under T but not under T^p, $p > 1$, then there is a divisor $m > 1$ of p and pairwise disjoint open subsets $S_0, S_1, \ldots, S_{m-1}$ of X with the following properties:
(i) $S := \bigcup_{j=0}^{m-1} S_j$ is dense in X;
(ii) $T(S_j) \subset S_{j+1}$, $j = 0, \ldots, m-2$, and $T(S_{m-1}) \subset S_0 \cup (X \setminus S)$;
(iii) $X \setminus S$ is invariant under T;
(iv) for $j = 0, \ldots, m-1$, the orbit of $T^j x$ under T^p is contained and dense in S_j;

see Figure 6.3.
(*Hint*: Consider the sets D_j of the previous exercise; set $S_{m-1} = X \setminus \bigcup_{j=0}^{m-2} \overline{D_j}$, with closure in X, and $S_j = T^{-m+j+1}(S_{m-1})$; show first that $T^{-m}(S_{m-1}) \subset S_{m-1}$, $D_j \subset S_j$, and that $T^n x \in S_j$ if and only if $n = j (\mathrm{mod}\, m)$.)

Exercise 6.4.8. Verify the results of the previous two exercises in the case of the map of Exercise 1.2.11.

Sources and comments

The results in this chapter have in common that their proofs use connectedness arguments. But their relationship runs deeper than that. As we have seen, Ansari's theorem is a consequence of the Costakis–Peris theorem, which in turn follows from the Bourdon–Feldman theorem. Moreover, the theorems of León–Müller and Conejero–Müller–Peris are proved by a common approach. Recently, Shkarin [284] was able to unify the latter two theorems with Ansari's theorem by deriving them as consequences of a single, quite general result. An alternative common framework was developed by Bayart and Matheron [44], which was further generalized by Matheron [235] to include Shkarin's result.

Section 6.1. Ansari [9] showed that powers of hypercyclic operators are hypercyclic. Independently, Banks [28] proved a more general result: any power of a minimal map on a connected topological space is also minimal (see Exercise 6.1.7). We combine ideas from Banks [28] and Peris [254] for the proof of Theorem 6.2. Lemma 6.1 is from Peris [254].

Section 6.2. Theorem 6.5 is due to Bourdon and Feldman [93], answering a question from Peris [254]. The corresponding result for semigroups of operators (see the next chapter for this notion) is due to Costakis and Peris [121]. It is interesting to note that for a weighted backward shift on ℓ^p, $1 \leq p < \infty$, to be hypercyclic it already suffices to have an orbit with a nonzero limit point, as was shown by Chan and Seceleanu [105]; such an orbit though, need not be dense.

Section 6.3. The fact that multi-hypercyclic operators are hypercyclic was independently proved by Costakis [117] and Peris [254], answering a question raised by Herrero [195]. The original proofs motivated the question leading to the Bourdon–Feldman theorem.

Section 6.4. Theorem 6.7 on rotations of hypercyclic operators is due to León and Müller [222]. Bayart and Bermúdez [37] show that the corresponding result for chaos fails. Badea, Grivaux and Müller [20] characterize the subsets of $\mathbb{C}$ that can appear as $\{\lambda \in \mathbb{C}\ ;\ \lambda T \text{ hypercyclic}\}$ for invertible operators T on a complex Hilbert space.

Theorem 6.8 on discretizations of hypercyclic C_0-semigroups is due to Conejero, Müller and Peris [110]. Exercise 6.4.4 shows that the result fails for semigroups indexed over $\mathbb{N}_0 \times \mathbb{R}_+$; see also Shkarin [284] and Exercise 7.3.1. Bayart [36] shows that it even fails for holomorphic groups over $\mathbb{C}$. And by Bayart and Bermúdez [37] there are chaotic C_0-semigroups on a Hilbert space for which no individual operator is chaotic.

The unified proof of Theorem 6.10 essentially follows the argument of [110]. The related approach to Ansari's theorem in Exercise 6.4.5 is due to Grosse-Erdmann, León and Piqueras [183]. As mentioned above, Shkarin [284], Bayart and Matheron [44] and Matheron [235] obtain much more general results that contain the theorems of Ansari, León–Müller and Conejero–Müller–Peris as special cases. Shkarin and Matheron point out that the main common idea in all these proofs can already be found in a paper by Furstenberg [156].

Exercises. Exercise 6.1.6 is taken from Montes and Salas [243], Exercise 6.1.7 from Banks [28]. Exercise 6.1.8 outlines essentially the original proof of Ansari [9]. Exercises 6.2.2 and 6.2.3 are taken from Bourdon and Feldman [93], while the result of Exercise 6.2.6 is due to Peris and Saldivia [257]. The result of Exercise 6.3.1 is due to Herrero and Kitai [196]. The notion of a countably hypercyclic operator (see Exercise 6.3.3), as well as the results of Exercises 6.3.4 and 6.3.5 are due to Feldman [149]. Exercise 6.4.1 is taken from León and Müller [222], Exercise 6.4.2 from Shkarin [284]. Exercises 6.4.6 and 6.4.7 are due to Grosse-Erdmann, León and Piqueras [183] (see also Marano and

Salas [226]); the case of $p = 2$ was previously obtained by Bourdon [91].

Extensions. We will show in Chapter 12 that the theorems of Ansari, Bourdon–Feldman, Costakis–Peris and León–Müller continue to hold in arbitrary topological vector spaces.

Part II
Selected topics

Chapter 7
Dynamics of semigroups, with applications to differential equations

In this chapter we study the dynamical properties of strongly continuous semigroups of operators on Banach spaces, that is, of C_0-semigroups. They can be viewed as the continuous-time analogue of the discrete-time case of iterates of a single operator.

Specifically, we introduce and study the notions of hypercyclicity, (weak) mixing and chaos for C_0-semigroups. We then develop corresponding criteria such as a Hypercyclicity Criterion and an eigenvalue criterion.

For discrete linear dynamical systems the shift operators on sequence spaces constitute one of the most important test classes. In the continuous case this role is played by the translation semigroups. We will obtain, in particular, a characterization of translation semigroups that are hypercyclic, mixing or chaotic.

We also investigate the relationship between the dynamical properties of a semigroup and those of its various discretizations.

C_0-semigroups describe the solutions of so-called abstract Cauchy problems. This makes the semigroup formulation of chaos applicable to linear differential equations. Some applications of the chaotic behaviour of C_0-semigroups to partial differential equations and to infinite linear systems of ordinary differential equations will be discussed in the last section.

Throughout this chapter, X will denote a separable Banach space.

7.1 Semigroups of operators

A one-parameter family $(T_t)_{t\geq 0}$ of operators on X is called a *strongly continuous semigroup of operators* if the following three conditions are satisfied:

(i) $T_0 = I$;
(ii) $T_{t+s} = T_t T_s$ for all $s, t \geq 0$;
(iii) $\lim_{s\to t} T_s x = T_t x$ for all $x \in X$ and $t \geq 0$.

One also refers to it as a *C_0-semigroup*.

K.-G. Grosse-Erdmann, A. Peris Manguillot, *Linear Chaos*, Universitext,
DOI 10.1007/978-1-4471-2170-1_7,

Condition (iii) expresses the pointwise continuity of the semigroup. The Banach–Steinhaus theorem (see Appendix A) yields that the family $(T_t)_{t\geq 0}$ is then *locally equicontinuous*, that is, for any $b > 0$ we have that

$$\sup_{t\in[0,b]} \|T_t\| < \infty,$$

or equivalently, there exists some $M > 0$ such that

$$\|T_t x\| \leq M\|x\| \quad \text{for all } t \in [0,b],\, x \in X. \tag{7.1}$$

We can express condition (iii) in some useful equivalent ways.

Lemma 7.1. *Let $(T_t)_{t\geq 0}$ be a family of operators on X. Then the following assertions are equivalent:*

(i) $\lim_{s\to t} T_s x = T_t x$ *for all $x \in X$ and $t \geq 0$;*

(ii) $(T_t)_{t\geq 0}$ *is locally equicontinuous and there is a dense subset X_0 of X such that* $\lim_{s\to t} T_s x = T_t x$ *for all $x \in X_0$ and $t \geq 0$;*

(iii) *the map*

$$\mathbb{R}_+ \times X \to X, \quad (t,x) \to T_t x$$

is continuous.

Proof. It suffices to show that (ii) implies (iii). This implication follows immediately from the identity

$$T_t x - T_s y = (T_t - T_s)(x - x_0) + (T_t - T_s)x_0 + T_s(x - y)$$

for $x, y \in X$, $x_0 \in X_0$ and $s, t \geq 0$. $\square$

Remark 7.2. Local equicontinuity implies, via (7.1), that $T_{t_n} x_n \to 0$ whenever $(t_n)_n$ is bounded and $x_n \to 0$. This observation will be used repeatedly.

Moreover, one can establish an exponential bound for the operator norm of the semigroup.

Proposition 7.3. *If $(T_t)_{t\geq 0}$ is a C_0-semigroup, then there exist $M \geq 1$ and $w \in \mathbb{R}$ such that $\|T_t\| \leq Me^{wt}$ for all $t \geq 0$.*

Proof. Let $M = \sup_{t\in[0,1]} \|T_t\|$, which is finite and at least 1. Setting $w = \log M$ and writing $t \geq 0$ as $t = n + s$ with $n \in \mathbb{N}_0$ and $s \in [0,1[$, we obtain that

$$\|T_t\| = \|T_s T_1^n\| \leq M\|T_1\|^n \leq Me^{nw} \leq Me^{tw},$$

where we have used the semigroup property. $\square$

We introduce a standard example that we will revisit throughout this chapter.

Example 7.4. **(Translation semigroups)** Let $1 \leq p < \infty$ and let $v : \mathbb{R}_+ \to \mathbb{R}$ be a strictly positive locally integrable function, that is, v is measurable with $\int_0^b v(x)\,dx < \infty$ for all $b > 0$. We consider the space of weighted p-integrable functions defined as

$$X = L^p_v(\mathbb{R}_+) = \{f : \mathbb{R}_+ \to \mathbb{K}\ ;\ f \text{ is measurable and } \|f\| < \infty\},$$

where

$$\|f\| = \Big(\int_0^\infty |f(x)|^p v(x)\,dx\Big)^{1/p}.$$

The *translation semigroup* is then given by

$$T_t f(x) = f(x+t), \quad t, x \geq 0.$$

We claim that this defines a C_0-semigroup on $L^p_v(\mathbb{R}_+)$ if and only if there exist $M \geq 1$ and $w \in \mathbb{R}$ such that, for all $t \geq 0$,

$$v(x) \leq M e^{wt} v(x+t) \ \text{ for almost all } x \geq 0. \tag{7.2}$$

Indeed, suppose that (7.2) is satisfied. If $f \in X$, we first observe that

$$\begin{aligned}\int_0^\infty |f(x+t)|^p v(x)\,dx &\leq M e^{wt} \int_0^\infty |f(x+t)|^p v(x+t)\,dx \\ &\leq M e^{wt} \int_0^\infty |f(x)|^p v(x)\,dx.\end{aligned}$$

This shows that $T_t : X \to X$, $t \geq 0$, is well defined and continuous, and that the family of translation operators is locally equicontinuous on X. Since the semigroup property is obviously satisfied, it only remains to show that $\lim_{s\to t} T_s = T_t$, for every $t \geq 0$, pointwise on a dense subset of X; see Lemma 7.1. We can check it easily on the dense subspace of continuous functions of compact support, using the fact that v is locally integrable.

Conversely, suppose that the translation semigroup $(T_t)_{t\geq 0}$ is a C_0-semigroup on X. Let M and w be given by Proposition 7.3. We will prove that, for any $t \geq 0$,

$$v(x) \leq 2M^p e^{pwt} v(x+t) \ \text{ for almost all } x \geq 0.$$

If this is not true, then there is some $t_0 > 0$ such that

$$B := \{x \geq 0\ ;\ v(x) > 2M^p e^{pwt_0} v(x+t_0)\}$$

has Lebesgue measure $\lambda(B) > 0$. Let $b > 0$ be such that $\lambda(B \cap [0,b]) > 0$, and define $f(x) = \frac{1}{v(x)^{1/p}}$ if $x \in t_0 + (B \cap [0,b])$, and $f(x) = 0$ otherwise. It is clear that $f \in X$ and $\|f\| > 0$. On the other hand

$$\|T_{t_0}f\|^p = \int_{B\cap[0,b]} |f(x+t_0)|^p v(x)\, dx$$
$$\geq 2M^p e^{pwt_0} \int_{B\cap[0,b]} |f(x+t_0)|^p v(x+t_0)\, dx = 2M^p e^{pwt_0}\|f\|^p,$$

a contradiction with the choice of M and w. Hence (7.2) holds.

Now, in order to avoid some technical problems we will demand in the sequel that the weight v satisfies (7.2) for any $x \geq 0$. Equivalently, there are constants $M \geq 1$ and $w \in \mathbb{R}$ such that

$$v(x) \leq Me^{w(y-x)}v(y) \quad \text{whenever } y \geq x \geq 0. \tag{7.3}$$

In that case, v is called an *admissible weight function.*

We collect here some properties of C_0-semigroups. The easiest general construction of C_0-semigroups is via an operator A on X. Since $\sum_{n=0}^{\infty} \frac{t^n}{n!}\|A\|^n < \infty$ for any $t \geq 0$,

$$T_t = e^{tA} = \sum_{n=0}^{\infty} \frac{t^n}{n!}A^n, \quad t \geq 0,$$

defines operators on X, and it is easily seen that $(T_t)_{t\geq 0}$ is a C_0-semigroup; we even have that, for any $t \geq 0$, $\lim_{s\to t} T_s = T_t$ in the operator norm topology. The semigroup is then called *uniformly continuous.* Moreover, for any $x \in X$, $Ax = \lim_{t\to 0} \frac{1}{t}(T_t x - x)$. For all these statements see Exercise 7.1.6.

Now let $(T_t)_{t\geq 0}$ be an arbitrary C_0-semigroup on X. It can be shown that

$$Ax := \lim_{t\to 0} \frac{1}{t}(T_t x - x)$$

exists on a dense subspace of X; the set of these x, the domain of A, is denoted by $D(A)$. Then A, or rather $(A, D(A))$, is called the *(infinitesimal) generator* of the semigroup. It turns out that $A : D(A) \to X$ is a linear map with *closed graph*, that is, for any sequence $(x_n)_n$ in $D(A)$, if $\lim_{n\to\infty} x_n =: x$ and $\lim_{n\to\infty} Ax_n =: y$ exist in X, then $x \in D(A)$ and $Ax = y$. Moreover, $T_t(D(A)) \subset D(A)$ with $AT_t x = T_t Ax$ for every $t \geq 0$ and $x \in D(A)$. See Exercise 7.1.7 for these statements. It can also be shown that the generator determines the semigroup uniquely.

Another important property is provided by the point spectral mapping theorem for semigroups (see Appendix B): if X is a complex Banach space then, for every $x \in X$ and $\lambda \in \mathbb{C}$,

$$Ax = \lambda x \implies T_t x = e^{\lambda t}x \tag{7.4}$$

for every $t \geq 0$.

As an example, the generator $(A, D(A))$ of the translation semigroup on $X = L^p_v(\mathbb{R}_+)$, $1 \leq p < \infty$, can be shown to be

$$D(A) = \{f \in X \; ; \; f \text{ is absolutely continuous and } f' \in X\}, \quad Af = f'.$$

The case when A is defined on all of X, that is, when A is an operator on X, occurs precisely when $(T_t)_{t\geq 0}$ is uniformly continuous; see Exercise 7.1.8.

Proposition 7.5. *Let $(T_t)_{t\geq 0}$ be a C_0-semigroup on X. Then the following assertions are equivalent:*

(i) *the semigroup is uniformly continuous;*
(ii) *the generator A of the semigroup is defined everywhere;*
(iii) *there is an operator A on X such that $T_t = e^{tA}$, $t \geq 0$.*

We finally mention an interpretation of the generator that is the starting point of the applications of semigroup theory to linear differential equations, as will be pursued in Section 7.5. Let $(T_t)_{t\geq 0}$ be a C_0-semigroup on X, $(A, D(A))$ its generator and, for any $x \in X$,

$$u(\cdot, x) : \mathbb{R}_+ \to X, \quad u(t, x) := T_t x,$$

the function that describes the orbit of x under the semigroup. Whenever $x \in D(A)$, $u(\cdot, x)$ is the unique solution of the *abstract Cauchy problem*

$$\begin{cases} \frac{d}{dt}u(t) = Au(t) & \text{for } t \geq 0, \\ u(0) = x. \end{cases} \tag{ACP}$$

And, for arbitrary $x \in X$, $u(\cdot, x)$ is the unique solution of the corresponding integral equation

$$u(t) = A \int_0^t u(s)\, ds + x, \quad t \geq 0.$$

In that sense, $u(\cdot, x)$ is called a *classical solution* of (ACP) if $x \in D(A)$ and a *mild solution* of (ACP) if $x \in X$; $(T_t)_{t\geq 0}$ is called the *solution semigroup* of (ACP).

Conversely, suppose that A is a linear map with closed graph defined on a dense subspace $D(A) \subset X$ such that (ACP) has a unique (classical) solution $u(\cdot, x)$ for every $x \in D(A)$. If, moreover, for any sequence $(x_n)_n$ in $D(A)$ with $x_n \to 0$ and any $b > 0$, $u(t, x_n) \to 0$ uniformly on $[0, b]$, then A is the infinitesimal generator of a C_0-semigroup.

7.2 Hypercyclic and chaotic C_0-semigroups

We now begin our investigation of the dynamical properties of C_0-semigroups. The concepts of hypercyclicity, topological transitivity, mixing, weak mixing and chaos for operators all have a natural continuous analogue.

Definition 7.6. Let $(T_t)_{t\geq 0}$ be a C_0-semigroup on X.

(a) For any $x \in X$ we call

$$\operatorname{orb}(x, (T_t)) = \{T_t x \ ; \ t \geq 0\}$$

the *orbit* of x under $(T_t)_{t\geq 0}$.

(b) The semigroup is called *hypercyclic* if there is some $x \in X$ whose orbit under $(T_t)_{t\geq 0}$ is dense in X. In such a case, x is called a *hypercyclic vector* for $(T_t)_{t\geq 0}$.

Definition 7.7. A C_0-semigroup $(T_t)_{t\geq 0}$ on X is called *topologically transitive* if, for any pair U, V of nonempty open subsets of X, there exists some $t \geq 0$ such that $T_t(U) \cap V \neq \varnothing$.

Using the fact that X is separable, we easily deduce from the definitions that hypercyclicity of a C_0-semigroup is equivalent to hypercyclicity of $(T_{t_n})_n$ for some positive sequence $(t_n)_n$ with $t_n \to \infty$; see Definition 3.23. Moreover, hypercyclicity and topological transitivity for C_0-semigroups are equivalent notions; see Exercise 7.2.1.

Definition 7.8. Let $(T_t)_{t\geq 0}$ be a C_0-semigroup on X.

(a) The semigroup is called *mixing* if, for any pair U, V of nonempty open subsets of X, there exists some $t_0 \geq 0$ such that $T_t(U) \cap V \neq \varnothing$ for all $t \geq t_0$.

(b) The semigroup is called *weakly mixing* if $(T_t \oplus T_t)_{t\geq 0}$ is topologically transitive on $X \oplus X$.

One should note that, for any C_0-semigroups $(T_t)_{t\geq 0}$ and $(S_t)_{t\geq 0}$ on Banach spaces X and Y, respectively, $(T_t \oplus S_t)_{t\geq 0}$ is a C_0-semigroup on $X \oplus Y$. As in the discrete case, the direct sum of a mixing semigroup with a hypercyclic semigroup is hypercyclic.

Finally, we also have a concept of chaos.

Definition 7.9. Let $(T_t)_{t\geq 0}$ be a C_0-semigroup on X.

(a) A point $x \in X$ is called a *periodic point* of $(T_t)_{t\geq 0}$ if there is some $t > 0$ such that $T_t x = x$. The set of periodic points for $(T_t)_{t\geq 0}$ is denoted by $\operatorname{Per}((T_t))$.

(b) The semigroup is said to be *chaotic* if it is hypercyclic and its set of periodic points is dense in X.

Example 7.10. (a) We consider the translation semigroup on the space $X = L^p_v(\mathbb{R}_+)$, where $1 \leq p < \infty$ and $v : \mathbb{R}_+ \to \mathbb{R}$ is an admissible weight function. We claim that the following assertions are equivalent:

(i) the translation semigroup is hypercyclic;
(ii) the translation semigroup is weakly mixing;
(iii) $\liminf_{x\to\infty} v(x) = 0$.

We first show that (i) implies (iii). Suppose that v is bounded away from zero. Since, by (7.3), v is bounded on $[0,1]$ there is then some $C>0$ such that $v(x) \leq C^p v(x+t)$ for all $x \in [0,1]$, $t \geq 0$. Let g be defined by $g(x) = v(x)^{-1/p}$

on $[0,1]$, and 0 otherwise. Then we have for any $f \in X$ with $\|f\| < (2C)^{-1}$ and any $t \geq 0$ that

$$\begin{aligned}\|T_t f - g\| &\geq \Big(\int_0^1 |f(x+t) - g(x)|^p v(x)\,dx\Big)^{1/p}\\ &\geq \Big(\int_0^1 |g(x)|^p v(x)\,dx\Big)^{1/p} - \Big(\int_0^1 |f(x+t)|^p v(x)\,dx\Big)^{1/p}\\ &\geq 1 - M\Big(\int_0^1 |f(x+t)|^p v(x+t)\,dx\Big)^{1/p} \geq 1 - C\|f\| > \frac{1}{2}.\end{aligned}$$

Consequently, $(T_t)_{t\geq 0}$ cannot be hypercyclic.

Next we will show that the semigroup is topologically transitive under condition (iii). Let X_0 be the dense subspace of continuous functions of compact support. It suffices to show that, for any $f_1, f_2 \in X_0$ and $\varepsilon > 0$, there is some $t \geq 0$ and some $g \in X$ such that

$$\|f_1 - g\| < \varepsilon \quad \text{and} \quad T_t g = f_2.$$

To this end, let

$$t_0 = \max\big(\operatorname{supp}(f_1) \cup \operatorname{supp}(f_2)\big).$$

Then, for any $t > t_0$, we define $g_t(x) = f_1(x)$ for $x < t$ and $g_t(x) = f_2(x-t)$ for $x \geq t$. We clearly have that $T_t g_t = f_2$. By (iii) there is some $t \geq t_0$ such that

$$M^2 e^{wt_0}\|f_2\|^p v(t_0 + t) < v(0)\varepsilon,$$

where M and w are the parameters associated with the admissibility of v; see (7.3). Therefore,

$$\begin{aligned}\|f_1 - g_t\|^p &= \int_t^\infty |f_2(x-t)|^p v(x)\,dx = \int_0^\infty |f_2(x)|^p v(x+t)\,dx\\ &\leq Mv(t_0+t)\int_0^\infty |f_2(x)|^p e^{w(t_0-x)}\,dx\\ &\leq M^2 e^{wt_0}\frac{v(t_0+t)}{v(0)}\int_0^\infty |f_2(x)|^p e^{wx} v(x) e^{-wx}\,dx\\ &= M^2 e^{wt_0}\frac{v(t_0+t)}{v(0)}\|f_2\|^p < \varepsilon,\end{aligned}$$

which had to be shown.

Indeed, by the same argument, the direct sum $(T_t \oplus T_t)_{t\geq 0}$ is topologically transitive, which shows (ii).

(b) Continuing the study of the translation semigroup of (a), we claim that the following assertions are equivalent:

(i) the translation semigroup is mixing;

(ii) $\lim_{x\to\infty} v(x) = 0$.

We need only modify the arguments in (a) slightly. Thus, suppose that (ii) fails, that is, there is some sequence $(t_n)_n$ with $t_n \to \infty$ such that $(v(t_n))_n$ is bounded away from zero. The admissibility condition (7.3) implies that v is bounded away from zero on $\bigcup_{n=1}^{\infty}[t_n, t_n + 1]$. But then there is some $C > 0$ such that $v(x) \leq C^p v(x + t_n)$ for all $x \in [0, 1]$, $n \geq 1$, and the argument in (a) shows that the sequence of operators $(T_{t_n})_n$ cannot be hypercyclic, which contradicts (i). The proof that (ii) implies (i) is exactly as in (a).

(c) Let $X = C_0(\mathbb{R}_+)$ be the space of continuous functions $f : \mathbb{R}_+ \to \mathbb{K}$ with $\lim_{x\to\infty} f(x) = 0$, endowed with the sup-norm $\|f\| = \sup_{x\in\mathbb{R}_+} |f(x)|$. Given a constant $w > 0$, we consider the family of operators defined by $(T_t f)(x) = e^{wt} f(x+t)$, $x \in \mathbb{R}_+$, which is easily seen to define a C_0-semigroup on X.

We claim that this semigroup is mixing and chaotic. To this end, let X_0 be the dense subspace of continuous functions with compact support, and let $f_1 \in X_0$, $f_2 \in X$ and $\varepsilon > 0$. For any $t > t_0 := \max \operatorname{supp}(f_1)$ we define $g_t \in X$ by

$$g_t(x) = \begin{cases} f_1(x) & \text{if } x \leq t_0, \\ e^{-wt} f_2(0) \frac{x-t_0}{t-t_0} & \text{if } t_0 < x < t, \\ e^{-wt} f_2(x - t) & \text{if } x \geq t. \end{cases}$$

Then we have that $T_t g_t = f_2$ and

$$\|f_1 - g_t\| = e^{-wt} \|f_2\| < \varepsilon$$

for any $t > t_0$ sufficiently large. As in (a), this yields the mixing property for the semigroup.

In order to show that the semigroup is even chaotic, it suffices to prove that, for any $f \in X_0$ and $\varepsilon > 0$, there is a periodic point $h \in X$ with $\|f - h\| < \varepsilon$. To this end, for any $t > t_0 := \max \operatorname{supp}(f)$, we define the function $h_t \in X$ by

$$h_t(x) = \begin{cases} e^{-wkt} f(x - kt) & \text{if } kt \leq x \leq kt + t_0,\ k \in \mathbb{N}_0, \\ e^{-w(k+1)t} f(0) \frac{x-kt-t_0}{t-t_0} & \text{if } kt + t_0 < x < (k+1)t,\ k \in \mathbb{N}_0. \end{cases}$$

Then $T_t h_t = h_t$, so that h_t is a periodic point. On the other hand,

$$\|f - h_t\| = \sup_{x \geq t_0} |h_t(x)| \leq \max_{k\geq 1} e^{-wkt} \|f\| < \varepsilon$$

for any $t > t_0$ sufficiently large, which had to be shown.

In the previous chapters we realized how important quasiconjugacies are for discrete dynamical systems. The same is true in the continuous case.

Definition 7.11. Let $(T_t)_{t\geq 0}$ and $(S_t)_{t\geq 0}$ be C_0-semigroups on separable Banach spaces X and Y, respectively.

(a) Then $(T_t)_{t\geq 0}$ is called *quasiconjugate* to $(S_t)_{t\geq 0}$ if there exists a continuous map $\phi : Y \to X$ with dense range such that $T_t \circ \phi = \phi \circ S_t$ for all $t \geq 0$. If ϕ can be chosen to be a homeomorphism then $(T_t)_{t\geq 0}$ and $(S_t)_{t\geq 0}$ are called *conjugate*.

(b) A property $\mathcal{P}$ of C_0-semigroups is said to be *preserved under (quasi) conjugacy* if any C_0-semigroup that is (quasi)conjugate to a C_0-semigroup with property $\mathcal{P}$ also possesses property $\mathcal{P}$.

Proposition 7.12. *Hypercyclicity, mixing, weak mixing and chaos for a C_0-semigroup are preserved under quasiconjugacy.*

As a special case we consider again the notion of complexification; see the discussion before Proposition 2.26. If X is a real Banach space and $\widetilde{X}$ is its complexification then the family $(\widetilde{T}_t)_{t\geq 0}$ of complexifications of a C_0-semigroup $(T_t)_{t\geq 0}$ on X is a C_0-semigroup on $\widetilde{X}$. Since $(T_t)_{t\geq 0}$ is quasiconjugate to $(\widetilde{T}_t)_{t\geq 0}$ via the canonical projection of $\widetilde{X}$ onto X, we obtain the following.

Corollary 7.13. *If the complexification $(\widetilde{T}_t)_{t\geq 0}$ is hypercyclic (or weakly mixing, mixing or chaotic, respectively) then so is $(T_t)_{t\geq 0}$.*

In the discrete case we saw that adjoints of hypercyclic operators have an empty point spectrum. The same happens for the infinitesimal generator A of a hypercyclic C_0-semigroup. In this context we call $\lambda \in \mathbb{K}$ an *eigenvalue* of A^* if there is some $x^* \in X^*$, $x^* \neq 0$, such that $\langle Ax, x^* \rangle = \lambda \langle x, x^* \rangle$ for all $x \in D(A)$. Note that we do not need to define A^*.

Lemma 7.14. *Let $(T_t)_{t\geq 0}$ be a hypercyclic C_0-semigroup on X with infinitesimal generator $(A, D(A))$.*

(a) *Then, for every $t > 0$, the adjoint T_t^* has no eigenvalues. Equivalently, every operator $T_t - \lambda I$, $\lambda \in \mathbb{K}$, has dense range. Moreover, A^* has no eigenvalues.*

(b) *If X is a real Banach space, then, for every $t > 0$, the adjoint $\widetilde{T}_t^*$ has no eigenvalues. Equivalently, every operator $\widetilde{T}_t - \lambda I$, $\lambda \in \mathbb{C}$, has dense range.*

Proof. (a) Proceeding by contradiction, we suppose that $\mathrm{orb}(x, (T_t))$ is dense in X for some $x \in X$ and that there are $t > 0$, $\lambda \in \mathbb{K}$, and $x^* \in X^* \setminus \{0\}$ such that $T_t^* x^* = \lambda x^*$. We will distinguish two cases:

Case 1: $|\lambda| < 1$. In this case we fix $y \in X$ such that $\langle y, x^* \rangle = 1$. Then there is a sequence $(t_n)_n$ that tends to infinity such that $\lim_{n\to\infty} T_{t_n} x = y$. We write $t_n = m_n t + s_n$ with $m_n \in \mathbb{N}_0$ and $s_n \in [0, t[$, $n \in \mathbb{N}$, and we obtain that

$$1 = \lim_{n\to\infty} \langle T_{t_n} x, x^* \rangle = \lim_{n\to\infty} \langle T_{s_n} x, (T_t^*)^{m_n} x^* \rangle = \lim_{n\to\infty} \lambda^{m_n} \langle T_{s_n} x, x^* \rangle = 0,$$

since the sequence $(T_{s_n} x)_n$ is bounded in X. This is a contradiction.

Case 2: $|\lambda| \geq 1$. On the one hand, let $r > 0$ be such that $|\langle T_r x, x^* \rangle| > 1$. On the other hand, by equicontinuity of $(T_s)_{s\in[0,t]}$, there is some $\varepsilon > 0$ such that $|\langle T_s y, x^* \rangle| < 1$ if $s \in [0,t]$ and $\|y\| < \varepsilon$. We can also find $t' > r$ such that $\|T_{t'} x\| < \varepsilon$. If we write $t' = mt - s + r$ for some $m \in \mathbb{N}$ and $s \in [0,t]$, then, putting all these facts together, we obtain that

$$1 > |\langle T_s(T_{t'}x), x^* \rangle| = |\langle T_r x, (T_t^*)^m x^* \rangle| = |\lambda|^m |\langle T_r x, x^* \rangle| > 1,$$

which is a contradiction.

Next, let $(A, D(A))$ be the generator of $(T_t)_{t\geq 0}$, and suppose that A^* has an eigenvalue $\lambda \in \mathbb{K}$. Let $x^* \in X^* \setminus \{0\}$ be such that $\langle Ax, x^* \rangle = \lambda \langle x, x^* \rangle$ for all $x \in D(A)$. Given $x \in D(A)$, we define $h(t) = \langle T_t x, x^* \rangle$, $t \geq 0$. Then $T_t x \in D(A)$ and $h'(t) = \langle AT_t x, x^* \rangle = \lambda \langle T_t x, x^* \rangle = \lambda h(t)$ for all $t \geq 0$; see Exercise 7.1.7. Therefore,

$$\langle x, T_t^* x^* \rangle = h(t) = h(0)e^{\lambda t} = e^{\lambda t} \langle x, x^* \rangle, \quad t \geq 0.$$

Since $x \in D(A)$ is arbitrary and $D(A)$ is dense in X, we conclude that x^* is an eigenvector of every T_t^*, and we know that this is impossible.

The remaining claim follows from the Hahn–Banach theorem; see the proof of Lemma 2.53.

(b) It is easy to adapt the above argument. One need only note that, for any $\widetilde{x}^* \in \widetilde{X}^* \setminus \{0\}$ there is some $y \in X$ such that $|\langle y, \widetilde{x}^* \rangle| > 0$; see the proof of Lemma 2.53(b). □

Since all operators on $\mathbb{C}^N$ have eigenvalues, we deduce the following as in the discrete case; see Corollary 2.59.

Theorem 7.15. *There are no hypercyclic C_0-semigroups on a finite-dimensional Banach space.*

Furthermore, we obtain the analogue of Bourdon's theorem for C_0-semigroups of operators. The proof is identical.

Theorem 7.16. *If $(T_t)_{t\geq 0}$ is a hypercyclic C_0-semigroup and p a nonzero polynomial, then every operator $p(T_t)$, $t > 0$, has dense range.*

From this we can deduce the continuous analogue of the Herrero–Bourdon theorem. The proof is the same as in the discrete case. However, one needs to add that, by Theorem 6.8, any vector that is hypercyclic for the semigroup $(T_t)_{t\geq 0}$ is also hypercyclic for every operator T_t, $t > 0$. It is important to realize that we have no vicious circle here: the proof of Theorem 6.8 used Theorem 7.16, but not Theorem 7.17.

Theorem 7.17. *If $(T_t)_{t\geq 0}$ is a C_0-semigroup and x is a hypercyclic vector, then, for any $t > 0$,*

$$\{p(T_t)x \;;\; p \text{ is a polynomial}\} \setminus \{0\}$$

is a dense set of hypercyclic vectors.

In particular, any hypercyclic C_0-semigroup admits a dense subspace of vectors consisting, except for zero, of hypercyclic vectors.

We can also apply Theorem 7.16 to obtain a necessary condition for chaos of a C_0-semigroup on complex Banach spaces; compare with Proposition 5.7.

Proposition 7.18. *If $(A, D(A))$ is the generator of a chaotic C_0-semigroup on a complex Banach space X, then*

$$\sigma_p(A) \cap i\mathbb{R}$$

is infinite and, moreover,

$$X = \overline{\text{span}} \bigcup_{\lambda \in i\mathbb{R}} \ker(\lambda I - A).$$

Proof. By hypothesis, the set $\text{Per}((T_t))$ of periodic points of the semigroup is dense in X. By the point spectral mapping theorem for semigroups (see Appendix B), we know that

$$\begin{aligned} \text{Per}((T_t)) &= \bigcup_{t>0} \ker(I - T_t) = \bigcup_{t>0} \overline{\text{span}} \bigcup_{n\in\mathbb{Z}} \ker\left(\frac{2\pi n i}{t} I - A\right) \\ &\subset \overline{\text{span}} \bigcup_{t>0} \bigcup_{n\in\mathbb{Z}} \ker\left(\frac{2\pi n i}{t} I - A\right). \end{aligned}$$

The density of periodic points implies that $X = \overline{\text{span}} \bigcup_{\lambda\in i\mathbb{R}} \ker(\lambda I - A)$.

On the other hand, if $\sigma_p(A) \cap i\mathbb{R}$ was finite, then there would exist finite collections of $n_k \in \mathbb{Z}$, $t_k > 0$, $k = 1, \ldots, N$, such that

$$\text{Per}((T_t)) \subset \overline{\text{span}} \bigcup_{k=1}^{N} \ker\left(\frac{2\pi n_k i}{t_k} I - A\right). \tag{7.5}$$

Moreover, by the point spectral mapping theorem for semigroups, we have that, for $k = 1, \ldots, N$,

$$\ker\left(\frac{2\pi n_k i}{t_k} I - A\right) \subset \ker(I - T_{t_k}). \tag{7.6}$$

It follows from (7.5) and (7.6) that the operator $T := (I - T_{t_1}) \cdots (I - T_{t_N})$ vanishes on $\text{Per}((T_t))$. Since this set is dense we have that $T = 0$. But this contradicts Theorem 7.16 by which T has dense range. □

Example 7.19. We consider again the translation semigroup on the space $X = L^p_v(\mathbb{R}_+)$, $1 \le p < \infty$, for an admissible weight v. Then the following assertions are equivalent:

(i) the translation semigroup is chaotic;
(ii) $\int_0^\infty v(x)\,dx < \infty$.

We will see below that a semigroup on a real Banach space is chaotic if and only if its complexification is; see Corollary 7.24. Therefore we may assume that we are dealing with the complex space X. Thus, if the semigroup is chaotic, Proposition 7.18 implies that its generator has an eigenvalue on the imaginary axis. Since the generator is the derivative, we can find some $s \in \mathbb{R}$ such that the function $x \to e^{isx}$, $x \in \mathbb{R}_+$, belongs to X. This yields (ii).

Conversely, suppose that v is integrable on $\mathbb{R}_+$. By Example 7.10(a), the semigroup is then hypercyclic. Now let f be a continuous function of compact support on $\mathbb{R}_+$ and $\varepsilon > 0$. Setting $K = 1 + \max_{x\in\mathbb{R}_+} |f(x)|$, we can find some $t > 0$ such that $\int_t^\infty v(x)\,dx < (\varepsilon/2K)^p$. We define $g(x+nt) = f(x)$, $x \in [0,t[$, $n \in \mathbb{N}_0$. The function g is bounded, thus $g \in X$ by the integrability of the weight. Moreover, $T_t g = g$ and

$$\begin{aligned}\|f-g\| &\leq \Big(\int_t^\infty |f(x)|^p v(x)\,dx\Big)^{1/p} + \Big(\int_t^\infty |g(x)|^p v(x)\,dx\Big)^{1/p}\\ &\leq 2K\Big(\int_t^\infty v(x)\,dx\Big)^{1/p} < \varepsilon.\end{aligned}$$

Since the continuous functions of compact support are dense in X, the semigroup has a dense set of periodic points and is therefore chaotic.

7.3 Discretizations of C_0-semigroups

In this section we study the relationship between hypercyclicity, mixing and weak mixing of a semigroup and the corresponding properties of its discretizations. In this way, we will establish a direct link between the continuous and the discrete case.

A *discretization* of a semigroup $(T_t)_{t\geq 0}$ is a sequence of operators $(T_{t_n})_n$ with $t_n \to \infty$. If there is some $t_0 > 0$ such that $t_n = nt_0$ for $n \in \mathbb{N}$, then $(T_{t_n})_n = (T_{t_0}^n)_n$ is called an *autonomous discretization* of $(T_t)_{t\geq 0}$.

As we have already observed, a C_0-semigroup is hypercyclic if and only if it admits a hypercyclic discretization $(T_{t_n})_n$. The following results provide characterizations of mixing and weakly mixing semigroups in terms of discretizations.

Proposition 7.20. *Let $(T_t)_{t\geq 0}$ be a C_0-semigroup on X. Then the following assertions are equivalent:*
(i) *$(T_t)_{t\geq 0}$ is weakly mixing;*
(ii) *some discretization of $(T_t)_{t\geq 0}$ is mixing;*
(iii) *some discretization of $(T_t)_{t\geq 0}$ is weakly mixing.*

Proof. If the semigroup is weakly mixing, then there exists a hypercyclic discretization $(T_{t_n} \oplus T_{t_n})_n$ of $(T_t \oplus T_t)_{t\geq 0}$. Since the operators T_{t_n}, $n \in \mathbb{N}$, commute, Theorem 3.25 (see also Exercise 1.4.2) implies that there exists a mixing subsequence $(T_{t_{n_k}})_k$ of $(T_{t_n})_n$. The other implications are trivial. □

Proposition 7.21. *Let $(T_t)_{t\geq 0}$ be a C_0-semigroup on X. Then the following assertions are equivalent:*

(i) *$(T_t)_{t\geq 0}$ is mixing;*
(ii) *every discretization of $(T_t)_{t\geq 0}$ is mixing;*
(iii) *every discretization of $(T_t)_{t\geq 0}$ is weakly mixing;*
(iv) *every discretization of $(T_t)_{t\geq 0}$ is hypercyclic;*
(v) *every autonomous discretization of $(T_t)_{t\geq 0}$ is mixing;*
(vi) *some autonomous discretization of $(T_t)_{t\geq 0}$ is mixing.*

Proof. The implications (i) $\Longrightarrow$ (ii) $\Longrightarrow$ (iii) $\Longrightarrow$ (iv) and (ii) $\Longrightarrow$ (v) $\Longrightarrow$ (vi) are trivial.

(iv) $\Longrightarrow$ (i). Suppose that $(T_t)_{t\geq 0}$ is not mixing. Then there exists a pair of nonempty open subsets U, V of X and a sequence $(t_n)_n$ tending to ∞ such that $T_{t_n}(U) \cap V = \varnothing$ for all $n \geq 1$. Then $(T_{t_n})_n$ cannot be topologically transitive.

(vi) $\Longrightarrow$ (i). Let $(T_{nt_0})_n$ be a mixing autonomous discretization of $(T_t)_{t\geq 0}$. We fix arbitrary nonempty open sets $U, V \subset X$. By Lemma 2.36 there are nonempty open sets $U_1 \subset U$, $V_1 \subset V$ and a 0-neighbourhood W such that $U_1 + W \subset U$ and $V_1 + W \subset V$. By local equicontinuity of the semigroup there is a 0-neighbourhood W_1 such that $T_s(W_1) \subset W$ for all $s \in [0, t_0]$. The hypothesis then implies the existence of some $N \in \mathbb{N}$ such that

$$T_{nt_0}(U_1) \cap W_1 \neq \varnothing \quad \text{and} \quad T_{nt_0}(W_1) \cap V_1 \neq \varnothing$$

for every $n \geq N$. Given any $t \geq Nt_0$, we write $t = nt_0 + s$ with $n \in \mathbb{N}$, $n \geq N$, and $s \in [0, t_0]$. We can then find $u_1 \in U_1$ and $w_1 \in W_1$ such that $T_{nt_0}u_1 \in W_1$ and $T_{(n+1)t_0}w_1 \in V_1$. Thus

$$T_t u_1 = T_s T_{nt_0} u_1 \in T_s(W_1) \subset W,$$

and for $w := T_{t_0-s}w_1 \in W$ we have that

$$T_t w = T_{(n+1)t_0} w_1 \in V_1,$$

that is, $T_t(u_1 + w) \in V$ with $u_1 + w \in U$. This shows (i). □

In this book we are mainly interested in autonomous discretizations. As we have noted, hypercyclicity is inherited by the semigroup if some autonomous discretization has this property, and the same is true for weak mixing, mixing and chaos. The natural question is whether, conversely, we can deduce hypercyclicity of autonomous discretizations from hypercyclicity of the semigroup. The Conejero–Müller–Peris theorem (see Chapter 6) gives a positive answer.

But since its proof is highly nontrivial we will provide here a simpler proof of a classical result that is somewhat weaker but still strong enough for many applications.

Theorem 7.22 (Oxtoby–Ulam). *If $(T_t)_{t\geq 0}$ is a C_0-semigroup on X and $x \in X$ is a hypercyclic vector, then there is a dense G_δ-set $J \subset]0,\infty[$ such that, for every $t \in J$, x is hypercyclic for T_t.*

Proof. Let $x \in X$ be a hypercyclic vector for the C_0-semigroup. We fix a countable base of nonempty open sets $(U_k)_k$ in X and set

$$J_k = \{t \in]0,\infty[\ ;\ \ T_{nt}x \in U_k \text{ for some } n \in \mathbb{N}\}.$$

Each J_k is an open subset of $]0,\infty[$. We will see that it is also dense. Indeed, if $0 < a < b < \infty$, there is some $n_0 \in \mathbb{N}$ such that $n_0 b > (n_0+1)a$. Then

$$\bigcup_{n\geq n_0}]na, nb[=]n_0 a, \infty[.$$

Let $s > n_0 a$ be such that $T_s x \in U_k$. Then there exists some $n \geq n_0$ with $s \in]na, nb[$. If we define $t = s/n \in]a,b[$, then we get that $T_{nt}x \in U_k$, so that $t \in J_k \cap]a,b[$. Thus, each set J_k is dense in $]0,\infty[$.

To finish the proof we consider

$$J = \bigcap_{k=1}^{\infty} J_k,$$

which is a dense G_δ-subset of $]0,\infty[$; note that the Baire category theorem is applicable since $]0,\infty[$ is isomorphic to $\mathbb{R}$. Let $t \in J$. Then, for every $k \in \mathbb{N}$, there exists some $n \in \mathbb{N}$ such that $T_{nt}x \in U_k$. That is, x is a hypercyclic vector for T_t. □

The theorem yields that, as in the discrete case, chaos implies weak mixing.

Theorem 7.23. *Let $(T_t)_{t\geq 0}$ be a hypercyclic C_0-semigroup on X. If there exists a dense subset X_0 of X such that the orbit of each $x \in X_0$ is bounded, then $(T_t)_{t\geq 0}$ is weakly mixing.*

In particular, every chaotic C_0-semigroup is weakly mixing.

Proof. By the Oxtoby–Ulam theorem, some operator T_s, $s > 0$, is hypercyclic. On the other hand, the orbit of every $x \in X_0$ under T_s is contained in the orbit of x under $(T_t)_{t\geq 0}$, which is a bounded set. Thus, by Theorem 2.48, T_s is weakly mixing, which implies the same property for the semigroup. □

Corollary 7.24. *A C_0-semigroup on a real Banach space is chaotic if and only if its complexification is.*

Proof. Suppose that the semigroup $(T_t)_{t\geq 0}$ on a real Banach space X is chaotic. Since its complexification $(\widetilde{T}_t)_{t\geq 0}$ can be identified with the direct sum $(T_t \oplus T_t)_{t\geq 0}$ it is hypercyclic by the preceding result. Moreover, if U and V are nonempty open subsets of X then, by topological transitivity and continuity, there is a nonempty open subset $U_1 \subset U$ and some $t \geq 0$ such that $T_t(U_1) \subset V$. We can then find some periodic point $x \in U_1$. Let $T_s x = x$, $s > 0$. Then $y := T_t x \in V$ and $T_s y = T_t T_s x = y$. Therefore, $(x, y) \in U \times V$ is a periodic point for $(T_t \oplus T_t)_{t\geq 0}$, which concludes the proof that $(\widetilde{T}_t)_{t\geq 0}$ is chaotic. The converse implication was obtained in Corollary 7.13. □

The Oxtoby–Ulam theorem also implies that weakly mixing C_0-semigroups contain weakly mixing autonomous discretizations. However, a stronger result can be derived.

Proposition 7.25. *Let $(T_t)_{t\geq 0}$ be a C_0-semigroup on X. Then the following assertions are equivalent:*

(i) *$(T_t)_{t\geq 0}$ is weakly mixing;*
(ii) *some autonomous discretization of $(T_t)_{t\geq 0}$ is weakly mixing;*
(iii) *every autonomous discretization of $(T_t)_{t\geq 0}$ is weakly mixing.*

Proof. It suffices to show that (ii) implies (iii). In view of the Bès–Peris theorem we have to show that if some T_{t_0}, $t_0 > 0$, satisfies the Hypercyclicity Criterion, then so does every T_{t_1}, $t_1 > 0$. Since $T_{t_1} = T^p_{t_1/p}$, $p \in \mathbb{N}$, we may assume that $t_1 < t_0$; see Exercise 6.1.3.

Now, Theorem 3.22 tells us that T_{t_0} even satisfies the Gethner–Shapiro criterion; that is, there are dense subsets X_0, Y_0 of X, an increasing sequence $(n_k)_k$ of positive integers and a mapping $S_{t_0} : Y_0 \to Y_0$ such that, for any $x \in X_0$ and $y \in Y_0$, $T^{n_k}_{t_0} x \to 0$, $S^{n_k}_{t_0} y \to 0$, and $T_{t_0} S_{t_0} y = y$. As the proof of Theorem 3.22 shows, Y_0 can be taken in the form $Y_0 = \{y_n \,;\, n \in \mathbb{N}\}$, where the y_n are pairwise distinct and $S_{t_0} y_n = y_{n+1}$ for all $n \in \mathbb{N}$.

Since $t_1 < t_0$ there are $m_k \in \mathbb{N}_0$ and $s_k \in [0, t_0[$, $k \in \mathbb{N}$, such that

$$m_k t_1 = n_k t_0 + s_k, \quad k \in \mathbb{N}.$$

We claim that also T_{t_1} satisfies the Hypercyclicity Criterion, with respect to X_0, Y_0, and the sequence $(m_k)_k$.

Indeed, for every $x \in X_0$ we have that

$$T^{m_k}_{t_1} x = T_{s_k} T^{n_k}_{t_0} x \to 0$$

as $k \to \infty$, where we use the fact that the family $(T_s)_{s\in[0,t_0]}$ is equicontinuous.

Next we define a family $(S_t)_{t\geq 0}$ of maps on Y_0 by setting $S_t y_n = T_s y_{n+k}$ if $t = kt_0 - s$ for some $k \in \mathbb{N}$ and $0 < s \leq t_0$; note that this is consistent with S_{t_0}. Then we have for every $y_n \in Y_0$, $n \in \mathbb{N}$, that

$$S_{m_k t_1} y_n = S_{(n_k+1)t_0-(t_0-s_k)} y_n = T_{t_0-s_k} y_{n+n_k+1} = T_{t_0-s_k} S^{n_k}_{t_0} y_{n+1} \to 0$$

as $k \to \infty$, again by equicontinuity.

Finally, for every $y_n \in Y_0$, $n \in \mathbb{N}$, we obtain that

$$T_{t_1}^{m_k} S_{m_k t_1} y_n = T_{m_k t_1} T_{t_0 - s_k} S_{t_0}^{n_k} y_{n+1} = T_{t_0}^{n_k+1} S_{t_0}^{n_k+1} y_n = y_n.$$

Altogether we find that T_{t_1} satisfies the Hypercyclicity Criterion. □

To finish this section we reformulate the Conejero–Müller–Peris theorem in the language of discretizations. The theorem contains both the Oxtoby–Ulam theorem and Proposition 7.25.

Theorem 7.26. *Let $(T_t)_{t\geq 0}$ be a C_0-semigroup on X and $x \in X$. Then the following assertions are equivalent:*
(i) *x is hypercyclic for $(T_t)_{t\geq 0}$;*
(ii) *x is hypercyclic for some discretization of $(T_t)_{t\geq 0}$;*
(iii) *x is hypercyclic for some autonomous discretization of $(T_t)_{t\geq 0}$;*
(iv) *x is hypercyclic for every autonomous discretization of $(T_t)_{t\geq 0}$.*

7.4 Criteria for the hypercyclicity and chaos of C_0-semigroups

In this section we derive, in analogy to the discrete case, the Hypercyclicity Criterion and various other criteria for weak mixing, mixing and chaos of semigroups of operators.

Theorem 7.27 (Hypercyclicity Criterion for semigroups). *Let $(T_t)_{t\geq 0}$ be a C_0-semigroup on X. If there are dense subsets $X_0, Y_0 \subset X$, a sequence $(t_n)_n$ in $\mathbb{R}_+$ with $t_n \to \infty$, and maps $S_{t_n} : Y_0 \to X$, $n \in \mathbb{N}$, such that, for any $x \in X_0$, $y \in Y_0$,*
(i) $T_{t_n} x \to 0$,
(ii) $S_{t_n} y \to 0$,
(iii) $T_{t_n} S_{t_n} y \to y$,
then $(T_t)_{t\geq 0}$ is weakly mixing, and in particular hypercyclic.

Proof. This is an immediate consequence of the Hypercyclicity Criterion for sequences of operators, Theorem 3.24, applied to the sequence $(T_{t_n})_n$. □

Calling a semigroup $(T_t)_{t\geq 0}$ *hereditarily hypercyclic* if there is a sequence $(t_n)_n$ in $\mathbb{R}_+$ with $t_n \to \infty$ such that, for every subsequence $(t_{n_k})_k$, the sequence of operators $(T_{t_{n_k}})_k$ admits a dense orbit, we have the following analogue of the Bès–Peris theorem.

Theorem 7.28. *Let $(T_t)_{t\geq 0}$ be a C_0-semigroup on X. Then the following assertions are equivalent:*
(i) *$(T_t)_{t\geq 0}$ satisfies the Hypercyclicity Criterion;*

(ii) $(T_t)_{t\geq 0}$ *is weakly mixing;*
(iii) $(T_t)_{t\geq 0}$ *is hereditarily hypercyclic.*

Proof. Suppose that $(T_t)_{t\geq 0}$ is weakly mixing. Then, by Proposition 7.25, the operator T_1 is weakly mixing and therefore hereditarily hypercyclic, so that the same holds for $(T_t)_{t\geq 0}$.

On the other hand, if $(T_t)_{t\geq 0}$ is hereditarily hypercyclic then there is some sequence $(t_n)_n$ in $\mathbb{R}_+$ with $t_n \to \infty$ such that $(T_{t_n})_n$ is a hereditarily transitive sequence of operators. By Theorem 3.25, $(T_n)_n$ satisfies the Hypercyclicity Criterion for sequences of operators (Theorem 3.24), which implies that the semigroup $(T_t)_{t\geq 0}$ satisfies the Hypercyclicity Criterion. □

If, in the Hypercyclicity Criterion, one has convergence along the whole real line then we obtain a criterion for mixing.

Theorem 7.29. *Let $(T_t)_{t\geq 0}$ be a C_0-semigroup on X. If there are dense subsets $X_0, Y_0 \subset X$, and maps $S_t : Y_0 \to X$, $t \geq 0$, such that, for any $x \in X_0$, $y \in Y_0$,*
(i) $T_t x \to 0$,
(ii) $S_t y \to 0$,
(iii) $T_t S_t y \to y$,
then $(T_t)_{t\geq 0}$ is mixing.

Proof. By the Hypercyclicity Criterion for sequences of operators (Theorem 3.24), every discretization $(T_{t_n})_n$ of $(T_t)_{t\geq 0}$ is weakly mixing. It then suffices to apply Proposition 7.21. Of course, a simple direct proof can also be given; see the proof of Kitai's theorem. □

We will now consider complex Banach spaces. In the discrete case, the Godefroy–Shapiro criterion tells us that a large supply of eigenvectors of an operator T to eigenvalues in an open set intersecting the unit circle leads to mixing and chaos of T. An analogous result is true when one studies eigenvectors of the generator A to eigenvalues in an open set that intersects the imaginary axis.

By a *weakly holomorphic* function $f : U \to X$ on an open set $U \subset \mathbb{C}$ we understand an X-valued function such that, for every $x^* \in X^*$, the complex-valued function $z \to \langle f(z), x^*\rangle$ is holomorphic on U. In the sequel, J is a nonempty index set.

Theorem 7.30. *Let X be a complex separable Banach space, and $(T_t)_{t\geq 0}$ a C_0-semigroup on X with generator $(A, D(A))$. Assume that there exists an open connected subset U and weakly holomorphic functions $f_j : U \to X$, $j \in J$, such that*
(i) $U \cap i\mathbb{R} \neq \varnothing$,
(ii) $f_j(\lambda) \in \ker(\lambda I - A)$ *for every* $\lambda \in U$, $j \in J$,
(iii) *for any* $x^* \in X^*$, *if* $\langle f_j(\lambda), x^*\rangle = 0$ *for all* $\lambda \in U$ *and* $j \in J$ *then* $x^* = 0$.
Then the semigroup $(T_t)_{t\geq 0}$ is mixing and chaotic.

Proof. We will use the Hahn–Banach theorem to prove the density of certain subspaces generated by eigenvectors, and then apply the Godefroy–Shapiro criterion.

For a fixed $t > 0$, we consider the subspaces

$$X_0 = \text{span}\{f_j(\lambda)\ ;\ j \in J,\ \lambda \in U \text{ with } \operatorname{Re}\lambda > 0\},$$

$$Y_0 = \text{span}\{f_j(\lambda)\ ;\ j \in J,\ \lambda \in U \text{ with } \operatorname{Re}\lambda < 0\},$$

and

$$Z_0 = \text{span}\{f_j(\lambda)\ ;\ j \in J,\ \lambda \in U \text{ with } \lambda t = \alpha\pi i \text{ for some } \alpha \in \mathbb{Q}\}.$$

We first show that these spaces are dense in X. To this end, suppose that $x^* \in X^*$ is identically 0 on X_0. For any $j \in J$, the function $\lambda \to \langle f_j(\lambda), x^*\rangle$ is holomorphic on U, and, in view of (i), it vanishes on some nonempty open subset of U. Therefore it vanishes on U, for any $j \in J$. By (iii) we have that $x^* = 0$. The Hahn–Banach theorem then implies that X_0 is dense in X. In the same way it follows that also Y_0 and Z_0 are dense.

Now, by (7.4), we have that

$$T_t x = e^{\lambda t} x$$

for all $x \in \ker(\lambda I - A)$. Therefore, by (ii), X_0 and Y_0 are contained in the span of the eigenvectors of T_t to eigenvalues of modulus greater than 1 and smaller than 1, respectively, and Z_0 is contained in the span of the eigenvectors of T_t to eigenvalues that are roots of unity. The Godefroy–Shapiro criterion yields that T_t is mixing and chaotic, properties that are inherited by the C_0-semigroup; see Proposition 7.21. □

One can strengthen the above criterion in the sense that the selection map f need not be defined on an open subset of $\mathbb{C}$, and we can replace weak holomorphy by continuity, as we will see next.

First of all, for any continuous vector-valued function $f : [a, b] \to X$ one can define the Riemann integral

$$\int_a^b f(t)\, dt$$

in the usual way by its Riemann sums; see Appendix A for more details and basic properties.

Lemma 7.31 (Riemann–Lebesgue lemma). *Let X be a complex Banach space and $f : [a, b] \to X$ a continuous function. Then $\int_a^b e^{irt} f(t)\, dt \to 0$ as $r \to \pm\infty$.*

Proof. Let $\varepsilon > 0$. We choose a partition $a = t_0 < t_1 < \ldots < t_N = b$ of $[a, b]$ such that $\|f(t) - f(t_j)\| < \varepsilon$ whenever $t_{j-1} \le t \le t_j$, $j = 1, \ldots, N$. Let g be the step function that takes the value $f(t_j)$ on $]t_{j-1}, t_j]$, $j = 1, \ldots, N$. Then we find that

$$\Big\| \int_a^b e^{irt} f(t)\, dt \Big\| \le \Big\| \int_a^b e^{irt}(f(t) - g(t))\, dt \Big\| + \Big\| \int_a^b e^{irt} g(t)\, dt \Big\|$$

$$\le (b-a)\varepsilon + \sum_{j=1}^{N} \|f(t_j)\| \Big| \int_{t_{j-1}}^{t_j} e^{irt}\, dt \Big|.$$

Now, since $\int_{t_{j-1}}^{t_j} e^{irt}\, dt \to 0$ as $r \to \pm\infty$, the claim follows. □

Theorem 7.32. *Let X be a complex separable Banach space, and $(T_t)_{t\ge 0}$ a C_0-semigroup on X with generator $(A, D(A))$. Assume that there are $a < b$ and continuous functions $f_j : [a,b] \to X$, $j \in J$, such that*
(i) *$f_j(s) \in \ker(isI - A)$ for every $s \in [a,b]$, $j \in J$,*
(ii) *$\operatorname{span}\{f_j(s)\ ;\ s \in [a,b], j \in J\}$ is dense in X.*
Then the semigroup $(T_t)_{t\ge 0}$ is mixing and chaotic.

Proof. We apply Theorem 7.29 to show that the semigroup is mixing. For any $r \in \mathbb{R}$, $j \in J$, we define $x_{r,j} = \int_a^b e^{irs} f_j(s)\, ds$, and we set

$$X_0 = Y_0 = \operatorname{span}\{x_{r,j}\ ;\ r \in \mathbb{R}, j \in J\}.$$

To prove that this subspace is dense in X we will again use the Hahn–Banach theorem. Thus let x^* be a continuous linear functional on X such that, for all $r \in \mathbb{R}$, $j \in J$,

$$\langle x_{r,j}, x^* \rangle = \int_a^b e^{irs} \langle f_j(s), x^* \rangle\, ds = 0.$$

The functions $s \to \langle f_j(s), x^* \rangle$ are continuous on $[a,b]$ and therefore belong to $L^2[a,b]$. Since $\big(\frac{1}{\sqrt{b-a}} \exp\big(\frac{2\pi}{b-a} ikt\big)\big)_{k\in\mathbb{Z}}$ is an orthonormal basis in this Hilbert space we deduce that, in view of continuity,

$$\langle f_j(s), x^* \rangle = 0 \ \text{ for all } s \in [a,b],\ j \in J.$$

Hence x^* vanishes on $\operatorname{span}\{f_j(s)\ ;\ s \in [a,b], j \in J\}$, which is dense in X by (ii). Thus $x^* = 0$, which implies that $X_0 = Y_0$ is dense.

Now, by (i) and (7.4) we have that, for any $t \ge 0$,

$$T_t f_j(s) = e^{ist} f_j(s) \ \text{ for } s \in [a,b],\ j \in J, \tag{7.7}$$

so that

$$T_t x_{r,j} = \int_a^b e^{irs} T_t f_j(s)\, ds = \int_a^b e^{i(t+r)s} f_j(s)\, ds = x_{t+r,j}. \tag{7.8}$$

The Riemann–Lebesgue lemma then implies that $T_t x \to 0$ for every $x \in X_0$.

Next we would like to define maps S_t on $Y_0 = X_0$ by demanding that $S_t x_{r,j} = x_{r-t,j}$, $r \in \mathbb{R}$, $j \in J$, and then extending linearly. However this will run into difficulties if the $x_{r,j}$ are not linearly independent. We therefore observe, as we have already done in a similar situation (see Exercise 3.1.1), that instead of verifying conditions (ii) and (iii) in Theorem 7.29 it suffices to produce, for any $y \in Y_0$, a family $(u_t)_{t\geq 0}$ in X such that $u_t \to 0$ and $T_t u_t \to y$ as $t \to \infty$. To this end, for any $y \in Y_0 = X_0$ we fix a representation $y = \sum_{k=1}^m a_k x_{r_k,j_k}$ and define

$$u_t = \sum_{k=1}^{m} a_k x_{r_k - t, j_k}.$$

By (7.8) we then have that $T_t u_t = y$ for all $t \geq 0$ and, again by the Riemann–Lebesgue lemma, that $u_t \to 0$ as $t \to \infty$.

Thus, $(T_t)_{t\geq 0}$ is a mixing semigroup.

Finally, by continuity of the functions f_j, also

$$Z_0 := \operatorname{span}\{f_j(s) \ ; \ s \in [a,b] \cap \mathbb{Q}, s \neq 0, j \in J\}$$

is dense in X. By (7.7), every vector $f_j(s)$, $s \neq 0$, is a fixed point of $T_{2\pi/s}$, and thus any point in Z_0 is a periodic point for $(T_t)_t$; note that any finitely many rational numbers are integer multiples of a common rational number. Hence the semigroup is also chaotic. □

Example 7.33. Let $(T_t)_{t\geq 0}$ be the translation semigroup on the complex Banach space $X = L^p_v(\mathbb{R}_+)$, $1 \leq p < \infty$, where $v(x) = e^{-x}$. The dual of X is given by the space $L^q_w(\mathbb{R}_+)$, where $1 < q \leq \infty$ satisfies $\frac{1}{p} + \frac{1}{q} = 1$, with $w = v^{-q/p}$ for $p > 1$ and $w = v^{-1}$ for $p = 1$ (see Example A.3(c)); note that $L^\infty_w(\mathbb{R}_+)$ is the space of measurable functions g on $\mathbb{R}_+$ such that gw is essentially bounded.

We define $f : \mathbb{D} \to X$ by $f(\lambda)(x) = e^{\lambda x}$, $\lambda \in \mathbb{D}$, $x \in \mathbb{R}_+$. Then f is well defined and weakly holomorphic on $\mathbb{D}$. Clearly $Af(\lambda) = \frac{d}{dx} f(\lambda) = \lambda f(\lambda)$. If a continuous linear functional on X^*, represented by $g \in L^q_w(\mathbb{R}_+)$, satisfies

$$\langle f(\lambda), g\rangle = \int_0^\infty g(x) e^{\lambda x}\, dx = 0$$

for all $\lambda \in \mathbb{D}$, then we obtain, by taking the nth derivative with respect to λ,

$$\int_0^\infty x^n g(x) e^{\lambda x}\, dx = 0, \quad n \in \mathbb{N}_0, \ \lambda \in \mathbb{D}.$$

Therefore, letting $\lambda = 0$ and taking linear combinations, we obtain that

$$\int_0^\infty g(x) p(x)\, dx = 0$$

for every polynomial p. Since the polynomials form a dense subspace of X, we conclude that $g = 0$.

Thus, Theorem 7.30 shows that the translation semigroup is mixing and chaotic, confirming our earlier findings in Examples 7.10 and 7.19.

We note that for C_0-semigroups on real Banach spaces one can still apply the above eigenvalue criteria. Indeed, if they are satisfied for the complexification of the semigroup then the original semigroup is also chaotic and mixing by Corollary 7.13.

7.5 Applications of C_0-semigroups to differential equations

The purpose of this section is to present some applications of C_0-semigroups to the study of the asymptotic behaviour of solutions of linear partial differential equations (PDEs) or of infinite linear systems of ordinary differential equations (ODEs). We do not pretend to obtain a full description of all possible situations, but we want to show the main techniques in certain representative cases.

A first-order PDE. We start with the following first-order abstract Cauchy problem on the space $X = L^1(\mathbb{R}_+)$:

$$\begin{cases} \dfrac{\partial u}{\partial t} = \dfrac{\partial u}{\partial x} + \dfrac{2x}{1+x^2}u, \\ u(0,x) = \varphi(x), \quad x \in \mathbb{R}_+. \end{cases} \tag{7.9}$$

The solution C_0-semigroup is given by

$$T_t\varphi(x) = \frac{1+(x+t)^2}{1+x^2}\varphi(x+t), \quad x,t \in \mathbb{R}_+.$$

That is, the action of the semigroup is a translation, together with multiplication by the function $h_t(x) = \frac{1+(x+t)^2}{1+x^2}$. The fact that $h_t(x) \to \infty$ as $t \to \infty$ opens up the possibility of approximating arbitrary functions by small functions under the action of T_t, and thus of obtaining the mixing property for the semigroup. On the other hand, the fact that $h_t(x) \to 1$ as $x \to \infty$ prevents eigenvectors of T_t to eigenvalues of modulus greater than 1. Therefore the eigenvalue criterion of Theorem 7.30 cannot be applied.

Proposition 7.34. *The solution semigroup $(T_t)_{t\geq 0}$ of (7.9) is mixing and chaotic on $L^1(\mathbb{R}_+)$.*

Proof. We first apply the mixing criterion, Theorem 7.29. We take for X_0 the dense subspace of $X = L^1(\mathbb{R}_+)$ of functions with compact support. If the support of $\varphi \in X_0$ is contained in $[0, b]$, then $T_t\varphi = 0$ for every $t \geq b$, and condition (i) of the criterion is satisfied. For conditions (ii) and (iii) we take $Y_0 = X$ and define the maps $S_t : X \to X$ by

$$S_t\varphi(x) = \frac{1+(x-t)^2}{1+x^2}\varphi(x-t), \quad \text{for } x \geq t \geq 0,$$

and $S_t(x) = 0$ for $0 \leq x < t$. Then $T_t S_t \varphi = \varphi$ for every $\varphi \in X$. Moreover,

$$\|S_t\varphi\| = \int_t^\infty \frac{1+(x-t)^2}{1+x^2}|\varphi(x-t)|\,dx = \int_0^\infty \frac{1+x^2}{1+(x+t)^2}|\varphi(x)|\,dx \to 0$$

as $t \to \infty$ by the dominated convergence theorem. Thus the semigroup is mixing.

As for chaos, let $\varphi \in X_0$ and $\varepsilon > 0$. Let the support of φ be contained in $[0, b]$, and fix $t \geq b$ such that $\sum_{n=1}^\infty \frac{1+b^2}{n^2t^2}\|\varphi\| < \varepsilon$. We define $\psi = \sum_{n=0}^\infty S_{nt}\varphi$. Since

$$\sum_{n=1}^\infty \|S_{nt}\varphi\| = \sum_{n=1}^\infty \int_0^b \frac{1+x^2}{1+(x+nt)^2}|\varphi(x)|\,dx \leq \sum_{n=1}^\infty \frac{1+b^2}{n^2t^2}\|\varphi\| < \varepsilon,$$

ψ is well defined and $\|\varphi - \psi\| < \varepsilon$. Moreover, ψ is periodic for T_t, so that the semigroup has a dense set of periodic points, implying that it is chaotic. □

A second-order PDE. In our first example we obtained mixing and chaos for a first-order PDE by analysing the corresponding solution C_0-semigroup. A similar treatment is possible for higher-order PDEs.

As a specific example we study the chaotic behaviour of the solutions of an abstract Cauchy problem that is given by the hyperbolic heat transfer equation in the absence of internal heat sources:

$$\begin{cases} \tau \dfrac{\partial^2 u}{\partial t^2} + \dfrac{\partial u}{\partial t} = \alpha \dfrac{\partial^2 u}{\partial x^2}, \\ u(0,x) = \varphi_1(x), \quad x \in \mathbb{R}, \\ \dfrac{\partial u}{\partial t}(0,x) = \varphi_2(x), \quad x \in \mathbb{R}, \end{cases} \tag{HHTE}$$

where φ_1 and φ_2 represent the initial temperature and the initial variation of temperature, respectively, $\alpha > 0$ is the thermal diffusivity, and $\tau > 0$ is the thermal relaxation time.

We will express this system as a first-order equation by representing it as a C_0-semigroup on the product of a certain function space with itself. To do this we set $u_1 = u$ and $u_2 = \frac{\partial u}{\partial t}$. Then the associated first-order equation is

$$\begin{cases} \dfrac{\partial}{\partial t}\begin{pmatrix} u_1 \\ u_2 \end{pmatrix} = \begin{pmatrix} 0 & I \\ \dfrac{\alpha}{\tau}\dfrac{\partial^2}{\partial x^2} & -\dfrac{1}{\tau}I \end{pmatrix}\begin{pmatrix} u_1 \\ u_2 \end{pmatrix}, \\ \begin{pmatrix} u_1(0,x) \\ u_2(0,x) \end{pmatrix} = \begin{pmatrix} \varphi_1(x) \\ \varphi_2(x) \end{pmatrix}, \quad x \in \mathbb{R}. \end{cases} \tag{7.10}$$

We fix $\rho > 0$ and consider the space

$$X_\rho = \Big\{ f : \mathbb{R} \to \mathbb{C}\ ;\ f(x) = \sum_{n=0}^{\infty} \frac{a_n \rho^n}{n!} x^n, (a_n)_{n\geq 0} \in c_0 \Big\}, \tag{7.11}$$

endowed with the norm $\|f\| = \sup_{n\geq 0}|a_n|$, where c_0 is the Banach space of complex sequences tending to 0. Then X_ρ is a Banach space of analytic functions with a certain growth control. By its definition it is isometrically isomorphic to c_0.

Since

$$A := \begin{pmatrix} 0 & I \\ \dfrac{\alpha}{\tau}\dfrac{\partial^2}{\partial x^2} & -\dfrac{1}{\tau}I \end{pmatrix}$$

is easily seen to be an operator on $X := X_\rho \oplus X_\rho$, we have that $(e^{tA})_{t\geq 0}$ is a C_0-semigroup on X, which is the solution semigroup of (7.10) on X.

Proposition 7.35. *Let $\rho > 0$ be such that $\alpha\tau\rho^2 > 2$. Then the solution semigroup $(e^{tA})_{t\geq 0}$ of (7.10) is mixing and chaotic on $X_\rho \oplus X_\rho$.*

Proof. Given $\lambda \in \mathbb{C}$, $z_0, z_1 \in \mathbb{R}$, we set $R_\lambda = (\tau\lambda^2 + \lambda)/\alpha$ and define

$$\varphi_{\lambda,z_0,z_1}(x) = z_0 \sum_{n=0}^{\infty} \frac{R_\lambda^n x^{2n}}{(2n)!} + z_1 \sum_{n=0}^{\infty} \frac{R_\lambda^n x^{2n+1}}{(2n+1)!}, \qquad x \in \mathbb{R}.$$

Let U be the open disk of radius $r = \sqrt{\alpha\rho^2/2\tau} > 0$ centred at zero. If $\lambda \in U$ then

$$\frac{|R_\lambda|}{\rho^2} < \frac{\tau\frac{\alpha\rho^2}{2\tau} + \sqrt{\frac{\alpha\rho^2}{2\tau}}}{\alpha\rho^2} = \frac{1}{2} + \frac{1}{2}\sqrt{\frac{2}{\alpha\tau\rho^2}} < 1, \tag{7.12}$$

so that $\varphi_{\lambda,z_0,z_1} \in X_\rho$. We now consider the functions $f_{z_0,z_1} : U \to X$, $z_0, z_1 \in \mathbb{R}$, given by

$$f_{z_0,z_1}(\lambda) = \begin{pmatrix} \varphi_{\lambda,z_0,z_1} \\ \lambda\varphi_{\lambda,z_0,z_1} \end{pmatrix}.$$

Then one finds that

$$f_{z_0,z_1}(\lambda) \in \ker(\lambda I - A), \quad \lambda \in U, z_0, z_1 \in \mathbb{R}.$$

According to Theorem 7.30 it remains to prove that, for any $x^* \in X^*_\rho \oplus X^*_\rho$, the functions $\lambda \to \langle f_{z_0,z_1}(\lambda), x^*\rangle$, $z_0, z_1 \in \mathbb{R}$, are holomorphic on U, and that if they all vanish on U then $x^* = 0$.

Thus, let $x^* \in X^*_\rho \oplus X^*_\rho$. By the isomorphism of X_ρ with c_0, x^* can be represented in a canonical way by a pair $((b_n)_{n\geq 0}, (c_n)_{n\geq 0}) \in \ell^1 \oplus \ell^1$. We then have that, for $\lambda \in U$,

$$\langle f_{z_0,z_1}(\lambda), x^*\rangle =$$
$$z_0 \sum_{n=0}^{\infty} R_\lambda^n \frac{b_{2n}}{\rho^{2n}} + z_1 \sum_{n=0}^{\infty} R_\lambda^n \frac{b_{2n+1}}{\rho^{2n+1}} + \lambda z_0 \sum_{n=0}^{\infty} R_\lambda^n \frac{c_{2n}}{\rho^{2n}} + \lambda z_1 \sum_{n=0}^{\infty} R_\lambda^n \frac{c_{2n+1}}{\rho^{2n+1}}.$$

By (7.12), these series converge uniformly on U, which implies that each function f_{z_0,z_1} is weakly holomorphic on U.

Finally, suppose that all the functions $\lambda \to \langle f_{z_0,z_1}(\lambda), x^*\rangle$, $z_0, z_1 \in \mathbb{R}$, vanish on U. Considering, in particular, the values $\lambda = 0$ and $\lambda = -1/\tau$ in U we obtain that $z_0 b_0 + z_1 b_1/\rho = 0$ and $z_0 c_0 + z_1 c_1/\rho = 0$ for all $z_0, z_1 \in \mathbb{R}$, hence $b_0 = b_1 = c_0 = c_1 = 0$. As a consequence, we may divide out R_λ and still retain holomorphic functions that vanish on U. Considering again $\lambda = 0$ and $\lambda = -1/\tau$ we obtain in the same way that $b_2 = b_3 = c_2 = c_3 = 0$. Proceeding inductively we deduce that $b_n = c_n = 0$ for all $n \geq 0$, and hence that $x^* = 0$. This had to be shown. □

An infinite system of ODEs. We conclude this section by providing an application of C_0-semigroups to an infinite linear system of ordinary differential equations.

Let us consider the following infinite system of ODEs associated with a linear kinetic model:

$$\begin{cases} \dfrac{df_n}{dt} = -\alpha_n f_n + \beta_n f_{n+1}, & n \geq 1, \\ f_n(0) = a_n, & n \geq 1, \end{cases} \tag{7.13}$$

where $(\alpha_n)_n, (\beta_n)_n$ are bounded positive sequences and $(a_n)_n \in \ell^1$ is a real sequence.

In this case we consider the real Banach space $X = \ell^1$ and the map A given by

$$Af = (-\alpha_n f_n + \beta_n f_{n+1})_n \quad \text{for } f = (f_n)_n.$$

Since A is an operator on ℓ^1, it generates a C_0-semigroup $(T_t)_{t\geq 0}$, which is then the solution semigroup of the abstract Cuachy problem (7.13).

Proposition 7.36. *Let $\alpha_n > 0$ and $\beta_n \in \mathbb{R}$, $n \in \mathbb{N}$, be such that*

$$\sup_{n\geq 1} \alpha_n < \liminf_{n\to\infty} \beta_n. \tag{7.14}$$

Then the solution semigroup $(T_t)_{t\geq 0}$ of (7.13) is mixing and chaotic on ℓ^1.

Proof. Let $\alpha = \sup_{n\geq 1} \alpha_n$, $\beta = \liminf_{n\to\infty} \beta_n$ and $\alpha/2 < r < \beta/2$. We fix $U \subset \mathbb{C}$ as the open disk of radius r centred at $-\alpha/2$, which intersects the imaginary axis $i\mathbb{R}$. We can calculate the eigenvectors of A to the eigenvalues $\lambda \in U$. Indeed, suppose that $f = (f_n)_n$ satisfies $Af = \lambda f$, $\lambda \in U$. Then

$$\lambda f_n = -\alpha_n f_n + \beta_n f_{n+1}, \quad n \geq 1,$$

and therefore

$$f_n = \gamma_n f_1 \;\text{ with }\; \gamma_n = \prod_{k=1}^{n-1} \frac{\lambda + \alpha_k}{\beta_k}, \; n \geq 1,$$

where $\gamma_1 = 1$. Conversely, $f(\lambda) := (\gamma_n)_n$ satisfies $Af(\lambda) = \lambda f(\lambda)$; we still need to verify that $f(\lambda) \in \ell^1$.

To this end let $\delta \in]2r, \beta[$. Then there is some $n_0 \in \mathbb{N}$ such that $\beta_n > \delta$ for all $n \geq n_0$. Thus, since our assumptions imply that $-\alpha_n \in U$ for all $n \in \mathbb{N}$, we have that

$$\frac{|\lambda + \alpha_n|}{\beta_n} \leq \frac{2r}{\delta} < 1 \tag{7.15}$$

for all $n \geq n_0$, which implies that $f(\lambda) \in X = \ell^1$. We can therefore consider the function

$$f : U \to X, \quad \lambda \to f(\lambda).$$

Let $x^* \in X^* = \ell^\infty$ be given by some bounded sequence $(b_n)_n$. Then

$$\langle f(\lambda), x^* \rangle = \sum_{n=1}^{\infty} \Big(\prod_{k=1}^{n-1} \frac{\lambda + \alpha_k}{\beta_k} \Big) b_n$$

converges uniformly for $\lambda \in U$ by (7.15), which implies that f is weakly holomorphic on U.

Finally, let $x^* \in X^*$, given by $(b_n)_n \in \ell^\infty$, be such that $\langle f(\lambda), x^* \rangle = 0$ for all $\lambda \in U$. Considering, in particular, $\lambda = -\alpha_1 \in U$ we obtain that $b_1 = 0$. As a consequence we may divide out $\lambda + \alpha_1$ and still retain a holomorphic function on U. Then taking $\lambda = -\alpha_2 \in U$ we obtain that $b_2 = 0$. Proceeding inductively we deduce that $b_n = 0$ for all $n \geq 1$, and therefore $x^* = 0$.

We have shown that all the hypotheses of Theorem 7.30 are satisfied, which implies the result. □

Exercises

Exercise 7.1.1. Let $\Delta = \Delta(\alpha) := \{re^{i\theta}\ ;\ r \geq 0,\ |\theta| \leq \alpha\}$, $0 < \alpha \leq \pi/2$, be a sector in the complex plane, and $v : \Delta \to \mathbb{R}$ a strictly positive measurable function for which there exist constants $M \geq 1$ and $w \in \mathbb{R}$ such that $v(z_1) \leq Me^{w|z_2|}v(z_1 + z_2)$ for all $z_1, z_2 \in \Delta$, called an *admissible weight function* on Δ. For $1 \leq p < \infty$ we define the space

$$L^p_v(\Delta) = \{f : \Delta \to \mathbb{K}\ ;\ f \text{ is measurable and } \|f\| < \infty\}$$

with $\|f\| = (\int_\Delta |f(z)|^p v(z)\, d\lambda(z))^{1/p}$, where λ denotes two-dimensional Lebesgue measure. Show that, for every $z \in \Delta$, the translation semigroup $(T_{tz})_{t\geq 0}$ given by $T_{tz}f(\zeta) = f(\zeta + tz)$ is a C_0-semigroup on $L^p_v(\Delta)$.

Exercise 7.1.2. Let $v : \mathbb{R}_+ \to \mathbb{R}$ be a strictly positive continuous function, and define the space $X = C_{0,v}(\mathbb{R}_+)$ by

$$C_{0,v}(\mathbb{R}_+) = \{f : \mathbb{R}_+ \to \mathbb{K}\ ;\ f \text{ is continuous and } \lim_{x\to\infty} f(x)v(x) = 0\},$$

endowed with the norm $\|f\| = \sup_{x\geq 0} |f(x)|v(x)$. Consider the translation semigroup given by $T_t f(x) = f(x + t)$, $t, x \geq 0$. Show that this defines a C_0-semigroup on X if and only if v is admissible in the sense of (7.3).

Exercise 7.1.3. For the operator A on $\mathbb{R}^2$ given by the matrix $\left(\begin{smallmatrix} 0 & -1 \\ 1 & 0 \end{smallmatrix}\right)$ determine the semigroup $(e^{tA})_{t\geq 0}$, and verify that A is its infinitesimal generator. Describe the orbits $\{e^{tA}x\ ;\ t \geq 0\}$, $x \in \mathbb{R}^2$. Do the same for $\left(\begin{smallmatrix} 1 & 1 \\ -1 & -1 \end{smallmatrix}\right)$.

Exercise 7.1.4. Let φ be a bounded holomorphic function on the unit disk $\mathbb{D}$. Then $T_t f = e^{t\varphi} f$, $t \geq 0$, defines multiplication operators on the Hardy space H^2; see Section 4.4. Show that $(T_t)_{t\geq 0}$ is a uniformly continuous semigroup on H^2, and identify its generator. Do the same for the family $(T_t^*)_{t\geq 0}$ of Hilbert space adjoints.

Exercise 7.1.5. Let T be an operator on a Banach space X such that $\|T\| < 1$. Then the series $R := \sum_{n=1}^\infty \frac{(-1)^{n+1}}{n} T^n$ converges in the operator norm, so that $R \in L(X)$. Prove that $e^R = I + T$.

Exercise 7.1.6. Let A be an operator on a Banach space X. Show that the operators $T_t = e^{tA}$, $t \geq 0$, define a C_0-semigroup on X. Moreover, show that $\lim_{s\to t} \|T_s - T_t\| = 0$ and $Ax = \lim_{t\to 0} \frac{1}{t}(T_t x - x)$ for all $x \in X$.

Exercise 7.1.7. Let $(T_t)_{t\geq 0}$ be a C_0-semigroup on a Banach space X and $(A, D(A))$ its infinitesimal generator. Show the following for any $t \geq 0$:
(i) for any $x \in D(A)$, $T_t x \in D(A)$ and $AT_t x = T_t Ax$;
(ii) for any $x \in D(A)$, $\lim_{s\to t} \frac{T_s x - T_t x}{s - t} = AT_t x$;
(iii) for any $x \in X$, $\int_0^t T_s x\, ds \in D(A)$ and $A \int_0^t T_s x\, ds = T_t x - x$;
(iv) for any $x \in D(A)$, $\int_0^t T_s Ax\, ds = T_t x - x$,
where the integral is the vector-valued Riemann integral; see Appendix B. Deduce that $D(A)$ is dense and that A has closed graph. (*Hint* for (ii): $\int_h^{t+h} T_s x\, ds - \int_0^t T_s x\, ds = \int_t^{t+h} T_s x\, ds - \int_0^h T_s x\, ds$.)

Exercise 7.1.8. Prove Proposition 7.5. (*Hint*: For (i) $\Longrightarrow$ (ii), use Lemma 8.20 to show that $\frac{1}{t}\int_0^t T_s\, ds$ is an invertible operator if $t > 0$ is sufficiently small, and use Exercise 7.1.7(ii). For (ii) $\Longrightarrow$ (iii), use the closed graph theorem and the fact that the infinitesimal generator determines the C_0-semigroup.)

Exercise 7.2.1. Let $(T_t)_{t\geq 0}$ be a C_0-semigroup on a separable Banach space X.

(a) If $x \in X$ is hypercyclic for $(T_t)_{t\geq 0}$ and U a nonempty open subset of X then, for any $t_0 \geq 0$, there is some $t \geq t_0$ such that $T_t x \in U$. (*Hint*: In an infinite-dimensional space, the unit ball is not relatively compact.)

(b) Prove the *Birkhoff transitivity theorem* for C_0-semigroups, that is: $(T_t)_{t\geq 0}$ is hypercyclic if and only if it is topologically transitive. In that case, the set of hypercyclic vectors for $(T_t)_{t\geq 0}$ is a dense G_δ-set.

Exercise 7.2.2. A one-parameter family $(T_t)_{t\in\mathbb{R}}$ of operators on a Banach space X is called a C_0*-group* of operators if $T_0 = I$, $T_{t+s} = T_tT_s$ for all $s, t \in \mathbb{R}$, and $\lim_{s\to t} T_sx = T_tx$ for all $x \in X$ and $t \in \mathbb{R}$. The *orbit* of a vector $x \in X$ is the set $\{T_tx \; ; \; t \in \mathbb{R}\}$ and the C_0-group is called *hypercyclic* if it admits a dense orbit. Prove that if a C_0-group $(T_t)_{t\in\mathbb{R}}$ of operators is hypercyclic, then so are the C_0-semigroups $(T_t)_{t\geq 0}$ and $(T_{-t})_{t\geq 0}$. (*Hint*: Proceed as in Exercise 6.3.1(ii).)

Exercise 7.2.3. Let $(T_t)_{t\geq 0}$ be the translation semigroup on $X = L^p_v(\mathbb{R}_+)$ for an admissible weight v. Establish the equivalence of the following assertions:
(i) $(T_t)_{t\geq 0}$ is chaotic;
(ii) $\sum_{n=1}^\infty v(n) < \infty$;
(iii) for all $\varepsilon > 0$ and $t > 0$, there is some $s > 0$ such that $\sum_{n=1}^\infty v(t + ns) < \varepsilon$.

Exercise 7.2.4. Let $(T_t)_{t\geq 0}$ be the translation semigroup on $X = C_{0,v}(\mathbb{R}_+)$ for an admissible continuous weight function v; see Exercise 7.1.2. Show that $(T_t)_{t\geq 0}$ is hypercyclic if and only if $\liminf_{x\to\infty} v(x)=0$. And that it is mixing, if and only if it is chaotic, and if and only if $\lim_{x\to\infty} v(x)=0$. (*Hint*: Use the fact that the generator A is given by $Af = f'$ on the subspace of continuously differentiable functions $f \in X$ with $f' \in X$.)

Exercise 7.2.5. Let $(T_t)_{t\geq 0}$ be the translation semigroup on the space $X = L^p_v(\Delta)$; see Exercise 7.1.1. Let $\Delta_m = \{z \in \Delta \; ; \; \mathrm{Im}(z) \in [-m, m]\}$, $m \in \mathbb{N}$. Demonstrate that the following are equivalent:
(i) the translation semigroup is chaotic on X;
(ii) $\int_{\Delta_m} v(z)\, d\lambda(z) < \infty$ for every $m \in \mathbb{N}$.

Exercise 7.3.1. Let $\Delta = \Delta(\pi/4)$, $v : \Delta \to \mathbb{R}$ the admissible weight function defined by $v(x+iy) = (x+y+1)^2(x-y+1)^{-2}$ if $x+y+1 \geq \sqrt{x-y+1}$, and $v(x+iy) := (x+y+1)^{-2}$ otherwise; see Figure 7.1. Let T_z, $z \in \Delta$, be the translation operators on the space $X = L^p_v(\Delta)$; see Exercise 7.1.1. The dynamical notions of C_0-semigroups introduced so far naturally extend to the semigroup $(T_z)_{z\in\Delta}$.

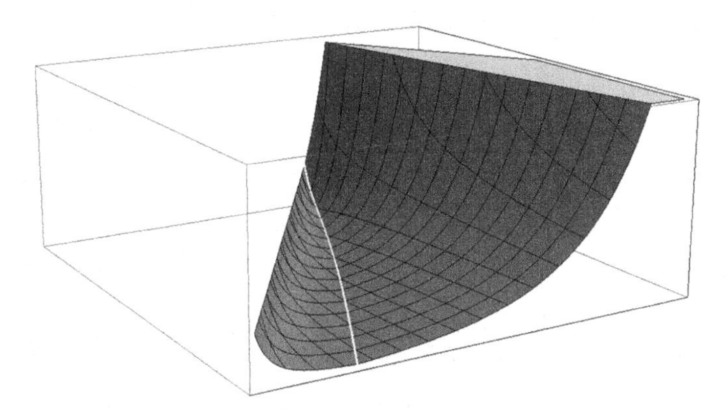

Fig. 7.1 The weight function on $\Delta(\pi/4)$

Show that there is a mixing nonautonomous discretization $(T_{z_n})_n$ of $(T_z)_{z\in\Delta}$, but that no C_0-subsemigroup $(T_{tz})_{t\geq 0}$, $z \in \Delta$, is hypercyclic. In particular, no autonomous discretization of $(T_z)_{z\in\Delta}$ is hypercyclic. (*Hint*: Take a discretization within the curve $x+y+1=\sqrt{x-y+1}$. On the other hand, observe that the weight v is asymptotically bounded below by 1 on any ray in Δ passing through 0.)

Exercise 7.3.2. Given a C_0-semigroup $(T_t)_{t\geq 0}$ and $z_1, z_2 \in \mathbb{K}$ with $\operatorname{Re} z_1 \leq \operatorname{Re} z_2$, we consider the rescaled semigroups $(e^{z_1 t}T_t)_{t\geq 0}$ and $(e^{z_2 t}T_t)_{t\geq 0}$. Show that, if the direct sum semigroup $(e^{z_1 t}T_t \oplus e^{z_2 t}T_t)_{t\geq 0}$ is hypercyclic, then the rescaled semigroup $(e^{zt}T_t)_{t\geq 0}$ is weakly mixing for every $z \in \mathbb{K}$ such that $\operatorname{Re} z_1 \leq \operatorname{Re} z \leq \operatorname{Re} z_2$. (*Hint*: Use the Oxtoby–Ulam theorem and Exercise 2.5.9.)

Exercise 7.3.3. Let $(T_t)_{t\geq 0}$ be a C_0-semigroup on X. Given a pair $U, V \subset X$ of open sets, we define

$$R(U,V) = \{t \geq 0 \; ; \; T_t(U) \cap V \neq \varnothing\}.$$

Prove that a C_0-semigroup is weakly mixing if and only if, for any 0-neighbourhood W and for any pair $U, V \subset X$ of nonempty open sets, we have that $R(U,W) \cap R(W,V) \neq \varnothing$. (*Hint*: One implication is trivial; for the other one use the Oxtoby–Ulam theorem and Theorem 2.45.)

Exercise 7.3.4. Prove that a C_0-semigroup $(T_t)_{t\geq 0}$ is weakly mixing if and only if, for every pair $U, V \subset X$ of nonempty open sets, $R(U,V)$ contains arbitrarily long intervals. (*Hint*: For one implication take any autonomous discretization and apply Theorem 1.54; the other implication is based on the local equicontinuity of the semigroup and, again, on Theorem 1.54.)

Exercise 7.3.5. Prove that a C_0-semigroup $(T_t)_{t\geq 0}$ is weakly mixing if and only if there exists some $\varepsilon > 0$ such that, for any pair $U, V \subset X$ of nonempty open sets, $R(U,V)$ contains some interval of length ε. (*Hint*: Exercise 7.3.4, and Exercise 1.5.8 for T_{t_0} with $t_0 \in \,]0, \varepsilon/2[$.)

Exercise 7.3.6. A C_0-semigroup $(T_t)_{t\geq 0}$ of operators is *topologically ergodic* if, for every pair $U, V \subset X$ of nonempty open sets, $R(U,V)$ is *syndetic*, that is, $\mathbb{R}_+ \setminus R(U,V)$ does not contain arbitrarily long intervals. Show that every autonomous discretization of a topologically ergodic C_0-semigroup is topologically ergodic. (*Hint*: Given $t > 0$, nonempty open sets $U, V \subset X$, and an arbitrary 0-neighbourhood W, use local equicontinuity to prove that $N(U,W) = \{n \in \mathbb{N}_0 \; ; \; nt \in R(U,W)\}$ and $N(W,V) = \{n \in \mathbb{N}_0 \; ; \; nt \in R(W,V)\}$ are syndetic. Then show that $N(U,W) \cap N(W,V)$ is (nonempty and) syndetic.)

Exercise 7.3.7. Prove that, if $(T_t)_{t\geq 0}$ and $(S_t)_{t\geq 0}$ are topologically ergodic C_0-semigroups, then $(T_t \oplus S_t)_{t\geq 0}$ is also topologically ergodic. (*Hint*: Exercises 7.3.6 and 2.5.5.)

Exercise 7.3.8. Show the following:

(i) every chaotic C_0-semigroup is topologically ergodic;
(ii) if $(T_t)_{t\geq 0}$ is a topologically ergodic C_0-semigroup and $(S_t)_{t\geq 0}$ is weakly mixing, then $(T_t \oplus S_t)_{t\geq 0}$ is weakly mixing.

(*Hint* to (ii): Proposition 7.25 and Exercises 7.3.6, 2.5.5 and 1.5.6.)

Exercise 7.4.1. Show directly that under the hypotheses of Theorem 7.27 the semigroup is topologically transitive. Prove Theorem 7.29 in the same way.

Exercise 7.4.2. By using a direct argument, prove that if a C_0-semigroup $(T_t)_{t\geq 0}$ satisfies the Hypercyclicity Criterion then the operator T_1 satisfies the Hypercyclicity Criterion. (*Hint*: See the proof of Proposition 7.25.)

Exercise 7.4.3. Let $(T_t)_{t\geq 0}$ be a C_0-semigroup. Show that the following assertions are equivalent:

(i) $(T_t)_{t\geq 0}$ satisfies the criterion for mixing of Theorem 7.29;
(ii) every discretization of $(T_t)_{t\geq 0}$ satisfies the Hypercyclicity Criterion for sequences of operators for the full sequence (n);
(iii) some autonomous discretization of $(T_t)_{t\geq 0}$ satisfies the Hypercyclicity Criterion for sequences of operators for the full sequence (n).

Exercise 7.4.4. Suppose that a C_0-semigroup $(T_t)_{t\geq 0}$ admits a discretization $(T_{t_n})_{n\in\mathbb{N}}$ with $\sup_{n\geq 1}(t_{n+1}-t_n)<\infty$ satisfying the Hypercyclicity Criterion for sequences of operators for the full sequence (n). Prove that $(T_t)_{t\geq 0}$ satisfies the criterion for mixing of Theorem 7.29.

Exercise 7.4.5. Show the following generalization of Theorem 7.32. Let X be a complex separable Banach space, and $(T_t)_{t\geq 0}$ a C_0-semigroup on X with generator $(A, D(A))$. Assume that there are compact intervals $I_j=[a_j,b_j]$ with $a_j<b_j$ and continuous functions $f_j: I_j\to X$, $j\in J$, such that
(i) $f_j(s)\in\ker(isI-A)$ for every $s\in I_j$, $j\in J$,
(ii) $\operatorname{span}\{f_j(s)\ ;\ s\in I_j, j\in J\}$ is dense in X.
Then the semigroup $(T_t)_{t\geq 0}$ is mixing and chaotic.

Exercise 7.5.1. Provide an alternative proof of Proposition 7.34 for complex functions by applying Exercise 7.4.5 to $f_j:[-j,j]\to X$, $f_j(s)(x)=\frac{\exp(isx)}{1+x^2}$, $j\in\mathbb{N}$. (*Hint*: For condition (ii), use the Hahn–Banach theorem and the fact that if the Fourier transform of an integrable function vanishes identically then so does the function itself.)

Exercise 7.5.2. Consider the following abstract Cauchy problem on $X=L^p(\mathbb{R}_+)$, $1\leq p<\infty$, or on $X=C_0(\mathbb{R}_+)$, that generalizes (7.9):

$$\begin{cases} \dfrac{\partial u}{\partial t}=\dfrac{\partial u}{\partial x}+h(x)u, \\ u(0,x)=\varphi(x), \quad x\in\mathbb{R}_+, \end{cases}$$

where $h:\mathbb{R}_+\to\mathbb{R}$ is a bounded continuous function. The solution semigroup is given by

$$T_t\varphi(x)=e^{\int_x^{x+t}h(s)\,ds}\varphi(x+t), \quad x,t\in\mathbb{R}_+.$$

Show that, on $L^p(\mathbb{R}_+)$, the semigroup is chaotic if and only if $\int_0^\infty \exp\left(-p\int_0^x h(s)\,ds\right)dx<\infty$, mixing if and only if $\lim_{x\to\infty}\int_0^x h(s)\,ds=\infty$, and hypercyclic if and only if $\sup_{x\geq 0}\int_0^x h(s)\,ds=\infty$. Find the corresponding characterizations on $C_0(\mathbb{R}_+)$. (*Hint*: Determine an admissible weight function v so that the semigroup is conjugate to the translation semigroup on $L^p_v(\mathbb{R}_+)$ or on $C_{0,v}(\mathbb{R}_+)$; see Exercise 7.2.4.)

Exercise 7.5.3. Consider the following wave equation:

$$\begin{cases} \dfrac{\partial^2 u}{\partial t^2}=\alpha\dfrac{\partial^2 u}{\partial x^2}, \\ u(0,x)=\varphi_1(x), \quad x\in\mathbb{R}, \\ \dfrac{\partial u}{\partial t}(0,x)=\varphi_2(x), \quad x\in\mathbb{R}, \end{cases}$$

where $\alpha>0$ is the square of the speed of wave propagation. As for the hyperbolic heat transfer equation (HHTE), by setting $u_1=u$ and $u_2=\frac{\partial u}{\partial t}$, one can reduce it to a first-order equation. Show that the solution semigroup to the corresponding first-order equation is chaotic on $X_\rho\oplus X_\rho$, for any $\rho>0$, where X_ρ is the space of analytic functions defined in (7.11).

Exercise 7.5.4. The classical heat equation

$$\begin{cases} \dfrac{\partial u}{\partial t} = \dfrac{\partial^2 u}{\partial x^2}, \\ u(0,x) = \varphi(x), \quad x \in \mathbb{R}, \end{cases}$$

on the real line admits a solution semigroup given by

$$T_t\varphi(x) = \frac{1}{2\sqrt{\pi t}} \int_{-\infty}^{\infty} \exp\left(-\frac{(x-s)^2}{4t}\right)\varphi(s)\,ds, \quad t \in \mathbb{R}_+,\ x \in \mathbb{R}.$$

If we consider the space X_ρ, $\rho > 0$, defined in (7.11), then $D^2 = \frac{\partial^2}{\partial x^2}$ is an operator on X_ρ and $T_t = e^{tD^2}$. Show that T_t is chaotic on X_ρ for any $t > 0$.

Exercise 7.5.5. For the abstract Cauchy problem associated with the linear kinetic model (7.13), let $\alpha_1 = 3$, $\alpha_n = 1$, $n \geq 2$, and $\beta_n = 2$, $n \in \mathbb{N}$. Prove that the solution semigroup is not hypercyclic on ℓ^1. This shows that condition (7.14) cannot be replaced by $\limsup_{n\to\infty} \alpha_n < \liminf_{n\to\infty} \beta_n$.

Sources and comments

Section 7.1. For the theory of C_0-semigroups we refer to the books by Engel and Nagel [143], [144]. All the results mentioned in this section can be found there.

Section 7.2. Rolewicz [268] was the first to observe the existence of a dense orbit under a C_0-semigroup. A systematic study of the dynamical properties of semigroups, however, was only started by Desch, Schappacher and Webb [131]. In particular, they introduced the notions of hypercyclicity and chaos for semigroups. Several other results in this chapter are due to them, for example the characterization of hypercyclic translation semigroups on $L^p_v(\mathbb{R}_+)$. The characterization of the corresponding mixing property was added by Bermúdez, Bonilla, Conejero, and Peris [49].

Lemma 7.14 on the necessity of empty point spectrum for the adjoint operators and for the adjoint of the generator of a hypercyclic semigroup was obtained by Costakis and Peris [121] and by Desch, Schappacher and Webb [131], respectively. Theorem 7.16 is due to Costakis and Peris [121].

Chaos for the translation semigroup was characterized by deLaubenfels and Emamirad [128]. Kalmes [209], [210] has undertaken a systematic study of hypercyclicity, weak mixing, mixing and chaos of C_0-semigroups that are induced by solution semiflows of differential equations on open subsets of $\mathbb{R}^d$.

Section 7.3. Theorem 7.22 is due to Oxtoby and Ulam [251]. As for the discretization results, Proposition 7.20 is due to Conejero and Peris [113], Proposition 7.21 to Bermúdez, Bonilla, Conejero, and Peris [49] and Conejero and Peris [113], and Proposition 7.25 to Desch and Schappacher [129]; see also Conejero and Peris [113]. Theorem 7.23 can be found in Bernal and Grosse-Erdmann [63]. As we already mentioned in Chapter 6, the discretization results fail for chaotic semigroups; in fact, Bayart and Bermúdez [37] have constructed a chaotic semigroup on a Hilbert space for which no autonomous discretization is chaotic.

Section 7.4. The first criteria for hypercyclicity of C_0-semigroups were found by Desch, Schappacher and Webb [131]. In the form given here, the Hypercyclicity Criterion, Theorem 7.27, is due to Conejero and Peris [111] and El Mourchid [139], while the criterion for mixing, Theorem 7.29, is due to Bermúdez, Bonilla, Conejero, and Peris [49]. See also Conejero and Peris [113].

As for the eigenvalue criteria, Theorem 7.30 is due to Desch, Schappacher and Webb [131], while Theorem 7.32 is due to El Mourchid [140]; see also Conejero and Mangino [109]. We mention that the latter result was motivated by a similar criterion for frequent hypercyclicity; see Section 9.3.

Theorem 7.28, together with further equivalent forms of the Hypercyclicity Criterion, can be found in Conejero and Peris [113]. Incidentally, Herrero's problem if hypercyclicity implies weak mixing (see Section 2.5), also has a negative answer for semigroups, as was proved by Shkarin [292].

Section 7.5. The three applications treated in this section are, in order, from El Mourchid [140], Conejero, Peris and Trujillo [114], and Banasiak and Lachowicz [24].

The first results on the hypercyclic behaviour of the heat equation were given by Herzog [198]. Generalizations of Herzog's results were accomplished by Betancor and Bonilla [73]. Some recent developments on the heat dynamics are due to Ji and Weber [207], [208].

The method presented here for the hyperbolic heat transfer equation is based on the conversion of the second-order Cauchy problem to first order so that the theory of C_0-semigroups can be applied. Alternatively, it is possible to keep the second-order problem and solve it directly, by applying the theory of hypercyclic cosine operator functions; this theory was initiated by Bonilla and Miana [89], and thoroughly developed by Kalmes [211].

The chaotic dynamics associated with infinite systems of differential equations modelling birth-and-death processes (evolution of cell population, etc.) have been systematically studied by Banasiak et al. in [24], [25], [22], [23], [26], [27].

We also mention an interesting recent application of the methods in this chapter to the Black–Scholes formula. Emamirad, Goldstein and Goldstein [141] have shown that the Black–Scholes semigroup is chaotic for certain choices of the parameters.

Exercises. The weight condition (iii) in Exercise 7.2.3 is due to Matsui, Yamada and Takeo [236]. For Exercise 7.2.5 we refer to Conejero and Peris [112]. The results of Exercise 7.2.4 are due to Desch, Schappacher and Webb [131] (hypercyclicity), Bermúdez, Bonilla, Conejero and Peris [49] (mixing) and Matsui, Yamada and Takeo [236] (chaos). Exercise 7.3.1 can be found in Conejero and Peris [113]. Exercises 7.3.3 and 7.3.4 are taken from Bernal and Grosse-Erdmann [63]. Topologically ergodic semigroups were considered by Desch and Schappacher in [129] under the name of semigroups satisfying the Recurrent Hypercyclicity Criterion. Exercises 7.3.6, 7.3.7 and 7.3.8 are extracted from their paper. Exercise 7.4.3 is taken from Conejero and Peris [113], Exercise 7.4.4 from Bermúdez, Bonilla, Conejero and Peris [49], and Exercise 7.4.5 from Conejero and Mangino [109]. Exercise 7.5.1 is extracted from El Mourchid [140], Exercise 7.5.2 is taken from, and improves on, Takeo [297], while Exercises 7.5.3 and 7.5.4 are taken from Conejero, Peris and Trujillo [114] and Herzog [198], respectively.

Chapter 8
Existence of hypercyclic operators

In the previous chapters we saw that there exist hypercyclic operators on a large variety of spaces. This suggests the question of whether every Fréchet space supports a hypercyclic operator. We will show here that this is indeed the case when one takes account of two natural restrictions: the space has to be separable and, by a result in Chapter 2, also infinite dimensional. In contrast, we will see that not every such Banach space supports a chaotic operator. We also study the corresponding questions for semigroups of operators.

We then ask how large the set of hypercyclic operators is. The answer depends on the underlying notion of size. We will show that if there exist hypercyclic operators on a given space then, under a suitable topology, they form a dense set.

In the final section, the flexibility of hypercyclic operators will be demonstrated in yet another way. We know that the orbit of a hypercyclic vector is a (dense) linearly independent set. Conversely, it will be shown that every dense linearly independent sequence in a Banach space is the orbit of a (necessarily) hypercyclic vector under a certain operator.

8.1 Mixing perturbations of the identity

In this section we prepare the ground for the proofs of the existence of hypercyclic operators and C_0-semigroups on arbitrary infinite-dimensional separable Fréchet and Banach spaces, respectively. What we will need are hypercyclic operators on ℓ^1 that are "small" perturbations of the identity or, equivalently, perturbations of the identity on weighted ℓ^1-spaces with rapidly growing weights. Such operators are also interesting in their own right.

Let $v = (v_n)_n$ be a positive weight sequence, that is, a sequence of strictly positive numbers, and let X be one of the weighted spaces

K.-G. Grosse-Erdmann, A. Peris Manguillot, *Linear Chaos*, Universitext,
DOI 10.1007/978-1-4471-2170-1_8,

$$\ell^p(v) = \Big\{(x_n)_{n\geq 1}\ ;\ \sum_{n=1}^{\infty} |x_n|^p v_n < \infty\Big\}, \quad 1 \leq p < \infty,$$
$$c_0(v) = \big\{(x_n)_{n\geq 1}\ ;\ \lim_{n\to\infty} |x_n| v_n = 0\big\}.$$

By Example 4.4(a), the backward shift B defines an operator on X if and only if $\sup_{n\in\mathbb{N}} \frac{v_n}{v_{n+1}} < \infty$. Under this assumption,

$$e^B = \sum_{k=0}^{\infty} \frac{1}{k!} B^k$$

is a well-defined operator on X; see the discussion before Proposition 4.41 or Appendix B. Moreover, its powers are given by

$$(e^B)^n = e^{nB} = \sum_{k=0}^{\infty} \frac{n^k}{k!} B^k, \quad n \geq 1.$$

Theorem 8.1. *Let X be one of the spaces $\ell^p(v)$, $1 \leq p < \infty$, or $c_0(v)$. If $\sup_{n\in\mathbb{N}} \frac{v_n}{v_{n+1}} < \infty$, then the operator $T := e^B$ is mixing on X.*

Proof. Since $\ell^1(v)$ is densely and continuously contained in $c_0(v)$ and in $\ell^p((v_n^p)_n)$, the hypercyclic comparison principle (see Exercise 2.2.6) tells us that it suffices to show the result for $\ell^1(v)$.

Let X_0 be the subspace of finite sequences, which is dense in $\ell^1(v)$. We fix arbitrary vectors $x = (x_k)_k$, $y = (y_k)_k \in X_0$ and $\varepsilon > 0$, and we choose $m \in \mathbb{N}$ such that $x_k = y_k = 0$ for $k > m$. Then the matrix

$$W = \begin{pmatrix} \frac{1}{m!} & \frac{1}{(m+1)!} & \cdots & \frac{1}{(2m-1)!} \\ \frac{1}{(m-1)!} & \frac{1}{m!} & \cdots & \frac{1}{(2m-2)!} \\ \vdots & \vdots & \ddots & \vdots \\ \frac{1}{1!} & \frac{1}{2!} & \cdots & \frac{1}{m!} \end{pmatrix}$$

is easily seen to be invertible; see Exercise 8.1.1. We let

$$C = \|W^{-1}\| \Big(\sum_{k=1}^{m} |y_k| + m \sum_{k=1}^{m} |x_k| \Big), \tag{8.1}$$

where $\|W^{-1}\|$ denotes the norm of W^{-1} as an operator from $(\mathbb{K}^m, \|\cdot\|_1)$ to $(\mathbb{K}^m, \|\cdot\|_\infty)$. Finally, let $N \in \mathbb{N}$ be such that

$$\sum_{k=m+1}^{2m} N^{m-k} v_k < \frac{\varepsilon}{eC}. \tag{8.2}$$

We will show that, for any $n \geq N$, there is some $z \in \ell^1(v)$ such that

$$\|x - z\| < \varepsilon \quad \text{and} \quad \|y - T^n z\| < \varepsilon,$$

which will imply that T is mixing.

Indeed, fix $n \geq N$, and let z be a finite sequence such that $z_k = x_k$ for $k = 1, \ldots, m$ and $z_k = 0$ for $k \geq 2m+1$. We claim that the missing m entries of z can be chosen in such a way that the first m entries of $T^n z = \sum_{j=0}^{\infty} \frac{n^j}{j!} B^j z$ coincide with those of y, that is

$$y_k = \sum_{j=0}^{\infty} \frac{n^j}{j!} z_{k+j} = \sum_{j=k}^{2m} \frac{n^{j-k}}{(j-k)!} z_j, \quad k = 1, \ldots, m.$$

This is equivalent to solving the system

$$A \begin{pmatrix} x_1 \\ \vdots \\ x_m \end{pmatrix} + V \begin{pmatrix} z_{m+1} \\ \vdots \\ z_{2m} \end{pmatrix} = \begin{pmatrix} y_1 \\ \vdots \\ y_m \end{pmatrix}, \tag{8.3}$$

where

$$A = \begin{pmatrix} 1 & \frac{n}{1!} & \cdots & \frac{n^{m-1}}{(m-1)!} \\ 0 & 1 & \cdots & \frac{n^{m-2}}{(m-2)!} \\ \vdots & \vdots & \ddots & \vdots \\ 0 & 0 & \cdots & 1 \end{pmatrix}, \quad V = \begin{pmatrix} \frac{n^m}{m!} & \frac{n^{m+1}}{(m+1)!} & \cdots & \frac{n^{2m-1}}{(2m-1)!} \\ \frac{n^{m-1}}{(m-1)!} & \frac{n^m}{m!} & \cdots & \frac{n^{2m-2}}{(2m-2)!} \\ \vdots & \vdots & \ddots & \vdots \\ \frac{n}{1!} & \frac{n^2}{2!} & \cdots & \frac{n^m}{m!} \end{pmatrix}.$$

Observe that $V = D_1 W D_2$, where

$$D_1 = \begin{pmatrix} n^{m-1} & 0 & \cdots & 0 \\ 0 & n^{m-2} & \cdots & 0 \\ \vdots & \vdots & \ddots & \vdots \\ 0 & 0 & \cdots & 1 \end{pmatrix}, \quad D_2 = \begin{pmatrix} n & 0 & \cdots & 0 \\ 0 & n^2 & \cdots & 0 \\ \vdots & \vdots & \ddots & \vdots \\ 0 & 0 & \cdots & n^m \end{pmatrix}.$$

Thus the solution of system (8.3) is given by

$$\begin{pmatrix} z_{m+1} \\ \vdots \\ z_{2m} \end{pmatrix} = D_2^{-1} W^{-1} D_1^{-1} \left[\begin{pmatrix} y_1 \\ \vdots \\ y_m \end{pmatrix} - A \begin{pmatrix} x_1 \\ \vdots \\ x_m \end{pmatrix} \right], \tag{8.4}$$

where

$$D_2^{-1} = \begin{pmatrix} n^{-1} & 0 & \cdots & 0 \\ 0 & n^{-2} & \cdots & 0 \\ \vdots & \vdots & \ddots & \vdots \\ 0 & 0 & \cdots & n^{-m} \end{pmatrix}, \quad D_1^{-1} = \begin{pmatrix} n^{-m+1} & 0 & \cdots & 0 \\ 0 & n^{-m+2} & \cdots & 0 \\ \vdots & \vdots & \ddots & \vdots \\ 0 & 0 & \cdots & 1 \end{pmatrix}.$$

This shows that the sequence z with the stated properties exists. Moreover, since the entries of D_1^{-1} and of $D_1^{-1}A$ are bounded by 1, we have that

$$\|D_1^{-1}[(y_1,\dots,y_m)^t - A(x_1,\dots,x_m)^t]\|_1 \leq \sum_{k=1}^{m}|y_k| + m\sum_{k=1}^{m}|x_k|.$$

With (8.1) and (8.4) we deduce that

$$|z_k| \leq Cn^{m-k}, \quad k = m+1,\dots,2m. \tag{8.5}$$

Thus (8.2) and (8.5) imply that

$$\|x - z\| = \sum_{k=m+1}^{2m}|z_k|v_k \leq C\sum_{k=m+1}^{2m} n^{m-k}v_k < \varepsilon.$$

On the other hand, the entries of $T^n z$ of index $k \geq 2m+1$ vanish, so that, again by (8.2) and (8.5),

$$\begin{aligned}\|y - T^n z\| &= \sum_{k=m+1}^{2m}\Big|\sum_{j=k}^{2m}\frac{n^{j-k}}{(j-k)!}z_j\Big|v_k \leq C\sum_{k=m+1}^{2m}\Big(\sum_{j=k}^{2m}\frac{n^{m-k}}{(j-k)!}\Big)v_k\\ &\leq eC\Big(\sum_{k=m+1}^{2m} n^{m-k}v_k\Big) < \varepsilon.\end{aligned}$$

This had to be shown. □

As a first application we also obtain that the operator $I + B$ is always mixing.

Theorem 8.2. *Let X be one of the spaces $\ell^p(v)$, $1 \leq p < \infty$, or $c_0(v)$. If $\sup_{n\in\mathbb{N}} \frac{v_n}{v_{n+1}} < \infty$, then the operator $T := I + B$ is mixing on X.*

Proof. Again, it suffices to prove the result for $\ell^1(v)$. We claim that there is a positive weight sequence $w = (w_n)_n$ and an operator $\phi : \ell^1(w) \to \ell^1(v)$ with dense range such that the following diagram commutes:

$$\begin{array}{ccc}\ell^1(w) & \xrightarrow{\;e^B\;} & \ell^1(w)\\ \phi\big\downarrow & & \big\downarrow\phi\\ \ell^1(v) & \xrightarrow{\;I+B\;} & \ell^1(v).\end{array}$$

To this end, let $A = (a_{nk})_{n\geq 1;k\leq n}$ be an infinite lower triangular matrix with $a_{nn} \neq 0$ for $n \geq 1$, and define ϕ by $\phi(e_n) = \sum_{k=1}^{n} a_{nk}e_k$, $n \geq 1$, where e_n denotes the canonical unit sequence, $n \geq 1$. We want to choose A in such a way that $(I + B)\phi(e_n) = \phi(e^B e_n)$ for $n \geq 1$. On the one hand we have that

$$(I+B)\phi(e_n) = \sum_{k=1}^{n} a_{nk}(I+B)e_k = a_{nn}e_n + \sum_{k=1}^{n-1}(a_{nk}+a_{n,k+1})e_k.$$

On the other hand,

$$\phi(e^B e_n) = \phi\Big(\sum_{k=0}^{n-1}\frac{1}{k!}e_{n-k}\Big) = \phi\Big(\sum_{j=1}^{n}\frac{1}{(n-j)!}e_j\Big)$$
$$= \sum_{j=1}^{n}\frac{1}{(n-j)!}\Big(\sum_{k=1}^{j}a_{jk}e_k\Big) = \sum_{k=1}^{n}\Big(\sum_{j=k}^{n}\frac{a_{jk}}{(n-j)!}\Big)e_k.$$

Setting these expressions equal and comparing the coefficients of e_{n-1} yields that $a_{n,n-1}+a_{nn} = a_{n-1,n-1}+a_{n,n-1}$ for $n\geq 2$, so that the elements on the diagonal must coincide; comparing the remaining coefficients implies that, for $k=1,\ldots,n-2$,

$$a_{nk}+a_{n,k+1} = \sum_{j=k}^{n}\frac{a_{jk}}{(n-j)!},$$

hence

$$a_{n,k+1} = \sum_{j=k}^{n-1}\frac{a_{jk}}{(n-j)!},$$

a system that we can solve inductively for $k=1,\ldots,n-2$, where $n\geq 3$. Observe that the elements a_{n1}, $n\geq 2$, are free, since no restriction on them is given. We can set them, for example, equal to 0. Similarly, the common value on the diagonal is free, which we may set equal to 1.

Once the matrix $A=(a_{nk})_{n\geq 1;k\leq n}$ is fixed, we can find a sufficiently rapidly increasing sequence $w=(w_n)_n$ with $\sup_{n\in\mathbb{N}}\frac{w_n}{w_{n+1}}<\infty$ such that $\phi:\ell^1(w)\to\ell^1(v)$ is a well-defined operator. By construction, $(I+B)\circ\phi = \phi\circ e^B$; moreover, the range of ϕ contains any finite sequence and is therefore dense. The result is now a consequence of Theorem 8.1 and Proposition 1.40.
□

We note that, conversely, Theorem 8.2 can also be derived from Theorem 8.1 (see Exercise 8.1.5), so that the two results are in fact equivalent. Another equivalent statement will be discussed in Exercise 8.1.2.

Using a suitable conjugacy one can reformulate the previous two results in terms of weighted shifts on unweighted spaces. We recall that, given a weight sequence $w=(w_n)_n$, the weighted shift B_w is a well-defined operator on ℓ^p, $1\leq p<\infty$, or on c_0 if and only if $\sup_{n\in\mathbb{N}}|w_n|<\infty$; see Example 4.9(a).

Corollary 8.3. *Let X be one of the spaces ℓ^p, $1\leq p<\infty$, or c_0, and let $w=(w_n)_n$ be a weight sequence such that $\sup_{n\in\mathbb{N}}|w_n|<\infty$. Then $T=I+B_w$ and $T=e^{B_w}$ are mixing operators on X.*

Proof. Let $X = \ell^p$, $1 \le p < \infty$, and set $u_n = (\prod_{\nu=1}^n w_\nu)^{-1}$, $v_n = |u_n|^p$. Then $\sup_{n\in\mathbb{N}} \frac{v_n}{v_{n+1}} < \infty$, and $\phi : \ell^p(v) \to \ell^p$, $(x_n)_n \to (x_n u_n)_n$ defines a conjugacy between $B : \ell^p(v) \to \ell^p(v)$ and $B_w : \ell^p \to \ell^p$. Linearity and continuity of ϕ imply that the operators $I + B_w$ and e^{B_w} on ℓ^p are also conjugate to $I + B$ and to e^B respectively, on $\ell^p(v)$. We then conclude with Theorems 8.2 and 8.1. The proof for $X = c_0$ is similar. □

The corollary furnishes an interesting example in connection with Kitai's theorem. Kitai tells us that, for any hypercyclic operator T on a complex Banach space, every connected component of the spectrum of T meets the unit circle. We will now see that, on the other hand, the spectrum may consist of a single point on the unit circle.

Example 8.4. Let B_w be the weighted shift with $w = (1/n)_n$ on a complex space ℓ^p or c_0. Since

$$(B_w)^n x = \big(\tfrac{1}{2\cdot 3\cdot \ldots \cdot (n+1)} x_{n+1}, \tfrac{1}{3\cdot 4\cdot \ldots \cdot (n+2)} x_{n+2}, \ldots\big),$$

we have that $\|(B_w)^n\|^{1/n} \le (n!)^{-1/n} \to 0$ as $n \to \infty$. Hence B_w is quasinilpotent, so that, by the spectral radius formula, $\sigma(B_w) = \{0\}$. Thus, $T = I + B_w$ is a mixing operator with $\sigma(T) = \{1\}$. Note also that B_w is compact; see Exercise 5.2.10.

We also consider a variant of the operator $I + B_w$ that will be of interest later on; see Theorem 8.13.

Proposition 8.5. *Let X be one of the spaces ℓ^p, $1 \le p < \infty$, or c_0. Let $w = (w_n)_n$ be a weight sequence such that $\sup_{n\in\mathbb{N}} |w_n| < \infty$ and D_w the operator on X given by*

$$D_w (x_n)_n = (w_2 x_2, w_4 x_4, w_6 x_6, \ldots).$$

Then the operator $T := I + D_w$ is mixing on X.

Proof. Once more it suffices to perform the proof for $X = \ell^1$. The operator D_w is best understood by its action on the unit sequences e_n, $n \ge 1$. If we write $n \ge 1$ as

$$n = m2^{k-1} \quad \text{with } m \ge 1 \text{ odd and } k \ge 1,$$

then

$$Te_n = 0 \ \text{ if } k = 1, \quad \text{and } Te_n = w_{m2^{k-1}} e_{m2^{k-2}} \ \text{ if } k \ge 2.$$

In this way it becomes obvious that D_w is conjugate to a countable direct sum of weighted shifts; see the explanations before Proposition 2.41. More precisely, if we define weights $w^{(m)} = (w_{m2^{k-1}})_{k\ge 1}$ for $m \ge 1$ odd, and the map

$$\phi : \Big(\bigoplus_{m \text{ odd}} \ell^1\Big)_{\ell^1} \to \ell^1, \quad \big((x_{m,k})_{k\geq 1}\big)_{m \text{ odd}} \to (y_n)_n,$$

where $y_n = x_{m,k}$ if $n = m2^{k-1}$, then ϕ is an isometric isomorphism, and the following diagram commutes:

$$\begin{array}{ccc} \Big(\bigoplus_{m \text{ odd}} \ell^1\Big)_{\ell^1} & \xrightarrow{\bigoplus_{m \text{ odd}} B_{w^{(m)}}} & \Big(\bigoplus_{m \text{ odd}} \ell^1\Big)_{\ell^1} \\ \phi \downarrow & & \downarrow \phi \\ \ell^1 & \xrightarrow{D_w} & \ell^1. \end{array}$$

Then also $I + D_w$ and $I + \bigoplus_{m \text{ odd}} B_{w^{(m)}} = \bigoplus_{m \text{ odd}} (I + B_{w^{(m)}})$ are conjugate via ϕ. Since, by Corollary 8.3, each operator $I + B_{w^{(m)}}$ is mixing on ℓ^1, Proposition 2.41 implies that $\bigoplus_{m \text{ odd}} (I + B_{w^{(m)}})$ is mixing, and therefore also $I + D_w$. □

To conclude this section we mention that all the results obtained so far are special cases of a single very general theorem. Its proof uses, in essence, the main argument of the proof of Theorem 8.1. To formulate the result we need the notion of an *extended backward shift*, that is, an operator T on a Banach space for which

$$\operatorname{span}\Big(\bigcup_{n=0}^{\infty} (\ker T^n \cap \operatorname{ran} T^n)\Big)$$

is dense. If T is surjective then the condition simply says that T has dense generalized kernel. Moreover, any weighted backward shift on a Banach sequence space in which the finite sequences are dense is an extended backward shift.

Theorem 8.6. *Let A be an extended backward shift on a Banach space X. Then the operators e^A and $I + A$ are mixing.*

Since this result will not be needed in the sequel we omit the proof; but see Exercise 8.1.9.

8.2 Existence of mixing operators and semigroups

In this section we will prove the existence of mixing, and therefore hypercyclic, operators on arbitrary infinite-dimensional separable Fréchet spaces. For the particular case of Banach spaces we even show the existence of mixing C_0-semigroups.

We first need two auxiliary results for Fréchet spaces. The first one is beyond the scope of this book and is therefore stated without proof.

Lemma 8.7. *Every separable Fréchet space that is not isomorphic to ω has a dense subspace that admits a continuous norm.*

Here, a continuous norm is exactly what it says: a norm functional that is continuous on the underlying space.

The second lemma is the crucial tool for transferring the mixing operators of the previous section to more general Fréchet spaces.

Lemma 8.8. *Let X be an infinite-dimensional separable Fréchet space that is not isomorphic to ω. Then there are sequences $(x_n)_n$ in X and $(x_n^*)_n$ in X^* such that*

(i) *$(x_n)_n$ converges to 0, and $\operatorname{span}\{x_n \ ; \ n \in \mathbb{N}\}$ is dense in X,*
(ii) *$(x_n^*)_n$ is equicontinuous,*
(iii) *$\langle x_k, x_n^* \rangle = 0$ if $k \neq n$, and $0 < \langle x_n, x_n^* \rangle \leq 1$, $n \geq 1$.*

Proof. By Lemma 8.7 there exists a dense subspace M of X that admits a continuous norm $|||\cdot|||$. We select a linearly independent sequence $(z_n)_n$ in M whose linear span is dense in M, and therefore in X. By the Hahn–Banach theorem (see Appendix A) there are functionals $y_n^* \in (M, |||\cdot|||)^*$ such that, for all $n \geq 1$, $\langle z_n, y_n^* \rangle = 1$ and $\langle z_k, y_n^* \rangle = 0$ for $k < n$. Following a Gram–Schmidt procedure in setting $y_1 = z_1$ and $y_n = z_n - \sum_{k=1}^{n-1} \langle z_n, y_k^* \rangle y_k$ for $n > 1$, we obtain a sequence $(y_n)_n$ in M such that $\operatorname{span}\{y_n \ ; \ n \in \mathbb{N}\} = \operatorname{span}\{z_n \ ; \ n \in \mathbb{N}\}$ is dense in X and $\langle y_k, y_n^* \rangle = \delta_{k,n}$, $k, n \in \mathbb{N}$.

By continuity there are $K_n \geq 1$, $n \geq 1$, such that $|\langle x, y_n^* \rangle| \leq K_n |||x|||$ for all $x \in M$. Since M is dense in X, there is a (unique) continuous seminorm p on X whose restriction to M coincides with $|||\cdot|||$, and each y_n^* has a unique continuous linear extension to X, which we denote again by y_n^*. Clearly, $|\langle x, y_n^* \rangle| \leq K_n p(x)$ for $x \in X$, $n \geq 1$. This implies that $\{K_n^{-1} y_n^* \ ; \ n \in \mathbb{N}\}$ is equicontinuous. Since X is metrizable, there are numbers $\alpha_n \in \left]0, 1\right]$ such that $(\alpha_n y_n)_n$ converges to 0 in X. Setting $x_n = \alpha_n y_n$ and $x_n^* = K_n^{-1} y_n^*$, $n \geq 1$, the claim follows. □

We are now ready for the main result of this section.

Theorem 8.9 (Ansari–Bernal). *Every infinite-dimensional separable Fréchet space supports a mixing, and therefore hypercyclic, operator.*

Proof. On $X = \omega$, the backward shift is a mixing operator; see Example 4.9(c). We may therefore suppose that X is an infinite-dimensional separable Fréchet space that is not isomorphic to ω. Applying Lemma 8.8 we find sequences $(x_n)_n$ in X and $(x_n^*)_n$ in X^* with the specified properties. Let us consider $T : X \to X$,

$$Tx = x + \sum_{n=1}^{\infty} 2^{-n} \langle x, x_{n+1}^* \rangle x_n, \quad x \in X.$$

The equicontinuity of $(x_n^*)_n$ and the fact that $(x_n)_n$ tends to 0 easily imply that T is a well-defined operator on X.

On the other hand, the operator

$$S : \ell^1 \to \ell^1, \quad S((\alpha_n)_n) = \Big(\alpha_1 + \frac{\langle x_2, x_2^* \rangle}{2}\alpha_2, \alpha_2 + \frac{\langle x_3, x_3^* \rangle}{2^2}\alpha_3, \ldots\Big)$$

is a perturbation of the identity by a weighted backward shift and therefore mixing by Corollary 8.3. The map $\phi : \ell^1 \to X$ given by $\phi((\alpha_n)_n) = \sum_{n=1}^{\infty} \alpha_n x_n$ is continuous, and it has dense range since the linear span of the x_n, $n \geq 1$, is dense in X. Since $T \circ \phi = \phi \circ S$, T is quasiconjugate to S, and the conclusion follows from Proposition 1.40. □

Remark 8.10. We note that in the case that X is a Banach space, the mixing operator T constructed above is of the form

$$T = I + K,$$

where K is a compact operator (see Exercise 5.2.4), and one may make $\|K\|$ arbitrarily small; see also Exercise 8.2.2.

Having answered the question of the existence of mixing operators one might wonder if every infinite-dimensional separable Fréchet space admits a chaotic operator. In fact, this is not even the case when we restrict ourselves to Banach spaces, as will follow from a deep recent result in Banach space theory.

Theorem 8.11 (Argyros–Haydon). *Let* $\mathbb{K} = \mathbb{R}$ *or* $\mathbb{C}$. *Then there exists an infinite-dimensional separable Banach space* X *over* $\mathbb{K}$ *on which all operators are of the form*

$$T = \lambda I + K,$$

where $\lambda \in \mathbb{K}$ *and* K *is a compact operator on* X.

In view of Proposition 5.20 we have the following.

Theorem 8.12. *Let* $\mathbb{K} = \mathbb{R}$ *or* $\mathbb{C}$. *Then there exists an infinite-dimensional separable Banach space over* $\mathbb{K}$ *that supports no chaotic operator.*

While we cannot always demand hypercyclic operators with a dense set of periodic points, there are, nonetheless, always hypercyclic operators with an infinite-dimensional closed subspace of such points.

Theorem 8.13. *Every infinite-dimensional separable Fréchet space supports a mixing, and therefore hypercyclic, operator with an infinite-dimensional closed subspace of fixed points.*

Proof. The proof is a variant of that of Theorem 8.9.

We first suppose that the Fréchet space X is not isomorphic to ω. Applying Lemma 8.8 we again find sequences $(x_n)_n$ in X and $(x_n^*)_n$ in X^* with the specified properties, and we define an operator T on X by

$$Tx = x + \sum_{n=1}^{\infty} 2^{-n} \langle x, x_{2n}^* \rangle x_n, \quad x \in X.$$

It follows, as in the proof of Theorem 8.9, that T is quasiconjugate to the operator S on ℓ^1 given by

$$S((\alpha_n)_n) = \Big(\alpha_1 + \frac{\langle x_2, x_2^* \rangle}{2}\alpha_2, \alpha_2 + \frac{\langle x_4, x_4^* \rangle}{2^2}\alpha_4, \alpha_3 + \frac{\langle x_6, x_6^* \rangle}{2^3}\alpha_6, \ldots\Big).$$

By Proposition 8.5, S is mixing on ℓ^1, and hence so is T on X. Moreover, every vector in the infinite-dimensional closed subspace $\overline{\text{span}}\{x_n \,;\, n \geq 1 \text{ odd}\}$ of X is a fixed point of T.

Finally, for the case when $X = \omega$, we see directly from Proposition 8.5 that $(\alpha_n)_n \to (\alpha_n + \alpha_{2n})_n$ defines a mixing operator on ℓ^1 and therefore, by continuous and dense inclusion of ℓ^1 into ω, also a mixing operator on ω, and it has an infinite-dimensional closed subspace of fixed points. □

In Chapter 7 we considered C_0-semigroups of operators on Banach spaces X. In particular, every operator $A : X \to X$ defines a C_0-semigroup $(T_t)_{t\geq 0}$ on X given by $T_t = e^{tA}$, and A is called the infinitesimal generator. In a way that is analogous to the existence of mixing operators on Fréchet spaces, we can establish the existence of mixing C_0-semigroups on Banach spaces.

Theorem 8.14. *Every infinite-dimensional separable Banach space supports a mixing, and therefore hypercyclic, C_0-semigroup $(T_t)_{t\geq 0}$.*

Proof. As in the Fréchet case we find sequences $(x_n)_n$ in X and $(x_n^*)_n$ in X^* satisfying the hypotheses of Lemma 8.8, and we consider the operator A on X given by

$$Ax = \sum_{n=1}^{\infty} 2^{-n} \langle x, x_{n+1}^* \rangle x_n, \quad x \in X.$$

Then A is quasiconjugate to a weighted shift B_w on ℓ^1 via the continuous linear map $\phi : \ell^1 \to X$ given in the proof of Theorem 8.9. Consequently, e^A is quasiconjugate to e^{B_w}, and the latter is a mixing operator by Corollary 8.3. It then follows from Propositions 1.40 and 7.21 that $(e^{tA})_{t\geq 0}$ is a mixing C_0-semigroup on X. □

We also have an analogue of Theorem 8.12.

Theorem 8.15. *Let $\mathbb{K} = \mathbb{R}$ or $\mathbb{C}$. Then there exists an infinite-dimensional separable Banach space over $\mathbb{K}$ that supports no chaotic C_0-semigroup.*

Proof. Let X be one of the spaces constructed by Argyros and Haydon (see Theorem 8.11), and suppose that $(T_t)_{t\geq 0}$ is a chaotic C_0-semigroup on X. By the Conejero–Müller–Peris theorem, every operator T_t, $t > 0$, is hypercyclic. Now, if $\mathbb{K} = \mathbb{C}$, then Lemma 5.19 implies that $\sigma(T_t)$ is a singleton contained in $\mathbb{T}$, for every $t > 0$. We deduce from the point spectral mapping theorem for semigroups (see Appendix B) that $e^{t\sigma_p(A)}$ contains at most one element, for every $t > 0$, where A is the infinitesimal generator of $(T_t)_{t\geq 0}$. This is only possible if $\sigma_p(A)$ itself contains at most one element, which contradicts Proposition 7.18. In the case when $\mathbb{K} = \mathbb{R}$ we consider the complexification of the semigroup, which, by Corollary 7.24, is again a chaotic semigroup consisting of compact perturbations of multiples of the identity, and we argue as before. □

8.3 Density of hypercyclic operators

For vectors we have the phenomenon that as soon as one hypercyclic vector (for a given operator) exists there is then automatically a large supply of them. Could something similar be true for operators? That is, does the existence of one hypercyclic operator on a given space imply that there must then be many of them, in a certain sense?

The most natural interpretation of this question leads to a negative answer. Indeed, no operator T on a Banach space X with $\|T\| \leq 1$ can be hypercyclic, because all of its orbits are bounded. Thus the set of hypercyclic operators on X is not dense in the space $L(X)$ of all operators, when we endow it with the operator norm topology. One can even show that in this topology the set of hypercyclic operators is nowhere dense.

But there is another well-known, weaker topology on $L(X)$, the *strong operator topology*, abbreviated SOT; it is the topology of pointwise convergence of operators. We will show in this section that the set of hypercyclic operators is indeed dense in this topology.

In order to prepare the ground, let X be any Fréchet space and $L(X)$ the space of operators on X. The strong operator topology is then defined in the following way: for $T \in L(X)$, a base of neighbourhoods is given by

$$U_{x_1,\dots,x_n}(T,\varepsilon) = \{S \in L(X) \;;\; \|Tx_k - Sx_k\| < \varepsilon \text{ for } k = 1,\dots,n\},$$

where $x_1,\dots,x_n$, $n \geq 1$, is an arbitrary collection of linearly independent vectors of X, $\|\cdot\|$ is an F-norm defining the topology of X, and $\varepsilon > 0$. Thus, a sequence (more generally, a net) $(T_\alpha)_\alpha$ is SOT-convergent to an operator T if and only if, for every $x \in X$, $T_\alpha x \to Tx$.

Starting from a single hypercyclic operator T on a given Fréchet space X it is very easy to find additional ones: if $A : X \to X$ is an invertible operator then $A^{-1}TA$ is conjugate to T and therefore also hypercyclic. The fundamental, and perhaps surprising result in this context says that, under a very weak condition on an arbitrary operator T, the *similarity orbit* of T,

$$\mathcal{S}(T) = \{A^{-1}TA \;;\; A : X \to X \text{ invertible}\},$$

is SOT-dense in $L(X)$.

Proposition 8.16. *Let X be an infinite-dimensional Fréchet space. Suppose that, for any $n \geq 1$, there are n vectors $x_1, \ldots, x_n$ in X such that*

$$x_1, \ldots, x_n, Tx_1, \ldots, Tx_n$$

are linearly independent. Then $\mathcal{S}(T)$ is SOT-dense in $L(X)$.

For the proof we need the following.

Lemma 8.17. *Let $x_1, \ldots, x_n$ and $y_1, \ldots, y_n$ be two sets of linearly independent vectors from X. Then there exists an invertible operator A of X such that $Ax_k = y_k$, $k = 1, \ldots, n$.*

Proof. Let $M = \text{span}\{x_1, \ldots, x_n, y_1, \ldots, y_n\}$, which is a space of dimension $m \leq 2n$. We can find vectors $x_{n+1}, \ldots, x_m$ and $y_{n+1}, \ldots, y_m$ such that both $x_1, \ldots, x_m$ and $y_1, \ldots, y_m$ are bases of M. Since finite-dimensional subspaces are closed, the Hahn–Banach theorem implies the existence of functionals $x_k^* \in X^*$ such that $\langle x_j, x_k^* \rangle = \delta_{j,k}$ for $j, k = 1, \ldots, m$.

We then define $A : X \to X$ by

$$Ax = \sum_{k=1}^{m} \langle x, x_k^* \rangle y_k + x - \sum_{k=1}^{m} \langle x, x_k^* \rangle x_k,$$

which is clearly continuous. Since $Ax_k = y_k$ for $k = 1, \ldots, m$, $A : M \to M$ is a bijection. Moreover, A is the identity on $M' := \bigcap_{k=1}^{m} \ker x_k^*$. Since, algebraically, $X = M \oplus M'$, A is a bijection (on X). By the inverse mapping theorem, A^{-1} is continuous so that A is invertible. □

Proof of Proposition 8.16. In order to show that $\mathcal{S}(T)$ is SOT-dense we fix an operator $S \in L(X)$, linearly independent vectors $x_1, \ldots, x_n$, $n \geq 1$, of X, and $\varepsilon > 0$. We then define inductively vectors $y_1, \ldots, y_n$ such that $x_1, \ldots, x_n, y_1, \ldots, y_n$ are linearly independent and $\|Sx_k - y_k\| < \varepsilon$, $k = 1, \ldots, n$; in fact, if $x_1, \ldots, x_n, y_1, \ldots, y_{k-1}, Sx_k$ are linearly independent we take $y_k = Sx_k$, and if not we choose a suitable small perturbation of Sx_k.

On the other hand, by hypothesis, there are $z_1, \ldots, z_n$ in X such that $z_1, \ldots, z_n, Tz_1, \ldots, Tz_n$ are linearly independent. By Lemma 8.17 there is an invertible operator A on X such that $Ax_k = z_k$ and $Ay_k = Tz_k$ for $k = 1, \ldots, n$. Thus $A^{-1}TAx_k = y_k$, so that

$$\|Sx_k - A^{-1}TAx_k\| < \varepsilon$$

for $k = 1, \ldots, n$. In other words, $A^{-1}TA \in U_{x_1,\ldots,x_n}(S, \varepsilon)$, as desired. □

We are now in a position to prove the announced density result for hypercyclic operators.

Theorem 8.18. *Let X be an infinite-dimensional separable Fréchet space. Then the set of hypercyclic operators on X is SOT-dense in $L(X)$.*

Proof. We know from the Ansari–Bernal theorem that there exists a hypercyclic operator T on X. By conjugacy, the similarity orbit $\mathcal{S}(T)$ consists entirely of hypercyclic operators. Thus the result follows from Proposition 8.16 once we know that, for any $n \geq 1$, there are n vectors $x_1, \ldots, x_n$ in X such that $x_1, \ldots, x_n, Tx_1, \ldots, Tx_n$ are linearly independent. But these are easily found: the orbit of any hypercyclic vector x for T is linearly independent and hence so is

$$x, T^2x, T^4x, \ldots, T^{2n-2}x, Tx, T^3x, T^5x, \ldots, T^{2n-1}x;$$

see Proposition 2.60. □

The proof tells us much more: the same result will be true for any nonempty class of hypercyclic operators that is invariant under conjugacies. In particular, we obtain the following.

Theorem 8.19. *Let X be an infinite-dimensional separable Fréchet space.*

(a) *The set of mixing operators on X is SOT-dense in $L(X)$.*

(b) *The set of chaotic operators on X is either empty or SOT-dense in $L(X)$; in particular, it is SOT-dense if X is a Hilbert space.*

(c) *The set of mixing operators on X with an infinite-dimensional closed subspace of fixed points is SOT-dense in $L(X)$.*

8.4 Existence of hypercyclic operators with prescribed orbits

There is another sense in which one can measure the richness of the set of hypercyclic operators. By Proposition 2.60 we know that a dense orbit is necessarily a linearly independent set. Therefore one may wonder if, conversely, every countable dense linearly independent set arises as an orbit of an operator; this operator is then automatically hypercyclic. We will show here that the answer is positive, at least in the setting of Banach spaces. The idea of the proof is not dissimilar to the approach in the last section: we construct an invertible operator that transforms a known dense orbit into an arbitrary countable dense linearly independent set.

We first need a property of perturbations of invertible operators.

Lemma 8.20. *Let X be a Banach space, A an invertible operator on X. If B is an operator on X such that $\|A-B\| < 1/\|A^{-1}\|$ then B is also invertible. Moreover, in that case,*

$$\|A^{-1} - B^{-1}\| \leq \frac{\|A^{-1}\|^2\,\|A-B\|}{1-\|A^{-1}\|\,\|A-B\|}.$$

Proof. First, if T is an operator on X with $\|T\| < 1$ then the series $C := \sum_{n=0}^{\infty} T^n$ converges in $L(X)$ under the operator norm, so that C defines an operator on X. Moreover, since

$$(I - T)C = (I + T + T^2 + \ldots) - (T + T^2 + \ldots) = I,$$

and similarly $C(I - T) = I$, the operator $I - T$ is invertible with inverse C.

Now, if $\|A - B\| < 1/\|A^{-1}\|$ then $\|I - A^{-1}B\| = \|A^{-1}(A - B)\| \leq \|A^{-1}\|\,\|A - B\| < 1$, so that $A^{-1}B = I - (I - A^{-1}B)$ is invertible and therefore also B. In addition,

$$A^{-1} - B^{-1} = (I - B^{-1}A)A^{-1} = -\sum_{n=1}^{\infty}(A^{-1}(A - B))^n A^{-1}.$$

Applying the operator norm and the sum formula of the geometric series gives the desired estimate on $\|A^{-1} - B^{-1}\|$. □

The next step is to prescribe values of an invertible operator.

Lemma 8.21. *Let X be an infinite-dimensional Banach space and X_0 and Y_0 dense subsets of X. Let E and F be subspaces of X with $\dim(E) = \dim(F) < \infty$ and A an invertible operator on X such that $A(E) = F$.*

Then, for any $x_0 \notin E$, $y_0 \notin F$ and $\varepsilon > 0$, there exists an invertible operator B on X such that $B|_E = A|_E$, $Bx_0 \in Y_0$, $B^{-1}y_0 \in X_0$ and $\|A - B\| < \varepsilon$.

Proof. Since E is closed as a finite-dimensional subspace, the Hahn–Banach theorem implies that there exists $x^* \in X^*$ such that $\langle x, x^* \rangle = 0$ for all $x \in E$ and $\langle x_0, x^* \rangle = 1$. By density of Y_0 there is some $y_1 \in Y_0$ with $\|x^*\|\|y_1 - Ax_0\| < \min(\varepsilon/2, 1/\|A^{-1}\|)$; perturbing y_1 if necessary we can assume that it does not belong to $F \oplus \text{span}\{y_0\}$. Then we define the operator C by

$$Cx = Ax + \langle x, x^* \rangle (y_1 - Ax_0), \quad x \in X.$$

Consequently, $C|_E = A|_E$, $Cx_0 = y_1$ and $\|A - C\| \leq \|x^*\|\|y_1 - Ax_0\| < \min(\varepsilon/2, 1/\|A^{-1}\|)$. By Lemma 8.20, C is invertible.

Now we repeat the procedure, with $F \oplus \text{span}\{y_1\}$, E, X_0, y_0, x_0, and C^{-1} taking the roles of E, F, Y_0, x_0, y_0, and A, respectively; note that $y_0 \notin F \oplus \text{span}\{y_1\}$. Thus, for $\eta > 0$, there is some $x_1 \in X_0$ and an invertible operator D on X such that $D|_F = C^{-1}|_F$, $Dy_1 = C^{-1}y_1 = x_0$, $Dy_0 = x_1$ and $\|C^{-1} - D\| < \eta$; moreover, by Lemma 8.20, if η is small enough then $\|C - D^{-1}\| < \varepsilon/2$. Then $B := D^{-1}$ is the desired invertible operator with $\|A - B\| \leq \|A - C\| + \|C - D^{-1}\| < \varepsilon$. □

This lemma implies that any two dense linearly independent sequences are isomorphically equivalent, in the following sense.

Lemma 8.22. *Let X be a Banach space, let $X_0 = \{x_n \,;\, n \in \mathbb{N}\}$ and $Y_0 = \{y_n \,;\, n \in \mathbb{N}\}$ be dense linearly independent sets in X, and $\varepsilon > 0$. Then there exists an invertible operator A on X such that $A(X_0) = Y_0$ and $\|I - A\| < \varepsilon$.*

Proof. We will show by induction that there are finite sets $V_n \subset X_0$, $W_n \subset Y_0$ and invertible operators A_n on X such that, for any $n \geq 1$, $V_{n-1} \subset V_n$, $\{x_1, \ldots, x_n\} \subset V_n$, $W_{n-1} \subset W_n$, $\{y_1, \ldots, y_n\} \subset W_n$, $A_n(V_n) = W_n$, $A_n|_{V_{n-1}} = A_{n-1}|_{V_{n-1}}$ and $\|A_n - A_{n-1}\| < \frac{\varepsilon}{2^n}$, where $A_0 = I$ and $V_0 = W_0 = \varnothing$.

For $n = 1$ we apply Lemma 8.21 to $E = F = \{0\}$ and $A = I$ to obtain an invertible operator A_1 with $\|A_1 - I\| < \frac{\varepsilon}{2}$ such that there are $p_1, q_1 \geq 1$ with $A_1 x_1 = y_{p_1}$ and $A_1 x_{q_1} = y_1$. We then set $V_1 = \{x_1, x_{q_1}\}$ and $W_1 = \{y_1, y_{p_1}\}$; note that both sets may be singletons.

Suppose that $V_1, \ldots, V_{n-1}, W_1, \ldots, W_{n-1}$ and $A_1, \ldots, A_{n-1}$ have been defined, $n \geq 2$. Since $\operatorname{card}(V_{n-1}) = \operatorname{card}(W_{n-1})$, we can apply Lemma 8.21 to $E = \operatorname{span} V_{n-1}$, $F = \operatorname{span} W_{n-1}$ and $A = A_{n-1}$; for x_0 we choose the vector $x_{k_n} \in V$ of smallest index that does not belong to E, and similarly a vector y_{m_n} for y_0. Then we obtain an invertible operator A_n and $p_n, q_n \geq 1$ such that $A_n|_E = A_{n-1}|_E$, $A_n x_{k_n} = y_{p_n}$, $A_n x_{q_n} = y_{m_n}$, and $\|A_n - A_{n-1}\| < \frac{\varepsilon}{2^n}$. We set $V_n = V_{n-1} \cup \{x_{k_n}, x_{q_n}\}$ and $W_n = W_{n-1} \cup \{y_{m_n}, y_{p_n}\}$. Using the linear independence of X_0 and Y_0, one deduces easily from the construction and the induction hypothesis that $\{x_1, \ldots, x_n\} \subset V_n$, $\{y_1, \ldots, y_n\} \subset W_n$ and $A_n(V_n) = W_n$.

Having obtained the operators A_n we note that $\sum_{n=1}^{\infty} \|A_n - A_{n-1}\| < \infty$. Thus

$$A := I + \sum_{n=1}^{\infty} (A_n - A_{n-1})$$

defines an operator on X. Moreover, $\|I - A\| \leq \sum_{n=1}^{\infty} \|A_n - A_{n-1}\| < \varepsilon$; choosing $\varepsilon < 1$, Lemma 8.20 implies that A is invertible. Moreover, since $A = \lim_{n\to\infty} A_n$ we deduce easily that $A(X_0) = Y_0$. □

We will need the following variant of this lemma.

Lemma 8.23. *Let X be a Banach space, let $X_0 = \{x_n \; ; \; n \in \mathbb{N}\}$ and $Y_0 = \{y_n \; ; \; n \in \mathbb{N}\}$ be dense linearly independent sets in X, B an invertible operator with $Bx_1 = y_1$, and $\varepsilon > 0$. Then there exists an invertible operator A on X such that $A(X_0) = Y_0$, $Ax_1 = y_1$ and $\|B - A\| < \varepsilon$.*

Proof. The proof is identical to that of the previous lemma if one starts with $A_1 = B$, $V_1 = \{x_1\}$, $W_1 = \{y_1\}$, and then proceeds with the induction. □

We can now show that dense orbits can be prescribed.

Theorem 8.24. *Let X be a Banach space, and let $\{x_n \; ; \; n \in \mathbb{N}\}$ be a dense linearly independent set in X. Then there exists an operator T on X, necessarily hypercyclic, such that* $\operatorname{orb}(x_1, T) = \{x_n \; ; \; n \in \mathbb{N}\}$.

Proof. Since X is an infinite-dimensional separable Banach space, the Ansari–Bernal theorem implies that there is a hypercyclic operator S on X. Applying Lemma 8.21 to $E = F = \{0\}$, $X_0 = \{x_n \; ; \; n \in \mathbb{N}\}$, $Y_0 = HC(S)$ and $A = I$ we obtain an invertible operator B with $y_1 := Bx_1 \in HC(S)$.

By Proposition 2.60, $\mathrm{orb}(y_1, S) =: \{y_n \; ; \; n \in \mathbb{N}\}$ is linearly independent. Now applying Lemma 8.23 to $X_0 = \{x_n \; ; \; n \in \mathbb{N}\}$ and $Y_0 = \{y_n \; ; \; n \in \mathbb{N}\}$ we obtain an invertible operator A on X with $Ax_1 = y_1$ and $A(X_0) = Y_0$. By conjugacy, the operator $T = A^{-1}SA$ is hypercyclic, and we have that

$$\mathrm{orb}(x_1, T) = \{T^n x_1 \; ; \; n \in \mathbb{N}_0\} = A^{-1}\{S^n y_1 \; ; \; n \in \mathbb{N}_0\} = A^{-1}(Y_0) = X_0,$$

which had to be shown. □

A more careful analysis of the proof shows that, for every $\varepsilon > 0$, the operator T can be of the form $T = I + K$, where K is a compact operator with $\|K\| < \varepsilon$; see Exercise 8.4.3.

Exercises

Exercise 8.1.1. Given $m, n \in \mathbb{N}$ with $m \geq n$, verify the identity

$$\det \begin{pmatrix} \frac{1}{m!} & \frac{1}{(m+1)!} & \cdots & \frac{1}{(m+n)!} \\ \frac{1}{(m-1)!} & \frac{1}{m!} & \cdots & \frac{1}{(m+n-1)!} \\ \vdots & \vdots & \ddots & \vdots \\ \frac{1}{(m-n)!} & \frac{1}{(m-n+1)!} & \cdots & \frac{1}{m!} \end{pmatrix} = \prod_{k=0}^{n} \frac{(n-k)!}{(m+k)!}.$$

Exercise 8.1.2. An entire function $\gamma(z) = \sum_{n=0}^{\infty} \gamma_n z^n$ is called an *admissible comparison function* provided that $\gamma_n > 0$ for $n \geq 0$ and the sequence $((n+1)\gamma_{n+1}/\gamma_n)_{n\geq 0}$ is bounded. Each admissible comparison function induces Banach spaces of entire functions

$$E^p(\gamma) = \Big\{ f \in H(\mathbb{C}) \; ; \; f(z) = \sum_{n=0}^{\infty} a_n z^n, \; \|f\|_{p,\gamma}^p = \sum_{n=0}^{\infty} \gamma_n^{-p} |a_n|^p < \infty \Big\}, \quad 1 \leq p < \infty;$$

see also Exercise 4.2.4. Show the following:

(i) any function $f \in E^p(\gamma)$ is of *exponential type* $\tau := \limsup_{n\to\infty} (n+1)\gamma_{n+1}/\gamma_n$, that is, for every $\varepsilon > 0$ there is some $M > 0$ such that $|f(z)| \leq Me^{(\tau+\varepsilon)|z|}$ for $z \in \mathbb{C}$;

(ii) $D : f \to f'$ defines an operator on $E^p(\gamma)$;

(iii) deduce from Theorem 8.1 that the translation operators $T_a f(z) = f(z+a)$, $a \in \mathbb{C}$, $a \neq 0$, are mixing on $E^p(\gamma)$;

(iv) conversely, show that the result of (iii) implies Theorem 8.1.

(*Hint*: Show that aD on $E^p(\gamma)$ is conjugate to B on $\ell^p(v)$ for a suitable sequence of weights $v = (v_n)_n$, and observe that $T_a = e^{aD}$.)

Exercise 8.1.3. Use the previous exercise to prove Theorem 4.23. (*Hint*: Note that $r^n \leq M_n \phi(r)$; then find γ_n such that any $f \in E^1(\gamma)$ satisfies the required growth condition, and apply the hypercyclic comparison principle.)

Exercise 8.1.4. Let $v = (v_n)_n$ be a positive weight sequence with $\sup_{n\in\mathbb{N}} \frac{v_{n+1}}{v_n} < \infty$. Show directly that $I + B$ is mixing on $\ell^p(v)$, $1 \leq p < \infty$, or on $c_0(v)$. (*Hint*: Following the proof of Theorem 8.1, use the same argument substituting the matrix W by a matrix W_n such that $\lim_n W_n = W$.)

Exercise 8.1.5. Deduce Theorem 8.1 from Theorem 8.2.

Exercise 8.1.6. An operator T on an infinite-dimensional Banach space X is called a *generalized backward shift* if $\dim\ker T = 1$ and the generalized kernel $\bigcup_{n=0}^{\infty}\ker T^n$ is dense in X. Show that there is a sequence $(e_n)_n$ of nonzero vectors in X such that $Te_1 = 0$, $Te_{n+1} = e_n$, $n \geq 1$, and $\operatorname{span}\{e_n \ ; \ n \in \mathbb{N}\}$ is dense in X. Deduce that T is an extended backward shift. (*Hint*: For $X_n = \ker T^n$, show that $X_{n-1} \subsetneq X_n$, and use the rank-nullity theorem to show inductively that $T(X_n) = X_{n-1}$ and $\dim X_n = n$.)

Exercise 8.1.7. Let A be an operator on a Banach space X for which there is a sequence $(e_n)_n$ of nonzero vectors such that $Ae_1 = 0$, $Ae_{n+1} = e_n$, $n \geq 1$, and $\operatorname{span}\{e_n \ ; \ n \in \mathbb{N}\}$ is dense in X. Prove that $I+A$ is mixing on X. (*Hint*: Construct a suitable quasiconjugacy to use Theorem 8.2.)

Exercise 8.1.8. Let A be an operator on a Banach space X that satisfies the hypothesis of the previous exercise. Let $\varphi(z) = \sum_{n=0}^{\infty} a_n z^n$ be a nonconstant formal power series with $|a_0| = 1$ such that $T := \varphi(A) = \sum_{n=0}^{\infty} a_n A^n$ converges pointwise on X; by the Banach–Steinhaus theorem, T is then an operator on X. Show that T is mixing. (*Hint*: One may assume that $a_0 = 1$. Then $T - I = \sum_{n=k}^{\infty} a_n A^n$ with $k \geq 1$, $a_k \neq 0$. For $j = 1, \ldots, k$, take $f_j := e_j$, and for $n \geq 1$ let $f_{k+n} := \sum_{j=1}^{n} w_{n,j} e_{k+j}$, where the $w_{n,j}$ have to be determined by induction to satisfy $(T-I)f_{k+n} = f_n$. Now, for $j = 1, \ldots, k$ define $X_j := \overline{\operatorname{span}\{f_{j+nk} \ ; \ n \geq 0\}}$. Then each X_j is $(T-I)$-invariant, and one may apply the previous exercise to each operator $A_j := (T-I)|_{X_j}$ on X_j. To finish, construct a commutative diagram

$$\begin{array}{ccc} X_1 \oplus \cdots \oplus X_k & \xrightarrow{\bigoplus_{j=1}^{k}(I+A_j)} & X_1 \oplus \cdots \oplus X_k \\ \phi\big\downarrow & & \phi\big\downarrow \\ X & \xrightarrow{T=\varphi(A)} & X \end{array}$$

and apply Proposition 1.40.)

Exercise 8.1.9. Prove Theorem 8.6. (*Hint*: By Proposition 2.37 it suffices to show that for any $x \in X$, $\varepsilon > 0$, and for any sufficiently large n there are $z, z' \in X$ (depending on n) such that $\|x - z\| < \varepsilon$ and $\|T^n z\| < \varepsilon$, $\|z'\| < \varepsilon$ and $\|x - T^n z'\| < \varepsilon$. It suffices to consider $x \neq 0$ from $\bigcup_{n=0}^{\infty}(\ker T^n \cap \operatorname{ran} T^n)$. Choose $m \geq 1$ minimal and $y \in X$ such that $T^m x = 0$ and $x = T^m y$; then construct e_k, $k = 1, \ldots, 2m$, such that $Te_1 = 0$, $Te_{k+1} = e_k$ for $k \geq 1$, and $e_m = x$. By minimality of m, the e_k are linearly independent. Now, for $T = e^A$, use the argument in the proof of Theorem 8.1 to show that there is some $C > 0$ such that, for any $n \geq 1$, there is $z \in X$ (depending on n) of the form $z = e_m + \sum_{k=m+1}^{2m} z_k e_k$ that satisfies $T^n z = \sum_{k=m+1}^{2m}(\sum_{j=k}^{2m} \frac{n^{j-k}}{(j-k)!} z_j) e_k$ and $|z_k| \leq C n^{m-k}$, $k = m+1, \ldots, 2m$. Deduce that, for large n, $\|x - z\| < \varepsilon$ and $\|T^n z\| < \varepsilon$. Construct z' in a similar way. For the proof of $T = I + A$ replace W by W_n as in Exercise 8.1.4.)

Exercise 8.2.1. Show that Lemma 8.8 fails for $X = \omega$.

Exercise 8.2.2. Let T be an operator on a Banach space X. For $n \geq 1$ the nth *approximation number* $\alpha_n(T)$ is defined as

$$\alpha_n(T) = \inf\{\|T - F\| \ ; \ F \in L(X), \ \dim(\operatorname{ran} F) \leq n\}.$$

By Exercise 5.2.4, if $\alpha_n(T) \to 0$ as $n \to \infty$ then T is compact. Let $\varepsilon > 0$, and let $\varepsilon_n > 0$, $n \geq 1$, be such that $\varepsilon_n \to 0$ as $n \to \infty$. Show that, on any infinite-dimensional separable

Banach space X, there exists a mixing operator of the form $T = I + K$, where K is a compact operator with $\|K\| < \varepsilon$ and $\alpha_n(K) \leq \varepsilon_n$ for any $n \geq 1$. (*Hint*: Modify the proof of Theorem 8.9.)

Exercise 8.2.3. Show the following:

(i) every infinite-dimensional separable Banach space supports an invertible mixing operator;

(ii) the operator constructed in the proof of Theorem 8.9 is not chaotic.

(*Hint*: See the previous exercise, Lemma 8.20 and Proposition 5.20.)

Exercise 8.2.4. Let X be an infinite-dimensional separable Fréchet space. Show that there are operators T_1, T_2 on X that are disjoint hypercyclic; see Exercise 3.4.1. (*Hint*: First note that when $x, y \neq 0$ then there is an invertible operator A such that $Ax = y$; indeed, if x, y are linearly independent then find x^*, $y^* \in X^*$ with $\langle x, x^* \rangle = 1$, $\langle y, y^* \rangle = 1$, $\langle y, x^* \rangle = 0$, $\langle x, y^* \rangle = 0$, and set $Az = z + \langle z, y^* - x^* \rangle (x - y)$. Now, let T be a mixing operator on X and (x_1, x_2) hypercyclic for $T \oplus T$. Let $x \neq 0$, and A_1, A_2 invertible operators such that $A_j x_j = x$, $j = 1, 2$. Then consider $T_j = A_j T A_j^{-1}$, $j = 1, 2$.)

Exercise 8.2.5. Let X be a separable Banach space and M a complemented subspace of X of infinite codimension; that is, M is a closed subspace for which there is an infinite-dimensional closed subspace M' of X such that, algebraically, $X = M \oplus M'$. Show that there exists a hypercyclic operator T on X that is the identity on M. (*Hint*: Apply Lemma 8.8 separately to both subspaces and follow the proof of Theorem 8.13.)

Exercise 8.2.6. Give an alternative proof of Theorem 8.14 as follows. Let T be the mixing operator on an infinite-dimensional separable Banach space X provided by the proof of Theorem 8.9 of the form $T = I + S$, where we can even have that $\|S\| < 1$. Then find an operator A such that $e^A = T$ to deduce that $(e^{tA})_{t \geq 0}$ is a mixing C_0-semigroup on X. (*Hint*: Use Exercise 7.1.5.)

Exercise 8.3.1. Even though every hypercyclic operator on ℓ^2 has norm bigger than one, Theorem 8.18 tells us that one can approximate the zero operator in the SOT-topology by such operators. Show that, concretely, if B is the backward shift, then $\frac{n+1}{n} B^n$, $n \geq 1$, are hypercyclic operators that converge to 0 in SOT as $n \to \infty$.

Exercise 8.3.2. Show that the class of all operators on an infinite-dimensional separable Fréchet space that have no nontrivial closed invariant subset is either empty or SOT-dense.

Exercise 8.3.3. Show that on any infinite-dimensional separable Fréchet space that is not isomorphic to ω the set of hypercyclic non-chaotic operators is SOT-dense. (*Hint*: Exercise 8.2.3.)

Exercise 8.3.4. Let X be an infinite-dimensional separable Fréchet space. Show the following:

(i) the operators in the similarity orbit $\mathcal{S}(T)$ of an operator T on X either have no common hypercyclic vector, or else every nonzero vector in X is hypercyclic for every operator in $\mathcal{S}(T)$;

(ii) there is no vector that is hypercyclic for every hypercyclic operator on X.

Exercise 8.3.5. Let X be an infinite-dimensional separable Fréchet space and $x \in X$, $x \neq 0$. Show that the set of hypercyclic operators on X having x as hypercyclic vector is SOT-dense in $L(X)$. (*Hint*: Consider $\{A^{-1}TA\ ;\ A \text{ invertible and } Ax \in HC(T)\}$.)

Exercise 8.3.6. Let X be a complex Banach space with the properties of Theorem 8.11. Show that if an operator T on X is the sum of two hypercyclic operators then it is of the form $T = \lambda I + K$, where $|\lambda| \leq 2$ and K is a compact operator. In particular, not every operator on X can be written as the sum of two hypercyclic operators. (*Hint*: Lemma 5.19.)

Exercise 8.4.1. Let X be a Banach space and M a dense subspace of X of countably infinite dimension. Show that for any nonzero vector $x \in M$ there exists an operator T on X such that $M = \operatorname{span} \operatorname{orb}(x, T)$; moreover, x may be hypercyclic for T. (*Hint*: Starting with x, construct a dense linearly independent sequence in M, complete to an algebraic basis of M, and apply Theorem 8.24.)

Exercise 8.4.2. Show that every normed space of countably infinite dimension supports an operator without nontrivial closed invariant subsets. (*Hint*: Apply the previous exercise to the completion of X.)

Exercise 8.4.3. Show that, for any $\varepsilon > 0$, the operator T in Theorem 8.24 can be of the form $T = I + K$, where K is a compact operator with $\|K\| < \varepsilon$. (*Hint*: Remark 8.10 and Exercise 5.2.3.)

Exercise 8.4.4. Show directly that the operator C defined in the proof of Lemma 8.21 is invertible if and only if it is injective, and if and only if $\langle A^{-1}y_1 - x_0, x^* \rangle \neq -1$. In that case, calculate the inverse.

Exercise 8.4.5. Show that Theorem 8.24 is false in ω. Indeed, there exists a dense linearly independent sequence in ω that cannot be the orbit of any operator on ω. To see this, let $X_1 = \{x_n \ ; \ n \in \mathbb{N}\}$ be a dense linearly independent set in ω such that, for $n \geq 1$, $x_n = (x_{n,k})_k$ is a finite sequence with $x_{n,1} \neq 0$. We consider $X_2 = \{y_n \ ; \ n \in \mathbb{N}\}$ defined by $y_{n,k} = 0$ if $k \leq n$, and $y_{n,k} = n^k$ if $k > n$. Then $X_0 = X_1 \cup X_2$ is dense and linearly independent in ω, but there is no operator T on ω such that X_0 coincides with an orbit under T. (*Hint*: Otherwise there would be infinitely many elements in X_2 whose image under T belonged to X_1; derive that then the composition of T with the projection onto the first coordinate would not be continuous.)

Exercise 8.4.6. Prove the following extension of Theorem 8.24 to C_0-semigroups. For any dense linearly independent set $\{x_n \ ; \ n \in \mathbb{N}\}$ in a Banach space X, there is a C_0-semigroup $(T_t)_{t\geq 0}$ on X such that $\operatorname{orb}(x_1, T_1) = \{x_n \ ; \ n \in \mathbb{N}\}$. (*Hint*: Combine the proof of Theorem 8.24 with Exercises 8.2.6 and 8.4.3.)

Sources and comments

Section 8.1. The main result of this section comes in three equivalent forms. Salas [274] showed that any perturbation of the identity by a weighted backward shift is hypercyclic on ℓ^p or c_0; see Corollary 8.3. Following his arguments, Desch, Schappacher and Webb [131] showed that the exponential of the backward shift is hypercyclic on the corresponding weighted spaces; see Theorem 8.1. Some years earlier, Chan and Shapiro [106] had obtained the hypercyclicity of the translation operator T_a on Hilbert spaces of entire functions of slow growth; see Exercise 8.1.2. The fact that the equivalence of these results was only noticed much later is, perhaps, due to the fact that the authors worked in very different contexts. While Salas regarded iterates of a function of the backward shift, Desch, Schappacher and Webb were interested in C_0-semigroups generated by shifts, and Chan and Shapiro studied translation operators on spaces of entire functions. In addition, the proof by Chan and Shapiro was very different from the one by Salas by using, in a crucial way, tools from complex analysis. Incidentally, the quasiconjugacies between $I + B$ and e^B were first observed in Martínez and Peris [232].

León and Montes [220] showed that Salas' operators satisfy the Hypercyclicity Criterion, while Grivaux [172] showed that they are even mixing.

Proposition 8.5 is due to Grivaux [171]. Theorem 8.6 was obtained by Grivaux and Shkarin [176, 286]; see also Bayart and Matheron [44]. Hypercyclicity of functions $\varphi(T)$ of rather general operators T have been studied by Herzog and Schmoeger [200], Miller and Miller [239], Bermúdez and Miller [53], Martínez and Peris [232], and Müller [248].

Section 8.2. The existence of hypercyclic operators on any infinite-dimensional separable Banach space was independently obtained by Ansari [10] and Bernal [54], solving a problem from Rolewicz [268]. Later, it was generalized to Fréchet spaces by Bonet and Peris [85]. The construction heavily depends on the hypercyclicity of perturbations of the identity by weighted shifts proved by Salas [274]. Grivaux [172] noticed that the operators of Ansari and Bernal are even mixing. It is also fair to say that, on Banach spaces, these operators are perturbations of the identity by the supercyclic operators constructed earlier by Herzog [197]. Lemma 8.7 was obtained by Metafune and Moscatelli [238].

Theorem 8.11 is due to Argyros and Haydon [12]. Bonet, Martínez and Peris [82] obtained Theorem 8.12; the deep result by Argyros and Haydon allowed us to simplify the proof. Theorem 8.13 is due to Grivaux [171]. De la Rosa, Frerick, Grivaux and Peris [127] showed that on every infinite-dimensional complex Fréchet space with an unconditional basis there is a chaotic operator, and that there is an infinite-dimensional real Banach space with an unconditional basis without chaotic operators.

Theorems 8.14 and 8.15 are due to Bermúdez and Bonilla with Conejero and Peris [49] and with Martinón [50].

Further existence results are due to Grivaux [172] (every infinite-dimensional separable Banach space supports a mixing operator that fails Kitai's criterion), Shkarin [285] (every infinite-dimensional separable Banach space supports an operator that satisfies Kitai's criterion), Grivaux and Shkarin [176, 286] (every infinite-dimensional separable Fréchet space not isomorphic to ω supports a hypercyclic non-mixing operator; on ω, every hypercyclic operator is mixing) and to Salas [276], see also Grivaux and Shkarin [176, 286] (every infinite-dimensional separable Banach space with separable dual supports a hypercyclic operator whose adjoint is also hypercyclic).

Hájek and Vivi [190] have applied the Ansari–Bernal theorem to limit sets of solutions of ordinary differential equations in Banach spaces.

Section 8.3. Chan [100] showed that the set of hypercyclic operators on an infinite-dimensional separable Hilbert space is SOT-dense in the space of all operators, which was extended by Bès and Chan [68] to Fréchet spaces. The simplified proofs using Proposition 8.16 by Hadwin, Nordgren, Radjavi, and Rosenthal [189] are due to Bès and Chan [67] and Prăjitură [260]. In the same vein many similar results are possible; see Bès and Chan [67].

The fact that hypercyclic operators are nowhere dense under the operator norm topology was observed by Wu [302], who even showed that the set of cyclic operators has this property. Herrero [194] has obtained a spectral description of the operator norm closure of the set of hypercyclic operators on an infinite-dimensional separable complex Hilbert space; see also Müller [248]. Herrero then deduced in [195] that the chaotic operators are norm dense in the set of hypercyclic operators.

As the set of hypercyclic operators is not dense in the operator norm topology one might wonder if its linear hull is dense. First, Bès and Chan [67] proved even more: on an infinite-dimensional separable complex Hilbert space the set of sums of two hypercyclic (or even chaotic) operators is dense in the operator norm. This motivated the following deep result.

Theorem 8.25 (Grivaux [170]). *Let X be an infinite-dimensional separable complex Hilbert space. Then every operator is the sum of two hypercyclic operators.*

In fact, both operators can even be chosen to be chaotic. But Grivaux also showed that, even for hypercyclicity, the result cannot be extended to all Banach spaces; see Exercise 8.3.6.

Section 8.4. The problem of whether every linearly independent sequence in a separable Banach space is contained in the orbit of a hypercyclic vector was posed by Halperin, Kitai and Rosenthal [192], who had obtained the result for Hilbert spaces. Theorem 8.24, which solves the problem, is due to Grivaux [169]. Albanese [5] has extended the theorem to Fréchet spaces with a continuous norm.

Exercises. Exercise 8.1.2 generalizes Exercise 4.2.4 for Birkhoff's operators. It was proved (with the exception of (iv)) by Chan and Shapiro [106] with completely different techniques based on complex analysis. Exercise 8.1.3 was observed by Chan and Shapiro [106]. Exercise 8.1.4 essentially asks for the original proof of Corollary 8.3 by Salas [274]. The notion of a generalized backward shift was introduced by Godefroy and Shapiro [165], where one also finds Exercise 8.1.6. Exercises 8.1.7 and 8.1.8 are from Martínez and Peris [229], and [230, 232], respectively. Exercise 8.2.1 is taken from Bonet and Peris [85], Exercise 8.2.2 generalizes a result by Chan and Shapiro [106]. The result of Exercise 8.2.4 is due to Bès, Martin, Peris, Salas and Shkarin [70, 277, 291]. Exercise 8.2.6 is taken from Bernal and Grosse-Erdmann [63], Exercise 8.3.1 from Chan [100], and Exercises 8.3.4 and 8.3.5 from Chan and Sanders [103]. Concerning Exercise 8.3.3, De la Rosa, Frerick, Grivaux and Peris [127] have shown that also on ω there are hypercyclic non-chaotic operators, so that the result of the exercise extends to all infinite-dimensional separable Fréchet spaces. Exercise 8.3.6 is due to Grivaux [170]; the use of the Argyros–Haydon spaces simplifies the proof. Exercises 8.4.1, 8.4.2, 8.4.3 are taken from Grivaux [169], and Exercises 8.4.4, 8.4.5 and 8.4.6 from Albanese [5], Bonet, Frerick, Peris and Wengenroth [81] and Bernal and Grosse-Erdmann [63], respectively.

Chapter 9
Frequently hypercyclic operators

The theory of linear dynamical systems has its roots in topological dynamics. But there is also a parallel theory of measurable dynamics, which is better known under the name of ergodic theory. In this chapter we show how concepts and results from that theory lead to a deepened understanding of linear dynamics. More specifically, we will see how the celebrated Birkhoff ergodic theorem suggests an interesting and rather strong variant of hypercyclicity, that of frequently hypercyclic operators. We point out that, while ergodic theory has turned out to be a most powerful tool in linear dynamics, we will use it here only for motivating the new concept.

Having introduced frequently hypercyclic operators, we then derive a Frequent Hypercyclicity Criterion and an eigenvalue criterion that allow us to show that, quite surprisingly, many of the hypercyclic operators met so far are in fact frequently hypercyclic. In the final section we revisit several of the structural properties of hypercyclicity within the new framework.

9.1 Frequently recurrent orbits

Let T be an operator on a separable Fréchet space X. In order to look at T from the point of view of ergodic theory we need to have a probability measure μ on X. Since we are in a topological situation it is natural to assume that μ is defined on the Borel σ-algebra $\mathcal{B}(X)$, that is, the smallest σ-algebra containing the open subsets of X; the elements of $\mathcal{B}(X)$ are called the Borel sets of X. Since T is continuous, it is then also measurable. We assume that T satisfies the minimum requirement in ergodic theory, namely, that it is μ-invariant, that is, $\mu(T^{-1}(A)) = \mu(A)$ for every Borel set A.

Of course, μ-invariance alone does not yet give us interesting dynamics since, for example, the identity operator is automatically μ-invariant for any measure μ. This changes when we inject ergodicity: one way of defining this notion is by demanding that, for any Borel sets A and B with $\mu(A) > 0$

K.-G. Grosse-Erdmann, A. Peris Manguillot, *Linear Chaos*, Universitext,
DOI 10.1007/978-1-4471-2170-1_9, © Springer-Verlag London Limited 2011

and $\mu(B) > 0$, there is some $n \in \mathbb{N}_0$ such that $\mu(T^{-n}(A) \cap B) > 0$. This notion is not only formally similar to topological transitivity. Suppose that the measure μ has the additional property that $\mu(U) > 0$ for any nonempty open set U; μ is then said to be of full (topological) support. Under this assumption, ergodicity obviously implies topological transitivity.

But there is an added bonus in the form of the Birkhoff ergodic theorem. It tells us that if T is ergodic with respect to μ then, for any μ-integrable function f on X, its time average with respect to T coincides with its space average; more precisely we have that

$$\frac{1}{N+1}\sum_{n=0}^{N} f(T^n x) \to \int_X f\,d\mu, \quad \text{for } \mu\text{-almost all } x \in X, \tag{9.1}$$

as $N \to \infty$. This then implies an interesting topological property for T. Indeed, since X is separable, its topology has a countable base $(U_k)_k$. When we apply (9.1) to the indicator functions $\mathbf{1}_{U_k}$, $k \geq 1$, the left-hand side turns out to be

$$\frac{1}{N+1}\sum_{n=0}^{N} \mathbf{1}_{U_k}(T^n x) = \frac{\operatorname{card}\{0 \leq n \leq N\ ;\ T^n x \in U_k\}}{N+1},$$

while the right-hand side is simply $\int_X \mathbf{1}_{U_k}\,d\mu = \mu(U_k) > 0$, where we have assumed again that μ is of full support. Thus there are subsets $A_k \subset X$, $k \geq 1$, of full measure such that, for any $x \in A_k$,

$$\lim_{N\to\infty} \frac{\operatorname{card}\{0 \leq n \leq N\ ;\ T^n x \in U_k\}}{N+1} > 0.$$

Since every nonempty open set contains some U_k and since $\bigcap_{k\geq 1} A_k$ has full measure we obtain that, for μ-almost all $x \in X$ and every nonempty open subset U of X,

$$\liminf_{N\to\infty} \frac{\operatorname{card}\{0 \leq n \leq N\ ;\ T^n x \in U\}}{N+1} > 0.$$

What we have found here is that, under the mentioned assumptions, the operator T has a property that is much stronger than hypercyclicity. There must even be an $x \in X$ whose orbit meets every nonempty open set very often, in the sense given above. Let us recall here the following.

Definition 9.1. The *lower density* of a subset $A \subset \mathbb{N}_0$ is defined as

$$\underline{\operatorname{dens}}(A) = \liminf_{N\to\infty} \frac{\operatorname{card}\{0 \leq n \leq N\ ;\ n \in A\}}{N+1}.$$

Our discussion so far leads us to the following concept.

Definition 9.2. An operator T on a Fréchet space X is called *frequently hypercyclic* if there is some $x \in X$ such that, for any nonempty open subset U of X,

$$\underline{\text{dens}}\,\{n \in \mathbb{N}_0 \;;\; T^n x \in U\} > 0.$$

In this case, x is called a *frequently hypercyclic vector* for T. The set of frequently hypercyclic vectors for T is denoted by $FHC(T)$.

The orbit of a frequently hypercyclic vector is therefore, in the specified sense, frequently recurrent. Obviously, frequent hypercyclicity is a stronger notion than hypercyclicity.

There is an equivalent formulation of frequent hypercyclicity that nicely differentiates it from hypercyclicity. Let A be a subset of $\mathbb{N}_0$; if $(n_k)_{k\geq 1}$ is the increasing sequence of integers forming A and $n_k \leq N < n_{k+1}$ then

$$\frac{k}{n_{k+1}} \leq \frac{\text{card}\{0 \leq n \leq N \;;\; n \in A\}}{N+1} \leq \frac{k}{n_k},$$

which implies that $\underline{\text{dens}}\,(A) = \liminf_{k\to\infty} \frac{k}{n_k}$. Thus A has positive lower density if and only if $(\frac{n_k}{k})_k$ is bounded; in other words, if $n_k = O(k)$.

Proposition 9.3. *A vector $x \in X$ is frequently hypercyclic for T if and only if, for any nonempty open subset U of X, there is a strictly increasing sequence $(n_k)_k$ of positive integers such that*

$$T^{n_k} x \in U \quad \textit{for all } k \in \mathbb{N}, \quad \textit{and } n_k = O(k).$$

By contrast, T is hypercyclic if and only if the same is true *for some* $(n_k)_k$, not necessarily of order $O(k)$. This seems to indicate that our new notion requires much more than mere hypercyclicity.

We have the usual behaviour under quasiconjugacies, which can be proved as in Proposition 1.19.

Proposition 9.4. *Frequent hypercyclicity is preserved under quasiconjugacy.*

Our first task will be to show that frequently hypercyclic operators exist. We saw above that an operator T on a separable Fréchet space X is frequently hypercyclic if one can find a Borel probability measure μ of full support on X with respect to which T is ergodic. However, in order to keep our introduction to frequent hypercyclicity simple we will not pursue this circle of ideas any further. Instead, we will favour a constructive approach to frequent hypercyclicity.

So, what does it take for a vector x to be frequently hypercyclic for an operator T? Let $\|\cdot\|$ denote an F-norm defining the topology of X, and let $(y_l)_l$ be a dense sequence in X. Then there are subsets $A(l,\nu)$, $l,\nu \geq 1$, of $\mathbb{N}_0$ of positive lower density such that, for any $n \in A(l,\nu)$,

$$\|T^n x - y_l\| < \frac{1}{\nu}.$$

Moreover, if $y_l \neq y_k$ then the sets $A(l,\nu)$ and $A(k,\mu)$ are disjoint if ν and μ are big. In fact, in the sequel we will need the existence of sets $A(l,\nu)$ with a stronger separation property.

Lemma 9.5. *There exist pairwise disjoint subsets $A(l,\nu)$, $l,\nu \geq 1$, of $\mathbb{N}_0$ of positive lower density such that, for any $n \in A(l,\nu)$ and $m \in A(k,\mu)$, we have that $n \geq \nu$ and*

$$|n-m| \geq \nu + \mu \quad \textit{if } n \neq m.$$

Proof. We start by partitioning $\mathbb{N}$ in a very natural fashion by using the dyadic representation

$$n = \sum_{j=0}^{\infty} a_j 2^j =: (a_0, a_1, a_2, \ldots)$$

of any positive integer n. We define $I(l,\nu)$, $l,\nu \geq 1$, as the set of all $n \in \mathbb{N}$ whose dyadic representation has the form

$$n = (0,\ldots,0,1,\ldots,1,0,*)$$

with $l-1$ leading zeros, followed by ν ones, then one zero, followed by an arbitrary tail. It is clear that the sets $I(l,\nu)$ form a partition of $\mathbb{N}$, but they do not satisfy the required separation property. To achieve this we let $\delta_k = \nu$ if $k \in I(l,\nu)$ for some $l \geq 1$, and we define

$$n_k = 2\sum_{i=1}^{k-1} \delta_i + \delta_k, \quad k \geq 1,$$

which is a strictly increasing sequence. We claim that

$$A(l,\nu) = \{n_k \ ; \ k \in I(l,\nu)\}, \quad l,\nu \geq 1$$

has the desired properties. First, these sets are pairwise disjoint. Moreover, if $n_k \in A(l,\nu)$ then $n_k \geq \delta_k = \nu$; and if $n_j \in A(l,\nu)$, $n_m \in A(k,\mu)$ with $n_j \neq n_m$, where we can assume that $j > m$, then

$$n_j - n_m = \delta_m + 2\sum_{i=m+1}^{j-1} \delta_i + \delta_j \geq \mu + \nu.$$

It remains to show that each set $A(l,\nu)$ has positive lower density. We begin by proving that there is some $M > 0$ such that

$$n_k \leq Mk, \quad k \geq 1. \tag{9.2}$$

It suffices to do this for $k = 2^N$, $N \geq 1$, because we then have for $2^{N-1} \leq k < 2^N$ that

$$n_k \leq n_{2^N} \leq M 2^N \leq 2Mk.$$

Thus let $k = 2^N$. A simple but tedious enumeration shows that, if $l + \nu \leq N + 2$, then $I(l, \nu)$ contains at most $2^{N+2-l-\nu}$ elements that do not exceed 2^N, and none if $l + \nu > N + 2$. Hence we have that

$$n_{2^N} \leq 2 \sum_{i=1}^{2^N} \delta_i \leq 2 \sum_{l+\nu \leq N+2} 2^{N+2-l-\nu} \nu \leq \Big(8 \sum_{l,\nu \geq 1} \frac{\nu}{2^{l+\nu}} \Big) 2^N,$$

so that (9.2) holds for some $M > 0$.

Now let $l, \nu \geq 1$. Let $(k_j)_j$ be the increasing sequence of elements of $I(l, \nu)$. Since the latter set has positive lower density, the argument leading up to Proposition 9.3 shows that there is some constant $K > 0$ such that

$$k_j \leq Kj, \quad j \geq 1.$$

It then follows that $A(l, \nu) = \{n_{k_j} \; ; \; j \geq 1\}$ and

$$n_{k_j} \leq M k_j \leq MKj, \quad j \geq 1.$$

Hence each set $A(l, \nu)$ has positive lower density. □

This result allows us to obtain a first example of a frequently hypercyclic operator.

Example 9.6. **(Birkhoff's operators)** The translation operators $T_a : f \to f(\cdot + a)$, $a \neq 0$, on the space $H(\mathbb{C})$ of entire functions are frequently hypercyclic. By Proposition 9.4 and Example 4.26 it suffices to consider $a = 1$.

Thus, let $A(l, \nu)$, $l, \nu \geq 1$, be subsets of $\mathbb{N}_0$ as given by Lemma 9.5, and let $(P_l)_l$ be a dense sequence of polynomials. Let $(n_k)_k$ be the increasing sequence of elements of $\bigcup_{l,\nu \geq 1} A(l, \nu)$. If $n_k \in A(l, \nu)$ then we define B_k as the closed ball around n_k of radius $r_k := \nu/2$, and on this ball we consider the function $g_k := P_l(z - n_k)$; see Figure 9.1. It follows from the lemma that

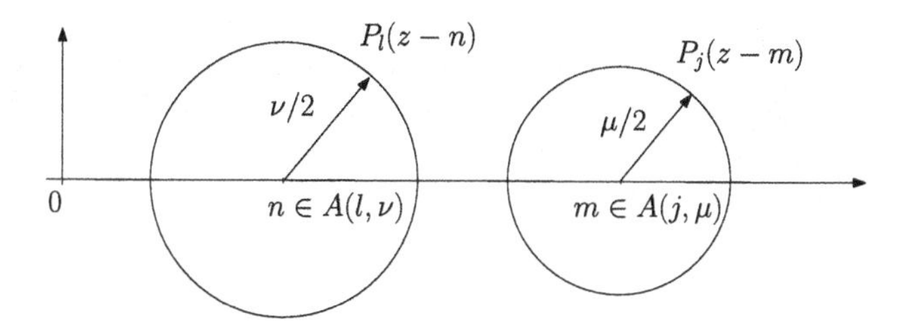

Fig. 9.1 Constructing Birkhoff frequently hypercyclic functions

the B_k are pairwise disjoint. We now apply Runge's theorem recursively. We start with $f_1 = g_1$. If entire functions $f_1, \ldots, f_k$, $k \geq 1$, have been constructed then we consider the function that is defined as f_k on $|z| \leq n_k + r_k$ and as g_{k+1} on B_{k+1}. Let $\varepsilon_k > 0$ be numbers that will be specified later. By Runge's approximation theorem there is an entire function f_{k+1} such that

$$\sup_{|z|\leq n_k+r_k} |f_{k+1}(z) - f_k(z)| < \varepsilon_k \quad \text{and} \quad \sup_{z\in B_{k+1}} |f_{k+1}(z) - g_{k+1}(z)| < \varepsilon_k.$$

If $\sum_{k=1}^{\infty} \varepsilon_k < \infty$ then it follows from the first inequality and the fact that $n_k \to \infty$ that

$$f(z) := f_1(z) + \sum_{k=1}^{\infty}(f_{k+1}(z) - f_k(z)) = \lim_{k\to\infty} f_k(z)$$

defines an entire function. Moreover we have with $\varepsilon_0 = 0$ that

$$\begin{aligned}\sup_{z\in B_k} |f(z) - g_k(z)| &\leq \sup_{z\in B_k} |f_k(z) - g_k(z)| + \sum_{j=k}^{\infty} \sup_{z\in B_k} |f_{j+1}(z) - f_j(z)| \\ &\leq \sum_{j=k-1}^{\infty} \varepsilon_j,\end{aligned}$$

that is,

$$\sup_{|z-n_k|\leq\nu/2} |f(z) - P_l(z - n_k)| \leq \sum_{j=k-1}^{\infty} \varepsilon_j$$

for $n_k \in A(l, \nu)$. It is easy to see that we can choose the ε_j in such a way that $\sum_{j=k-1}^{\infty} \varepsilon_j < \frac{1}{\nu}$ whenever $n_k \in A(l, \nu)$. We therefore have that

$$\sup_{|z|\leq\nu/2} |T_1^{n_k} f(z) - P_l(z)| < \frac{1}{\nu}$$

for $n_k \in A(l, \nu)$. Since the sets $\{g \in H(\mathbb{C})\ ;\ \sup_{|z|\leq\nu/2} |g(z) - P_l(z)| < \frac{1}{\nu}\}$, $l, \nu \geq 1$, form a basis of the topology of $H(\mathbb{C})$ and since each set $A(l, \nu)$ has positive lower density, it follows that T_1 is frequently hypercyclic.

It is of interest to compare the new, and strong, form of hypercyclicity with other strong forms such as weak mixing, mixing and chaos. We start here by showing that every frequently hypercyclic operator is weakly mixing. For the proof we need a property of sets of positive lower density. For any subset A of $\mathbb{N}_0$ its *difference set* is defined as

$$A - A = \{n - m\ ;\ n, m \in A, n \geq m\};$$

it should be noted that we consider here only nonnegative differences. We recall that a subset B of $\mathbb{N}_0$ is called syndetic if its complement does not contain intervals of arbitrary length; one also says that it has bounded gaps.

Theorem 9.7 (Erdős–Sárközy). *Let $A \subset \mathbb{N}_0$ be a set of positive lower density. Then the difference set $A - A$ is syndetic.*

Proof. Suppose that the difference set D of A is not syndetic. In particular, there exists some $n_1 \notin D$. Moreover, since $\mathbb{N}_0 \setminus D$ contains intervals of arbitrary length, there is some $n_2 \notin D$ such that also $n_2 + n_1 \notin D$. Hence $\{n_1, n_2, n_1 + n_2\} \subset \mathbb{N}_0 \setminus D$. Similarly, there is some $n_3 \notin D$ such that $n_3+n_1, n_3+n_2 \notin D$, which implies that $\{n_1, n_2, n_3, n_1+n_2, n_1+n_3, n_2+n_3\} \subset \mathbb{N}_0 \setminus D$. Continuing in this way we obtain a sequence $(n_k)_k$ in $\mathbb{N}_0$ such that any finite sum of elements in the sequence belongs to $\mathbb{N}_0 \setminus D$.

We now fix a positive integer m such that $\underline{\mathrm{dens}}(A) > \frac{1}{m}$, and we consider the sets

$$A_k = A + (n_1 + \ldots + n_k), \quad k \in \mathbb{N}.$$

Since each set A_k also has lower density larger than $\frac{1}{m}$, there is some $N \geq 1$ such that, for any $k \leq m$,

$$\mathrm{card}\{n \leq N \; ; \; n \in A_k\} > \frac{N+1}{m}.$$

If the A_k, $k = 1, \ldots, m$, were pairwise disjoint, we would have that

$$\mathrm{card}\{n \leq N \; ; \; n \in A_1 \cup \ldots \cup A_m\} > m\frac{N+1}{m} = N + 1,$$

which is impossible. Hence there are $j < k$ with $A_j \cap A_k \neq \varnothing$, which implies that

$$n_{j+1} + \ldots + n_k \in A - A = D.$$

This contradicts the construction of the n_k. $\square$

With this we can prove the announced result.

Theorem 9.8. *Any frequently hypercyclic operator on a Fréchet space is weakly mixing.*

Proof. Let T be a frequently hypercyclic operator on a Fréchet space X. We want to show that the condition of Theorem 2.47 is satisfied. Thus, let W be a 0-neighbourhood and U and V nonempty open subsets of X.

First, since T is hypercyclic and therefore topologically transitive, there is some $n_0 \geq 0$ such that $T^{n_0}(U) \cap W \neq \varnothing$. By continuity there is a nonempty open subset U_0 of U such that $T^{n_0}(U_0) \subset W$. Now let x be an arbitrary frequently hypercyclic vector for T. Then there is a set $A \subset \mathbb{N}_0$ of positive lower density such that

$$T^n x \in U_0 \quad \text{for any } n \in A.$$

For $m, n \in A$, $m \geq n$, we then find that

$$T^{n_0+m-n}(T^n x) = T^{n_0}(T^m x) \in W.$$

We thus have that

$$n_0 + (A - A) \subset N(U_0, W) \subset N(U, W).$$

It follows from Theorem 9.7 that $N(U, W)$ is syndetic.

Secondly, by continuity and linearity, each set $T^{-k}(W)$ is a 0-neighbourhood. Thus, given any $m \geq 1$, there is a 0-neighbourhood W_0 such that $T^k(W_0) \subset W$ for $k = 1, \ldots, m$. Again by topological transitivity there is some $K > m$ and some $y \in W_0$ such that $T^K y \in V$ and hence, for $1 \leq k \leq m$,

$$T^{K-k}(T^k y) \in T^{K-k}(W) \cap V.$$

This shows that, for any $m \geq 1$, $N(W, V)$ contains m consecutive integers.

Our two conclusions imply that $N(U, W) \cap N(W, V) \neq \varnothing$, so that, by Theorem 2.47, T is weakly mixing. □

9.2 The Frequent Hypercyclicity Criterion

In order to obtain further examples of frequently hypercyclic operators, we derive here a sufficient condition for frequent hypercyclicity that resembles the Hypercyclicity Criterion. Its proof is inspired by Kitai's constructive approach to that criterion; see the alternative proof of Theorem 3.12. However, at a crucial point we have to depart from that proof: since we require approximation on sets of positive lower density we can no longer define the k_j inductively. On the other hand, fixing the k_j in advance is no option either as they will necessarily depend on the chosen operator. Lemma 9.5 provides us with exactly the right tool for the construction.

We refer to Appendix A for the notion of unconditionally convergent series.

Theorem 9.9 (Frequent Hypercyclicity Criterion). *Let T be an operator on a separable Fréchet space X. If there is a dense subset X_0 of X and a map $S : X_0 \to X_0$ such that, for any $x \in X_0$,*

(i) $\sum_{n=0}^{\infty} T^n x$ *converges unconditionally,*

(ii) $\sum_{n=0}^{\infty} S^n x$ *converges unconditionally,*

(iii) $TSx = x$,

then T is frequently hypercyclic.

Proof. Since X is separable we can choose a sequence $(y_j)_j$ from X_0 that is dense in X. Let $\|\cdot\|$ denote an F-norm that defines the topology of X. Conditions (i) and (ii) imply that there are $N_l \in \mathbb{N}$, $l \geq 1$, such that, for any $j \leq l$ and any finite set $F \subset \{N_l, N_l+1, N_l+2, \ldots\}$ we have that

$$\Big\| \sum_{n\in F} T^n y_j \Big\| < \frac{1}{l2^l}, \tag{9.3}$$

$$\Big\| \sum_{n\in F} S^n y_j \Big\| < \frac{1}{l2^l}. \tag{9.4}$$

Now let $A(l,\nu)$, $l,\nu \geq 1$, be subsets of $\mathbb{N}_0$ as given by Lemma 9.5. We set

$$A = \bigcup_{l=1}^{\infty} A(l, N_l)$$

and

$$z_n = y_l \quad \text{if} \quad n \in A(l, N_l).$$

We then consider

$$x = \sum_{n\in A} S^n z_n. \tag{9.5}$$

First we want to verify that this series converges unconditionally. Let us fix $l \geq 1$. For any finite set $F \subset \mathbb{N}_0$ we have that

$$\sum_{\substack{n\in A\\ n\in F}} S^n z_n = \sum_{j=1}^{\infty} \sum_{\substack{n\in A(j,N_j)\\ n\in F}} S^n y_j = \sum_{j=1}^{l} \sum_{\substack{n\in A(j,N_j)\\ n\in F}} S^n y_j + \sum_{j=l+1}^{\infty} \sum_{\substack{n\in A(j,N_j)\\ n\in F}} S^n y_j.$$

It follows from (9.4) that, for $j \leq l$ and $F \subset \{N_l, N_l+1, N_l+2, \ldots\}$ finite,

$$\Big\| \sum_{\substack{n\in A(j,N_j)\\ n\in F}} S^n y_j \Big\| < \frac{1}{l2^l};$$

moreover, since $n \geq N_j$ for any $n \in A(j, N_j)$ by Lemma 9.5, we also have by (9.4) that, for any $j \geq 1$ and any finite set F,

$$\Big\| \sum_{\substack{n\in A(j,N_j)\\ n\in F}} S^n y_j \Big\| < \frac{1}{j2^j} \leq \frac{1}{2^j}.$$

Altogether we have that, for any finite set $F \subset \{N_l, N_l+1, N_l+2, \ldots\}$,

$$\Big\| \sum_{\substack{n\in A\\ n\in F}} S^n z_n \Big\| < \sum_{j=1}^{l} \frac{1}{l2^l} + \sum_{j=l+1}^{\infty} \frac{1}{2^j} = \frac{2}{2^l}.$$

Since l was arbitrary we have proved that the series (9.5) converges unconditionally.

We now show that x is frequently hypercyclic for T. To this end, fix $l \geq 1$. Then, for $n \in A(l, N_l)$,

$$T^n x - y_l = \sum_{\substack{k \in A \\ k<n}} T^n S^k z_k + \sum_{\substack{k \in A \\ k>n}} T^n S^k z_k + T^n S^n z_n - y_l.$$

For the second sum we have, for any $m \geq n$, using condition (iii),

$$\sum_{\substack{k \in A \\ n<k\leq m}} T^n S^k z_k = \sum_{j=1}^{l} \sum_{\substack{k \in A(j,N_j) \\ n<k\leq m}} S^{k-n} y_j + \sum_{j=l+1}^{\infty} \sum_{\substack{k \in A(j,N_j) \\ n<k\leq m}} S^{k-n} y_j.$$

Note that, by Lemma 9.5, $k-n \geq N_l$ in the first sum and $k-n \geq N_j$ in the second sum. Therefore, the same argument as above shows that

$$\Big\| \sum_{\substack{k \in A \\ n<k\leq m}} T^n S^k z_k \Big\| < \sum_{j=1}^{l} \frac{1}{l2^l} + \sum_{j=l+1}^{\infty} \frac{1}{2^j} = \frac{2}{2^l},$$

hence

$$\Big\| \sum_{\substack{k \in A \\ k>n}} T^n S^k z_k \Big\| \leq \frac{2}{2^l}.$$

In the same way, but using (9.3) instead of (9.4), we obtain that also

$$\Big\| \sum_{\substack{k \in A \\ k<n}} T^n S^k z_k \Big\| \leq \frac{2}{2^l}.$$

Finally, since $n \in A(l, N_l)$ we have that

$$T^n S^n z_n = y_l.$$

Altogether we find that for all $n \in A(l, N_l)$

$$\|T^n x - y_l\| \leq \frac{4}{2^l}.$$

Since the y_l form a dense set in X and since each set $A(l, N_l)$ is of positive lower density we conclude that x is frequently hypercyclic for T. □

Remark 9.10. For a later application we note that the same proof works when we replace conditions (ii) and (iii) by the following:

For any $x \in X_0$ there is a sequence $(u_n)_{n\geq 0}$ in X with $u_0 = x$ such that $\sum_{n=0}^{\infty} u_n$ converges unconditionally and $T^n u_k = u_{k-n}$ if $n \leq k$.

Let us also note here that the Frequent Hypercyclicity Criterion not only implies frequent hypercyclicity but also two other strong forms of hypercyclicity.

Proposition 9.11. *An operator on a separable Fréchet space that satisfies the Frequent Hypercyclicity Criterion is also chaotic and mixing.*

Proof. The mixing property follows immediately from Kitai's criterion.

As for chaos, we have from conditions (i) and (ii) that, for any $x \in X_0$ and $N \geq 1$,

$$y_{x,N} := \sum_{j=1}^{\infty} S^{jN} x + x + \sum_{j=1}^{\infty} T^{jN} x$$

converges in X. Moreover, by condition (iii), $T^N y_{x,N} = y_{x,N}$, and by (i) and (ii) we have that $y_{x,N} \to x$ as $N \to \infty$. Since X_0 is dense, the $y_{x,N}$ therefore form a dense set of periodic points for T. Knowing already that T is hypercyclic we deduce that T is even chaotic. □

In the last section we saw that Birkhoff's operators are frequently hypercyclic. The Frequent Hypercyclicity Criterion allows us to show that also the other two classical hypercyclic operators, the operators of MacLane and Rolewicz, are in fact frequently hypercyclic; see also Exercise 9.2.2.

Example 9.12. **(MacLane's operator)** The differentiation operator D on $H(\mathbb{C})$ is frequently hypercyclic. To see this we proceed as in Example 3.7. Let X_0 be the set of polynomials and S the operator $Sf(z) = \int_0^z f(\zeta)\, d\zeta$. Condition (i) of the Frequent Hypercyclicity Criterion is satisfied since any finite series converges unconditionally, and (iii) is trivial. For (ii) we need only consider the monomials, for which we find that $\sum_{n=0}^{\infty} S^n(z^k) = k! \sum_{n=0}^{\infty} \frac{1}{(k+n)!} z^{k+n}$, which converges uniformly and unconditionally on any compact set.

We study Rolewicz's operators in the broader context of general weighted shifts. We will use the notation and terminology of Section 4.1. In particular, we consider weighted (backward) shifts

$$B_w : (x_1, x_2, x_3, \ldots) \to (w_2 x_2, w_3 x_3, w_4 x_4, \ldots),$$

where $w = (w_n)_n$ is a weight sequence. By e_n, $n \geq 1$, we denote the canonical unit sequences.

Proposition 9.13. *Let B_w be a weighted shift on a Fréchet sequence space X in which* $\operatorname{span}\{e_n \; ; \; n \geq 1\}$ *is dense. If the series*

$$\sum_{n=1}^{\infty}\Big(\prod_{\nu=1}^{n} w_\nu\Big)^{-1} e_n$$

converges unconditionally in X then B_w is frequently hypercyclic.

Proof. We apply the Frequent Hypercyclicity Criterion. We choose X_0 as the set of finite sequences, which is dense by assumption, and for S we consider the weighted forward shift $(x_1, x_2, x_3, \ldots) \to (0, x_1/w_2, x_2/w_3, \ldots)$. Then condition (i) holds because any finite series converges unconditionally, and condition (iii) is obvious. By linearity, we need to confirm (ii) only for the sequences e_k, $k \geq 1$. But then

$$\sum_{n=0}^{\infty} S^n e_k = \sum_{n=0}^{\infty} \frac{e_{k+n}}{w_{k+1}\cdots w_{k+n}} = \Big(\prod_{\nu=1}^{k} w_\nu\Big) \sum_{n=0}^{\infty}\Big(\prod_{\nu=1}^{k+n} w_\nu\Big)^{-1} e_{k+n},$$

which converges unconditionally by hypothesis. □

In particular, by Theorem 4.8 we have the following.

Corollary 9.14. *On a Fréchet sequence space X in which $(e_n)_n$ is an unconditional basis, every chaotic weighted shift is frequently hypercyclic.*

The result covers some interesting special cases.

Example 9.15. **(Rolewicz's operators)** For $\lambda \in \mathbb{K}$ we consider the multiples $T = \lambda B$ of the shift operator B with $|\lambda| > 1$. Then T is frequently hypercyclic on any Fréchet sequence space on which it is defined, in which $(e_n)_n$ is an unconditional basis and that contains the sequence $(1/\lambda^n)_n$. This includes, in particular, the spaces ℓ^p, $1 \leq p < \infty$, and c_0.

Example 9.16. (a) We follow Example 4.9(b) and consider weighted shifts $T = B_w$ on $H(\mathbb{C})$, or rather its corresponding sequence space. Since the sequences e_n, $n \geq 0$, correspond to the monomials z^n, B_w turns out to be frequently hypercyclic if $\sum_{n=1}^{\infty}(\prod_{\nu=1}^{n} w_\nu)^{-1} z^n$ converges unconditionally in $H(\mathbb{C})$, which is equivalent to

$$\lim_{n\to\infty}\Big(\prod_{\nu=1}^{n} |w_\nu|\Big)^{1/n} = \infty.$$

Since the differentiation operator D corresponds to the weights $w_n = n$, we also get a new proof that D is frequently hypercyclic.

(b) In the space $\omega = \mathbb{K}^{\mathbb{N}}$, every series $\sum_{n=1}^{\infty} a_n e_n$ converges unconditionally. As a consequence, every weighted shift is frequently hypercyclic on ω.

One is still far away from a characterization of frequently hypercyclic weighted shifts. We complement here the sufficient condition derived above by a necessary condition.

Proposition 9.17. *Let B_w be a weighted shift on a Fréchet sequence space X in which $(e_n)_n$ is an unconditional basis. If B_w is frequently hypercyclic then there exists a subset $A \subset \mathbb{N}_0$ of positive lower density such that*

$$\sum_{n\in A} \Big(\prod_{\nu=1}^{n} w_\nu\Big)^{-1} e_n \quad \textit{converges.}$$

Proof. Let x be a frequently hypercyclic vector for $T = B_w$. Since

$$T^n x = (w_2 w_3 \cdots w_{n+1} x_{n+1}, \ldots)$$

and since the projection onto the first coordinate is continuous on X, there is a set $B \subset \mathbb{N}_0$ of positive lower density such that, for any $n \in B$,

$$|w_2 w_3 \cdots w_{n+1} x_{n+1} - 2| < 1,$$

hence

$$|x_{n+1}| > \frac{1}{|w_2 w_3 \cdots w_{n+1}|}.$$

Together with the unconditional convergence of $\sum_{n=1}^{\infty} x_n e_n$ this implies that

$$\sum_{n\in B} \frac{1}{w_1 w_2 \cdots w_{n+1}} e_{n+1}$$

converges; see Theorem A.16. This proves the claim for $A = \{n+1\ ;\ n \in B\}$. □

We note that this condition is not, in general, a sufficient condition; see Exercise 9.2.5.

While at the outset it was not even clear if frequently hypercyclic operators exist, we have now actually seen that all the classical hypercyclic operators and many others have this strong form of hypercyclicity. Although it is not to be expected that hypercyclicity and frequent hypercyclicity coincide, we are also in the position to give an example that differentiates the two concepts.

Example 9.18. On $X = \ell^2$ we consider the weighted shift B_w with weights $w_n = (\frac{n+1}{n})^{1/2}$. It follows from Example 4.9(a) that B_w is hypercyclic and even mixing. However, if B_w were frequently hypercyclic then by the previous proposition we could find a set $A = \{n_k\ ;\ k \geq 1\}$ of positive lower density such that $\sum_{k=1}^{\infty} \frac{1}{n_k+1} < \infty$, which is impossible since $n_k = O(k)$; see the discussion before Proposition 9.3. Note that B_w is conjugate to the shift operator B on the Bergman space A^2; see Example 4.4(b).

This example takes us back to the problem of comparing frequent hypercyclicity with other forms of hypercyclicity. We saw in Theorem 9.8 that every frequently hypercyclic operator is weakly mixing. Some other implications have turned out to be false. We have just seen that the mixing property does not imply frequent hypercyclicity. But there is also a frequently hypercyclic operator on c_0 that is neither mixing nor chaotic. In particular, by Proposition 9.11, this operator does not satisfy the Frequent Hypercyclicity Criterion. The construction of this example goes well beyond the scope of this book.

The reader may have noticed that in frequent hypercyclicity so far we have not made use of the Baire category theorem. This is unlike the situation in hypercyclicity where the existence of a hypercyclic vector was deduced from the fact that, in the sense of Baire category, there must be many of them; see the proof of the Birkhoff transitivity theorem. In fact, this procedure is ruled out in frequent hypercyclicity because, in general, the set $FHC(T)$ of frequently hypercyclic vectors for an operator T is only of first Baire category.

Proposition 9.19. *Let T be an operator on a Fréchet space X. If there is a dense set X_0 such that $T^n x \to 0$ for all $x \in X_0$ then $FHC(T)$ is of first Baire category. This is true, in particular, for all operators satisfying the Frequent Hypercyclicity Criterion.*

Proof. Let $\|\cdot\|$ be an F-norm defining the topology of X, and choose $\delta > 0$ such that $\{x \in X \; ; \; \|x\| > \delta\}$ is nonempty. Then every frequently hypercyclic vector for T belongs to the set

$$E := \{x \in X \; ; \; \underline{\text{dens}}\{n \in \mathbb{N}_0 \; ; \; \|T^n x\| \geq \delta\} > 0\}.$$

We have that

$$E = \bigcup_{k\geq 1} \bigcup_{M \geq 1} E_{k,M},$$

where

$$E_{k,M} = \bigcap_{N \geq M} \left\{x \in X \; ; \; \text{card}\{n \leq N \; ; \; \|T^n x\| \geq \delta\} \geq \tfrac{N+1}{k}\right\}.$$

The continuity of T implies that the complement of $E_{k,M}$,

$$X \setminus E_{k,M} = \bigcup_{N \geq M} \left\{x \in X \; ; \; \text{card}\{n \leq N \; ; \; \|T^n x\| < \delta\} > (N+1)(1 - \tfrac{1}{k})\right\},$$

is open, and it contains the dense set X_0. Hence each set $E_{k,M}$ is nowhere dense, so that E is of first Baire category. □

Thus one cannot argue as in the case of hypercyclicity (see Proposition 2.52), that any vector in the underlying space is the sum of two frequently hypercyclic vectors. Indeed, there are frequently hypercyclic operators T for which $X \neq FHC(T) + FHC(T)$; see Exercise 9.2.6. On the other hand, there are operators T for which the set $FHC(T)$ is sufficiently large to ensure that $X = FHC(T) + FHC(T)$; see Exercise 9.1.4.

We end this section with another interesting phenomenon. Chapter 11 will be devoted to the question of whether an uncountable family of hypercyclic operators on a given space can have a common hypercyclic vector. The answer is positive, for example, for the Rolewicz operators λB, $\lambda > 1$, on any of the spaces $X = \ell^p$, $1 \leq p < \infty$, or c_0; see Example 11.11. The corresponding result is false, however, for frequent hypercyclicity.

Example 9.20. Let X be one of the spaces ℓ^p, $1 \leq p < \infty$, or c_0. Then the Rolewicz operators λB, $\lambda > 1$, on X have no common frequently hypercyclic vector. Indeed, suppose that x was such a vector. By the proof of Proposition 9.17 it then follows that, for any $\lambda > 1$,

$$\delta_\lambda := \underline{\text{dens}}\{n \in \mathbb{N}_0 \ ; \ |\lambda^n x_{n+1} - 2| < 1\} > 0.$$

Since there are uncountably many λ, one can find a finite subset, $\lambda_1 < \lambda_2 < \ldots < \lambda_K$ say, such that

$$\sum_{k=1}^{K} \delta_{\lambda_k} > 2.$$

Let $\rho = \min_{1\leq k<K} \frac{\lambda_{k+1}}{\lambda_k}$, and choose $M \in \mathbb{N}$ such that $\rho^M \geq 3$. We then have for N sufficiently large that, for any $k = 1, \ldots, K$,

$$\text{card}\{M \leq n \leq N \ ; \ |\lambda_k^n x_{n+1} - 2| < 1\} \geq \tfrac{1}{2}\delta_{\lambda_k} N.$$

Since $\sum_{k=1}^{K} \frac{1}{2}\delta_{\lambda_k} N > N$, the corresponding sets cannot be pairwise disjoint. Hence there are $1 \leq k < l \leq K$ and $n \geq M$ such that $|\lambda_k^n x_{n+1} - 2| < 1$ and $|\lambda_l^n x_{n+1} - 2| < 1$. Thus, $\lambda_k^n |x_{n+1}| > 1$ and $\lambda_l^n |x_{n+1}| < 3$, which implies that $\rho^M \leq (\lambda_l/\lambda_k)^n < 3$, a contradiction.

9.3 An eigenvalue criterion for frequent hypercyclicity

By the Godefroy–Shapiro criterion, eigenvalues inside and outside the unit disk with many associated eigenvectors are useful for proving an operator to be hypercyclic. Additional eigenvectors to certain unimodular eigenvalues are responsible for chaos. We recall that an eigenvalue λ is called *unimodular* if $|\lambda| = 1$.

In this section we will see that, rather surprisingly, a large supply of eigenvectors to unimodular eigenvalues by itself may lead to hypercyclicity, and in some cases to frequent hypercyclicity. Let us only mention that the correct interpretation of largeness in this context was again motivated by ergodic theoretic considerations.

Suppose for the moment that T is an operator on a complex Fréchet space X whose eigenspaces to unimodular eigenvalues all have dimension at most one. One can then define an eigenvector field $E : \mathbb{T} \to X$ so that, for any

$\lambda \in \mathbb{T}$, $E(\lambda)$ is either an eigenvector to the eigenvalue λ, or 0. Since we want a large supply of eigenvectors we would demand that $\text{span}\{E(\lambda)\ ;\ \lambda \in \mathbb{T}\}$ is dense in X, in which case E is called spanning. In order to capture the situation where eigenspaces are higher-dimensional one has to allow for a collection of eigenvector fields. In the sequel, J is a nonempty index set.

Definition 9.21. Let T be an operator on a complex Fréchet space X. Then a collection of functions $E_j : \mathbb{T} \to X$, $j \in J$, is called a *spanning eigenvector field associated to unimodular eigenvalues* if $E_j(\lambda) \in \ker(\lambda I - T)$ for any $\lambda \in \mathbb{T}$, $j \in J$, and

$$\text{span}\{E_j(\lambda)\ ;\ \lambda \in \mathbb{T}, j \in J\} \quad \text{is dense in } X.$$

In addition, the vector field is said to be *continuous* (or C^2) if each function $E_j : \mathbb{T} \to X$, $j \in J$, is continuous (or C^2, respectively).

As usual, a function $E : \mathbb{T} \to X$, is called C^2 if it is twice continuously differentiable, where differentiation is defined as in the scalar-valued case.

We now have the announced eigenvalue criterion.

Theorem 9.22. *Let T be an operator on a complex separable Fréchet space.*

(a) *If T has a spanning continuous eigenvector field associated to unimodular eigenvalues then it is mixing and chaotic.*

(b) *If T has a spanning C^2-eigenvector field associated to unimodular eigenvalues then it is frequently hypercyclic.*

The proof is similar to that of Theorem 7.32. We will need the Riemann integral

$$\int_0^{2\pi} f(t)\, dt$$

for a continuous function $f : [0, 2\pi] \to X$; see Appendix A for details and basic properties.

Lemma 9.23. *Let X be a complex Fréchet space and $f : [0, 2\pi] \to X$ a continuous function.*

(a) **(Riemann–Lebesgue lemma)** *Then $\int_0^{2\pi} e^{int} f(t)\, dt \to 0$ as $n \to \pm\infty$.*

(b) *If f is twice continuously differentiable with $f(0) = f(2\pi)$ and $f'(0) = f'(2\pi)$ then $\sum_{n=0}^{\infty} \int_0^{2\pi} e^{int} f(t)\, dt$ and $\sum_{n=0}^{\infty} \int_0^{2\pi} e^{-int} f(t)\, dt$ converge unconditionally.*

Proof. Let $(p_k)_k$ be an increasing sequence of seminorms defining the topology of X.

(a) As in the proof of Lemma 7.31 one shows that, for any $k \geq 1$, $p_k\big(\int_0^{2\pi} e^{int} f(t)\, dt\big) \to 0$ as $n \to \pm\infty$. This implies the claim.

(b) Upon integrating by parts twice we obtain that, for any $k \geq 1$, $n \neq 0$,

$$p_k\left(\int_0^{2\pi} e^{int} f(t)\,dt\right) = p_k\left(-\frac{1}{n^2}\int_0^{2\pi} e^{int} f''(t)\,dt\right) \leq \frac{1}{n^2}\int_0^{2\pi} p_k(f''(t))\,dt,$$

which implies the claim. □

We are now in a position to prove the eigenvalue criterion.

Proof of Theorem 9.22. (a) Let $(E_j)_{j\in J}$ be the given eigenvector field of T. Since each $E_j : \mathbb{T} \to X$ is continuous the integrals

$$x_{k,j} := \int_0^{2\pi} e^{ikt} E_j(e^{it})\,dt \in X, \quad k \in \mathbb{Z}, j \in J,$$

are defined. In order to apply Kitai's criterion we set

$$X_0 = Y_0 = \mathrm{span}\{x_{k,j}\ ;\ k \in \mathbb{Z}, j \in J\}.$$

We will use the Hahn–Banach theorem to show that this set is dense. Thus, let x^* be a continuous linear functional on X so that, for all $k \in \mathbb{Z}$, $j \in J$,

$$\langle x_{k,j}, x^*\rangle = \int_0^{2\pi} e^{ikt}\langle E_j(e^{it}), x^*\rangle\,dt = 0.$$

The functions $t \to \langle E_j(e^{it}), x^*\rangle$ are continuous and therefore belong to $L^2[0,2\pi]$. Since $(\frac{1}{\sqrt{2\pi}} e^{ikt})_{k\in\mathbb{Z}}$ is an orthonormal basis in this Hilbert space, we deduce that, by continuity,

$$\langle E_j(e^{it}), x^*\rangle = 0 \quad \text{for all } t \in [0,2\pi],\ j \in J.$$

Hence x^* vanishes on the set $\mathrm{span}\{E_j(\lambda)\ ;\ \lambda \in \mathbb{T}, j \in J\}$, which is dense by assumption, so that x^* itself must vanish. Thus $X_0 = Y_0$ is dense.

Now, since each $E_j(\lambda)$ is in the eigenspace of λ we have for any $k \in \mathbb{Z}$ and $j \in J$ that

$$T^n x_{k,j} = \int_0^{2\pi} e^{ikt} T^n E_j(e^{it})\,dt = \int_0^{2\pi} e^{i(k+n)t} E_j(e^{it})\,dt \to 0$$

as $n \to \infty$, as a result of the Riemann–Lebesgue lemma. By linearity, we conclude that $T_n x \to 0$ for all $x \in X_0$.

It would seem natural to define the mapping $S : Y_0 \to Y_0$ by

$$x_{k,j} = \int_0^{2\pi} e^{ikt} E_j(e^{it})\,dt \to \int_0^{2\pi} e^{i(k-1)t} E_j(e^{it})\,dt = x_{k-1,j},$$

followed by linear extension to Y_0. Since this may lead to a conflict if the $x_{k,j}$ are not linearly independent we apply, instead, the variant, Exercise 3.1.1, of Kitai's criterion. Thus, for any $y \in Y_0$, we consider a representation $y = \sum_{l=1}^m a_l x_{k_l,j_l}$ and define

$$u_n = \sum_{l=1}^{m} a_l x_{k_l - n, j_l}, \quad n \geq 0.$$

We then have $T^n u_n = y$ and, again by the Riemann–Lebesgue lemma, that $x_{k_l-n,j} \to 0$ as $n \to \infty$, so that $u_n \to 0$. We can therefore conclude that T is mixing.

Moreover, by continuity of the eigenvector field, also

$$\text{span}\{E_j(\lambda) \ ; \ j \in J, \lambda = e^{\alpha\pi i} \text{ for some } \alpha \in \mathbb{Q}\}$$

is dense in X, and each vector in this span is a periodic point for T. Consequently, T is chaotic.

(b) The proof follows the same lines, this time using Lemma 9.23(b) and the Frequent Hypercyclicity Criterion in the form of Remark 9.10. □

The eigenvalue criterion provides a new proof that the three classical hypercyclic operators are even frequently hypercyclic.

Example 9.24. **(Rolewicz' operators)** We consider the Rolewicz operators $T = \mu B$, $|\mu| > 1$, on one of the complex spaces $X = \ell^p$, $1 \leq p < \infty$, or c_0. Then

$$E : \mathbb{T} \to X, \quad \lambda \to (\lambda^n/\mu^n)_n$$

is an eigenvector field associated to unimodular eigenvalues. An elementary but tedious calculation shows that the field is C^2 (see Exercise 9.3.2), while the spanning property was proved in Example 3.2.

Concerning MacLane's and Birkhoff's operators we will show a much more general result, namely that Theorem 4.21 by Godefroy and Shapiro also holds for frequent hypercyclicity.

Theorem 9.25. *Suppose that $T : H(\mathbb{C}) \to H(\mathbb{C})$, $T \neq \lambda I$, is an operator that commutes with D, that is,*

$$TD = DT.$$

Then T is frequently hypercyclic.

Proof. Following the proof of Theorem 4.21 we can write $T = \varphi(D)$ with a nonconstant entire function φ of exponential type, which also implies that every function $e_\lambda(z) = e^{\lambda z}$, $\lambda \in \mathbb{C}$, is an eigenvector of T to the eigenvalue $\varphi(\lambda)$. Since $\varphi(\mathbb{C})$ is connected and dense (see Appendix A), there is a point $z \in \mathbb{C}$ with $w := \varphi(z) \in \mathbb{T}$; and since $\varphi(\mathbb{C})$ is open and the zeros of φ' are isolated points we can also achieve that $\varphi'(z) \neq 0$. Thus φ maps a neighbourhood of z conformally onto a neighbourhood U of w; let ψ be the inverse map, which is holomorphic. Fix a nontrivial closed subarc $\gamma \subset U$ of $\mathbb{T}$ containing w and a C^2-function $f : \mathbb{T} \to \mathbb{C}$ with $f(w) \neq 0$ that vanishes outside γ. It follows that $E : \mathbb{T} \to H(\mathbb{C})$ with $E(\lambda) = f(\lambda)e_{\psi(\lambda)}$ if $\lambda \in \gamma$ and $E(\lambda) = 0$, else, defines an eigenvector field associated to unimodular eigenvalues for T. It was shown in the proof of Lemma 2.34 that the function $\mathbb{C} \to H(\mathbb{C})$, $\lambda \to e_\lambda$ is, in fact, infinitely differentiable, so that E is a C^2-field. Finally, E is spanning by Lemma 2.34. Now the eigenvalue criterion for frequent hypercyclicity implies the result. □

As in the case of hypercyclicity one may ask how slowly a frequently hypercyclic entire function can grow at infinity. The eigenvalue criterion allows us to deduce corresponding results for any operator $T = \varphi(D)$; see Exercise 9.3.3. Here we consider only the special case of Birkhoff's operators $T_a f(z) = f(z+a)$, $a \neq 0$. The theorem of Duyos-Ruiz tells us that corresponding hypercyclic functions can grow arbitrarily slowly. This is no longer true in the frequent context.

Theorem 9.26. *Let $a \neq 0$.*

(a) *Let $\varepsilon > 0$. Then there exists an entire function f that is frequently hypercyclic for T_a and that satisfies*

$$|f(z)| \leq M e^{\varepsilon r} \quad \textit{for } |z| = r > 0$$

with some $M > 0$.

(b) *Let $\varepsilon : \mathbb{R}_+ \to \mathbb{R}_+$ be a function with $\liminf_{r\to\infty} \varepsilon(r) = 0$. Then there is no entire function f that is frequently hypercyclic for T_a and that satisfies*

$$|f(z)| \leq M e^{\varepsilon(r) r} \quad \textit{for all } |z| = r > 0 \textit{ sufficiently large}$$

with some $M > 0$.

Proof. (a) This result follows from a general growth result for all operators that commute with D (see Exercise 9.3.3) because $T_a = e^{aD}$ and $e^{az} = 1$ for $z = 0$.

(b) We will assume that $a = 1$; see Example 4.26. Suppose, on the contrary, that f is a frequently hypercyclic entire function with the stated growth condition; by adding a constant, if necessary, we can assume that $f(0) = 1$. Then there is a strictly increasing sequence $(n_k)_k$ of positive integers with $n_k = O(k)$ such that, for any $k \geq 1$,

$$|f(z+n_k) - z| < \frac{1}{2} \text{ for } |z| \leq \frac{1}{2}.$$

Thus, by Rouché's theorem (see Appendix A), f has a zero in $|z - n_k| < \frac{1}{2}$. If $N(r)$ denotes the number of zeros of f in $|z| < r$, counting multiplicity, then

$$N(n_k + 1) \geq k, \quad k \geq 1.$$

On the other hand, it follows from Jensen's formula (see Theorem A.23) and the growth assumption on f that

$$N(r)\log 2 \le \log M + \varepsilon(2r)2r$$

for r sufficiently large.

Now let $r_\nu \to \infty$ be such that $\varepsilon(2r_\nu) \to 0$ as $\nu \to \infty$. For sufficiently large ν choose k_ν such that $n_{k_\nu} + 1 \le r_\nu \le n_{k_\nu+1}$. Altogether we conclude that

$$\frac{k_\nu}{n_{k_\nu+1}} \le \frac{N(n_{k_\nu}+1)}{n_{k_\nu+1}} \le \frac{N(r_\nu)}{r_\nu} \le \frac{\log M + \varepsilon(2r_\nu)2r_\nu}{r_\nu \log 2} \to 0,$$

hence that $\frac{n_{k_\nu+1}}{k_\nu+1} = \frac{k_\nu}{k_\nu+1}\frac{n_{k_\nu+1}}{k_\nu} \to \infty$, which is a contradiction. □

9.4 Structural properties

In the previous chapters we derived various structural properties of hypercyclicity. In this section we revisit several of them in the context of frequent hypercyclicity.

We begin by looking at the main results of Chapter 6. Ansari's theorem says that every power T^p, $p \ge 1$, of a hypercyclic operator is again hypercyclic; in fact, T and T^p have the same hypercyclic vectors. For frequent hypercyclicity we have the corresponding property, but its proof relies on very different techniques than Ansari's theorem.

Theorem 9.27. *Let T be an operator on a Fréchet space. Then, for any $p \in \mathbb{N}$, $FHC(T) = FHC(T^p)$. In particular, if T is frequently hypercyclic then so is every power T^p.*

Proof. Since every orbit $\text{orb}(x, T^p)$ is obtained from the orbit $\text{orb}(x, T)$ by retaining only the powers $T^{np}x$, $n \ge 0$, it is clear that every frequently hypercyclic vector for T^p is also frequently hypercyclic for T.

Conversely, let $x \in X$ be a frequently hypercyclic vector for T and $p \ge 1$. In order to show that x is also frequently hypercyclic for T^p we fix a nonempty open subset U of X. Since the sequence $(kp-1)_{k\ge1}$ is syndetic, we can deduce from Theorems 9.8 and 1.54 that there is some $m_1 \ge 0$ of the form $m_1 = k_1p - 1$ such that $U_1 := U \cap T^{-m_1}(U) \ne \varnothing$. For the same reason, there is some $m_2 \ge 0$ of the form $m_2 = k_2p - 2$ such that $U_2 := U_1 \cap T^{-m_2}(U_1) \ne \varnothing$. Proceeding inductively we find, for $j = 1, \ldots, p-1$, integers $m_j \ge 0$ of the form $m_j = k_jp - j$ such that $U_j := U_{j-1} \cap T^{-m_j}(U_{j-1}) \ne \varnothing$, where $U_0 := U$. Moreover we set $k_0 = 0$.

Now let $V = U_{p-1}$, which clearly satisfies $V \subset U$ and $T^{k_jp-j}(V) \subset U$, for $j = 0, 1, \ldots, p-1$. Since x is frequently hypercyclic there is a subset $A \subset \mathbb{N}_0$ of positive lower density such that $T^n x \in V$ for all $n \in A$. We then define the function $f : \mathbb{N}_0 \to \mathbb{N}_0$ by $f(n) = \frac{n-j}{p} + k_j$ if $n = j \,(\text{mod}\, p)$, $j = 0, \ldots, p-1$; note that this is well defined.

We finally set $B = f(A)$. It is easy to show that $\underline{\text{dens}}(B) \ge \underline{\text{dens}}(A) > 0$; see Exercise 9.4.1. Moreover, if $m \in B$, then $m = \frac{n-j}{p} + k_j$ for some $n \in A$

with $n = j \pmod p$, and hence

$$(T^p)^m x = T^{n-j+k_j p} x = T^{k_j p - j}(T^n x) \in T^{k_j p - j}(V) \subset U.$$

This proves that x is frequently hypercyclic for T^p. □

We saw in Section 6.3 how Ansari's theorem follows from the fact that if the union of the orbits of finitely many vectors is dense then one of these orbits must already be dense. The corresponding result fails for frequent hypercyclicity.

Example 9.28. We consider the Rolewicz operator $T = 2B$ on ℓ^1. We claim that there are two vectors $v, w \in \ell^1$ such that, for any nonempty open subset $U \subset \ell^1$,

$$\underline{\text{dens}}\{n \in \mathbb{N}_0 \; ; \; T^n v \in U \text{ or } T^n w \in U\} > 0, \tag{9.6}$$

but neither v nor w is frequently hypercyclic for T.

To see this, let $(y_j)_j$ be a dense sequence in ℓ^1 consisting of finite sequences. The proof of the Frequent Hypercyclicity Criterion, with S being half the forward shift, then constructs a frequently hypercyclic vector x for T given by

$$x = \sum_{n \in A} S^n z_n,$$

where A is the union of certain pairwise disjoint sets $A(l, N_l)$, $l \geq 1$, of positive lower density, and $z_n = y_l$ for $n \in A(l, N_l)$. For an increasing sequence $(m_k)_{k \geq 0}$ of positive integers with $m_0 = 0$, that will be determined later, we split the set A into two subsets

$$B = \{n \in A \; ; \; \exists\, k \geq 0 : m_{2k} \leq n < m_{2k+1}\},$$
$$C = \{n \in A \; ; \; \exists\, k \geq 0 : m_{2k+1} \leq n < m_{2k+2}\}.$$

The proof of the Frequent Hypercyclicity Criterion then shows that the series

$$v := \sum_{n \in B} S^n z_n, \quad w := \sum_{n \in C} S^n z_n$$

converge and that, for any $n \in A(l, N_l)$,

$$\|T^n v - y_l\| \leq \frac{4}{2^l} \quad \text{or} \quad \|T^n w - y_l\| \leq \frac{4}{2^l}$$

depending on whether $n \in B$ or $n \in C$. This implies that the joint orbits of v and w are frequently recurrent, in the sense of (9.6).

On the other hand, let l_n be the length of the finite sequence z_n, $n \in A$. Let $k \geq 0$. If $n \in \mathbb{N}_0$ satisfies

$$M_{2k+1} := \max_{\nu \in B, \nu < m_{2k+1}} (\nu + l_\nu) \leq n < m_{2k+2}$$

then the sequence $T^n v$ starts with a 0 and hence $T^n v \notin U$, where $U = \{x \in \ell^1 \ ; \ \|x - e_1\| < 1\}$. Now, if we choose the m_{2k+2}, $k \geq 1$, such that

$$\frac{M_{2k+1}}{m_{2k+2}} \leq \frac{1}{k}$$

then $\underline{\text{dens}}\{n \geq 0 \ ; \ T^n v \in U\} = 0$, which shows that v is not frequently hypercyclic for T. Imposing, in addition, a similar condition on the m_{2k+1} one can also achieve that w is not frequently hypercyclic for T.

For a variant of the Bourdon–Feldman theorem for frequent hypercyclicity see Exercise 9.4.4.

We next turn to the results of Section 6.4. For this we need to define frequent hypercyclicity for C_0-semigroups. The *lower density* of a measurable subset $A \subset \mathbb{R}_+$ is given by

$$\underline{\text{dens}}(A) := \liminf_{T\to\infty} \frac{\lambda\{t \in [0,T] \ ; \ t \in A\}}{T},$$

where λ denotes the Lebesgue measure.

Definition 9.29. A C_0-semigroup $(T_t)_{t\geq 0}$ on a Banach space X is called *frequently hypercyclic* if there is a vector $x \in X$ such that, for any nonempty open subset U of X,

$$\underline{\text{dens}}\,\{t \in \mathbb{R}_+ \ ; \ T_t x \in U\} > 0.$$

In this case, x is called a *frequently hypercyclic vector* for $(T_t)_{t\geq 0}$.

As before we will treat the problems of unimodular multiples and of discretizations of semigroups within the common framework of semigroup actions; see Section 6.4. We recall that if T is an operator on a complex Fréchet space X then

$$\Psi(n,t) = e^{2\pi t i} T^n, \quad n \in \mathbb{N}_0, \ t \geq 0, \tag{9.7}$$

defines a semigroup action. Similarly, if $(T_t)_{t\geq 0}$ is a C_0-semigroup on a Banach space X then

$$\Psi(n,t) = T_t, \quad n \in \mathbb{N}_0, \ t \geq 0, \tag{9.8}$$

defines a semigroup action. In both cases, properties (α) and (β) of Section 6.4 are satisfied.

We then also need a concept of frequent hypercyclicity for semigroup actions. The natural notion of *lower density* on $G = \mathbb{N}_0 \times \mathbb{R}_+$ is given by

$$\underline{\text{dens}}(A) := \liminf_{N\to\infty} \frac{1}{N(N+1)} \sum_{n=0}^{N} \lambda\{t \in [0,N] \ ; \ (n,t) \in A\},$$

where $A \subset G$ is such that $\{t \geq 0 \ ; \ (n,t) \in A\}$ is measurable for each $n \geq 0$; λ denotes the Lebesgue measure.

Definition 9.30. A semigroup action $\Psi : G \to L(X)$ is called *frequently hypercyclic* if there is some $x \in X$ such that, for any nonempty open subset U of X,

$$\underline{\text{dens}}\,\{g \in G\ ;\ \Psi(g)x \in U\} > 0.$$

In this case, x is called a *frequently hypercyclic vector* for Ψ.

Now, frequent hypercyclicity of some operator $\Psi(n,t)$, $n,t > 0$, implies frequent hypercyclicity of Ψ.

Proposition 9.31. *Let Ψ be a semigroup action on a Fréchet space X satisfying property (α). If $x \in X$ is frequently hypercyclic for some operator $\Psi(n,t)$, $n,t > 0$, then it is frequently hypercyclic for Ψ.*

Proof. Let U be a nonempty open subset of X. Since $\Psi(0,0) = I$ and Ψ is continuous, there is a nonempty open subset V of U and some $\eta > 0$ such that $\Psi(0,s)V \subset U$ if $0 \le s < \eta$. By assumption, there is some $(n,t) \in G$ such that $\underline{\text{dens}}(A) = \delta > 0$, where $A = \{k \in \mathbb{N}_0\ ;\ \Psi(n,t)^k x \in V\}$. Now, if $k \in A$ and $0 \le s < \eta$ then $\Psi(kn, kt + s)x = \Psi(0,s)\Psi(n,t)^k x \in U$. In view of property (α), if $\Psi(1,0) = I$ then also $\Psi(kn + m, kt + s)x \in U$ for $m \in \mathbb{Z}$ with $kn + m \ge 0$; if $\Psi(0,1) = I$ then $\Psi(kn, kt + s + m)x \in U$ for $m \in \mathbb{Z}$ with $kt + s + m \ge 0$. In both cases a simple count reveals that $\underline{\text{dens}}\{(k,s) \in G\ ;\ \Psi(k,s)x \in U\} \ge \eta\delta/\max(n,t) > 0$. □

Our main aim is to prove the converse statement. The following will be crucial.

Lemma 9.32. *Let Ψ be a semigroup action on an infinite-dimensional Fréchet space X satisfying properties (α) and (β). If $x \in X$ is frequently hypercyclic for Ψ then, for any $k \in \mathbb{N}$ and any nonempty open subset U of X, we have that*

$$\underline{\text{dens}}\,\big\{(n,t) \in G\ ;\ \Psi(n,t)x \in U,\ t \in \textstyle\bigcup_{m=1}^{\infty}[m - \frac{1}{k}, m[\big\} > 0.$$

Proof. We fix $k \in \mathbb{N}$ and a nonempty open subset U of X. For $j = 1, \ldots, k$ we define the sets

$$I_j = \bigcup_{m=1}^{\infty} [m - \tfrac{j}{k}, m - \tfrac{j-1}{k}[.$$

By Theorem 6.10, x is hypercyclic for $\Psi(1,1)$. It follows from property (β) that, for any $t \ge 0$, also $\Psi(0,t)x$ is hypercyclic for $\Psi(1,1)$. Therefore there are $n_j \in \mathbb{N}_0$, $j = 1, \ldots, k$, such that

$$\Psi(n_j, n_j + \tfrac{j-1}{k})x = \Psi(1,1)^{n_j}\Psi(0, \tfrac{j-1}{k})x \in U.$$

By continuity there is a neighbourhood V of x such that

$$\Psi(n_j, n_j + \tfrac{j-1}{k})(V) \subset U, \quad j = 1, \ldots, k.$$

Let $N_0 = \max(n_1, \ldots, n_k) + 1$. It follows from frequent hypercyclicity of x for Ψ that there are $\delta > 0$ and $N_1 \geq N_0$ such that, if $N \geq N_1$, then

$$\frac{1}{N(N+1)} \sum_{n=0}^{N} \lambda\{t \in [0, N] \ ; \ \Psi(n,t)x \in V\} \geq \delta.$$

We now fix $N \geq N_1$. Since the I_j, $j = 1, \ldots, k$, form a partition of $\mathbb{R}_+$, there is some j such that

$$\sum_{n=0}^{N} \lambda\{t \in [0, N] \ ; \ \Psi(n,t)x \in V, \ t \in I_j\} \geq \frac{1}{k} \sum_{n=0}^{N} \lambda\{t \in [0, N] \ ; \ \Psi(n,t)x \in V\}.$$

We fix such a j. If $0 \leq n \leq N$, $0 \leq t \leq N$, $t \in I_j$, and $\Psi(n,t)x \in V$ then $\nu := n + n_j \leq 2N$, $\tau := t + n_j + \frac{j-1}{k} \leq 2N$, $\tau \in I_1$, and

$$\Psi(\nu, \tau)x = \Psi\left(n_j, n_j + \tfrac{j-1}{k}\right) \Psi(n,t)x \in \Psi\left(n_j, n_j + \tfrac{j-1}{k}\right)(V) \subset U.$$

We conclude that

$$\begin{aligned}\sum_{\nu=0}^{2N} \lambda\{\tau \in [0, 2N] \ ; \ \Psi(\nu,\tau)x \in U, \ \tau \in I_1\}& \\ \geq \frac{1}{k} \sum_{n=0}^{N} \lambda\{t \in [0, N] \ ; \ \Psi(n,t)x \in V\},&\end{aligned}$$

so that

$$\frac{1}{2N(2N+1)} \sum_{\nu=0}^{2N} \lambda\{\tau \in [0, 2N] \ ; \ \Psi(\nu,\tau)x \in U, \ \tau \in I_1\} \geq \frac{\delta}{4k}.$$

Since $N \geq N_1$ was arbitrary, the claim follows. □

We can now prove the analogue of Theorem 6.10 for frequent hypercyclicity.

Theorem 9.33. *Let Ψ be a semigroup action on an infinite-dimensional Fréchet space X satisfying properties (α) and (β). If $x \in X$ is frequently hypercyclic for Ψ then it is frequently hypercyclic for every operator $\Psi(1,t)$, $t > 0$.*

Proof. We first prove the case when $t = 1$. Thus, let U be a nonempty open subset of X. Since $\Psi(0,0) = I$, continuity of Ψ implies that there is a nonempty open subset V of U and some $\eta > 0$ such that $\Psi(0,s)V \subset U$ if $0 \leq s < \eta$. Let $k \in \mathbb{N}$ be such that $\frac{1}{k} < \eta$. Then, by Lemma 9.32, there are $\delta > 0$ and $N_0 \in \mathbb{N}$ such that, for any $N \geq N_0$,

$$r := \sum_{n=0}^{N} \lambda\{t \in [0, N] \ ; \ \Psi(n,t)x \in V, \ t \in \bigcup_{m=1}^{\infty} [m - \tfrac{1}{k}, m[\} \geq N(N+1)\delta.$$

Now, if $\Psi(n,t)x \in V$ and $t \in [m-\frac{1}{k}, m[$ then $\Psi(n,m)x = \Psi(0, m-t)\Psi(n,t)x \in U$. Thus, for

$$p := \mathrm{card}\{(n,m) \ ; \ 0 \leq n, m \leq N, \Psi(n,m)x \in U\}$$

we have that $p\frac{1}{k} \geq r$.

Next, let $\Psi(1,1)^n x \in U$. We distinguish the two cases described by (α). If $\Psi(1,0) = I$ then $\Psi(m,n)x = \Psi(n,n)x = \Psi(1,1)^n x \in U$ for any $m \in \mathbb{Z}$; and if $\Psi(0,1) = I$ then $\Psi(n,m)x = \Psi(n,n)x \in U$ for any $m \in \mathbb{Z}$. Thus, for

$$q := \mathrm{card}\{0 \leq n \leq N \ ; \ \Psi(1,1)^n x \in U\}$$

we have that $p = (N+1)q$.

Altogether we find that, for any $N \geq N_0$,

$$\frac{\mathrm{card}\{0 \leq n \leq N \ ; \ \Psi(1,1)^n x \in U\}}{N+1} = \frac{p}{(N+1)^2} \geq \frac{kr}{(N+1)^2} \geq \frac{kN}{N+1}\delta.$$

Hence x is frequently hypercyclic for $\Psi(1,1)$.

Now, if $t > 0$ is arbitrary then we rescale the semigroup action as in the proof of Theorem 6.10. It is then not difficult to see, using property (α), that x is also frequently hypercyclic for $\widetilde{\Psi}$ and thus frequently hypercyclic for $\widetilde{\Psi}(1,1) = \Psi(1,t)$. □

If we combine Theorem 9.33 with Theorem 9.27, noting that $\psi(n,t) = \Psi(1,t/n)^n$, we obtain the announced converse of Proposition 9.31.

Corollary 9.34. *Let Ψ be a semigroup action on a Fréchet space X satisfying properties (α) and (β). If $x \in X$ is frequently hypercyclic for Ψ then it is frequently hypercyclic for every operator $\Psi(n,t)$, $n, t > 0$.*

Proposition 9.31 and Theorem 9.33, applied to the semigroup action (9.7), immediately imply a version of the León–Müller theorem for frequent hypercyclicity.

Theorem 9.35. *Let T be an operator on a complex Fréchet space and $\lambda \in \mathbb{C}$ with $|\lambda| = 1$. Then T and λT have the same frequently hypercyclic vectors, that is, $FHC(T) = FHC(\lambda T)$.*

Similarly, applying Theorem 9.33 to the semigroup action (9.8) yields an analogue of the Conejero–Müller–Peris theorem.

Theorem 9.36. *Let $(T_t)_{t\geq 0}$ be a C_0-semigroup on a Banach space X. If $x \in X$ is frequently hypercyclic for $(T_t)_{t\geq 0}$, then it is frequently hypercyclic for every operator T_t, $t > 0$.*

Apart from being interesting in its own right, Theorem 9.35 has an important application. We saw in Chapter 5 that the spectrum of a hypercyclic operator on a complex Banach space has the property that each of its connected components meets the unit circle; this is the content of Kitai's theorem. In particular, the spectrum cannot have isolated points outside the unit circle. We will now show that the spectrum of frequently hypercyclic operators, just like that of chaotic operators (see Proposition 5.7), cannot even have isolated points on the unit circle.

We start with a crucial lemma whose proof uses complex analysis in a very clever way.

Lemma 9.37. *Let T be an operator on a real Fréchet space X. Let $x \in X$ and $x^* \in X^*$ with $\langle x, x^*\rangle \neq 0$ be such that*

$$|\langle (T-I)^n x, x^*\rangle|^{1/n} \to 0$$

as $n \to \infty$. Then x is not frequently hypercyclic for T.

Proof. First, we may assume that $\langle x, x^*\rangle = 1$. Suppose that x is frequently hypercyclic for T. Then $(\langle T^n x, x^*\rangle)_{n\geq 0}$ is dense in $\mathbb{R}$. Thus there must be some $n \geq 0$ such that $\langle T^n x, x^*\rangle \leq 0$ and $\langle T^{n+1}x, x^*\rangle > 0$. Then, for $\alpha > 0$ sufficiently small, $\langle T^n x - \alpha x, x^*\rangle < 0$ and $\langle T^{n+1}x - \alpha T x, x^*\rangle > 0$, so that the open set

$$U = \{y \in X \ ; \ \langle y, x^*\rangle < 0 \quad \text{and} \quad \langle Ty, x^*\rangle > 0\}$$

is nonempty.

We now consider the series

$$f(z) = \sum_{k=0}^{\infty} \langle (T-I)^k x, x^*\rangle \frac{z(z-1)\cdots(z-k+1)}{k!}, \qquad z \in \mathbb{C},$$

where we regard the quotient as 1 if $k = 0$. We claim that this defines an entire function. Indeed, it follows from the assumption that, for any $\varepsilon \in\,]0,1[$, there is some $M > 0$ such that

$$|\langle (T-I)^n x, x^*\rangle| \leq M\varepsilon^n, \quad n \geq 0,$$

so that, for any $R > 0$ and $|z| \leq R$,

$$\begin{aligned}\sum_{k=0}^{\infty} |\langle (T-I)^k x, x^*\rangle| \left|\frac{z(z-1)\cdots(z-k+1)}{k!}\right| \\ &\leq M \sum_{k=0}^{\infty} \varepsilon^k \frac{R(R+1)\cdots(R+k-1)}{k!} \\ &= M \sum_{k=0}^{\infty} \binom{-R}{k} (-\varepsilon)^k = \frac{M}{(1-\varepsilon)^R} < \infty,\end{aligned}$$

where we have used the binomial theorem. Moreover, setting $\eta = -\log(1-\varepsilon)$, this inequality implies that

$$|f(z)| \leq Me^{\eta|z|}, \quad z \in \mathbb{C}.$$

In addition, $f(0) = \langle x, x^* \rangle = 1$. It follows from Jensen's formula (see Theorem A.23) that if $N(r)$ denotes the number of zeros of f in $|z| < r$, counting multiplicity, then

$$N(r) \log 2 \leq \log M + 2r\eta, \quad r > 0. \tag{9.9}$$

On the other hand we have that for $n \in \mathbb{N}_0$,

$$\begin{aligned} f(n) &= \sum_{k=0}^{n} \langle (T-I)^k x, x^* \rangle \frac{n(n-1)\cdots(n-k+1)}{k!} \\ &= \Big\langle \sum_{k=0}^{n} \binom{n}{k} (T-I)^k x, x^* \Big\rangle = \langle T^n x, x^* \rangle. \end{aligned}$$

Thus, if $T^n x \in U$ then $f(n) < 0$ and $f(n+1) > 0$, so that f, being real on the real axis, has a zero in the interval $]n, n+1[$. It follows with (9.9) that

$$\frac{\operatorname{card}\{0 \leq n \leq m\ ;\ T^n x \in U\}}{m+1} \leq \frac{N(m+1)}{m+1} \leq \frac{\log M + 2(m+1)\eta}{(m+1)\log 2} \to \frac{2\eta}{\log 2}$$

as $m \to \infty$. Since $\eta > 0$ is arbitrary, we deduce that x is not frequently hypercyclic. □

As an immediate consequence we have the following. Recall that an operator T on a Banach space is called quasinilpotent if $\|T^n\|^{1/n} \to 0$ as $n \to \infty$.

Lemma 9.38. *Let T be an operator on a Banach space X of the form $T = \lambda I + S$ with $|\lambda| = 1$ and S quasinilpotent. Then T is not frequently hypercyclic.*

Proof. By Theorem 9.35 we may assume that $\lambda = 1$, so that $\|(T-I)^n\|^{1/n} \to 0$ as $n \to \infty$. Moreover, we can regard X as a real Banach space and $T - I$ as a (real-linear) operator on X. We then have that, for any $x \in X$ and any (real-linear) continuous linear functional x^* on X,

$$|\langle (T-I)^n x, x^* \rangle|^{1/n} \leq \big(\|(T-I)^n\| \|x\| \|x^*\| \big)^{1/n} \to 0.$$

By the previous lemma, T cannot be frequently hypercyclic on X; note that this notion does not depend on the scalar field. □

We can now prove the mentioned spectral property of frequently hypercyclic operators.

Theorem 9.39. *Let T be a frequently hypercyclic operator on a complex Banach space. Then its spectrum $\sigma(T)$ has no isolated points.*

Proof. Suppose that $\lambda \in \mathbb{C}$ is an isolated point of the spectrum. Then $\sigma(T)$ can be partitioned into some closed subset and the singleton $\{\lambda\}$. By the Riesz decomposition theorem (see Appendix B) there are nontrivial T-invariant closed subspaces M_1 and M_2 of X such that $X = M_1 \oplus M_2$ and $\sigma(T|_{M_2}) = \{\lambda\}$. By Exercise 2.2.8 and Proposition 9.4, $T|_{M_2}$ is frequently hypercyclic. By Kitai's theorem we have that $|\lambda| = 1$, and the spectral radius formula (see Appendix B) implies that $T|_{M_2} = \lambda I + S$ with a quasinilpotent operator S. This contradicts Lemma 9.38. □

Lemma 9.38 has another application. First, combining it with Lemma 5.19 yields the following.

Proposition 9.40. *No compact perturbation of a multiple of the identity on a Banach space is frequently hypercyclic.*

We can then apply the Argyros–Haydon theorem; see Theorem 8.11.

Corollary 9.41. *Let $\mathbb{K} = \mathbb{R}$ or $\mathbb{C}$. Then there exists an infinite-dimensional separable Banach space over $\mathbb{K}$ that supports no frequently hypercyclic operator.*

With this we end our introduction to frequent hypercyclicity.

Exercises

Exercise 9.1.1. Show that the Herrero–Bourdon theorem also holds for frequent hypercyclicity. In particular, every frequently hypercyclic operator on a Fréchet space admits a dense T-invariant subspace consisting, except for 0, of frequently hypercyclic vectors.

Exercise 9.1.2. Using Lemma 9.5, show that every weighted shift is frequently hypercyclic on the space $\omega = \mathbb{K}^{\mathbb{N}}$.

Exercise 9.1.3. Show that every frequently hypercyclic operator on a Fréchet space is topologically ergodic; see Exercise 1.5.6. Deduce that if T is a frequently hypercyclic operator on a Banach space then its adjoint T^* cannot be frequently hypercyclic. (*Hint*: Exercise 2.5.5, Remark 4.17.)

Exercise 9.1.4. Show that every entire function is the sum of two functions that are frequently hypercyclic for the translation operator $T_1 f(z) = f(z+1)$. (*Hint*: Use a variant of the construction in Example 9.6.)

Exercise 9.2.1. Let $T_n : X \to Y$, $n \geq 0$, be operators between separable Fréchet spaces X and Y. The definition of frequent hypercyclicity for the sequence $(T_n)_{n\geq 0}$ is obvious. Prove the following version of the Frequent Hypercyclicity Criterion for $(T_n)_n$. For the notion of uniformly unconditionally convergent series see Definition 11.7 below.

If there is a dense subset Y_0 of Y and maps $S_n : Y_0 \to X$, $n \geq 0$, such that, for any $y \in Y_0$,

(i) $\sum_{n=0}^{m} T_m S_{m-n} y$ converges unconditionally in Y, uniformly for $m \geq 0$,

(ii) $\sum_{n=0}^{\infty} T_m S_{m+n} y$ converges unconditionally in Y, uniformly for $m \geq 0$,

(iii) $\sum_{n=0}^{\infty} S_n y$ converges unconditionally in X,

(iv) $T_n S_n y \to y$, as $n \to \infty$,

then $(T_n)_n$ is frequently hypercyclic.

Note that, in (i), the finite sums can be understood as infinite series by adding 0 terms.

Exercise 9.2.2. Use the Frequent Hypercyclicity Criterion to give a new proof that Birkhoff's operators are frequently hypercyclic. (*Hint*: Example 3.8.)

Exercise 9.2.3. Formulate and prove an analogue of Proposition 9.13 for weighted bilateral shifts.

Exercise 9.2.4. Let B_w be a frequently hypercyclic weighted shift on ℓ^p, $1 \leq p < \infty$. Show that, for any $\varepsilon > 0$, there exists a subset $A \subset \mathbb{N}_0$ of positive lower density such that, for any $m \in A$,

$$\sum_{\substack{n \in A \\ n > m}} \frac{1}{|w_2 w_3 \cdots w_{n-m+1}|^p} < \varepsilon.$$

(*Hint*: Proceed as in the proof of Proposition 9.17 and consider the coordinates of index $n - m + 1$ in $B_w^m x - e_1$.)

Exercise 9.2.5. Let $N_j = 2j^2 - 2j + 1$, $j \geq 1$. Define w_n as $(\frac{n+1}{n})^2$ for $N_j \leq n < N_j + j$, as $(N_j + j)^{-2/j}$ for $N_j + j \leq n < N_j + 2j$, as 1 for $N_j + 2j \leq n < N_j + 3j$, and as $(N_{j+1})^{2/j}$ for $N_j + 3j \leq n < N_j + 4j = N_{j+1}$. Show that B_w is a weighted shift on ℓ^p, $1 \leq p < \infty$, that satisfies the condition given in Proposition 9.17 but not the condition in the previous exercise. Thus, the condition in Proposition 9.17 does not characterize frequent hypercyclicity of weighted shifts on ℓ^p. (*Hint*: Use the result by Erdős and Sárközy.)

Exercise 9.2.6. The aim of this exercise is to show that not every vector $x \in \ell^1$ is the sum of two frequently hypercyclic vectors for the Rolewicz operator $2B$. Suppose that $x = y + z$ with y and z frequently hypercyclic. Then there is an increasing sequence $(m_k)_k$ of positive integers such that $\|T^{m_k} y\| < 1$ for $k \geq 1$; further let $\underline{\text{dens}}\{n \in \mathbb{N}_0 \;;\; \|T^n z\| < 1\} =: 2\delta > 0$. Deduce that there are positive integers n_k such that $\|T^{n_k} z\| < 1$ and $\delta m_k \leq n_k \leq m_k$, $k \geq 1$ sufficiently large, and hence that $\|T^{m_k} x\| \leq 1 + 2^{(1-\delta)m_k}$. Finally find some $x \in \ell^1$ that fails this inequality for any $\delta > 0$ and any increasing sequence $(m_k)_k$ of positive integers.

Exercise 9.2.7. Generalize Example 9.20: let T be an operator on a Fréchet space X and $\Lambda \subset\,]0, \infty[$ an uncountable set such that λT is frequently hypercyclic for any $\lambda \in \Lambda$. Show that these operators have no common frequently hypercyclic vector. (*Hint*: Consider $U = \{x \in X \;;\; |\langle x, x^* \rangle - 2| < 1\}$.)

Exercise 9.3.1. Let $L^p(\mathbb{T})$, $1 \leq p < \infty$, be the space of all complex-valued functions f on $\mathbb{T}$ such that $\|f\|_p := (\int_0^{2\pi} |f(e^{it})|^p \, dt)^{1/p} < \infty$. Show that $Tf(\lambda) = \lambda f(\lambda) - \int_{(1,\lambda)} f(\zeta)\, d\zeta$ defines a mixing and chaotic operator on $L^p(\mathbb{T})$, where (λ_1, λ_2) denotes the positively oriented arc from λ_1 to λ_2. (*Hint*: Consider the indicator functions $f = \mathbf{1}_{(\lambda,1)}$.)

Exercise 9.3.2. Let X be one of the complex spaces ℓ^p, $1 \leq p < \infty$, or c_0. Show that the map $\mathbb{D} \to X$, $\lambda \to (\lambda^n)_n$, is infinitely differentiable. Deduce that also the maps $\mathbb{D} \to H^2$, $\lambda \to k_{\overline{\lambda}}$ (see Proposition 4.38) and $\mathbb{D}_\tau \to E_\tau^2$, $\lambda \to e_\lambda$ (see Exercise 4.2.4) are infinitely differentiable.

Exercise 9.3.3. Let φ be a nonconstant entire function of exponential type and $A = \min\{|z| \; ; \; z \in \mathbb{C}, |\varphi(z)| = 1\}$. Show that, for any $\varepsilon > 0$, there is an entire function f that is frequently hypercyclic for $\varphi(D)$ such that

$$|f(z)| \leq Me^{(A+\varepsilon)r} \quad \text{for } |z| = r > 0$$

with some $M > 0$. (*Hint*: Combine the ideas of Exercise 4.2.4 and the proof of Theorem 9.25.)

Exercise 9.3.4. Let D be the differentiation operator on $H(\mathbb{C})$. Let $\phi :]0, \infty[\to [1, \infty[$ be a function with $\phi(r) \to \infty$ as $r \to \infty$. Show that there exists an entire function f that is frequently hypercyclic for D and that satisfies

$$|f(z)| \leq M\phi(r)e^r \quad \text{for } |z| = r > 0$$

with some $M > 0$. (*Hint*: Look at the proof of Theorem 4.22, using the Frequent Hypercyclicity Criterion in the version of Exercise 9.2.1.)

Exercise 9.3.5. Let φ be a nonconstant bounded holomorphic function on $\mathbb{D}$ and let M_φ^* be the corresponding adjoint multiplication operator on H^2; see Section 4.4. Show that M_φ^* is frequently hypercyclic if and only if it is hypercyclic, that is, if $\varphi(\mathbb{D}) \cap \mathbb{T} \neq \varnothing$. (*Hint*: Look at the proofs of Theorems 4.42 and 9.25.)

Exercise 9.4.1. In the proof of Theorem 9.27, show that $\underline{\text{dens}}(B) \geq \underline{\text{dens}}(A)$.

Exercise 9.4.2. Let T be a frequently hypercyclic operator on a Fréchet space X. Show that then $T^p \oplus T^q$ is hypercyclic on $X \oplus X$ for any $p, q \in \mathbb{N}$. (*Hint*: Exercises 2.5.5 and 9.1.3.)

Exercise 9.4.3. Let T be a topologically ergodic operator on a separable Fréchet space; see Exercise 1.5.6. Show that T^p is then also topologically ergodic for any $p \geq 1$. (*Hint*: Follow the proof of Theorem 9.27, using Exercise 2.5.5; see Exercise 6.1.5 for an alternative proof.)

Exercise 9.4.4. Let T be an operator on a separable Fréchet space X. Suppose that there is a vector $x \in X$ and a nonempty open subset U of X such that $\underline{\text{dens}}\{n \in \mathbb{N}_0 \; ; \; T^n x \in V\} > 0$ for all nonempty open subsets V of U. Show that x is frequently hypercyclic for T. (*Hint*: Use the Bourdon–Feldman theorem.)

Exercise 9.4.5. Let T be an operator on a (real or complex) Banach space X. Show that if there is some $x^* \in X^*$, $x^* \neq 0$, and some λ with $|\lambda| = 1$ such that $\|(\lambda I - T^*)^n x^*\|^{1/n} \to 0$ as $n \to \infty$ then T is not frequently hypercyclic.

Sources and comments

Section 9.1. Frequently hypercyclic operators were introduced by Bayart and Grivaux [38], [40]. The idea of using ergodic theory to obtain the dynamical properties of linear operators seems to be due to Rudnicki [272] and Flytzanis [152, 153]. Bayart and Grivaux

[40] obtained Lemma 9.5 (see also Bonilla and Grosse-Erdmann [87]) as well as the frequent hypercyclicity of the Birkhoff operators. The theorem of Erdős and Sárközy can be found in [296]. Theorem 9.8 is due to Grosse-Erdmann and Peris [185]; Bayart and Matheron [45] show that this result is essentially optimal.

For an introduction to ergodic theory we refer to Walters [300].

Section 9.2. The Frequent Hypercyclicity Criterion was obtained by Bayart and Grivaux [38, 40]; the form given here is due to Bonilla and Grosse-Erdmann [87]. Grivaux [173] also provided a probabilistic version of it. Proposition 9.11 is due to Bonilla and Grosse-Erdmann [87]. The remaining results in this section can essentially be found in Bayart and Grivaux [40]; see also Bonilla and Grosse-Erdmann [87]. The latter paper also contains further conditions under which the set $FHC(T)$ of frequently hypercyclic operators is of first Baire category, or when $FHC(T)+FHC(T)$ does or does not coincide with the full space.

Bayart and Grivaux [41] constructed a weighted shift on c_0 that is frequently hypercyclic, but neither chaotic nor mixing; this also shows that not every frequently hypercyclic operator satisfies the Frequent Hypercyclicity Criterion, and that Proposition 9.13 does not characterize frequently hypercyclic weighted shifts on c_0. Badea and Grivaux [19] found operators on a Hilbert space that are frequently hypercyclic and chaotic but not mixing.

It remains an open problem whether every chaotic operator is frequently hypercyclic, and to find a characterization of frequently hypercyclic weighted shifts, even on ℓ^2 or on c_0.

Section 9.3. The proof of Theorem 9.22 follows Bayart and Grivaux [38]; see also [39]. Theorems 9.25 and 9.26 are due to Blasco, Bonilla and Grosse-Erdmann [86, 76]; these authors also show that the operators of differentiation and translation on the space of harmonic functions on $\mathbb{R}^N$ are frequently hypercyclic, and they obtain some related growth results.

In order to keep the presentation simple we have imposed rather strong assumptions on the eigenvector fields. A much deeper analysis leads to one of the most striking results in linear dynamics.

To be more specific, an operator T on a complex separable Banach space X is said to have a *perfectly spanning set of eigenvectors associated to unimodular eigenvalues* if one of the following two equivalent conditions holds:

(i) there exists an atomless probability measure σ on $\mathbb{T}$ such that, for any measurable set $A \subset \mathbb{T}$ with $\sigma(A) = 1$, $\operatorname{span}\{\ker(\lambda I - T)\ ;\ \lambda \in A\}$ is dense in X;

(ii) for any countable set $D \subset \mathbb{T}$, $\operatorname{span}\{\ker(\lambda I - T)\ ;\ \lambda \in \mathbb{T} \setminus D\}$ is dense in X.

These conditions were first introduced by Flytzanis [152, 153]. Their equivalence was shown by Grivaux [174], who also obtained the following fundamental principle.

Theorem 9.42. *Any operator on a complex separable Banach space with a perfectly spanning set of eigenvectors associated to unimodular eigenvalues is frequently hypercyclic.*

When the underlying space is even a Hilbert space then one can show that there exists a Borel probability measure of full support on X with respect to which T is ergodic (see Bayart and Grivaux [40]); as explained in Section 9.1, this immediately implies that T is frequently hypercyclic. The measure can even be a so-called Gaussian measure. A similar result for nuclear Fréchet spaces is due to Grosse-Erdmann [182]. For surveys on the application of ergodic theory to linear dynamics we refer to Godefroy [164] and Grosse-Erdmann [182]. A detailed treatment can be found in Bayart and Matheron [44].

Bayart and Grivaux [40, 41] have applied their results to various operators. In particular they have shown that if φ is an automorphism of the unit disk $\mathbb{D}$ then the corresponding composition operator C_φ (see Section 4.5) is frequently hypercyclic on the Hardy space H^2 if and only if it is hypercyclic, that is, if and only if φ is parabolic or hyperbolic.

The example of Bayart and Grivaux [41] of a frequently hypercyclic weighted shift on c_0, mentioned above, has no unimodular eigenvalues, so that the approach chosen in this section is not always possible. Moreover, their operator does not possess any invariant Gaussian measure of full support.

Section 9.4. Theorem 9.27 is due to Bayart and Grivaux [40], whose proof uses Ansari's theorem. The alternative proof given in Grosse-Erdmann and Peris [185] contains an error; in fact, Example 9.28 contradicts Theorem 1.4 in that paper. The proof given here is due to Grosse-Erdmann and Peris [186].

Theorem 9.33 provides a new common approach to Theorems 9.35 and Theorem 9.36 that were previously obtained by Bayart and Matheron [44] and by Conejero, Müller and Peris [110], respectively. The remainder of the section, including Theorem 9.39 and Corollary 9.41, is due to Shkarin [287]. Grivaux [174] has recently shown that the necessary spectral conditions of Theorems 5.6 and 9.39 actually characterize spectra of frequently hypercyclic operators on Hilbert spaces.

Theorem 9.43. *Let $K \subset \mathbb{C}$ be a nonempty compact set. There exists a frequently hypercyclic operator T on a complex Hilbert space such that $\sigma(T) = K$ if and only if K has no isolated points and each of its connected components meets the unit circle.*

Further interesting results on frequent hypercyclicity include the facts that every operator on an infinite-dimensional complex separable Hilbert space is the sum of two frequently hypercyclic operators (Bayart and Grivaux [40]) and that every infinite-dimensional complex Fréchet space with an unconditional basis supports a frequently hypercyclic and chaotic operator (De la Rosa, Frerick, Grivaux, and Peris [127]).

Many questions concerning frequently hypercyclic operators remain open. For example (see Bayart and Grivaux [40]), whether the frequent hypercyclicity of an operator T is inherited by its direct sum $T \oplus T$; and whether it is inherited by its inverse T^{-1}, if it exists.

Exercises. Exercise 9.1.1 is taken from Bayart and Grivaux [40], Exercises 9.1.4, 9.2.1 and 9.2.6 from Bonilla and Grosse-Erdmann [87], and Exercises 9.2.4 and 9.2.5 from Grosse-Erdmann and Peris [185]. For Exercise 9.3.1 we refer to Bayart and Grivaux [39], for Exercise 9.3.3 to Bonilla and Grosse-Erdmann [86]. Exercise 9.3.4 is taken from Blasco, Bonilla and Grosse-Erdmann [76] who also show that, in the converse direction, given any function $\phi : \mathbb{R}_+ \to \mathbb{R}_+$ with $\lim_{r\to\infty} \phi(r) = 0$ there is no entire function f that is frequently hypercyclic for D such that $|f(z)| \leq \phi(r)\frac{e^r}{r^{1/4}}$ for $|z| = r$ sufficiently large. Exercise 9.3.5 is taken from Bayart and Grivaux [40], Exercise 9.4.2 from Costakis and Ruzsa [122], Exercise 9.4.4 from Grosse-Erdmann and Peris [185], and Exercise 9.4.5 from Shkarin [287].

Chapter 10
Hypercyclic subspaces

By the Herrero–Bourdon theorem, every hypercyclic operator admits a dense subspace in which every nonzero vector is hypercyclic. In this chapter we ask for a large space of hypercyclic vectors in a different sense: does a given hypercyclic operator admit a closed and infinite-dimensional subspace in which every nonzero vector is hypercyclic? We will see that in many cases the answer is positive. However, in contrast to the problem of dense subspaces of hypercyclic vectors, we will also find counterexamples. This makes the notion considered here particularly interesting.

We point out that the two meanings of largeness are almost incompatible. The only dense and closed subspace is the whole space itself. And while there do exist operators for which every nonzero vector is hypercyclic, Read's operator being one of them, such examples are extremely rare and difficult to construct. Indeed, the commonly known hypercyclic operators all have large supplies of non-hypercyclic vectors.

Since several of the proofs in this chapter are technically more demanding in the Fréchet space setting, we present them first in Banach (or Hilbert) spaces; in the final section we then supply the proofs in the general case.

10.1 Operators with hypercyclic subspaces

In the present context the following terminology has been generally accepted.

Definition 10.1. Let T be an operator on a separable Fréchet space X. Then a *hypercyclic subspace* for T is an infinite-dimensional closed subspace M of X so that every nonzero vector in M is hypercyclic for T.

In this section we derive a useful sufficient condition for an operator to have a hypercyclic subspace.

Thus let T be a hypercyclic operator on X. Which additional assumption do we need in order for T to support a hypercyclic subspace? The following

K.-G. Grosse-Erdmann, A. Peris Manguillot, *Linear Chaos*, Universitext,
DOI 10.1007/978-1-4471-2170-1_10,

idea is at least plausible. Suppose that there is an infinite-dimensional closed subspace M_0 of X so that $T^n x \to 0$ for every $x \in M_0$, and suppose that M_0 has a basis $(e_n)_n$. By the density of hypercyclic vectors we can find, arbitrarily closely to each e_n, a hypercyclic vector f_n. Now, if the f_n are sufficiently close to the e_n then one may hope that the f_n, in turn, form a basis in their closed linear span M, and that each nonzero vector in M is also hypercyclic. This strategy does indeed work, at least in certain spaces.

Theorem 10.2 (Montes). *Let X be a separable Fréchet space with a continuous norm, and let T be an operator on X. Suppose that there exists an increasing sequence $(n_k)_k$ of positive integers such that*

(i) *T satisfies the Hypercyclicity Criterion for $(n_k)_k$,*
(ii) *there exists an infinite-dimensional closed subspace M_0 of X such that $T^{n_k} x \to 0$ for all $x \in M_0$.*

Then T has a hypercyclic subspace.

We have already met the notion of a continuous norm; see Lemma 8.7.

Example 10.3. Of course, the norm in any Banach space is continuous. The Fréchet space $H(\mathbb{C})$ of entire functions has a continuous norm. One may take, for example, $\|f\| = \sup_{|z|\leq 1} |f(z)|$, $f \in H(\mathbb{C})$. By contrast, the space $\omega = \mathbb{K}^{\mathbb{N}}$ of all sequences (see Example 2.2) has no continuous norm: it would have to be dominated by a multiple of some seminorm $p_n(x) = \sup_{1\leq k\leq n} |x_k|$ (see Exercise 2.1.7) and would thus assign the value 0 to some nonzero vector.

Remark 10.4. While Montes' theorem is usually applied in the form stated above, we will see that it remains true if condition (ii) is replaced by the following weaker condition:

(ii′) *there exists an infinite-dimensional closed subspace M_0 of X such that $(T^{n_k} x)_k$ converges for all $x \in M_0$.*

We will give two proofs of this result. The first one is straightforward but slightly technical, and it uses the notion of a basic sequence. The second one, which we describe in the next section, provides a very interesting link between the hypercyclicity of an operator T and the hypercyclicity of the corresponding left-multiplication operator $L_T : S \to TS$.

Let us turn to the first proof. As is the case for several results in this chapter, the proof of Montes' theorem is considerably more transparent when X is a Banach space. We will therefore restrict ourselves here to these spaces. The proof in the general case will be given in Section 10.5.

Definition 10.5. A (finite or infinite) sequence $(e_n)_n$ in a Banach space X is called a *basic sequence* if it is a basis in its closed linear span.

We have that $(e_n)_n$ is a basic sequence if and only if every vector $x \in M := \overline{\operatorname{span}}\{e_n \ ; \ n \geq 1\}$ has a unique representation

$$x = \sum_{n\geq 1} a_n e_n$$

with scalars $a_n \in \mathbb{K}$, $n \geq 1$. One then defines the coefficient functionals

$$e_n^* : M \to \mathbb{K}, \quad x \to a_n$$

for $n \geq 1$. Since M is a Banach space these functionals are continuous; see Appendix A. By $\|e_n^*\|$, $n \geq 1$, we denote the norm of e_n^* on M.

We will now show that a sufficiently small perturbation of a basic sequence remains basic.

Lemma 10.6. *Let $(e_n)_n$ be a basic sequence in a Banach space X with coefficient functionals e_n^*, $n \geq 1$. If $(f_n)_n$ is a sequence in X with*

$$\sum_{n\geq 1} \|e_n^*\| \, \|e_n - f_n\| = \delta < 1,$$

then $(f_n)_n$ is also a basic sequence.

Moreover, a series $\sum_{n\geq 1} a_n e_n$ converges if and only if $\sum_{n\geq 1} a_n f_n$ does, and

$$\|f_n^*\| \leq \frac{1}{1-\delta}\|e_n^*\|, \quad n \geq 1.$$

Proof. Let M denote the closed linear span of the e_n. We then consider the operator

$$T : M \to X, \quad x = \sum_{n\geq 1} a_n e_n \to \sum_{n\geq 1} a_n f_n,$$

that is, $Tx = \sum_{n\geq 1} \langle x, e_n^* \rangle f_n$. In order to see that this is well defined, we note that for $1 \leq n \leq m$,

$$\begin{aligned}\Big\| \sum_{k=n}^{m} a_k f_k \Big\| &\leq \Big\| \sum_{k=n}^{m} a_k (f_k - e_k) \Big\| + \Big\| \sum_{k=n}^{m} a_k e_k \Big\| \\ &\leq \Big(\sum_{k=n}^{m} \|e_k^*\| \|f_k - e_k\| \Big) \|x\| + \Big\| \sum_{k=n}^{m} a_k e_k \Big\|, \end{aligned} \tag{10.1}$$

and the right-hand side tends to 0 as $m \geq n \to \infty$. Thus Tx exists and, by (10.1),

$$\|Tx\| \leq \delta \|x\| + \|x\| = (1+\delta)\|x\|, \quad x \in M, \tag{10.2}$$

so that T is continuous.

Moreover, the second triangle inequality tells us that, for any $x \in M$,

$$\begin{aligned}\|Tx\| = \Big\| \sum_{n\geq 1} a_n f_n \Big\| &\geq \Big\| \sum_{n\geq 1} a_n e_n \Big\| - \Big\| \sum_{n\geq 1} a_n (e_n - f_n) \Big\| \\ &\geq \|x\| - \Big(\sum_{n\geq 1} \|e_n^*\| \, \|f_n - e_n\| \Big) \|x\| \end{aligned}$$

$$\geq \|x\| - \delta\|x\| = (1-\delta)\|x\|. \tag{10.3}$$

Now, (10.2) and (10.3) imply that T is an isomorphism of M onto $M' := \operatorname{ran} T$. Thus, $(f_n)_n = (Te_n)_n$, is a basis in M' and therefore a basic sequence in X. Moreover, a series $\sum_{n\geq 1} a_n e_n$ converges if and only if $\sum_{n\geq 1} a_n f_n$ does. Finally, we have by (10.3) that for any $y = Tx = \sum_{n\geq 1} a_n f_n \in M'$ and $n \geq 1$

$$|\langle y, f_n^* \rangle| = |a_n| = |\langle x, e_n^* \rangle| \leq \|e_n^*\| \|x\| \leq \frac{1}{1-\delta}\|e_n^*\| \|y\|,$$

which implies that $\|f_n^*\| \leq \frac{1}{1-\delta}\|e_n^*\|$. □

As a second tool we need the following fundamental result.

Theorem 10.7 (Mazur). *Every infinite-dimensional Banach space contains a basic sequence.*

We prepare its proof by a lemma.

Lemma 10.8. *Let X be an infinite-dimensional Banach space, E a finite-dimensional subspace of X and $\varepsilon > 0$. Then there exists an $x \in X$ with $\|x\| = 1$ such that, for any $\lambda \in \mathbb{K}$ and $y \in E$,*

$$\|y\| \leq (1+\varepsilon)\|\lambda x + y\|.$$

Since we will obtain a more general result later (see Lemma 10.39), we omit the proof.

Proof of Theorem 10.7. First, let $(\varepsilon_n)_n$ be a sequence of positive numbers such that $\prod_{n=1}^{\infty}(1+\varepsilon_n) \leq 2$. Choose $e_1 \in X$ with $\|e_1\| = 1$. By Lemma 10.8 we can inductively construct vectors $e_2, e_3, \ldots$ of norm 1 such that, for all $n \geq 1$ and $a_1, \ldots, a_{n+1} \in \mathbb{K}$,

$$\Big\| \sum_{k=1}^{n} a_k e_k \Big\| \leq (1+\varepsilon_n) \Big\| \sum_{k=1}^{n+1} a_k e_k \Big\|.$$

We then have for any $a_1, \ldots, a_n \in \mathbb{K}$ and any $m \leq n$ that

$$\begin{aligned}
\Big\| \sum_{k=1}^{m} a_k e_k \Big\| &\leq (1+\varepsilon_m) \Big\| \sum_{k=1}^{m+1} a_k e_k \Big\| \\
&\leq (1+\varepsilon_m)(1+\varepsilon_{m+1})\ldots(1+\varepsilon_{n-1}) \Big\| \sum_{k=1}^{n} a_k e_k \Big\| \\
&\leq 2 \Big\| \sum_{k=1}^{n} a_k e_k \Big\|,
\end{aligned} \tag{10.4}$$

which also implies that, for $k \leq n$,

$$|a_k| = \|a_k e_k\| \le \Big\|\sum_{j=1}^{k} a_j e_j\Big\| + \Big\|\sum_{j=1}^{k-1} a_j e_j\Big\| \le 4\Big\|\sum_{j=1}^{n} a_j e_j\Big\|. \qquad (10.5)$$

We claim that $(e_n)_n$ is a basic sequence in X. Indeed, let M be the closed linear span of these vectors, and let $x \in M$. Then there are vectors

$$x_\nu = \sum_{k=1}^{N_\nu} a_{\nu,k} e_k$$

that converge to x as $\nu \to \infty$. We let $a_{\nu,k} = 0$ for $k > N_\nu$. It then follows from (10.5) that, for $k \ge 1$,

$$|a_{\nu,k} - a_{\mu,k}| \le 4\|x_\nu - x_\mu\| \to 0$$

as $\mu, \nu \to \infty$, so that $a_k := \lim_{\nu\to\infty} a_{\nu,k}$ exists for all $k \ge 1$. Using (10.4) we deduce that, for any $n \ge 1$,

$$\begin{aligned}\Big\|\sum_{k=1}^{n} a_{\nu,k} e_k - \sum_{k=1}^{n} a_k e_k\Big\| &= \lim_{\mu\to\infty} \Big\|\sum_{k=1}^{n} a_{\nu,k} e_k - \sum_{k=1}^{n} a_{\mu,k} e_k\Big\| \\ &\le 2 \limsup_{\mu\to\infty} \|x_\nu - x_\mu\| = 2\|x_\nu - x\|\end{aligned}$$

and therefore, whenever $n \ge N_\nu$,

$$\Big\|x - \sum_{k=1}^{n} a_k e_k\Big\| \le \|x - x_\nu\| + \Big\|\sum_{k=1}^{N_\nu} a_{\nu,k} e_k - \sum_{k=1}^{n} a_k e_k\Big\| \le 3\|x_\nu - x\|,$$

so that $x = \sum_{k=1}^\infty a_k e_k$. This proves the claim. □

We are now in a position to prove Theorem 10.2 in a special case.

Proof of Theorem 10.2 *for Banach spaces.* The proof will be divided into three steps. To simplify notation we perform the proof in the case when $(n_k)_k$ is the full sequence of positive integers. The general case follows in exactly the same way.

Step 1. By Mazur's theorem, since M_0 is an infinite-dimensional Banach space in the induced topology, there exists a basic sequence $(e_n)_n$ in X of elements from M_0.

Step 2. We show that $(e_n)_n$ can be perturbed into a basic sequence $(f_n)_n$ of hypercyclic vectors. To this end, let $K_n = \max(1, \|e_n^*\|)$. Further let X_0 and Y_0 be the dense subsets of X appearing in the Hypercyclicity Criterion and let $(y_n)_n$ be a sequence in Y_0 that is dense in X. We claim that there then exist vectors $x_{j,k} \in X_0$ and positive integers $n(j,k)$ such that $(n(j,k))_{k\ge1}$ is increasing for each $j \ge 1$ and such that, for all $j, k, j', k' \ge 1$,

$$\|x_{j,k}\| \leq \frac{1}{2^{j+k+1}K_j}, \tag{10.6}$$

$$\|T^{n(j,k)}x_{j,k} - y_k\| \leq \frac{1}{2^k}, \tag{10.7}$$

$$\|T^{n(j',k')}x_{j,k}\| \leq \frac{1}{2^{j+k+k'}K_j} \quad \text{if } (j',k') \neq (j,k), \tag{10.8}$$

$$\|T^{n(j,k)}e_j\| \leq \frac{1}{2^k}. \tag{10.9}$$

This is easily seen by induction with respect to the strict order $<$ on $\mathbb{N}\times\mathbb{N}$ that is defined by $(1,1) < (1,2) < (2,1) < (1,3) < (2,2) < (3,1) < (1,4) < \ldots$. The existence of $x_{j,k}$ and $n(j,k)$ satisfying (10.6), (10.7) and (10.9) follows from the assumptions on Y_0 in the Hypercyclicity Criterion and the fact that $e_j \in M_0$; note that $x_{j,k} \in X_0$ can be achieved because X_0 is dense in X. Condition (10.8) can be rewritten as

$$\|T^{n(j,k)}x_{j',k'}\| \leq \frac{1}{2^{j'+k'+k}K_{j'}} \quad \text{and} \quad \|T^{n(j',k')}x_{j,k}\| \leq \frac{1}{2^{j+k+k'}K_j}$$

if $(j',k') < (j,k)$; this condition can therefore also be ensured when we use the fact that each $x_{j',k'}$ belongs to X_0 and that T is continuous.

We now define, for any $j \geq 1$,

$$f_j = e_j + \sum_{k=1}^{\infty} x_{j,k}.$$

By (10.6) these series converge, and simple calculations using (10.6)–(10.9) show that we have, for $j, k, j' \geq 1$,

$$\|e_j - f_j\| \leq \frac{1}{2^{j+1}K_j}, \tag{10.10}$$

$$\|T^{n(j,k)}f_j - y_k\| \leq \frac{3}{2^k}, \tag{10.11}$$

$$\|T^{n(j',k)}(e_j - f_j)\| \leq \frac{1}{2^{j+k}K_j} \quad \text{if } j' \neq j. \tag{10.12}$$

By (10.10) we have that $\sum_{j=1}^{\infty} \|e_j^*\|\,\|e_j - f_j\| < 1$, whence $(f_n)_n$ is a basic sequence by Lemma 10.6.

Step 3. We claim that the closed linear span M of the f_n, $n \geq 1$, is the desired hypercyclic subspace. Thus let $z \in M$, $z \neq 0$; we need to show that z is hypercyclic for T. Since $(f_n)_n$ is a basis of M we can write

$$z = \sum_{j=1}^{\infty} a_j f_j.$$

Since $z \neq 0$, one of the coefficients, a_m say, must be nonzero. We can assume that $a_m = 1$ because any nonzero multiple of a hypercyclic vector remains hypercyclic. Moreover, by Lemma 10.6,

$$w := \sum_{j \neq m} a_j e_j$$

exists, and w belongs to M_0.

We then have by (10.11) and (10.12) that

$$\begin{aligned}
&\|T^{n(m,k)} z - y_k\| \\
&\quad \leq \|T^{n(m,k)} f_m - y_k\| + \Big\| \sum_{j \neq m} a_j T^{n(m,k)} (f_j - e_j) \Big\| + \|T^{n(m,k)} w\| \\
&\quad \leq \frac{3}{2^k} + \sum_{j \neq m} |a_j| \|T^{n(m,k)} (f_j - e_j)\| + \|T^{n(m,k)} w\| \\
&\quad \leq \frac{3}{2^k} + \sum_{j \neq m} \|e_j^*\| \|w\| \frac{1}{2^{j+k} K_j} + \|T^{n(m,k)} w\| \\
&\quad \leq \frac{3}{2^k} + \frac{1}{2^k} \|w\| + \|T^{n(m,k)} w\| \to 0
\end{aligned}$$

as $k \to \infty$ because $w \in M_0$. Since the y_k, $k \geq 1$, are dense in X, z is hypercyclic for T. □

Remark 10.9. In order to see that Montes' theorem remains true under condition (ii′) of Remark 10.4 one need only weaken (10.9) to

$$\|T^{n(j,k)} e_j - v_j\| \leq \frac{1}{2^k}$$

with certain $v_j \in X, j \geq 1$, so that (10.11) has to be replaced by

$$\|T^{n(j,k)} f_j - v_j - y_k\| \leq \frac{3}{2^k}.$$

In addition, there is some $v \in X$ such that $T^n w \to v$. We can then conclude as before that $\|T^{n(m,k)} z - v_m - v - y_k\| \to 0$ as $k \to \infty$. Since the vectors $y_k + v_m + v$, $k \geq 1$, form a dense set in X, we have again that z is hypercyclic.

Our first application of Montes' theorem treats weighted shifts.

Example 10.10. Let $w = (w_n)_n$ be a bounded weight sequence and B_w the corresponding weighted backward shift on one of the spaces $X = \ell^p$, $1 \leq p < \infty$, or $X = c_0$; see Section 4.1. By Example 4.9(a), B_w satisfies the Hypercyclicity Criterion if (and only if)

$$\sup_{n \geq 1} \prod_{\nu=1}^{n} |w_\nu| = \infty.$$

We claim that B_w has a hypercyclic subspace if, in addition,

$$\sup_{n\geq 1}\limsup_{k\to\infty}\prod_{\nu=1}^{n}|w_{\nu+k}|<\infty.$$

For example, if $w_n = (\frac{n+1}{n})^\alpha$, $n \geq 1$, $\alpha > 0$, then B_w has a hypercyclic subspace. In particular, the backward shift on the Bergman space does; see Example 4.9(a).

In order to show the claim, let C denote the latter supremum, and suppose that the Hypercyclicity Criterion is satisfied for the sequence $(n_k)_k$. Setting $m_1 = n_1$, we can find a subsequence $(m_k)_k$ of $(n_k)_k$ such that, for $k \geq 1$,

$$\prod_{\nu=1}^{m_k}|w_{\nu+\mu}|\leq C+1 \quad \text{for } \mu\geq m_{k+1}-m_k.$$

This implies that, for $j > k \geq 1$,

$$\prod_{\nu=m_j-m_k+1}^{m_j}|w_\nu|\leq C+1. \tag{10.13}$$

If e_n, $n \geq 1$, denote the unit sequences then

$$M_0 := \Big\{\sum_{k=1}^{\infty}a_k e_{m_k}\ ;\ (a_k)_k\in X\Big\}$$

is an infinite-dimensional closed subspace of X. For any $x = \sum_{k=1}^{\infty} a_k e_{m_k} \in M_0$ we have, using (10.13), that

$$\begin{aligned}\|T^{m_k}x\| &= \Big\|\sum_{j=1}^{\infty}a_j T^{m_k}e_{m_j}\Big\| = \Big\|\sum_{j>k}a_j\Big(\prod_{\nu=m_j-m_k+1}^{m_j}w_\nu\Big)e_{m_j-m_k}\Big\|\\ &\leq (C+1)\Big\|\sum_{j>k}a_j e_{m_j-m_k}\Big\|\to 0\end{aligned}$$

as $k \to \infty$. Montes' theorem then implies the claim.

For the application to Birkhoff's operators we derive a simple but useful consequence from Montes' theorem.

Corollary 10.11. *Let X be a separable Fréchet space with a continuous norm, and let T be an operator on X that satisfies the Hypercyclicity Criterion. If $\ker(\lambda I - T)$ is infinite-dimensional for some λ with $|\lambda| < 1$, then T has a hypercyclic subspace.*

Proof. By continuity of T, $M_0 := \ker(\lambda I - T)$ is a closed subspace, and for any $x \in M_0$ we have that $T^n x = \lambda^n x \to 0$. □

We remark that this result remains true for any λ with $|\lambda| \leq 1$; see Exercise 10.1.3.

Example 10.12. **(Birkhoff's operators)** The Birkhoff operators $T_a : f \to f(\cdot + a)$, $a \neq 0$, on $H(\mathbb{C})$ have hypercyclic subspaces. This follows from the fact that the linearly independent entire functions

$$z \to \frac{1}{2^{z/a}} \exp\left(\frac{2k\pi i}{a} z\right), \quad k \in \mathbb{Z},$$

belong to $\ker(\frac{1}{2}I - T_a)$. The remaining hypotheses of Corollary 10.11 are also satisfied; see Examples 2.38 and 10.3.

More generally, if φ is an entire function of exponential type that is not a polynomial then the operator $\varphi(D)$ on $H(\mathbb{C})$ defined in Proposition 4.19 has a hypercyclic subspace. Indeed, by the proof of Theorem 4.21 we have that $\varphi(D)e_\mu = \varphi(\mu)e_\mu$, where $e_\mu(z) = e^{\mu z}$. The big Picard theorem tells us that for any $\lambda \in \mathbb{C}$ with at most one exception the equation $\varphi(\mu) = \lambda$ has infinitely many solutions. This implies that for such a λ, $\ker(\lambda I - \varphi(D))$ is infinite-dimensional, and we can choose $|\lambda| < 1$.

As our last example in this section we consider MacLane's operator.

Example 10.13. **(MacLane's operator)** The operator of differentiation on $H(\mathbb{C})$ has a hypercyclic subspace. Since D satisfies the Hypercyclicity Criterion for the full sequence (see Example 3.7), it suffices to exhibit an infinite-dimensional closed subspace M_0 of $H(\mathbb{C})$ on which suitable powers of D tend to 0.

To start with, let us note that, for any $n \geq 1$, there is some $C_n > 0$ such that

$$x^n \leq 2^x \quad \text{for all } x \geq C_n. \tag{10.14}$$

We choose a strictly increasing sequence of positive integers $(n_k)_k$ with $n_1 \geq 1$ such that, for all $k \geq 1$,

$$n_{k+1} \geq C_{n_k}.$$

For $j \geq k+1$ we then have that $n_j \geq n_{k+1} \geq C_{n_k}$ and hence, by (10.14),

$$n_j^{n_k} \leq 2^{n_j} \quad \text{for } j \geq k+1. \tag{10.15}$$

Now let M_0 be the closed subspace of $H(\mathbb{C})$ of all entire functions f of the form

$$f(z) = \sum_{k=1}^{\infty} a_k z^{n_k - 1}.$$

We claim that

$$D^{n_k} f \to 0 \quad \text{in } H(\mathbb{C}) \text{ as } k \to \infty.$$

Indeed, let $R \geq 1$. Then we have that

$$\begin{aligned}
\sup_{|z|\leq R} |D^{n_k} f(z)| &= \sup_{|z|\leq R} \Big| \sum_{j=k+1}^{\infty} a_j D^{n_k} z^{n_j-1} \Big| \\
&\leq \sum_{j=k+1}^{\infty} |a_j| \, (n_j-1)\cdots(n_j-n_k) R^{n_j-n_k-1} \\
&\leq \sum_{j=k+1}^{\infty} |a_j| \, n_j^{n_k} R^{n_j} \\
&\leq \sum_{j=k+1}^{\infty} |a_j| (2R)^{n_j} \to 0
\end{aligned}$$

as $k \to \infty$, where in the last inequality we have used (10.15). This had to be shown.

Further examples will be considered in Section 10.4.

10.2 Hypercyclic left-multiplication operators

In this section we outline an alternative approach to hypercyclic subspaces that is also interesting in its own right. In order to emphasize ideas over technical details we first restrict ourselves again to Banach spaces and postpone the general proofs to Section 10.5.

Thus, let T be an operator on a separable Banach space X. The principal idea, a priori unrelated to the topic of this chapter, is to study the dynamical properties of the operator T by way of its induced *left-multiplication operator*

$$L_T : L(X) \to L(X), \quad L_T S = TS,$$

on the space $L(X)$ of operators on X. The inequality

$$\|L_T S\| = \|TS\| \leq \|S\| \|T\|, \quad S \in L(X),$$

shows that L_T indeed defines an operator on $L(X)$ when the latter is endowed with the operator norm topology.

This new approach of studying the operator T is quite promising. For if we can describe properties of an orbit

$$\{(L_T)^n S \; ; \; n \geq 0\} = \{T^n S \; ; \; n \geq 0\} \subset L(X) \tag{10.16}$$

of some fixed operator S under L_T then we can deduce properties of the orbit

$$\{T^n S x \; ; \; n \geq 0\} \subset X \tag{10.17}$$

of Sx under T, for an arbitrary $x \in X$. In particular if the orbit (10.16) is dense in $L(X)$ then any orbit (10.17) is dense in X, unless for $x = 0$, giving us automatically a large set of hypercyclic vectors for T.

This striking idea, however, meets with an unexpected obstacle: for an orbit in $L(X)$ to be dense the space of operators would have to be separable, which is hardly ever the case under the operator norm topology; see Exercise 10.2.1 (but also Exercise 10.2.8). The only hope is to weaken the topology of $L(X)$. In Section 8.3 the strong operator topology (SOT) has already come to our rescue in another case where the operator norm topology proved to be too strong. It will save us again because $L(X)$ is indeed separable under this weaker topology. Here, and in the sequel, topological notions will be prefixed by SOT if they refer to the strong operator topology.

Proposition 10.14. *Let X be a separable Banach space. Then $L(X)$ is SOT-separable.*

For the proof we need a related result concerning the dual X^* of X. Note that the dual of a separable Banach space need not be separable under the usual operator norm; a simple example is provided by the sequence space ℓ^1 whose dual is ℓ^∞. Again we obtain separability under a weaker topology. Just like the strong operator topology on $L(X)$, the *weak-$*$-topology* on X^* is defined as the topology of pointwise convergence on X. A base of neighbourhoods of an element $x^* \in X^*$ is given by

$$U_{x_1,\dots,x_n}(x^*, \varepsilon) = \{y^* \in X^* \ ; \ |\langle x_k, x^*\rangle - \langle x_k, y^*\rangle| < \varepsilon \text{ for } k = 1, \dots, n\},$$

where $x_1, \dots, x_n$, $n \geq 1$, is an arbitrary collection of linearly independent vectors of X and $\varepsilon > 0$. Thus, a sequence (more generally, a net) $(x^*_\alpha)_\alpha$ is weak-$*$-convergent to x^* if and only if, for every $x \in X$, $\langle x, x^*_\alpha\rangle \to \langle x, x^*\rangle$.

For later use we consider general normed spaces.

Lemma 10.15. *Let X be a separable normed space. Then X^* is weak-$*$-separable.*

Proof. Let E be a countable dense subset of X. We then consider the set of all continuous linear functionals on X of the following form: there are linearly independent elements $e_1, \dots, e_m \in E$ and numbers $q_1, \dots, q_m$ in $\mathbb{Q}$ or in $\mathbb{Q} + i\mathbb{Q}$ such that

$$x^* = \sum_{j=1}^{m} q_j e^*_j,$$

where $e^*_1, \dots, e^*_m$ are the coordinate functionals corresponding to the (finite) basic sequence $(e_j)_{j=1,\dots,m}$ (see Definition 10.5 and the subsequent discussion); we assume that the e^*_j are extended continuously to all of X, with preservation of the norm.

We show that this countable set of functionals is weak-$*$-dense in X^*. To see this, let $x^* \in X^*$, let $x_1, \dots, x_m$, $m \geq 1$, be linearly independent vectors of X and $\varepsilon > 0$. Let $x_1^*, \ldots, x_m^*$ be the coordinate functionals corresponding to the (finite) basic sequence $(x_j)_{j=1,\ldots,m}$, extended to all of X, and let $M = \max_{j=1,\ldots,m} \|x_j^*\|$. We fix rational numbers $q_1, \ldots, q_m \in \mathbb{Q}(+i\mathbb{Q})$ such that, for $j = 1, \ldots, m$,

$$|\langle x_j, x^* \rangle - q_j| < \frac{\varepsilon}{2}.$$

Let $e_1, \ldots, e_m$ be vectors from E such that, for $j = 1, \ldots, m$,

$$\|e_j - x_j\| \leq \min\Big(\frac{1}{2mM}, \frac{\varepsilon}{4M(\sum_{k=1}^m |q_k| + 1)}\Big).$$

By Lemma 10.6, applied to the Banach space $\mathrm{span}\{x_1, \ldots, x_m, e_1, \ldots, e_m\}$, we have that $(e_j)_{j=1,\ldots,m}$ is a basic sequence in X with $\|e_j^*\| \leq 2M$, $j = 1, \ldots, m$.

Let

$$y^* = \sum_{j=1}^m q_j e_j^*$$

be the corresponding continuous linear functional. Since $\sum_{j=1}^m q_j \langle e_k, e_j^* \rangle = q_k$, we have for $k = 1, \ldots, m$ that

$$\begin{aligned}
|\langle x_k, x^* \rangle - \langle x_k, y^* \rangle| &\leq |\langle x_k, x^* \rangle - q_k| + \Big|\sum_{j=1}^m q_j \langle e_k, e_j^* \rangle - \sum_{j=1}^m q_j \langle x_k, e_j^* \rangle\Big| \\
&< \frac{\varepsilon}{2} + \sum_{j=1}^m |q_j| \|e_j^*\| \|e_k - x_k\| \\
&\leq \frac{\varepsilon}{2} + 2M \frac{\varepsilon}{4M(\sum_{j=1}^m |q_j| + 1)} \sum_{j=1}^m |q_j| \leq \varepsilon,
\end{aligned}$$

as had to be shown. □

Proof of Proposition 10.14. Let E be a countable dense subset of X. By Lemma 10.15 there is a countable weak-$*$-dense subset Φ of X^*. We claim that the countable set

$$\mathcal{F} = \mathcal{F}_{\Phi,E} = \Big\{ \sum_{j=1}^m \langle \cdot\,, y_j^* \rangle e_j \;;\; y_j^* \in \Phi, e_j \in E, 1 \leq j \leq m \Big\}$$

of operators on X is SOT-dense in $L(X)$. Indeed, let $T \in L(X)$, let $x_1, \ldots, x_m$, $m \geq 1$, be linearly independent vectors of X and $\varepsilon > 0$. Denote by $x_1^*, \ldots, x_m^*$ the coordinate functionals corresponding to the basic sequence $(x_j)_{j=1,\ldots,m}$, extended to X. Then, by density of E and weak-$*$-density of Φ, there are vectors $e_1, \ldots, e_m \in E$ and $y_1^*, \ldots, y_m^* \in \Phi$ such that for $j, k = 1, \ldots, m$,

$$\|Tx_k - e_k\| < \frac{\varepsilon}{2}, \quad \|(\langle x_j, x_k^*\rangle - \langle x_j, y_k^*\rangle)\, e_k\| < \frac{\varepsilon}{2m}.$$

For $S = \sum_{j=1}^m \langle \cdot\,, y_j^*\rangle e_j \in \mathcal{F}$ we then have that, for $k = 1, \ldots, m$,

$$\|Tx_k - Sx_k\| \le \|Tx_k - e_k\| + \|(1 - \langle x_k, y_k^*\rangle)e_k\| + \sum_{j\neq k} \|\langle x_k, y_j^*\rangle e_j\| < \varepsilon;$$

note that $\langle x_j, x_k^*\rangle = \delta_{j,k}$. This had to be shown. □

The SOT-separability of $L(X)$ allows us to speak of SOT-dense orbits. However, since $L(X)$ is not a Fréchet space under the strong operator topology, we are no longer in the framework of Chapter 2. Thus we need to define the following.

Definition 10.16. Let T be an operator on a Banach space X. Then the left-multiplication operator L_T is called *SOT-hypercyclic* if there is some operator $S \in L(X)$ whose orbit $\{T^n S\ ;\ n \ge 0\}$ under L_T is SOT-dense in $L(X)$; such an operator S is then called *SOT-hypercyclic* for L_T.

Remark 10.17. Under the strong operator topology, $L(X)$ is a topological vector space and L_T is an operator on $L(X)$. The dynamical properties of operators on topological vector spaces will be treated in detail in Chapter 12.

The following explains our interest in SOT-hypercyclicity of left-multiplication operators.

Proposition 10.18. *Let T be an operator on a separable Banach space X. If $S \in L(X)$ is SOT-hypercyclic for L_T then Sx is a hypercyclic vector for T for any $x \in X$, $x \neq 0$.*

Proof. By the Hahn–Banach theorem there is some $x^* \in X^*$ with $\langle x, x^*\rangle = 1$. Let $y \in X$ be arbitrary. Then the operator $R = \langle \cdot\,, x^*\rangle y$ maps x to y. Since S is SOT-hypercyclic there is an increasing sequence $(n_k)_k$ of positive integers such that $T^{n_k}Sx = (L_T)^{n_k}Sx \to Rx = y$. Hence Sx is hypercyclic for T. □

Remark 10.19. As an immediate consequence we have the following: if T itself is SOT-hypercyclic for L_T, or if some surjective operator S is hypercyclic for L_T then every nonzero vector $x \in X$ is hypercyclic for T.

The main result of this section provides us with a close link between the hypercyclicity properties of the operator T and those of the operator L_T.

Theorem 10.20. *Let T be an operator on a separable Banach space X. Then the following assertions are equivalent:*
(i) *T satisfies the Hypercyclicity Criterion;*
(ii) *L_T is SOT-hypercyclic.*

In order to prepare the proof we fix a countable weak-$*$-dense subset Φ of X^*. Then we define

$$\mathcal{K} = \mathcal{K}_\Phi = \overline{\operatorname{span}}\{\langle\,\cdot\,,y^*\rangle x\ ;\ y^* \in \Phi, x \in X\} \subset L(X),$$

where the closure is taken in the operator norm topology. Moreover, since

$$\|\langle\,\cdot\,,y^*\rangle x\| \le \|y^*\|\|x\|, \quad y^* \in X^*, x \in X, \tag{10.18}$$

the countable set $\mathcal{F}$ of operators considered in the proof of Proposition 10.14 is dense in $\mathcal{K}$. Consequently, $\mathcal{K}$ is a separable Banach space. Finally, L_T obviously maps $\mathcal{F}$, and therefore also $\mathcal{K}$, into $\mathcal{K}$; thus L_T defines an operator on $\mathcal{K}$.

Proof of Theorem 10.20. (i)$\Longrightarrow$(ii). We first note that, by the Hypercyclicity Criterion, any n-fold direct sum $T \oplus \ldots \oplus T$ of T is topologically transitive; see Remark 3.13(b).

We want to show that L_T is topologically transitive as an operator on $\mathcal{K}$. Thus, let U and V be nonempty open subsets of $\mathcal{K}$. By definition there are $m \ge 1$, $y_1^*, \ldots, y_m^* \in \Phi$ as well as $x_1, \ldots, x_m, z_1, \ldots, z_m \in X$ such that

$$\sum_{j=1}^m \langle\,\cdot\,,y_j^*\rangle x_j \in U \quad \text{and} \quad \sum_{j=1}^m \langle\,\cdot\,,y_j^*\rangle z_j \in V;$$

note that we can have the same elements $y_1^*, \ldots, y_m^* \in \Phi$ here by taking some x_j and z_j as zero. By (10.18) there are open neighbourhoods $U_1 \times \cdots \times U_m$ of $(x_1, \ldots, x_m)$ and $V_1 \times \cdots \times V_m$ of $(z_1, \ldots, z_m)$ such that, if $(x_1', \ldots, x_m') \in U_1 \times \cdots \times U_m$ and $(z_1', \ldots, z_m') \in V_1 \times \cdots \times V_m$, then

$$\sum_{j=1}^m \langle\,\cdot\,,y_j^*\rangle x_j' \in U \quad \text{and} \quad \sum_{j=1}^m \langle\,\cdot\,,y_j^*\rangle z_j' \in V.$$

Applying the topological transitivity of the m-fold direct sum $T \oplus \ldots \oplus T$ we find some $(x_1', \ldots, x_m') \in U_1 \times \cdots \times U_m$ and some $n \ge 1$ such that $(T^n x_1', \ldots, T^n x_m') \in V_1 \times \cdots \times V_m$. This implies that

$$S := \sum_{j=1}^m \langle\,\cdot\,,y_j^*\rangle x_j' \in U \quad \text{and} \quad (L_T)^n S = T^n S = \sum_{j=1}^m \langle\,\cdot\,,y_j^*\rangle T^n x_j' \in V,$$

which proves our initial claim.

By the Birkhoff transitivity theorem, the operator L_T is therefore hypercyclic on $\mathcal{K}$, that is, there is an operator $S \in \mathcal{K}$ whose orbit $\{(L_T)^n S\ ;\ n \ge 0\}$ is dense in $\mathcal{K}$ with respect to the operator norm topology. But then the orbit is also SOT-dense in $\mathcal{K}$; and since $\mathcal{K}$ contains $\mathcal{F}$, which is SOT-dense in $L(X)$ by the proof of Proposition 10.14, the orbit of S under L_T is SOT-dense in $L(X)$. This shows that L_T is SOT-hypercyclic on $L(X)$.

(ii)$\Longrightarrow$(i). By the Bès–Peris theorem it suffices to show that $T \oplus T$ is hypercyclic. Let S be an operator that is SOT-hypercyclic for L_T. Since X cannot be one-dimensional, there are two linearly independent vectors $x_1, x_2 \in X$, and we denote by x_1^*, x_2^* the coordinate functionals corresponding to the basic sequence (x_1, x_2), extended to X. Let $(y_1, y_2) \in X \oplus X$ be arbitrary. Then the operator $R := \langle \cdot, x_1^* \rangle y_1 + \langle \cdot, x_2^* \rangle y_2$ maps x_1 to y_1, x_2 to y_2. By SOT-hypercyclicity of S there is an increasing sequence $(n_k)_k$ of positive integers such that $T^{n_k} S x_1 \to R x_1 = y_1$ and $T^{n_k} S x_2 \to R x_2 = y_2$. This shows that (Sx_1, Sx_2) is hypercyclic for $T \oplus T$. □

The theorem allows us to deduce Montes' theorem for Banach spaces in the case when (n_k) is the full sequence (n). However, when we do not want to restrict the sequence $(n_k)_k$ we need to generalize Theorem 10.20 to sequences $(T_n)_{n\geq 0}$ of operators on a Banach space X.

Definition 10.21. Let $(T_n)_{n\geq 0}$ be a sequence of operators on a Banach space X. Then the sequence $(L_{T_n})_n$ of left-multiplication operators is called *SOT-hypercyclic* if there is some operator $S \in L(X)$ whose orbit $\{T_n S \; ; \; n \geq 0\}$ under $(L_{T_n})_n$ is SOT-dense in $L(X)$; such an operator S is then called *SOT-hypercyclic* for $(L_{T_n})_n$.

Then exactly as in the proof of Theorem 10.20, using Theorems 3.24 and 3.25, we obtain the following generalization.

Theorem 10.22. *Let $(T_n)_n$ be a commuting sequence of operators on a separable Banach space X with $\dim X \geq 2$. Then the following assertions are equivalent:*
(i) *$(T_n)_n$ satisfies the Hypercyclicity Criterion;*
(ii) *$(L_{T_n})_n$ is SOT-hypercyclic.*

From this we can deduce Montes' theorem for Banach spaces.

Second proof of Theorem 10.2 *for Banach spaces.* We suppose that T satisfies the Hypercyclicity Criterion for an increasing sequence $(n_k)_k$ of positive integers. Then also the sequence $(T^{n_k})_k$ satisfies the Hypercyclicity Criterion; see Theorem 3.24. By the preceding theorem, the sequence $(L_{T^{n_k}})_k$ admits an SOT-hypercyclic vector $S \in L(X)$. Since any nonzero multiple of an SOT-hypercyclic vector is SOT-hypercyclic we may assume that $\|S\| = \frac{1}{2}$. Then, for any $x \in X$,

$$\|(I+S)x\| \geq \|x\| - \|Sx\| \geq \frac{1}{2}\|x\|.$$

Hence $I+S$ is an injective operator, and it defines an isomorphism of X onto its range $\operatorname{ran}(I+S)$.

Now let M_0 be an infinite-dimensional closed subspace of X so that $T^{n_k}x \to 0$ for all $x \in M_0$. By the above, $M := (I+S)M_0$ is an infinite-dimensional closed subspace of X. We claim that every nonzero vector $x \in M$

is hypercyclic for T. Indeed, there is some $y \in M_0$, $y \neq 0$, such that $x = (I + S)y$. Then

$$T^{n_k}x = T^{n_k}y + T^{n_k}Sy, \quad k \geq 1,$$

is a dense sequence in X because $T^{n_k}y \to 0$ and Sy is hypercyclic for $(T^{n_k})_k$; Proposition 10.18 also holds for sequences of operators. □

For the deduction of Montes' theorem in the Fréchet space setting we refer to Section 10.5.

10.3 Operators without hypercyclic subspaces

In this section we want to show that not all hypercyclic operators have hypercyclic subspaces. To this end we derive a useful necessary condition for the existence of a hypercyclic subspace in which the behaviour of the iterates of T on subspaces of finite codimension plays an essential role. We recall that a subspace M of a vector space X is said to be of *finite codimension* if there is a finite-dimensional subspace E of X such that $X = M + E$. We collect here some rather obvious properties, whose proof we leave to the reader; see Exercise 10.3.1.

Lemma 10.23. *Let X be a vector space over $\mathbb{K}$. Then we have the following:*
- (i) *a subspace M is of finite codimension if and only if there is some $n \geq 1$ and a linear map $u : X \to \mathbb{K}^n$ such that $M = \ker u$;*
- (ii) *if $M_1, \ldots, M_n$ are subspaces of finite codimension then so is $\bigcap_{k=1}^n M_k$;*
- (iii) *if M is a subspace of finite codimension and $T : X \to X$ is a linear map then $T^{-1}(M)$ is of finite codimension;*
- (iv) *if M is an infinite-dimensional subspace and L is a subspace of finite codimension then $M \cap L$ is infinite-dimensional.*

We turn to the announced condition that prevents an operator from having hypercyclic subspaces. We first state a version for Banach spaces; later on we will present a more technical result for general Fréchet spaces.

Theorem 10.24. *Let T be an operator on a Banach space X. Suppose that there are subspaces $M_n \subset X$, $n \geq 1$, of finite codimension and positive numbers C_n, $n \geq 1$, with $C_n \to \infty$ as $n \to \infty$ such that*

$$\|T^n x\| \geq C_n \|x\| \quad \text{for any } x \in M_n,\ n \geq 1.$$

Then T does not possess any hypercyclic subspace.

Since the proof is more transparent for Hilbert spaces we consider this case first; the general case will be treated later within the context of Fréchet spaces.

Proof of Theorem 10.24 *for Hilbert spaces.* Thus, let X be a Hilbert space. We assume that T satisfies the hypotheses of the theorem and that M is an infinite-dimensional closed subspace of X; we need to show then that M contains a vector $x \neq 0$ that is not hypercyclic for T. First, it follows from the assumption that there is an increasing sequence $(k_n)_n$ of positive integers such that, for $n \geq 2$,

$$C_j \geq n^3 \quad \text{for } k_{n-1} < j \leq k_n.$$

We claim that there are points $x_n \in X$ and closed subspaces $L_n \subset X$ of finite codimension such that, for $n \geq 1$ and $1 \leq j \leq k_n$,

$$x_n \in M \cap M_j, \quad \|x_n\| = \frac{1}{n^2}, \tag{10.19}$$

$$T^j x_1, \ldots, T^j x_{n-1} \perp L_n \quad (n \geq 2), \tag{10.20}$$

$$T^j x_n \in L_1 \cap \ldots \cap L_n. \tag{10.21}$$

We show existence by induction on $n \geq 1$. For $n = 1$ one can take any vector $x_1 \in M \cap M_1 \cap \ldots \cap M_{k_1}$ of norm 1 and $L_1 = X$. Now, if $x_1, \ldots, x_{n-1}$ and $L_1, \ldots, L_{n-1}$, $n \geq 2$, have been constructed then we let

$$E_n = \operatorname{span}\{T^j x_1, \ldots, T^j x_{n-1} \; ; \; 1 \leq j \leq k_n\} \quad \text{and} \quad L_n = E_n^{\perp}.$$

Clearly, L_n is a closed subspace of finite codimension such that (10.20) holds. Moreover, by Lemma 10.23 we can find a nonzero vector

$$x_n \in M \cap \bigcap_{1 \leq j \leq k_n} \big(M_j \cap T^{-j}(L_1 \cap \ldots \cap L_n)\big),$$

which we can assume to have norm $\frac{1}{n^2}$; hence also (10.19) and (10.21) are satisfied.

Having constructed the x_n and L_n, $n \geq 1$, we set

$$x = \sum_{n=1}^{\infty} x_n;$$

by (10.19), the series converges in X, and x belongs to M because M is a closed subspace. Now let $n \geq 2$ and $k_{n-1} < j \leq k_n$. Then

$$T^j x = \sum_{\nu=1}^{n-1} T^j x_\nu + T^j x_n + \sum_{\nu=n+1}^{\infty} T^j x_\nu.$$

It follows from (10.21) that, for $\nu \geq n+1$,

$$T^j x_\nu \in L_{n+1}.$$

Since L_{n+1} is closed, $\sum_{\nu=n+1}^{\infty} T^j x_\nu$ belongs to L_{n+1} and hence is orthogonal to $\sum_{\nu=1}^{n} T^j x_\nu$ by (10.20). By Pythagoras' theorem we have for any orthogonal vectors $y, z \in X$ that $\|y+z\| = (\|y\|^2 + \|z\|^2)^{1/2} \geq \|y\|$. In our present situation we therefore have that

$$\|T^j x\| \geq \Big\| \sum_{\nu=1}^{n-1} T^j x_\nu + T^j x_n \Big\|.$$

Moreover, by (10.20) and (10.21) we have that $\sum_{\nu=1}^{n-1} T^j x_\nu$ is orthogonal to $T^j x_n$. Again by Pythagoras' theorem this yields that

$$\|T^j x\| \geq \|T^j x_n\|,$$

and therefore, using the fact that $x_n \in M_j$ and $k_{n-1} < j \leq k_n$,

$$\|T^j x\| \geq C_j \|x_n\| \geq \frac{n^3}{n^2} = n.$$

Thus $\|T^j x\| \to \infty$ as $j \to \infty$, which shows that $x \in M$, $x \neq 0$, is not a hypercyclic vector for T, as desired. □

For Banach spaces this proof breaks down because we no longer have a notion of orthogonality. In Section 10.5 we will prove the following generalization of Theorem 10.24, which then also contains the Banach space case.

Theorem 10.25. *Let T be an operator on a Fréchet space X with defining increasing sequence $(p_n)_n$ of seminorms. Suppose that there are subspaces $M_n \subset X$, $n \geq 1$, of finite codimension, positive numbers C_n, $n \geq 1$, with $C_n \to \infty$ as $n \to \infty$, and some $N \geq 1$ such that*
(i) *$p_N(x) > 0$ for every hypercyclic vector x for T;*
(ii) *$p_N(T^n x) \geq C_n p_n(x)$ for any $x \in M_n$, $n \geq 1$.*
Then T does not possess any hypercyclic subspace.

We remark that the additional assumption that $p_N(x) > 0$ for any hypercyclic vector is automatically satisfied if the defining seminorms can be chosen to be norms, that is, if the space admits a continuous norm.

Our first application uses the criterion in the case of Banach spaces.

Example 10.26. **(Rolewicz's operators)** We consider Rolewicz's operators $T = \lambda B$, $|\lambda| > 1$, on $X = \ell^p$, $1 \leq p < \infty$, or c_0. Theorem 10.24 implies immediately that they do not have hypercyclic subspaces. Indeed, for the subspaces $M_n = \{(x_k)_k \; ; \; x_1 = x_2 = \ldots = x_n = 0\}$, $n \geq 1$, of finite codimension we have that $\|T^n (x_k)_k\| = \|\lambda^n (x_{k+n})_k\| = |\lambda|^n \|(x_k)_k\|$ for all $(x_k)_k \in M_n$. Since $|\lambda|^n \to \infty$ as $n \to \infty$ the claim follows.

More generally, let $T = B_w$ be a weighted backward shift on one of these spaces X, with $w = (w_n)_n$ a bounded weight sequence; see Section 4.1. In this case we have for $n \geq 1$ that

$$(B_w)^n(x_k)_k = (w_2 \cdots w_{n+1}x_{n+1}, w_3 \cdots w_{n+2}x_{n+2}, w_4 \cdots w_{n+3}x_{n+3}, \ldots);$$

hence, taking the same spaces M_n as above, we obtain that

$$\|(B_w)^n x\| \geq \Big(\inf_{k\geq 1} \prod_{\nu=1}^{n} |w_{\nu+k}|\Big)\|x\| \quad \text{for } x \in M_n.$$

Consequently, by Theorem 10.24, if

$$\lim_{n\to\infty} \inf_{k\geq 1} \prod_{\nu=1}^{n} |w_{\nu+k}| = \infty,$$

then B_w has no hypercyclic subspace.

We note that, strangely enough, this condition implies that B_w is mixing; see Example 4.9(a).

Theorem 10.25 can also be applied in a different direction. In Montes' theorem, which gives a sufficient condition for an operator to have a hypercyclic subspace, we have assumed that the space possesses a continuous norm. This is not just a technical requirement, as the following example shows.

Example 10.27. There is an even mixing operator on a Fréchet space (without a continuous norm) that satisfies condition (ii) of Theorem 10.2 for the full sequence (n) but that does not have a hypercyclic subspace. Indeed, we consider the sequence space over $\mathbb{Z}$,

$$X = \Big\{(x_n)_{n\in\mathbb{Z}}\ ;\ \sum_{n=0}^{\infty} |x_n| < \infty\Big\},$$

endowed with the seminorms

$$p_n(x) = \sum_{k=-n}^{\infty} |x_k|, \quad n = 1, 2, 3, \ldots.$$

Clearly, X is isomorphic to $\mathbb{K}^{\mathbb{N}} \oplus \ell^1$ and therefore a separable Fréchet space. On X we consider the multiple $T = 2B$ of the bilateral backward shift, that is,

$$T(x_n)_{n\in\mathbb{Z}} = 2(x_{n+1})_{n\in\mathbb{Z}}.$$

Then Kitai's criterion, applied to the finite sequences and the inverse of T as S, shows that T is mixing (or one might apply Theorem 4.13). Moreover,

$$M_0 = \{(x_n)_{n\in\mathbb{Z}}\ ;\ x_n = 0 \text{ for } n \geq 0\}$$

is an infinite-dimensional closed subspace of X on which T^n tends pointwise to 0 as $n \to \infty$. Thus T satisfies conditions (i) and (ii) of Theorem 10.2 for the full sequence (n). But we will now show that T has no hypercyclic subspace.

For this we will apply Theorem 10.25. First, it is clear that every hypercyclic sequence x satisfies $p_1(x) > 0$. For M_n we consider the subspaces of finite codimension

$$M_n = \{(x_k)_{k\in\mathbb{Z}} \ ; \ x_k = 0 \text{ for } -n \leq k \leq n\}, \quad n \geq 1.$$

We then have for $x \in M_n$, $n \geq 1$, that

$$p_1(T^n x) = 2^n p_n(x).$$

Therefore, T has no hypercyclic subspace.

10.4 Further operators with hypercyclic subspaces

We first show that operators with hypercyclic subspaces exist in a large class of Fréchet spaces.

Theorem 10.28. *Let X be an infinite-dimensional separable Fréchet space with a continuous norm. Then there exists a mixing operator on X that possesses a hypercyclic subspace.*

Proof. By Theorem 8.13 there is a mixing operator T on X with an infinite-dimensional closed subspace M_0 of fixed points. Thus, T satisfies the Hypercyclicity Criterion, and every $x \in M_0$ trivially has a converging orbit. It follows from Montes' theorem, taking into account Remark 10.4, that T has a hypercyclic subspace. □

The usual techniques of Chapter 8 then also show that, in the setting of the theorem, the mixing operators with a hypercyclic subspace are SOT-dense in the space of operators on X; see Exercise 10.4.1.

Montes' theorem provides us with a powerful tool for verifying that an operator has hypercyclic subspaces. But it leaves us with the task of finding an infinite-dimensional closed subspace on which (sub)orbits converge to 0. In many cases, coming up with such a subspace is by no means easy or obvious. In the remainder of this section we address this problem in two particular cases. We will see that once again the notion of a basic sequence is crucial.

Our first result is in the spirit of Theorem 10.24.

Theorem 10.29. *Let T be an operator on a separable Banach space X. Suppose that there exists an increasing sequence $(n_k)_k$ of positive integers such that*

(i) *T satisfies the Hypercyclicity Criterion for $(n_k)_k$,*
(ii) *there is a decreasing sequence $(M_k)_k$ of infinite-dimensional closed subspaces of X such that $\sup_{k\geq 1} \|T^{n_k}|_{M_k}\| < \infty$.*

Then T has a hypercyclic subspace.

Proof. For simplicity we will perform the proof in the case when $(n_k)_k$ is the full sequence; the general case follows in the same way.

By (i) there is a dense subset X_0 of X such that, for all $x \in X_0$, $T^n x \to 0$ as $n \to \infty$. By (ii) there is some $C > 0$ such that $\|T^n|_{M_n}\| \le C$ for $n \ge 1$. And by a slight strengthening of Mazur's theorem (see Exercise 10.4.4), there exists a basic sequence $(e_n)_n$ in X with $e_n \in M_n$ for $n \ge 1$; let $K_n = \max(1, \|e_n^*\|)$.

By density of X_0 and continuity of T we can then find $f_n \in X_0$, $n \ge 1$, such that

$$\|T^n e_j - T^n f_j\| < \frac{1}{2^j K_j}, \quad j \ge 1, n = 0, 1, \ldots, j. \tag{10.22}$$

Since the f_n belong to X_0 one can construct inductively an increasing sequence $(n_k)_k$ of positive integers such that, for all $k \ge 1$,

$$\|T^{n_k} f_j\| \le \frac{1}{2^{j+k} K_j}, \quad j = 1, \ldots, n_{k-1}, \tag{10.23}$$

where $n_0 = 1$. Since also $(e_{n_k})_k$ is a basic sequence and

$$\sum_{k=1}^{\infty} \|e_{n_k}^*\| \|e_{n_k} - f_{n_k}\| < \sum_{k=1}^{\infty} K_{n_k} \frac{1}{2^{n_k} K_{n_k}} \le 1,$$

we have by Lemma 10.6 that $(f_{n_k})_k$ is a basic sequence. Let M_0 be its closed linear span.

We now claim that the powers T^{n_k} of T tend pointwise to 0 on M_0. Indeed, let $x \in M_0$, $x = \sum_{k=1}^{\infty} a_k f_{n_k}$. By Lemma 10.6, also $z := \sum_{k=1}^{\infty} a_k e_{n_k}$ converges. Moreover, since the subspaces M_k are closed and $(M_k)_k$ is decreasing, we have that

$$\sum_{j=k}^{\infty} a_j e_{n_j} \in M_{n_k}, \quad k \ge 1. \tag{10.24}$$

We then deduce from (10.22), (10.23) and (10.24) that

$$\begin{aligned}
\|T^{n_k} x\| &\le \sum_{j=1}^{k-1} |a_j| \|T^{n_k} f_{n_j}\| + \Big\| T^{n_k}\Big(\sum_{j=k}^{\infty} a_j e_{n_j}\Big)\Big\| + \sum_{j=k}^{\infty} |a_j| \|T^{n_k}(f_{n_j} - e_{n_j})\| \\
&\le \sum_{j=1}^{k-1} \|e_{n_j}^*\| \|z\| \frac{1}{2^{n_j+k} K_{n_j}} + C\Big\| \sum_{j=k}^{\infty} a_j e_{n_j}\Big\| + \sum_{j=k}^{\infty} \|e_{n_j}^*\| \|z\| \frac{1}{2^{n_j} K_{n_j}} \\
&\le \frac{1}{2^k}\|z\| + C\Big\| \sum_{j=k}^{\infty} a_j e_{n_j}\Big\| + \frac{2}{2^{n_k}}\|z\| \to 0
\end{aligned}$$

as $k \to \infty$, which proves the claim.

An application of Montes' theorem finishes the proof. □

As an application one obtains a more direct verification of Example 10.10; see Exercise 10.4.2.

We use the theorem here to obtain a new, large class of operators with hypercyclic subspaces. For this we need the following.

Lemma 10.30. *Let K be a compact operator on a Banach space X. Then there is a closed subspace M of finite codimension such that $\|K|_M\| \leq \frac{1}{2}$.*

Proof. Since the image under K of the closed unit sphere is relatively compact in X there are finitely many points $x_k \in X$ with $\|x_k\| = 1$ for $k = 1, \ldots, N$, such that, for any $x \in X$ with $\|x\| = 1$ there is some k with $\|Kx_k - Kx\| \leq \frac{1}{4}$. By the Hahn–Banach theorem (see Appendix A), there are continuous linear functionals y_k^*, $k = 1, \ldots, N$, on X such that $y_k^*(Kx_k) = \|Kx_k\|$ and $\|y_k^*\| = 1$. By Lemma 10.23,

$$M = \bigcap_{k=1}^{N} \ker(y_k^* \circ K)$$

is a closed subspace of finite codimension.

Now let $x \in M$ with $\|x\| = 1$, and let k be such that $\|Kx_k - Kx\| \leq \frac{1}{4}$. Then

$$\|Kx_k\| = y_k^*(Kx_k) = y_k^*(Kx_k - Kx) \leq \|y_k^*\| \, \|Kx_k - Kx\| \leq \frac{1}{4}$$

and therefore

$$\|Kx\| \leq \|Kx - Kx_k\| + \|Kx_k\| \leq \frac{1}{2}.$$

This implies that, for all $x \in M$, $\|Kx\| \leq \frac{1}{2}\|x\|$, and the claim follows. □

We can now present the announced class of operators with hypercyclic subspaces.

Corollary 10.31. *Let T be an operator on a separable Banach space of the form*

$$T = U + K,$$

where $\|U\| \leq 1$ and K is compact. If T satisfies the Hypercyclicity Criterion then it has a hypercyclic subspace.

Proof. For $n \geq 1$ we have that

$$T^n = (U + K)^n = U^n + K_n$$

with compact operators K_n; see Exercise 5.2.3. It then follows from Lemma 10.30 that there are closed subspaces M_n of finite codimension such that $\|K_n|_{M_n}\| \leq \frac{1}{2}$, hence

$$\|T^n|_{M_n}\| = \|(U^n + K_n)|_{M_n}\| \leq \frac{3}{2}.$$

Since X must be infinite dimensional, an application of Theorem 10.29, using the decreasing sequence $(M_1 \cap \ldots \cap M_n)_n$ of infinite-dimensional closed subspaces, yields the result. □

Example 10.32. (a) Let X be one of the spaces ℓ^p, $1 \le p < \infty$, or c_0, $w = (w_n)_n$ a weight sequence with $\lim_{n\to\infty} w_n = 0$, and B_w the corresponding weighted shift. Then $T = I + B_w$ has a hypercyclic subspace. This follows from the corollary with Corollary 8.3 and Exercise 5.2.10.

(b) In the Banach space setting the mixing operator constructed in the proof of Theorem 8.9 is a compact perturbation of the identity; see also Exercise 8.2.2. By the corollary it therefore has a hypercyclic subspace. This provides a new proof of Theorem 10.28 for Banach spaces.

We finish this section with another example where basic sequences help us to verify the subspace condition in Montes' theorem. By Example 10.12 we know already that Birkhoff's translation operators, $f \to f(\cdot + a)$ on $H(\mathbb{C})$, have hypercyclic subspaces. The proof was based on the largeness of some eigenspace. We will give here another proof in a more general setting. As in Section 4.3 we consider general composition operators $C_\varphi : f \to f \circ \varphi$ on arbitrary domains Ω in $\mathbb{C}$ with automorphisms φ of Ω.

Proposition 10.33. *Let Ω be a domain in $\mathbb{C}$. Then every hypercyclic composition operator C_φ, $\varphi \in \mathrm{Aut}(\Omega)$, has a hypercyclic subspace.*

Proof. Let C_φ be hypercyclic on $H(\Omega)$. By Proposition 4.31, Ω is either simply connected or infinitely connected. By Remark 4.35 and Theorem 3.25, there is a subsequence $(m_n)_n$ such that C_φ satisfies the Hypercyclicity Criterion for the sequence $(m_n)_n$ and such that $(\varphi^{m_n})_n$ is a run-away sequence. By Montes' theorem it then suffices to construct an infinite-dimensional closed subspace M_0 of $H(\Omega)$ and a subsequence $(n_k)_k$ of $(m_n)_n$ such that $(C_\varphi)^{n_k} g \to 0$ as $k \to \infty$, for all $g \in M_0$.

Since Ω is conformally equivalent to any domain $r\Omega + z_0$, $r > 0, z_0 \in \mathbb{C}$, we can assume by Proposition 4.25 that Ω contains the closed unit disk $\overline{\mathbb{D}}$. Let $(K_n)_n$ be an exhaustion of Ω by compact sets with $K_1 = \overline{\mathbb{D}}$. Then we construct inductively Ω-convex compact subsets L_k and a subsequence $(n_k)_k$ of $(m_n)_n$ such that, for $k \ge 1$,

$$\varphi^{n_k}(L_k) \cap L_k = \varnothing,\ \varphi^{n_k}(L_k) \cup L_k \text{ is } \Omega\text{-convex}, \tag{10.25}$$

$$K_k \cup L_{k-1} \cup \varphi^{n_{k-1}}(L_{k-1}) \subset L_k, \tag{10.26}$$

where $L_0 = \varnothing$ and $n_0 = 0$; in the simply connected case this follows immediately from the run-away property, in the infinitely connected case we use Lemma 4.33.

Since the unit circle $\mathbb{T}$ is contained in Ω, we can consider functions $f \in H(\Omega)$ as functions in the Hilbert space

$$L^2(\mathbb{T}) = \left\{ f : \mathbb{T} \to \mathbb{C}\ ;\ \int_0^{2\pi} |f(e^{it})|^2\, dt < \infty \right\}$$

with norm $\|f\|_2 = (\frac{1}{2\pi}\int_0^{2\pi} |f(e^{it})|^2\, dt)^{1/2}$; the embedding is obviously continuous. In particular, the functions $e_n(z) = z^n$, $n \ge 1$, form an orthonormal system in $L^2(\mathbb{T})$ and are therefore a basic sequence there; the norms of the corresponding coefficient functionals are 1.

By (10.25) we can apply Runge's theorem to obtain, for any $j \geq 1$, functions $f_{j,k} \in H(\Omega)$ such that

$$\sup_{z \in L_1} |e_j(z) - f_{j,1}(z)| < \frac{1}{2^{j+2}}, \qquad \sup_{z \in \varphi^{n_1}(L_1)} |f_{j,1}(z)| < \frac{1}{2^{j+2}} \tag{10.27}$$

and, for $k \geq 2$,

$$\sup_{z \in L_k} |f_{j,k-1}(z) - f_{j,k}(z)| < \frac{1}{2^{j+k+1}}, \qquad \sup_{z \in \varphi^{n_k}(L_k)} |f_{j,k}(z)| < \frac{1}{2^{j+k+1}}. \tag{10.28}$$

Since, by (10.26), $L_k \subset L_{k+1}$ for all $k \geq 1$, (10.28) implies that

$$f_j := f_{j,1} + \sum_{k=2}^{\infty} (f_{j,k} - f_{j,k-1}) = f_{j,m} + \sum_{k=m+1}^{\infty} (f_{j,k} - f_{j,k-1})$$

converges uniformly on any L_m and hence on any K_m, $m \geq 1$, so that $f_j \in H(\Omega)$, $j \geq 1$.

By (10.26) we also have that $\varphi^{n_k}(L_k) \subset L_{k+1} \subset L_{k+2} \subset \ldots$ for all $k \geq 1$. By (10.27) and (10.28) we then have, for any $m \geq 1$,

$$\begin{aligned} \sup_{z \in \varphi^{n_m}(L_m)} |f_j(z)| &\leq \sup_{z \in \varphi^{n_m}(L_m)} |f_{j,m}(z)| + \sum_{k=m+1}^{\infty} \sup_{z \in L_k} |f_{j,k}(z) - f_{j,k-1}(z)| \\ &\leq \sum_{k=m}^{\infty} \frac{1}{2^{j+k+1}} = \frac{1}{2^{j+m}}. \end{aligned} \tag{10.29}$$

In the same way we have, using that $\mathbb{T} \subset \overline{\mathbb{D}} \subset L_1$,

$$\|e_j - f_j\|_2 \leq \sup_{z \in L_1} |e_j(z) - f_j(z)| \leq \sum_{k=1}^{\infty} \frac{1}{2^{j+k+1}} = \frac{1}{2^{j+1}}$$

and therefore $\sum_{j=1}^{\infty} \|e_j - f_j\|_2 \leq \frac{1}{2}$. It follows from Lemma 10.6 that $(f_n)_n$ is a basic sequence in $L^2(\mathbb{T})$ with $\|f_n^*\| \leq 2$.

We now claim that the closed linear span M_0 of the f_j in $H(\Omega)$ is the desired subspace; note that it is infinite dimensional because the f_j are linearly independent. Thus, let $g \in M_0$; then there are linear combinations

$$g_\nu = \sum_{j=1}^{N_\nu} a_{\nu,j} f_j$$

that converge to g in $H(\Omega)$ and then also in $L^2(\mathbb{T})$. Fix $m \geq 1$; then we have for $k \geq m$ that $L_m \subset L_k$ and therefore, using (10.29),

$$\begin{aligned}\sup_{z\in L_m}|(C_\varphi)^{n_k}g(z)| &\leq \sup_{z\in L_m}|(C_\varphi)^{n_k}(g-g_\nu)(z)| + \sum_{j=1}^{N_\nu}|a_{\nu,j}|\sup_{z\in L_k}|f_j(\varphi^{n_k}(z))|\\ &\leq \sup_{z\in L_m}|(g-g_\nu)(\varphi^{n_k}(z))| + \sum_{j=1}^{\infty}\|f_j^*\|\,\|g_\nu\|_2\frac{1}{2^{j+k}}\\ &\leq \sup_{w\in\varphi^{n_k}(L_m)}|(g-g_\nu)(w)| + \frac{2}{2^k}\|g_\nu\|_2.\end{aligned}$$

Letting $\nu\to\infty$ we obtain that, for any $m\geq 1$ and $k\geq m$,

$$\sup_{z\in L_m}|(C_\varphi)^{n_k}g(z)| \leq \frac{2}{2^k}\|g\|_2.$$

Since L_m contains K_m we obtain that $(C_\varphi)^{n_k}g\to 0$ in $H(\Omega)$ for any $g\in M_0$, which had to be shown. □

10.5 The Fréchet space setting

In this section we discuss the extensions of the main results of the previous sections to Fréchet spaces.

Montes' theorem. We start with the proof of Montes' theorem.

Proof of Theorem 10.2. As in the Banach space case we assume that $(n_k)_k$ is the full sequence of positive integers.

Step 1. In order to construct a basic sequence $(e_n)_n$ again, we first need to specify a Banach space in which it lives. Thus, let $|||\cdot|||$ denote a continuous norm on X. Then $(X,|||\cdot|||)$ is a normed space and as such has a completion $(\widehat{X},|||\cdot|||)$, which is a Banach space that contains X densely. Let $\widehat{M_0}$ be the closure of M_0 in $\widehat{X}$. By Mazur's theorem, $(\widehat{M_0},|||\cdot|||)$ contains a basic sequence $(\widehat{e}_n)_n$. Approximating each $\widehat{e}_n$ sufficiently closely from inside M_0, Lemma 10.6 tells us that we can find a sequence $(e_n)_n$ from M_0 that is a basic sequence in $\widehat{X}$.

Step 2. We will again perturb $(e_n)_n$ into a basic sequence $(f_n)_n$ of hypercyclic vectors. To do this, let $K_n=\max(1,|||e_n^*|||)$, $n\geq 1$, where $|||e_n^*|||$ is the norm of the coefficient functionals e_n^* on the space $(\widehat{X},|||\cdot|||)$. Furthermore, let $\|\cdot\|$ denote an F-norm defining the topology of X; see Section 2.1. As in the Banach space case there is then a dense sequence $(y_n)_n$ in X, vectors $x_{j,k}\in X$ and positive integers $n(j,k)$ such that $(n(j,k))_{k\geq 1}$ is increasing for each $j\geq 1$ and such that, for all $j,k,j',k'\geq 1$,

$$\max(\|x_{j,k}\|, |||x_{j,k}|||) \leq \frac{1}{2^{j+k+1}K_j}, \tag{10.30}$$

$$\|T^{n(j,k)}x_{j,k} - y_k\| \leq \frac{1}{2^k},$$

$$\|T^{n(j',k')}x_{j,k}\| \leq \frac{1}{2^{j+k+k'}K_j} \quad \text{if } (j',k') \neq (j,k),$$

$$\|T^{n(j,k)}e_j\| \leq \frac{1}{2^k}.$$

In (10.30) we have used the continuity of the inclusion of X into $\widehat{X}$. Defining

$$f_j = e_j + \sum_{k=1}^{\infty} x_{j,k}, \quad j \geq 1,$$

we obtain as before that these series converge in X and that, for $j, k, j' \geq 1$,

$$\max(\|e_j - f_j\|, |||e_j - f_j|||) \leq \frac{1}{2^{j+1}K_j}, \tag{10.31}$$

$$\|T^{n(j,k)}f_j - y_k\| \leq \frac{3}{2^k}, \tag{10.32}$$

$$\|T^{n(j',k)}(e_j - f_j)\| \leq \frac{1}{2^{j+k}K_j} \quad \text{if } j' \neq j. \tag{10.33}$$

By (10.31), $\sum_{j=1}^{\infty} |||e_j^*|||\, |||e_j - f_j||| \leq \frac{1}{2} < 1$. Lemma 10.6 implies that $(f_n)_n$ is a basic sequence in $\widehat{X}$ with

$$|||f_j^*||| \leq 2K_j, \quad j \geq 1. \tag{10.34}$$

Step 3. We now define the desired hypercyclic subspace. We denote by M the closed linear span (in X) of the f_n, $n \geq 1$. Note that $(f_n)_n$ is not necessarily a basis in M, but since the f_n are linearly independent, M is an infinite-dimensional closed subspace of X. It remains to show that every nonzero vector in M is hypercyclic for T.

We first derive an estimate for finite linear combinations

$$z = \sum_{j=1}^{N} a_j f_j$$

with $a_m = 1$ for some m. To this end we define

$$w := \sum_{j \leq N,\, j \neq m} a_j e_j.$$

Let $k \geq 1$. Then by (2.3), (10.32), (10.33) and (10.34) we find that

$$
\begin{aligned}
&\left\|T^{n(m,k)}z - y_k\right\| \\
&\qquad \leq \|T^{n(m,k)}f_m - y_k\| + \Big\| \sum_{j\leq N, j\neq m} a_j T^{n(m,k)}(f_j - e_j)\Big\| + \|T^{n(m,k)}w\| \\
&\qquad \leq \frac{3}{2^k} + \sum_{j\leq N,\, j\neq m} (|a_j|+1)\|T^{n(m,k)}(f_j - e_j)\| + \|T^{n(m,k)}w\| \\
&\qquad \leq \frac{4}{2^k} + \sum_{j=1}^{\infty} |||e_j^*|||\; |||w|||\, \frac{1}{2^{j+k}K_j} + \|T^{n(m,k)}w\| \\
&\qquad \leq \frac{4}{2^k} + \frac{1}{2^k}|||w||| + \|T^{n(m,k)}w\|. \qquad (10.35)
\end{aligned}
$$

Now let $z \in M$, $z \neq 0$. We want to show that z is hypercyclic for T. As z also belongs to the closed linear span of the f_n when taken in $\widehat{X}$, we have a representation

$$z = \sum_{j=1}^{\infty} a_j f_j$$

with convergence in $\widehat{X}$. As in the Banach space case we can assume some a_m to be 1. By assumption there are vectors

$$z_\nu := \sum_{j=1}^{N_\nu} a_{\nu,j} f_j$$

converging to z in X. Since we also have convergence in $\widehat{X}$ and the coefficient functionals are continuous, we have that $a_{\nu,j} \to a_j$ for all $j \geq 1$. In particular, $a_{\nu,m} \to a_m$, and we can assume without loss of generality that $a_{\nu,m} = 1$ for all ν. Let us also consider the vectors

$$w_\nu := \sum_{j\leq N_\nu,\, j\neq m} a_{\nu,j} e_j \in M_0.$$

Setting $a_{\nu,j} = 0$ for $j > N_\nu$, we find for $\nu, \mu \geq 1$,

$$\|(w_\nu - z_\nu) - (w_\mu - z_\mu)\| \leq \sum_{j\neq m} \|(a_{\nu,j} - a_{\mu,j})(e_j - f_j)\|.$$

Since $(a_{\nu,j})_\nu$ converges for all $j \geq 1$ and since, by (2.3), (10.31) and (10.34),

$$
\begin{aligned}
\|(a_{\nu,j} - a_{\mu,j})(e_j - f_j)\| &\leq (|a_{\nu,j} - a_{\mu,j}| + 1)\|e_j - f_j\| \\
&\leq |||f_j^*|||\; |||z_\nu - z_\mu|||\; \|e_j - f_j\| + \frac{1}{2^{j+1}} \\
&\leq \frac{|||z_\nu - z_\mu|||}{2^j} + \frac{1}{2^{j+1}} \leq \frac{C}{2^j}
\end{aligned}
$$

with some $C > 0$, the dominated convergence theorem implies that $(w_\nu - z_\nu)_\nu$ is a Cauchy sequence in X; hence $(w_\nu)_\nu$ converges to some vector w, which necessarily belongs to M_0.

By our previous argument, z_ν and w_ν satisfy (10.35) for any $\nu \geq 1$. Letting $\nu \to \infty$ we have by continuity that, for any $k \geq 1$,

$$\|T^{n(m,k)} z - y_k\| \leq \frac{4}{2^k} + \frac{1}{2^k} |||w||| + \|T^{n(m,k)} w\|.$$

Since $w \in M_0$ we obtain that $\|T^{n(m,k)} z - y_k\| \to 0$ as $k \to \infty$, which implies that z is hypercyclic. □

Remark 10.34. Similar modifications to those in the Banach space case yield that condition (ii′) of Remark 10.4 suffices in Montes' theorem.

Left-multiplication operators. Next we turn to the generalization of Theorems 10.20 and 10.22 to separable Fréchet spaces X with a continuous norm $|||\cdot|||$. A crucial role in the proof will be played by operators from X under the norm topology induced by $|||\cdot|||$ to X under its original topology. For brevity we will denote $(X, |||\cdot|||)$ by $X_{|||\cdot|||}$.

Theorem 10.35. *Let $(T_n)_n$ be a commuting sequence of operators on a separable Fréchet space X with a continuous norm $|||\cdot|||$ and* $\dim X \geq 2$. *Then the following assertions are equivalent:*
(i) *$(T_n)_n$ satisfies the Hypercyclicity Criterion;*
(ii) *$(L_{T_n})_n$ is SOT-hypercyclic.*

Note that the strong operator topology on $L(X)$ is defined in Section 8.3, and the definition of SOT-hypercyclicity is the same as in Definitions 10.16 and 10.21. Proposition 10.18 extends as well, with unchanged proof.

We endow the space $L(X_{|||\cdot|||}, X)$ of operators $X_{|||\cdot|||} \to X$ with its natural operator topology. More precisely, if $(p_n)_n$ is an increasing sequence of seminorms defining the topology of X then we set

$$\|S\|_n = \sup_{|||x||| \leq 1} p_n(Sx), \quad n \geq 1. \tag{10.36}$$

It is a standard exercise to show that this defines a Fréchet space topology on $L(X_{|||\cdot|||}, X)$ and to show that the left-multiplication operator L_T induced by an operator T on X is an operator on $L(X_{|||\cdot|||}, X)$.

It is now not difficult to generalize the various preliminary results of Section 10.2 to the present setting. First, when we apply Lemma 10.15 to the separable normed space $X_{|||\cdot|||}$ we obtain a countable set Φ of continuous linear functionals on $X_{|||\cdot|||}$ that is weak-*-dense in $(X_{|||\cdot|||})^*$. Note that every continuous linear functional on $X_{|||\cdot|||}$ is also a continuous linear functional on X in its original topology.

Lemma 10.36. *Let X be a separable Fréchet space with a continuous norm $|||\cdot|||$. Let Φ be a countable set of continuous linear functionals on $X_{|||\cdot|||}$ that is weak-*-dense in $(X_{|||\cdot|||})^*$, and let E be a dense subset of X. We define*

$$\mathcal{F} = \mathcal{F}_{\Phi,E} = \Big\{ \sum_{j=1}^{m} \langle \cdot\,, y_j^* \rangle e_j \ ; \ y_j^* \in \Phi, e_j \in E, 1 \leq j \leq m \Big\},$$

$$\mathcal{K} = \mathcal{K}_{\Phi} = \overline{\operatorname{span}}\{ \langle \cdot\,, y^* \rangle x \ ; \ y^* \in \Phi, x \in X \},$$

where the closure is taken in $L(X_{|||\cdot|||}, X)$. Then:

(i) *$\mathcal{F}$ is SOT-dense in $L(X)$;*
(ii) *$L(X)$ is SOT-separable;*
(iii) *every $S \in \mathcal{K}$ is a compact operator $S : X_{|||\cdot|||} \to X$;*
(iv) *$\mathcal{K}$ is a separable Fréchet space, and $\mathcal{F}$ is dense in $\mathcal{K}$;*
(v) *for any operator T on X, L_T is an operator on $\mathcal{K}$.*

Proof. (i) Suppose that $\|\cdot\|$ is an F-norm that defines the topology of X. Let $T \in L(X)$, let $x_1, \ldots, x_m$, $m \geq 1$, be linearly independent vectors of X, and $\varepsilon > 0$. The coordinate functionals $x_1^*, \ldots, x_m^*$ corresponding to the basic sequence $(x_j)_{j=1,\ldots,m}$ are continuous with respect to $|||\cdot|||$ on $\operatorname{span}\{x_1, \ldots, x_m\}$ (because all norms on a finite-dimensional space are equivalent) and hence have continuous linear extensions to $X_{|||\cdot|||}$. One then concludes exactly as in the proof of Proposition 10.14 that there is some $S \in \mathcal{F}$ such that $\|Tx_k - Sx_k\| < \varepsilon$ for $k = 1, \ldots, m$.

(ii) Since X is separable we can choose E to be countable. Then $\mathcal{F}$ is countable, and the assertion follows from (i).

(iii) The proof of this assertion can be given as in Exercise 5.2.4.

(iv) Since, for any $y^* \in \Phi$ and $x \in X$,

$$\|\langle \cdot\,, y^* \rangle x\|_n = p_n(x) \sup_{|||y||| \leq 1} |\langle y, y^* \rangle|, \quad n \geq 1, \tag{10.37}$$

and since E is dense in X, $\mathcal{F}$ is dense in $\mathcal{K}$. Moreover, if E is countable then so is $\mathcal{F}$, which makes $\mathcal{K}$ separable.

(v) Clearly, L_T is an operator on $L(X_{|||\cdot|||}, X)$ that maps $\mathcal{F}$ into $\mathcal{K}$; by (iv) it then also maps $\mathcal{K}$ into $\mathcal{K}$. □

Lemma 10.37. *Under the assumptions of Theorem 10.35 and Lemma 10.36, the following assertions are equivalent:*

(i) *$(T_n)_n$ satisfies the Hypercyclicity Criterion;*
(ii) *$(L_{T_n})_n$ is hypercyclic on $\mathcal{K}$;*
(iii) *$(L_{T_n})_n$ is SOT-hypercyclic.*

Moreover, if $S \in \mathcal{K}$ is hypercyclic for $(L_{T_n})_n$ on $\mathcal{K}$, then it is SOT-hypercyclic for $(L_{T_n})_n$ on $L(X)$.

Proof. The proof follows the same lines as that of Theorem 10.20, using (10.37), Theorem 3.24 and Theorem 3.25. □

This lemma proves Theorem 10.35.

Lemma 10.38. *Under the assumptions of Theorem 10.35 and Lemma 10.36, let $S \in \mathcal{K}$ be hypercyclic for $(L_{T_n})_n$ on $\mathcal{K}$. Then there is some $\lambda \neq 0$ such that $|||\lambda Sx||| \leq \frac{1}{2}|||x|||$ for all $x \in X$. If M_0 is an infinite-dimensional closed subspace of X such that $T_n x \to 0$ for all $x \in M_0$ then $(I + \lambda S)M_0$ is a hypercyclic subspace for $(T_n)_n$.*

Proof. Since $S : X_{|||\cdot|||} \to X$ is continuous, S is also a continuous operator on $X_{|||\cdot|||}$. Hence there is some $\lambda \neq 0$ such that $|||\lambda Sx||| \leq \frac{1}{2}|||x|||$ for all $x \in X$, which implies that

$$|||(I + \lambda S)x||| \geq |||x||| - |||\lambda Sx||| \geq \frac{1}{2}|||x|||, \tag{10.38}$$

so that $I + \lambda S$ is an injective operator.

Moreover, under the usual topology, $I + \lambda S$ maps closed sets onto closed sets. Indeed, let $A \subset X$ be closed, $y_n = (I + \lambda S)x_n$, $x_n \in A$, and $y_n \to y$ in X. Since $(y_n)_n$ also converges in $X_{|||\cdot|||}$, it follows from (10.38) that $(x_n)_n$ is bounded in $X_{|||\cdot|||}$. By compactness of S, a subsequence of $(Sx_n)_n$ converges in X. Since $x_n = y_n - \lambda Sx_n$, a subsequence of $(x_n)_n$ converges, to $x \in X$ say. Then $x \in A$ and $y = (I + \lambda S)x$, so that $(I + \lambda S)(A)$ is closed.

Now let M_0 be an infinite-dimensional closed subspace of X so that $T_n x \to 0$ for all $x \in M_0$. Then $M := (I + \lambda S)M_0$ is an infinite-dimensional closed subspace of X. Let $x = (I + \lambda S)y$ with $y \in M_0$, $y \neq 0$. Then

$$T_n x = T_n y + \lambda T_n S y, \quad n \geq 1,$$

is a dense sequence in X because Sy is hypercyclic for $(T_n)_n$ by the extension of Proposition 10.18. Hence M is a hypercyclic subspace for $(T_n)_n$. □

The previous two lemmas, applied to the sequence $(T^{n_k})_k$, imply Montes' theorem in full generality.

Operators without hypercyclic subspaces. We give here the proof of Theorem 10.25, which will also prove Theorem 10.24 for general Banach spaces. For this we first need a substitute for the notion of orthogonality in Hilbert spaces. We refer to Exercise 2.1.7 for a characterization of continuous seminorms.

Lemma 10.39. *Let X be a Fréchet space, E a finite-dimensional subspace of X, p a continuous seminorm on X and $\varepsilon > 0$. Then there exists a closed subspace L of finite codimension such that, for any $x \in L$ and $y \in E$,*

$$p(x + y) \geq \max\left(\frac{p(x)}{2 + \varepsilon}, \frac{p(y)}{1 + \varepsilon}\right). \tag{10.39}$$

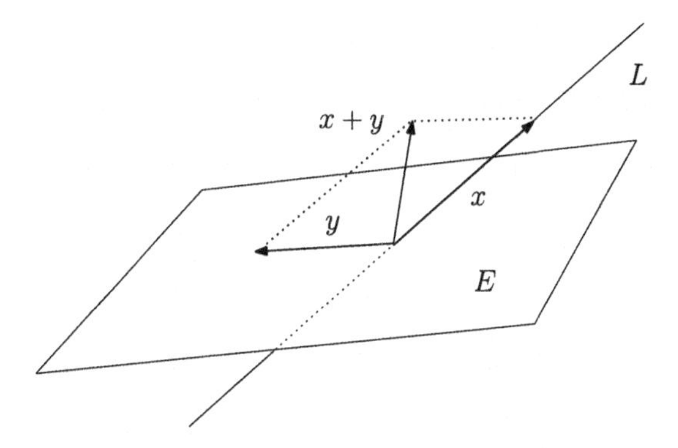

Fig. 10.1 A substitute for orthogonality

Proof. We first note that $\ker p$ is a subspace of X. There is then a subspace F of E such that $E = (\ker p \cap E) \oplus F$ algebraically. Clearly, p is a norm on F. On the finite-dimensional space F all norms define the same topology; see Appendix A. Hence the unit sphere with respect to p in F, being bounded and closed, is compact. Consequently there are finitely many points $y_k \in F$ with $p(y_k) = 1$, $k = 1, \ldots, N$, such that, for any $y \in F$ with $p(y) = 1$, there is some k with $p(y_k - y) \leq \frac{\varepsilon}{1+\varepsilon}$. More generally, if $y \in E$ with $p(y) = 1$ then $y = u + v$ with $p(u) = 0$ and $v \in F$, $p(v) = 1$. Hence there is some k with $p(y_k - y) = p(y_k - v) \leq \frac{\varepsilon}{1+\varepsilon}$.

By the Hahn–Banach theorem (see Appendix A), there are linear functionals y_k^*, $k = 1, \ldots, N$, on X such that $\langle y_k, y_k^* \rangle = p(y_k) = 1$ and $|\langle y, y_k^* \rangle| \leq p(y)$ for all $y \in X$. By continuity of p, these functionals are continuous. By Lemma 10.23,

$$L = \bigcap_{k=1}^{N} \ker y_k^*$$

is a closed subspace of finite codimension.

Now let $x \in L$ and $y \in E$; see Figure 10.1. If $p(y) = 0$ then $p(x + y) = p(x)$ and the claim is trivially true. Hence let $p(y) \neq 0$. Then $y' := y/p(y)$ satisfies $p(y') = 1$, so that there is some k with $p(y_k - y') \leq \frac{\varepsilon}{1+\varepsilon}$. Setting $x' = x/p(y) \in L$ we have that

$$p(x'+y') \geq p(y_k+x')-p(y_k-y') \geq |\langle y_k+x', y_k^* \rangle|-\frac{\varepsilon}{1+\varepsilon} = 1-\frac{\varepsilon}{1+\varepsilon} = \frac{1}{1+\varepsilon},$$

and therefore

$$p(x + y) \geq \frac{p(y)}{1+\varepsilon}.$$

In addition,

$$(2 + \varepsilon)p(x + y) = p(x + y) + (1 + \varepsilon)p(x + y) \geq p(x + y) + p(y) \geq p(x).$$

This proves the claim. □

The similarities between this proof and that of Lemma 10.30 are evident. In Exercise 10.5.4 we ask the reader to prove a common generalization of the two lemmas in the case of Banach spaces.

We can now prove the generalization of Theorem 10.24 to Fréchet spaces.

Proof of Theorem 10.25. Suppose that M is a hypercyclic subspace for T. Then, exactly as in the proof of Theorem 10.24, there is an increasing sequence $(k_n)_n$ of positive integers such that $C_j \geq n^3$ for $k_{n-1} < j \leq k_n$, $n \geq 2$, and with Lemma 10.39 we can construct points x_n in X, finite-dimensional subspaces E_n of X and corresponding closed subspaces L_n of finite codimension satisfying (10.39) for all $x \in L_n$, $y \in E_n$, with $p = p_N$, such that, for $n \geq 1$ and $1 \leq j \leq k_n$,

$$\begin{aligned} p_n(x_n) = \frac{1}{n^2} \quad (n \geq N), \quad x_n \in M \cap M_j, \\ T^j x_1, \ldots, T^j x_{n-1} \in E_n \quad (n \geq 2), \\ T^j x_n \in L_1 \cap \ldots \cap L_n, \end{aligned} \tag{10.40}$$

where we set $E_1 = \{0\}$ and $L_1 = X$. Note that $p_n(x_n) = \frac{1}{n^2}$, $n \geq N$, is possible because, by assumption, $p_n(x) \geq p_N(x) > 0$ for every hypercyclic vector x. Defining

$$x = \sum_{n=1}^{\infty} x_n,$$

it follows from (10.40) that the series converges in X, and we have that $x \in M$. Once more following the proof of Theorem 10.24, replacing orthogonality by condition (10.39), we find that, for $n \geq N$ and $k_{n-1} < j \leq k_n$,

$$\begin{aligned} p_N(T^j x) &\geq \frac{1}{1+\varepsilon} p_N\Big(\sum_{\nu=1}^{n} T^j x_\nu\Big) \geq \frac{p_N(T^j x_n)}{(1+\varepsilon)(2+\varepsilon)} \geq \frac{C_j p_j(x_n)}{(1+\varepsilon)(2+\varepsilon)} \\ &\geq \frac{n^3 p_n(x_n)}{(1+\varepsilon)(2+\varepsilon)} = \frac{n}{(1+\varepsilon)(2+\varepsilon)} \to \infty \end{aligned}$$

as $n \to \infty$; we have used here that $x_n \in M_j$ and $j \geq k_{n-1} + 1 \geq n$. This is a contradiction because every nonzero element of M is hypercyclic. □

Exercises

Exercise 10.1.1. Show that the following operators have hypercyclic subspaces:

(i) the operator T on $C_0(\mathbb{R}_+)$ given by $Tf(x) = \lambda f(x+a)$, $a > 0$, $\lambda > 1$ (see Exercise 2.2.1);

(ii) the operator T on $L^p_v(\mathbb{R}_+)$, $1 \leq p < \infty$, given by $Tf(x) = f(x+1)$, where v is an admissible weight function with $\liminf_{x\to\infty} v(x) = 0$; see Exercise 2.2.4.

Compare with Examples 10.10 and 10.26.

Exercise 10.1.2. Show that a Fréchet space has a continuous norm if and only if its topology can be defined by an increasing sequence $(p_n)_n$ of norms.

Exercise 10.1.3. Let X be a separable Fréchet space with a continuous norm, and let T be an operator on X that satisfies the Hypercyclicity Criterion. Show that if $\ker(\lambda I - T)$ is infinite-dimensional for some λ with $|\lambda| \leq 1$, then T has a hypercyclic subspace. (*Hint*: Use Remark 10.4 and the León–Müller theorem.)

Exercise 10.1.4. Let B_w be a weighted backward shift on $H(\mathbb{C})$; see Example 4.9(b). Show that if $\lim_{n\to\infty}(\prod_{\nu=1}^n |w_\nu|)^{1/n} = \infty$ and $\lim_{n\to\infty} |w_n|^{1/n} = 1$ then B_w has a hypercyclic subspace. See also Exercise 10.3.3.

Exercise 10.1.5. Let T be an operator on a separable Fréchet space. Show the following:

(i) T and T^n, $n \geq 1$, have the same hypercyclic subspaces;

(ii) if T is surjective and $M \neq X$ is a hypercyclic subspace then $M_n := T^{-n}(M)$, $n \geq 0$, are hypercyclic subspaces of T; moreover, $M_n \neq M_k$ for $n \neq k$;

(iii) if T is bijective and $M \neq X$ is a hypercyclic subspace then $M_n := T^{-n}(M)$, $n \in \mathbb{Z}$, are hypercyclic subspaces of T; moreover, $M_n \neq M_k$ for $n \neq k$.

(*Hint*: Ansari's theorem.)

Exercise 10.2.1. Let X be a Banach space with a basis $(e_n)_n$.

(a) Suppose that the basis is unconditional. Show that $L(X)$ is not separable in the operator norm topology. (*Hint*: Consider the operators $\sum_{k\in\mathbb{N}} a_k e_k \to \sum_{k\in A} a_k e_k$, $A \subset \mathbb{N}$.)

(b) Give a short proof that $L(X)$ is SOT-separable. (*Hint*: For $T \in L(X)$ consider $P_n T P_n$, where $P_n : \sum_{k=1}^\infty a_k e_k \to \sum_{k=1}^n a_k e_k$.)

Exercise 10.2.2. Let X be a Banach space with separable dual X^*. Show that X itself is separable. (*Hint*: If $\{x_n^* \ ; \ n \geq 1\}$ is dense in X^*, choose $x_n \in X$ with $\|x_n\| = 1$, $|x_n^*(x_n)| \geq \frac{1}{2}\|x_n^*\|$; show that $\operatorname{span}\{x_n \ ; \ n \geq 1\}$ is dense in X.)

Exercise 10.2.3. Let T be an operator on a separable Banach space X such that L_T is SOT-hypercyclic. Show the following:

(i) the set of SOT-hypercyclic operators for L_T is SOT-dense in $L(X)$;

(ii) the set of SOT-hypercyclic operators for L_T is not dense in $L(X)$ under the operator norm topology, unless all nonzero vectors are hypercyclic for T.

(*Hint*: Generalize Proposition 1.15; see Remark 10.19, Exercise 5.1.4.)

Exercise 10.2.4. Let X be a Banach separable space. Show that an operator T on X is chaotic if and only if L_T is SOT-chaotic, that is, L_T is SOT-hypercyclic and has an SOT-dense set of periodic points in $L(X)$.

Exercise 10.2.5. A Banach space X is said to have the *approximation property* if, for every Banach space Y, for every compact operator $T : Y \to X$ and for every $\varepsilon > 0$ there is a finite-rank operator $F : Y \to X$ such that $\|T - F\| < \varepsilon$; see Exercise 5.2.2. Let X be a Banach space with the approximation property and $K(X)$ the space of compact operators on X. Let $\Phi \subset X^*$ be norm-dense and $E \subset X$ be dense. Show that then the operators of the form $\sum_{k=1}^m \langle \cdot, y_k^*\rangle x_k$, $y_1^*, \ldots, y_m^* \in \Phi$, $x_1, \ldots, x_m \in E$, $m \geq 1$, constitute a dense subset of $K(X)$ under the operator norm topology. Deduce that if X has the approximation property and X^* is separable then $K(X)$ is separable. (*Hint*: Exercise 10.2.2.)

Exercise 10.2.6. Let X be a Banach space with the approximation property such that X^* is separable. By Exercises 5.2.4 and 10.2.5, the space $K(X)$ of compact operators is a separable Banach space under the operator norm topology. Show that $L_T : K(X) \to K(X)$, $S \to TS$, is a well-defined operator on $K(X)$. Show moreover that T satisfies the Hypercyclicity Criterion if and only if L_T is hypercyclic on $K(X)$.

Exercise 10.2.7. Let X be a Banach space and T an operator on X. Then the *right-multiplication operator* R_T on $L(X)$ is defined as $R_T S = ST$ for $S \in L(X)$.

(a) Let $x_j \in X$, $y_j \in X^*$, $j = 1, \ldots, m$. Show that $R_T(\sum_{j=1}^m \langle \cdot, y_j^* \rangle x_j) = \sum_{j=1}^m \langle \cdot, T^* y_j^* \rangle x_j$.

(b) Suppose that X and X^* are separable. Show that if $T^* : X^* \to X^*$ satisfies the Hypercyclicity Criterion then R_T is SOT-hypercyclic, that is, there is some $S \in L(X)$ such that $\{ST^n \; ; \; n \geq 0\}$ is SOT-dense in $L(X)$.

Exercise 10.2.8. By Theorem 8.11 there exists an infinite-dimensional separable Banach space X on which every operator is of the form $T = \lambda I + K$ with $\lambda \in \mathbb{K}$ and K a compact operator. In fact, the space constructed by Argyros and Haydon has the approximation property and a separable dual. For such a Banach space X, show the following:

(i) the space $L(X)$ of operators on X is separable under the operator norm topology;
(ii) the functional $u : L(X) \to \mathbb{K}$, $\lambda I + K \to \lambda$ is well defined and continuous;
(iii) the functional u is an eigenvector of $(L_T)^*$ for any operator T on X.

Deduce that, even for this Banach space X with separable operator space $L(X)$, no left-multiplication operator L_T is hypercyclic on $L(X)$ with respect to the operator norm. (*Hint*: Exercises 5.2.3, 5.2.4 and 10.2.5.)

Exercise 10.2.9. Verify the contents of Remark 10.4 for Banach spaces using the approach of this section.

Exercise 10.3.1. Prove Lemma 10.23.

Exercise 10.3.2. Let B_w be a hypercyclic weighted backward shift on one of the spaces $X = \ell^p$, $1 \leq p < \infty$, or c_0. Show the following:

(i) if $\liminf_{n\to\infty} |w_n| > 1$, then B_w has no hypercyclic subspace;
(ii) if $(|w_n|)_n$ is decreasing, then B_w has a hypercyclic subspace if and only if $\lim_{n\to\infty} |w_n| = 1$;
(iii) if $(|w_n|)_n$ is increasing, then B_w is even mixing, but it has no hypercyclic subspace.

Exercise 10.3.3. Let B_w be a weighted backward shift on $H(\mathbb{C})$; see Exercise 10.1.4. Show that if $\lim_{n\to\infty} |w_n|^{1/n} = \infty$ then B_w has no hypercyclic subspace. (*Hint*: The seminorms $p_n(\sum_{k=0}^\infty a_k z^k) = \sum_{k=0}^\infty |a_k| n^k$, $n \geq 1$, define the topology of $H(\mathbb{C})$.)

Exercise 10.3.4. Use Example 10.27 to show that the existence of hypercyclic subspaces is not preserved under quasiconjugacies.

Exercise 10.4.1. Using the methods of Section 8.3 show that in every Fréchet space X, the set of operators having hypercyclic subspaces is either empty or SOT-dense in $L(X)$. Moreover, if X is an infinite-dimensional separable Fréchet space with a continuous norm then the set of mixing operators having hypercyclic subspaces is SOT-dense in $L(X)$.

Exercise 10.4.2. Verify Example 10.10 using Theorem 10.29.

Exercise 10.4.3. Let B_w be a hypercyclic weighted backward shift on $X = \ell^p(\mathbb{Z})$, $1 \leq p < \infty$, or $X = c_0(\mathbb{Z})$. Suppose that there is some $N \in \mathbb{Z}$ such that $|w_n| \leq 1$ for all $n \leq N$. Show that B_w has a hypercyclic subspace.

Exercise 10.4.4. Prove the following strengthening of Mazur's theorem. Let X be a Banach space and M_n, $n \geq 1$, infinite-dimensional closed subspaces of X. Then there exists a basic sequence $(e_n)_n$ in X with $e_n \in M_n$ for all $n \geq 1$. (*Hint*: Use Lemma 10.39.)

Exercise 10.4.5. Show the following variant of Corollary 10.31. Let T be an operator on a separable Banach space of the form $T = S + K$ where $\ker(\lambda I - S)$ is infinite dimensional for some λ with $|\lambda| \leq 1$ and K is compact. If T satisfies the Hypercyclicity Criterion then it has a hypercyclic subspace.

Exercise 10.5.1. Let X be a separable Fréchet space with a continuous norm and T an operator on X that satisfies the Hypercyclicity Criterion. Let $\mathcal{K}$ be the space defined in Lemma 10.36.

(a) Show that the set of operators from $\mathcal{K}$ with dense range is a dense G_δ-set.

(b) Show that L_T has an SOT-hypercyclic vector S that is a dense range operator.

(c) Deduce that there is a dense subspace of X consisting, except for zero, of hypercyclic vectors for T. Compare with the Herrero–Bourdon theorem.

Exercise 10.5.2. Let Y be a separable normed space and X a separable Fréchet space. The space $L(Y, X)$ of operators from Y to X turns into a Fréchet space under seminorms as in (10.36). For an operator T on X define the operator L_T on $L(Y, X)$ by $L_T S = TS$. The definitions of the strong operator topology on $L(Y, X)$ and SOT-hypercyclicity of L_T are obvious. Show that the following assertions are equivalent when $\dim Y \geq 2$:

(i) T satisfies the Hypercyclicity Criterion;

(ii) L_T is SOT-hypercyclic.

In that case, L_T has an SOT-hypercyclic vector S that defines a compact operator $S : Y \to X$. Deduce Theorem 10.35 for iterates of an operator.

Exercise 10.5.3. Let T be the operator of Example 10.27. Show that it is even chaotic, but that the left-multiplication operator L_T is not SOT-hypercyclic. Hence, in Theorem 10.35, the assumption of existence of a continuous norm cannot be dropped, even for iterates of an operator. (*Hint*: Let $S \in L(X)$. If P denotes the canonical projection of X onto ℓ^1, then show that there exists some $N \geq 1$ such that $PSe_{-k} = 0$ for all $k \geq N$ and deduce that Se_{-N} is not hypercyclic for T.)

Exercise 10.5.4. Let X and Y be Banach spaces, $T : X \to Y$ an operator, E a subspace of X such that $T|_E : E \to Y$ is compact, and $\varepsilon > 0$. Then there is a closed subspace L of X of finite codimension such that for any $x \in L$ and $y \in E$

$$\|Tx + Ty\| - \|Ty\| \geq -\varepsilon\|y\|,$$

and for any $x \in L \cap E$

$$\|Tx\| \leq \varepsilon\|x\|.$$

Deduce Lemma 10.39 (for a Banach space X and p the norm on X) and Lemma 10.30.

Sources and comments

In view of the fact that every hypercyclic operator possesses a dense subspace all of whose nonzero vectors are hypercyclic, it seemed natural to reserve the term "hypercyclic subspace" for the more interesting case of infinite-dimensional closed subspaces, as was first suggested by Chan and Taylor [107]. An example of an operator for which every nonzero vector is due to Read [266]; see Theorem 2.64.

Section 10.1. For Banach spaces, Montes' theorem is due to Montes [240]; see also González, León and Montes [167]. Our proof is taken from Bonilla and Grosse-Erdmann [88]; for a similar proof see León and Müller [223]. For Fréchet spaces, the result is due independently to Bonet, Martínez and Peris [84] who use a tensor product technique, and to Petersson [258] who adapts Montes' technique. Bès and Conejero [69] have studied hypercyclic subspaces of operators on a Fréchet space without a continuous norm; they have shown that every nonconstant polynomial $p(B)$ of the backward shift has a hypercyclic subspace on $\omega = \mathbb{K}^{\mathbb{N}}$.

For more on basic sequences in Banach spaces we refer to Diestel [133].

In the Banach space setting, González, León and Montes [167] (see also León and Montes [221] and León and Müller [223]) have improved Montes' theorem to a characterization under the assumption that the operator satisfies the Hypercyclicity Criterion, or equivalently, that it is weakly mixing.

Theorem 10.40 (González–León–Montes). *Let T be a weakly mixing operator on a separable Banach space X. Then the following assertions are equivalent:*

(i) *T has a hypercyclic subspace;*

(ii) *there exists an increasing sequence $(n_k)_k$ of positive integers and an infinite-dimensional closed subspace M_0 of X such that $T^{n_k}x \to 0$ for all $x \in M_0$;*

(iii) *there exists an increasing sequence $(n_k)_k$ of positive integers and an infinite-dimensional closed subspace M_1 of X such that $\sup_{k\geq 1} \|T^{n_k}|_{M_1}\| < \infty$;*

(iv) *the essential spectrum $\sigma_e(T)$ of T meets the closed unit disk $\overline{\mathbb{D}}$.*

It is important to note that the sequence $(n_k)_k$ in (ii) need not be related to the sequence in the Hypercyclicity Criterion. One way of defining the essential spectrum of an operator T is that $\lambda \notin \sigma_e(T)$ if and only if $\lambda I - T$ is a Fredholm operator, that is, $\lambda I - T$ has finite-dimensional kernel and finite-codimensional closed range.

As an application, González, León and Montes [167], [221] identify the operators with hypercyclic subspaces among various classes of operators. For example, a hypercyclic weighted backward shift B_w on ℓ^2 has a hypercyclic subspace if and only if

$$\lim_{n\to\infty} \Big(\inf_{k\geq 0} \prod_{\nu=1}^{n} |w_{\nu+k}|\Big)^{1/n} \leq 1,$$

while any hypercyclic bilateral weighted backward shift on $\ell^2(\mathbb{Z})$ has a hypercyclic subspace. A hypercyclic adjoint multiplier M_φ^* on the Hardy space H^2, induced by an injective bounded holomorphic function φ on $\mathbb{D}$ (see Section 4.4), has a hypercyclic subspace if and only if the boundary of $\varphi(\mathbb{D})$ meets the closed unit disk. And for any automorphism φ of $\mathbb{D}$ the composition operator C_φ on H^2 (see Section 4.5) has a hypercyclic subspace whenever it is hypercyclic; see also Montes [240].

Remark 10.4 is due to Bernal [57]; in the case of Banach spaces it also follows from the sufficiency of condition (iii) in Theorem 10.40 and the Banach–Steinhaus theorem. Corollary 10.11 and Example 10.12 are due to Petersson [258]; hypercyclic subspaces for Birkhoff's operators were found earlier by Bernal and Montes [65]; see the remarks on Proposition 10.33 below. The fact that MacLane's operator has a hypercyclic subspace was only recently proved by Shkarin [290]; he also notes that its essential spectrum is empty, so that Theorem 10.40 breaks down for Fréchet spaces.

Frequently hypercyclic subspaces are introduced and studied in Bonilla and Grosse-Erdmann [88].

Section 10.2. The alternative approach to hypercyclic subspaces via left-multiplication operators is due to Chan [99], who considered Hilbert spaces. He also introduced the notion of SOT-hypercyclicity and observed Proposition 10.18. Chan and Taylor [107] and Montes and Romero [242] extended Chan's investigation to Banach spaces. In all three papers, the implication (i)$\Longrightarrow$(ii) in Theorem 10.20 is obtained by a construction. Martínez and Peris [231] show this implication by a tensor product technique; they also prove the converse implication as well as Theorem 10.22. The proof of Theorem 10.20 given here follows Martínez and Peris [231] but avoids the language of tensor products; see also Aron, Bès, León and Peris [14]. Remark 10.19 was made in Montes and Romero [242].

For more on the weak-$*$-topology we refer to Diestel [133].

Section 10.3. Montes [240] not only proved his sufficient condition for the existence of hypercyclic subspaces, he was also the first to come up with operators without such a subspace, namely Rolewicz's operators; see Example 10.26. Theorem 10.24 is due to León and Müller [223]; its extension to Fréchet spaces, Theorem 10.25, seems to be new. Example 10.27 provides a new proof of a result by Bonet, Martínez and Peris [84].

Section 10.4. Theorem 10.28 was obtained independently by Bernal [57] and Petersson [258]; our proof follows that of Bernal. For Banach spaces the result was obtained earlier by León and Montes [220]. The problem of whether Theorem 10.28 remains true for all infinite-dimensional separable Fréchet spaces remains open; in the case of $X = \omega$, a positive answer was given by Bès and Conejero [69], as already mentioned.

Theorem 10.29 is due to León and Müller [223], but it is also implicitly contained in León and Montes [220]. Corollary 10.31 is due to León and Montes [220] who deduced Theorem 10.28 for Banach spaces from it. In a related result, Petersson [258] has shown that any weakly mixing nuclear perturbation of the identity on a separable Fréchet space with a continuous norm has a hypercyclic subspace; he used it to deduce Theorem 10.28.

The result with which we close this section, Proposition 10.33, is due to Bernal and Montes [65]; historically, it was, in fact, the first result on hypercyclic subspaces after Read's theorem mentioned above.

Section 10.5. The proof of Montes' theorem via basic sequences in a containing Banach space, given here, is from Bonilla and Grosse-Erdmann [88].

Theorem 10.35 is due to Bonet, Martínez and Peris [84] who refine the tensor product technique of Martínez and Peris [231]. The proof given here uses similar ideas without relying on the theory of tensor products.

For Lemma 10.39 in the case of Banach spaces we refer to Müller [246].

Exercises. Exercises 10.1.3 and 10.1.5 are taken from Petersson [258], Exercise 10.2.3 from Chan [99], Exercise 10.2.4 from Martínez and Peris [231], Exercises 10.2.6 and 10.2.7 from Bonet, Martínez and Peris [84]. As for Exercise 10.2.8, the space constructed by Argyros and Haydon [12] has the approximation property because it has a basis, and its dual is separable because it is isomorphic to ℓ^1. For Exercise 10.4.1 we refer to Bès and Chan [100], [67], [68] and Bernal [57]. Exercises 10.4.5 and 10.5.1 are taken from León and Müller [223], and Exercises 10.5.2 and 10.5.3 from Bonet, Martínez and Peris [84].

Chapter 11
Common hypercyclic vectors

It is an immediate consequence of the Baire category theorem that whenever we have two, three, or even countably many hypercyclic operators on a given Fréchet space then there exists a vector that is simultaneously hypercyclic for each of these operators; such a vector is called a common hypercyclic vector. This is a purely structural property and has nothing to do with the nature of the operators under consideration.

However, the problem comes to life again when we allow uncountable families of operators. We show in this chapter that even some natural uncountable families do not admit common hypercyclic vectors. The main result will provide a sufficient condition for the existence of common hypercyclic vectors. This criterion will then be applied to the main families of hypercyclic operators. In the final section we will again take up the topic of the previous chapter by studying the existence of common hypercyclic subspaces.

11.1 The Common Hypercyclicity Criterion

We first fix our terminology.

Definition 11.1. Let $(T_\lambda)_{\lambda\in\Lambda}$ be a family of hypercyclic operators on a separable Fréchet space X. Then a vector $x \in X$ is called a *common hypercyclic vector* for this family if it is hypercyclic for each operator T_λ, $\lambda \in \Lambda$.

In other words, the common hypercyclic vectors are exactly the elements of

$$\bigcap_{\lambda\in\Lambda} HC(T_\lambda).$$

In view of the Baire category theorem, the following is an immediate consequence of Theorem 2.19.

K.-G. Grosse-Erdmann, A. Peris Manguillot, *Linear Chaos*, Universitext, 305
DOI 10.1007/978-1-4471-2170-1_11, © Springer-Verlag London Limited 2011

Proposition 11.2. *If Λ is countable, then the set of common hypercyclic vectors for $(T_\lambda)_{\lambda\in\Lambda}$ is a dense G_δ-set, and in particular, nonempty.*

The situation changes dramatically when we allow uncountable families of operators. The following example shows that in this case there may not exist any common hypercyclic vector; see also Exercise 11.1.1.

Example 11.3. Let B be the backward shift operator on ℓ^2. We consider the operators

$$T_{\lambda,\mu} := \lambda B \oplus \mu B, \quad \lambda, \mu > 1,$$

on $\ell^2 \oplus \ell^2$. We know that the Rolewicz operators λB, $\lambda > 1$, are mixing (see Example 2.38), so that each operator $T_{\lambda,\mu}$ is hypercyclic by Proposition 1.42. But they do not possess a common hypercyclic vector. Indeed, suppose that (x, y) is such a vector. Then, in particular, for every $\lambda, \mu > 1$, the pairs of real numbers $(\lambda^n\|B^n x\|, \mu^n\|B^n y\|)$, $n \geq 1$, form a dense set in $\mathbb{R}^2_+$. Let us fix $a, b > 0$ and $\delta > 1$. Then, for any $\lambda, \mu > 1$, there is some $n \geq 1$ such that

$$\delta^{-1}a < \lambda^n\|B^n x\| < \delta a \quad \text{and} \quad \delta^{-1}b < \mu^n\|B^n y\| < \delta b. \tag{11.1}$$

We set $c_n = \frac{1}{n}\log(a\,\|B^n x\|^{-1})$ and $d_n = \frac{1}{n}\log(b\,\|B^n y\|^{-1})$. Taking logarithms in (11.1) we find that, for any $\lambda, \mu > 1$ there is some $n \geq 1$ such that

$$-\tfrac{1}{n}\log\delta < \log\lambda - c_n < \tfrac{1}{n}\log\delta \quad \text{and} \quad -\tfrac{1}{n}\log\delta < \log\mu - d_n < \tfrac{1}{n}\log\delta.$$

Hence the squares of side length $\frac{2}{n}\log\delta$ centred at (c_n, d_n), $n \geq 1$, cover $]0,\infty[^2$. But this is impossible because the total area of these squares is finite due to the convergence of the series $\sum_{n=1}^\infty \frac{1}{n^2}$.

On the other hand, some uncountable families do have common hypercyclic vectors. Two results that we have already met provide us with a multitude of examples. By the León–Müller theorem, each family $(\lambda T)_{\lambda\in\mathbb{T}}$ has common hypercyclic vectors as soon as T is a hypercyclic operator on a complex Fréchet space. And by the Conejero–Müller–Peris theorem, if a C_0-semigroup $(T_t)_{t\geq 0}$ of operators on a Banach space is hypercyclic then the operators T_t, $t > 0$, have common hypercyclic vectors. However, it is important to realize that these results should be understood as structural properties of hypercyclicity; they do not represent interesting examples of the phenomenon of common hypercyclicity. Indeed, in both cases the operators in the family have exactly the same hypercyclic vectors, which will not be the case in general; see Exercise 11.1.4.

Suppose now that an uncountable family $(T_\lambda)_{\lambda\in\Lambda}$ of hypercyclic operators does not fall into one of these two categories. How would we prove the existence of a common hypercyclic vector? In view of uncountability, one might expect that we have to do without the Baire category theorem. This is true, but only partly so. Baire's theorem can still help us to simplify the search, and it serves to show that in many cases the set of common hypercyclic vectors is a dense G_δ-set.

To this end we suppose that the index set Λ is a metric space. Then a family $(T_\lambda)_{\lambda\in\Lambda}$ of operators on a Fréchet space X is called *continuous* if, for any $x \in X$, the map

$$\Lambda \to X, \quad \lambda \to T_\lambda x$$

is continuous. In the present setting, this implies joint continuity with respect to λ and x.

Proposition 11.4. *Let Λ be a metric space and $(T_\lambda)_{\lambda\in\Lambda}$ a continuous family of operators on a Fréchet space X. Then the map $(\lambda, x) \to T_\lambda x$ is continuous on $\Lambda \times X$.*

Proof. Let $x_n \to x$ in X and $\lambda_n \to \lambda$ in Λ. Then

$$T_\lambda x - T_{\lambda_n} x_n = (T_\lambda - T_{\lambda_n})x + T_{\lambda_n}(x - x_n).$$

Since $(T_{\lambda_n} z)_n$ converges for any $z \in X$, the Banach–Steinhaus theorem (Theorem A.10) implies that $(T_{\lambda_n})_n$ is equicontinuous. It then follows easily that $T_{\lambda_n} x_n \to T_\lambda x$ as $n \to \infty$. □

We need to add one more assumption, namely that Λ is *σ-compact*, that is, a countable union of compact sets.

Theorem 11.5. *Let X be a separable Fréchet space, Λ a σ-compact metric space and $(T_\lambda)_{\lambda\in\Lambda}$ a continuous family of operators on X.*
(a) *The set of common hypercyclic vectors is a G_δ-set.*
(b) *The following assertions are equivalent:*
 (i) *the set of common hypercyclic vectors is a dense G_δ-set;*
 (ii) *for any compact set $K \subset \Lambda$ and for any nonempty open subsets U and V of X there is some $x \in U$ such that, for any $\lambda \in K$, there is some $n \geq 0$ such that $T_\lambda^n x \in V$.*

Proof. (a) By separability, the topology of X has a countable base $(V_k)_k$. Let $(K_m)_m$ be a sequence of compact sets whose union is Λ. We then have that

$$\bigcap_{\lambda\in\Lambda} HC(T_\lambda) = \bigcap_{m\geq 1}\bigcap_{k\geq 1} E(K_m, V_k),$$

where we define

$$E(K, V) = \{x \in X \ ; \ \text{for all } \lambda \in K \text{ there is } n \geq 0 \text{ such that } T_\lambda^n x \in V\}$$

for any compact subset $K \subset \Lambda$ and any nonempty open subset $V \subset X$.

We show that each set $E(K, V)$ is open. Let $x_0 \in E(K, V)$. Then, for any $\lambda \in K$, there is some $n_\lambda \geq 0$ such that $T_\lambda^{n_\lambda} x_0 \in V$. By Proposition 11.4, the continuity of $(T_\mu)_{\mu\in\Lambda}$ implies the continuity of each family $(T_\mu^{n_\lambda})_{\mu\in\Lambda}$, so that there exist open neighbourhoods O_λ of λ and U_λ of x_0 such that, for any $\mu \in O_\lambda$ and $x \in U_\lambda$, $T_\mu^{n_\lambda} x \in V$. Since the sets O_λ, $\lambda \in K$, form an open cover of the compact set K there exists a finite subcover $O_{\lambda_1}, \ldots, O_{\lambda_J}$ of K. Then $U_0 := \bigcap_{j=1}^J U_{\lambda_j}$ is a neighbourhood of x_0, and if $x \in U_0$ and $\lambda \in K$ then there is some j such that $\lambda \in O_{\lambda_j}$, hence $T_\lambda^{n_{\lambda_j}} x \in V$, so that $x \in E(K, V)$. This shows that $E(K, V)$ is open, which implies (a).

(b) It now follows from the Baire category theorem that the set of common hypercyclic vectors is dense if and only if each set $E(K, V)$ is dense, that is, if it meets any nonempty open set U. But this is precisely condition (ii). □

Note that if Λ is a singleton, then the result reduces to the Birkhoff transitivity theorem; see Theorem 2.19.

Remark 11.6. (a) If, under the conditions of the theorem, one of the operators, T_{λ_0} say, commutes with all the others then the set of common hypercyclic vectors is either empty or a dense G_δ-set. Indeed, if x is a common hypercyclic vector then the vectors $T_{\lambda_0}^n x$, $n \geq 0$, form a dense set of common hypercyclic vectors; see Exercise 2.6.3. See also Exercise 11.1.5.

(b) Condition (ii) is deceivingly simple. As the proof shows, the following, formally stronger, condition also characterizes when the set of common hypercyclic vectors is a dense G_δ-set:

(ii′) *for any compact set $K \subset \Lambda$ and for any nonempty open subsets U and V of X there are $x \in U$ and $n_1, \ldots, n_J \geq 0$ such that, for any $\lambda \in K$ there is some j such that $T_\lambda^{n_j} x \in V$;*

see Exercise 11.1.6. In all the existence proofs in this chapter we will in fact implicitly verify condition (ii′) rather than condition (ii).

As for ordinary hypercyclicity, the Baire category theorem simplifies the task of finding common hypercyclic vectors considerably. But, as for frequent hypercyclicity, the conditions (ii) or (ii′) still demand the construction of a vector x having several approximation properties simultaneously.

The following criterion, the main result of this chapter, provides a sufficient condition that allows us to construct such a vector. It turns out that the criterion has some similarities with the Frequent Hypercyclicity Criterion; see Theorem 9.9 and Exercise 9.2.1.

Again the notion of unconditionally convergent series enters the picture, but convergence should be uniform over a family of series. As usual we will work with an F-norm $\|\cdot\|$ that induces the topology of the given Fréchet space.

Definition 11.7. Let X be a Fréchet space. Then the series $\sum_{n=1}^\infty x_{\lambda,n}$, $\lambda \in \Lambda$, in X are said to be *uniformly unconditionally convergent* if, for any $\varepsilon > 0$, there is some $N \in \mathbb{N}$ such that for any finite set $F \subset \{N, N+1, N+2, \ldots\}$ we have that

$$\Big\| \sum_{n \in F} x_{\lambda,n} \Big\| < \varepsilon \quad \text{for all } \lambda \in \Lambda.$$

Example 11.8. The following are sufficient conditions for uniform unconditional convergence of a family of series $\sum_{n=1}^{\infty} x_{\lambda,n}$, $\lambda \in \Lambda$:

(i) there is a convergent series $\sum_{n=1}^{\infty} c_n$ of positive numbers such that

$$\|x_{\lambda,n}\| \leq c_n \quad \text{for all } n \geq 1,\ \lambda \in \Lambda;$$

(ii) for any $k \geq 1$ and $\varepsilon > 0$ there is some $N \in \mathbb{N}$ such that

$$\sum_{n=N}^{\infty} p_k(x_{\lambda,n}) < \varepsilon \quad \text{for all } \lambda \in \Lambda,$$

where $(p_k)_k$ is an increasing sequence of seminorms defining the topology of X;

(iii) there are $e_n \in X$, $a_{\lambda,n} \in \mathbb{K}$ and $c_n > 0$ such that $\sum_{n=1}^{\infty} c_n e_n$ converges unconditionally and, for all $n \geq 1$ and $\lambda \in \Lambda$,

$$x_{\lambda,n} = a_{\lambda,n} e_n \quad \text{and} \quad |a_{\lambda,n}| \leq c_n.$$

Here, condition (ii) is sufficient because $\{x \in X\ ;\ \|x\| < \varepsilon\}$ is a neighbourhood of 0 and thus contains a set of the form $\{x \in X\ ;\ p_k(x) < \delta\}$, $k \geq 1$, $\delta > 0$. And condition (iii) is sufficient in view of condition (vi) in Theorem A.16.

We can now formulate the announced sufficient condition for common hypercyclicity. In order to increase readability we will use the notation

$$T_\lambda = T(\lambda) \quad \text{and} \quad T_\lambda^n = T^n(\lambda).$$

Let us also note that the criterion requires a one-dimensional parameter set, a restriction that we will discuss later.

Theorem 11.9 (Common Hypercyclicity Criterion). *Let $\Lambda \subset \mathbb{R}$ be an interval and $(T_\lambda)_{\lambda \in \Lambda} = (T(\lambda))_{\lambda \in \Lambda}$ a continuous family of operators on a separable Fréchet space X. Suppose that, for any compact subinterval $K \subset \Lambda$, there is a dense subset X_0 of X and maps $S_n(\lambda) : X_0 \to X$, $n \geq 0$, $\lambda \in K$, such that, for any $x \in X_0$,*

(i) $\sum_{n=0}^{m} T^m(\lambda) S_{m-n}(\mu_n) x$ *converges unconditionally, uniformly for $m \geq 0$ and $\lambda \geq \mu_0 \geq \ldots \geq \mu_m$ from K;*

(ii) $\sum_{n=0}^{\infty} T^m(\lambda) S_{m+n}(\mu_n) x$ *converges unconditionally, uniformly for $m \geq 0$ and $\lambda \leq \mu_0 \leq \mu_1 \leq \ldots$ from K;*

(iii) *for any $\varepsilon > 0$ there is some $\delta > 0$ such that, for any $n \geq 1$, $\lambda, \mu \in K$,*

$$\text{if} \quad 0 \leq \mu - \lambda < \tfrac{\delta}{n} \quad \text{then} \quad \|T^n(\lambda) S_n(\mu) x - x\| < \varepsilon;$$

(iv) *$T^n(\lambda)x \to 0$ as $n \to \infty$, uniformly for $\lambda \in K$.*

Then the set of common hypercyclic vectors of the family $(T_\lambda)_{\lambda\in\Lambda}$ is a dense G_δ-set, and in particular, nonempty.

In (i), we consider the finite sums as infinite series by adding 0 terms.

Proof. We will verify that the characterizing condition of Theorem 11.5 holds. Thus let $K \subset \Lambda$ be a compact set, which we can assume to be a subinterval $K = [a, b]$, and let U and V be nonempty open subsets of X. Then there are points $x_0, y_0 \in X_0$ and some $\varepsilon > 0$ such that, whenever $\|x - x_0\| < \varepsilon$ and $\|y - y_0\| < \varepsilon$ then $x \in U$ and $y \in V$.

We can deduce from conditions (i), (ii) and (iv) that there is some $N \geq 0$ such that, for any finite set $F \subset \{N, N+1, N+2, \ldots\}$, we have that

$$\Big\| \sum_{n\in F, n\leq m} T^m(\lambda)S_{m-n}(\mu_n)y_0 \Big\| < \frac{\varepsilon}{4} \quad \text{for } m \geq 0,\ \lambda \geq \mu_0 \geq \ldots \geq \mu_m, \tag{11.2}$$

$$\Big\| \sum_{n\in F} T^m(\lambda)S_{m+n}(\mu_n)y_0 \Big\| < \frac{\varepsilon}{4} \quad \text{for } m \geq 0,\ \lambda \leq \mu_0 \leq \mu_1 \leq \ldots, \tag{11.3}$$

$$\|T^n(\lambda)x_0\| < \frac{\varepsilon}{4} \quad \text{for } n \geq N,\ \lambda \in K. \tag{11.4}$$

Moreover, by condition (iii) there is some $\delta > 0$ such that

$$\|T^n(\lambda)S_n(\mu)y_0 - y_0\| < \frac{\varepsilon}{4} \quad \text{for } n \geq 1,\ 0 \leq \mu - \lambda < \tfrac{\delta}{n}. \tag{11.5}$$

By the divergence of the harmonic series there is some $J \geq 1$ such that

$$a + \sum_{\nu=1}^{J-1} \frac{\delta}{2\nu N} \leq b < a + \sum_{\nu=1}^{J} \frac{\delta}{2\nu N}.$$

We then set

$$\mu_0 = a, \quad \mu_j = a + \sum_{\nu=1}^{j} \frac{\delta}{2\nu N} \ (0 < j < J), \quad \mu_J = b,$$

so that $a = \mu_0 \leq \mu_1 \leq \ldots \leq \mu_J = b$.

After these preparations we define

$$x = x_0 + S_N(\mu_1)y_0 + S_{2N}(\mu_2)y_0 + \ldots + S_{JN}(\mu_J)y_0$$

and claim that it is a vector as required in condition (ii) of Theorem 11.5(b). Indeed, setting $m = 0$ and $\lambda = a$ in (11.3) we find that

$$\|x - x_0\| = \|S_N(\mu_1)y_0 + S_{2N}(\mu_2)y_0 + \ldots + S_{JN}(\mu_J)y_0\| < \varepsilon,$$

so that $x \in U$. Moreover, let $\lambda \in K$. Then there is some j with $1 \leq j \leq J$ such that $\mu_{j-1} \leq \lambda \leq \mu_j$. Since $0 \leq \mu_j - \lambda \leq \frac{\delta}{2jN} < \frac{\delta}{jN}$, we conclude that

$$\begin{aligned}
T^{jN}(\lambda)x - y_0 &= T^{jN}(\lambda)x_0 + \sum_{\nu=1}^{j-1} T^{jN}(\lambda)S_{\nu N}(\mu_\nu)y_0 \\
&\quad + \big(T^{jN}(\lambda)S_{jN}(\mu_j)y_0 - y_0\big) + \sum_{\nu=j+1}^{J} T^{jN}(\lambda)S_{\nu N}(\mu_\nu)y_0 \\
&= T^{jN}(\lambda)x_0 + \sum_{\alpha=1}^{j-1} T^{jN}(\lambda)S_{jN-\alpha N}(\mu_{j-\alpha})y_0 \\
&\quad + \big(T^{jN}(\lambda)S_{jN}(\mu_j)y_0 - y_0\big) + \sum_{\alpha=1}^{J-j} T^{jN}(\lambda)S_{jN+\alpha N}(\mu_{j+\alpha})y_0 \\
&< \frac{\varepsilon}{4} + \frac{\varepsilon}{4} + \frac{\varepsilon}{4} + \frac{\varepsilon}{4} = \varepsilon,
\end{aligned}$$

where we have applied, in turn, (11.4), (11.2), (11.5) and (11.3). This implies that

$$T^{jN}(\lambda)x \in V,$$

which proves the claim. □

Remark 11.10. (a) In applications one often has that $S_0(\lambda)x = x$ for all $\lambda \in \Lambda$ and $x \in X_0$. In such a case condition (iv) can be dropped because it follows from condition (i) by considering the index $n = m$. Incidentally, condition (iii) implies that $T^n(\lambda)S_n(\lambda)x = x$ for all $n \geq 1$, $\lambda \in K$ and $x \in X_0$.

(b) The conditions in the Common Hypercyclicity Criterion are rather strong. They imply that any operator T_λ, $\lambda \in \Lambda$, is frequently hypercyclic and mixing; see Exercise 9.2.1 and Remark 3.13(a). Moreover, if we can take, for any $\lambda \in \Lambda$, $S_n(\lambda) = S^n(\lambda)$, $n \geq 0$, with some map $S(\lambda) : X_0 \to X_0$, then each operator T_λ, $\lambda \in \Lambda$, is also chaotic; see Proposition 9.11.

(c) Condition (i) in particular is quite restrictive; see the discussion before Example 11.18.

(d) It is obvious from the proof that condition (iii) can be relaxed. Clearly, the only property of the sequence $(\delta/n)_n$ that we needed was that the series $\sum_{n=1}^\infty \frac{\delta}{nN}$ diverges for any $N \geq 1$. Thus, condition (iii) can be weakened to the following:

(iii′) *for any $\varepsilon > 0$ there is a decreasing sequence $(\delta_n)_n$ of positive numbers such that $\sum_{n=1}^\infty \delta_n$ diverges and such that, for all $n \geq 1$, $\lambda, \mu \in K$,*

$$\textit{if} \quad 0 \leq \mu - \lambda < \delta_n \quad \textit{then} \quad \|T^n(\lambda)S_n(\mu)x - x\| < \varepsilon.$$

The fact that $(\delta_n)_n$ is decreasing ensures that also $\sum_{n=1}^\infty \delta_{nN}$ diverges for any $N \geq 1$. See Exercise 11.3.1 for an application. As a concrete example, one may replace δ/n by $\delta/(n \log n)$ in condition (iii).

In many situations one faces families with a complex or (at least) two-dimensional real parameter set of positive Lebesgue measure. When one tries to extend the Common Hypercyclicity Criterion to such a setting, condition

(iii) seems to create a serious problem. The property that we needed in the proof was that, essentially, the balls of radius δ/n about the μ_n cover the compact part K of the parameter set. To have something similar in a complex setting, say, would require choosing larger balls, for example of radius $\delta/\sqrt{n}$; see also the argument in Example 11.3. But in this weakened form condition (iii) can usually no longer be satisfied.

Let us now see the Common Hypercyclicity Criterion at work.

Example 11.11. **(Rolewicz's operators)** The multiples λB, $|\lambda| > 1$, of the backward shift on any of the spaces ℓ^p, $1 \le p < \infty$, or c_0, share a common hypercyclic vector, and the set of common hypercyclic vectors is a dense G_δ-set.

Thanks to the León–Müller theorem, for any fixed $\lambda > 1$, the operators $z\lambda B$, $|z| = 1$, have the same hypercyclic vectors. Hence it suffices to prove the claim for $\Lambda =]1, \infty[$.

In order to apply the Common Hypercyclicity Criterion, let $K = [a, b]$ with $1 < a < b$. As is usual for backward shifts, we set $S(\lambda) = \frac{1}{\lambda}F$, with F the forward shift (see Example 3.6), and we let $S_n(\lambda) = S^n(\lambda)$, $n \ge 0$. For X_0 we choose the set of finite sequences.

It then suffices to verify conditions (i)–(iii) for the canonical unit sequences $x = e_k$, $k \ge 1$; see Remark 11.10(a). For any finite set $G \subset \{N, N+1, \ldots\}$ we have that

$$\sum_{n\in G, n\le m} T^m(\lambda) S_{m-n}(\mu_n) e_k = \sum_{n\in G, n\le m} \frac{\lambda^m}{\mu_n^{m-n}} B^n e_k,$$

which will vanish, irrespective of m, λ, and the μ_n, whenever $N \ge k$. This shows condition (i). Next, for any $m \ge 0$ and $\lambda \le \mu_0 \le \mu_1 \le \ldots$ in $[a, b]$ we have that $0 \le \lambda^m/\mu_n^{m+n} \le 1/a^n$ and hence

$$\Big\| \sum_{n\in G} T^m(\lambda) S_{m+n}(\mu_n) e_k \Big\| = \Big\| \sum_{n\in G} \frac{\lambda^m}{\mu_n^{m+n}} F^n e_k \Big\| \le \Big\| \sum_{n=N}^{\infty} \frac{1}{a^n} e_{k+n} \Big\| \to 0$$

as $N \to \infty$. This shows condition (ii). And finally,

$$\|T^n(\lambda) S_n(\mu) e_k - e_k\| = \Big| \Big(\frac{\lambda}{\mu}\Big)^n - 1 \Big|.$$

Therefore, if $0 \le \mu - \lambda < \frac{\delta}{n}$ and $\lambda, \mu \in K$, then

$$0 \le 1 - \Big(\frac{\lambda}{\mu}\Big)^n = n \int_{\lambda/\mu}^{1} t^{n-1}\, dt \le n\Big(1 - \frac{\lambda}{\mu}\Big) < \frac{\delta}{\mu} \le \frac{\delta}{a}, \tag{11.6}$$

so that also condition (iii) holds with $\delta = a\varepsilon$.

It is obvious from the argument that this result allows a far-reaching generalization. We will return to this in the next section.

11.2 Common hypercyclic vectors for multiples of an operator

One of the simplest parametrized families of operators is given by multiples

$$\lambda T, \quad \lambda \in \mathbb{K},$$

of a given operator, which always constitutes a continuous family. We start with a simple observation that is an immediate consequence of Remark 11.6(a) and Exercise 11.1.5.

Proposition 11.12. *Let T be an operator on a separable Fréchet space, $\Lambda \subset \mathbb{K}$ a σ-compact set. If the set*

$$\bigcap_{\lambda \in \Lambda} HC(\lambda T)$$

of common hypercyclic vectors is nonempty then it is a dense G_δ-set and it contains a dense subspace, except for 0.

Moreover, we know from the León–Müller theorem that any operator λT shares its hypercyclic vectors with each of the operators $z\lambda T$, $|z| = 1$. Thus we only have to search for common hypercyclic vectors when λ varies inside the positive real numbers.

We will consider here the special case of operators T whose generalized kernel

$$\bigcup_{n=0}^{\infty} \ker T^n$$

forms a dense set in X; we have already encountered this notion in Corollary 2.49. For example, if $T = B_w$ is a unilateral weighted backward shift on a sequence space X, then its generalized kernel is simply the set of finite sequences, which often forms a dense set.

In this context we have the following general result.

Theorem 11.13. *Let T be an operator on a separable Fréchet space X with dense generalized kernel X_0 and $\lambda_0 \geq 0$. Suppose that there is a map $S : X_0 \to X$ such that $TSx = x$ and*

$$\frac{1}{\lambda^n} S^n x \to 0$$

for all $x \in X_0$, $\lambda > \lambda_0$. Then the set of common hypercyclic vectors for the operators λT, $|\lambda| > \lambda_0$, is a dense G_δ-set, and in particular, nonempty.

Proof. We first note that, for any $x \in X_0$, $T^n Sx = T^{n-1}x = 0$ if n is sufficiently large, so that S maps X_0 into itself. Therefore the maps S^n are defined on X_0. Now, by the theorem of León and Müller it suffices to show that the operators λT, $\lambda > \lambda_0$, have a common hypercyclic vector. To this end we apply the Common Hypercyclicity Criterion with $\Lambda =]\lambda_0, \infty[$, where we define

$$S_n(\lambda)x = \frac{1}{\lambda^n} S^n x, \quad x \in X_0, \lambda \in \Lambda.$$

Now let $b > a > \lambda_0$. Then we have for $x \in X_0$ and $\lambda, \mu_0, \ldots, \mu_m \in [a, b]$ that

$$\sum_{n=0}^{m} T^m(\lambda) S_{m-n}(\mu_n) x = \sum_{n=0}^{m} \frac{\lambda^m}{\mu_n^{m-n}} T^n x,$$

which implies condition (i) of the Common Hypercyclicity Criterion because $T^n x = 0$ for all sufficiently large n.

Moreover, for any $x \in X_0$ and $\lambda \leq \mu_0 \leq \mu_1 \leq \ldots$ from $[a, b]$ we have that

$$\sum_{n=0}^{\infty} T^m(\lambda) S_{m+n}(\mu_n) x = \sum_{n=0}^{\infty} \frac{\lambda^m}{\mu_n^{m+n}} S^n x.$$

Let $(p_k)_k$ be an increasing sequence of seminorms defining the topology of X, and let $\lambda_0 < c < a$. Since

$$0 \leq \frac{\lambda^m}{\mu_n^{m+n}} \leq \frac{1}{\mu_n^n} \leq \frac{1}{a^n},$$

we then have for any $k \geq 1$,

$$\sum_{n=0}^{\infty} p_k(T^m(\lambda) S_{m+n}(\mu_n) x) \leq \sum_{n=0}^{\infty} \Big(\frac{c}{a}\Big)^n p_k\Big(\frac{1}{c^n} S^n x\Big) < \infty,$$

which implies condition (ii); see Example 11.8.

Next, for any $\varepsilon > 0$ and $x \in X$ there is some $\widetilde{\varepsilon} > 0$ such that $\|cx\| < \varepsilon$ whenever $|c| < \widetilde{\varepsilon}$. Now, if $0 \leq \mu - \lambda < a\widetilde{\varepsilon}/n$, then $|(\frac{\lambda}{\mu})^n - 1| < \widetilde{\varepsilon}$ (see (11.6)), and hence

$$\|T^n(\lambda) S_n(\mu) x - x\| = \Big\|\Big(\Big(\frac{\lambda}{\mu}\Big)^n - 1\Big) x\Big\| < \varepsilon,$$

which implies condition (iii).

Finally, condition (iv) is trivial. □

Remark 11.14. For some applications it is useful to allow arbitrary dense subsets X_0 of the generalized kernel $\bigcup_{n=0}^{\infty} \ker T^n$ of T; see Exercises 11.2.3–11.2.5. In that case we assume the existence of maps $S_n : X_0 \to X$, $n \geq 0$, such that, for all $x \in X_0$, $T^n S_n x = x$, $T^m S_{m+n} x = S_n x$ for $m, n \geq 0$, and $\frac{1}{\lambda^n} S_n x \to 0$, $\lambda > \lambda_0$. Then the result remains true with virtually the same proof.

In Banach spaces the theorem leads to a very general result under rather weak hypotheses.

Corollary 11.15. *Let T be a surjective operator on a separable Banach space X with dense generalized kernel. Then there exists some $\lambda_0 > 0$ such that the set of common hypercyclic vectors for the operators λT, $|\lambda| > \lambda_0$, is a dense G_δ-set, and in particular, nonempty.*

Proof. By the open mapping theorem, the image of the open unit ball under T contains the open ball of radius 2ε for some $\varepsilon > 0$. Hence for any $y \in X$ with $\|y\| = \varepsilon$ there is some $x \in X$ with $\|x\| < 1$, which we call Sy, such that $Tx = TSy = y$. For general $y \in X$, $y \neq 0$, we define

$$Sy = \frac{\|y\|}{\varepsilon} S\Big(\frac{\varepsilon}{\|y\|} y\Big),$$

with $S0 = 0$. Then S is well defined, and for any $y \in X$ we have that $TSy = y$ and $\|Sy\| \le \|y\|/\varepsilon$. Hence

$$\Big\| \frac{1}{\lambda^n} S^n y \Big\| \le \frac{1}{\lambda^n \varepsilon^n} \|y\| \to 0$$

for any $y \in X$ and $\lambda > 1/\varepsilon$. We then conclude with Theorem 11.13. □

In situations where Theorem 11.13 can be applied, the task that remains is that of determining a possibly small value of λ_0. We start with a generalization of Example 11.11; see Section 4.1 for the general framework of this result.

Example 11.16. Let X be a Fréchet sequence space in which $(e_n)_n$ is a basis and suppose that the (unweighted) backward shift B is an operator on X. If $\lambda_0 \ge 0$ is such that the sequence $(\lambda^{-n})_n$ belongs to X for any $\lambda > \lambda_0$ then the multiples λB, $|\lambda| > \lambda_0$, share a hypercyclic vector. In fact, the set of common hypercyclic vectors is a dense G_δ-set.

This follows from Theorem 11.13. Indeed, the generalized kernel of B is the set of finite sequences, which is dense, and the condition $(\lambda^{-n})_n \in X$ implies that $\sum_{n=1}^\infty \lambda^{-n} e_n$ converges and hence that $\lambda^{-n} F^n e_k = \lambda^k \lambda^{-(n+k)} e_{n+k} \to 0$ for all $k \ge 1$, where F is the forward shift.

Example 11.17. **(MacLane's operator)** Let D be the differentiation operator on the space $H(\mathbb{C})$ of entire functions.

(a) The family $(\lambda D)_{\lambda \neq 0}$ has a common hypercyclic vector. Indeed, the generalized kernel X_0 of D consists of the polynomials, which are dense in $H(\mathbb{C})$. The map $S : X_0 \to H(\mathbb{C})$ is defined, as usual, by $Sf(z) = \int_0^z f(\zeta)\, d\zeta$, $z \in \mathbb{C}$. Since, for every $\lambda \neq 0$ and $k \ge 0$, $\frac{1}{\lambda^n} S^n(z^k) = \frac{\lambda^k k!}{(k+n)!} (z/\lambda)^{k+n} \to 0$ as $n \to \infty$, uniformly on compact sets, Theorem 11.13 implies the claim.

(b) Let $T = \mu I + D$, where $\mu \in \mathbb{C}$. Then $(\lambda T)_{\lambda \neq 0}$ has a common hypercyclic vector. In this case, we consider the set X_0 of functions $f(z) = e^{-\mu z} p(z)$, where p is a polynomial; then X_0 is dense, and a subset of the generalized kernel of T (in fact, it is the generalized kernel). Moreover, $Sf(z) =$

$e^{-\mu z}\int_0^z e^{\mu\zeta} f(\zeta)\,d\zeta$ defines a map $S : X_0 \to X_0$ for which the maps S^n, $n \geq 1$, satisfy the assumptions of Remark 11.14; in particular, an argument as in (a) shows that, for any $f \in X_0$ and $\lambda \neq 0$, $\frac{1}{\lambda^n} S^n f \to 0$ as $n \to \infty$, uniformly on compact sets. Again, this implies the claim.

Having treated Rolewicz's and MacLane's operators it is natural to study common hypercyclic vectors for the multiples of Birkhoff's operators T_a on $H(\mathbb{C})$ given by $T_a f(z) = f(z+a)$, $a \neq 0$. However, since T_a is injective its generalized kernel is trivial so that we cannot apply the methods of this section. What is worse, even reverting directly to the Common Hypercyclicity Criterion does not work, at least when using the obvious choice for the maps $S_n(\lambda)$, namely $S_n(\lambda)f(z) = \frac{1}{\lambda^n} f(z - na)$. Indeed, condition (i) is then not satisfied; see Exercise 11.2.7. Thus we need to rely on what seems to be the canonical tool when dealing with Birkhoff's operators, the Runge approximation theorem; see Example 2.20.

Example 11.18. **(Birkhoff's operators, I)** Let $a \in \mathbb{C}$, $a \neq 0$. We claim that the multiples of the corresponding translation operator,

$$\lambda T_a f(z) = \lambda f(z+a), \quad \lambda \in \mathbb{C} \setminus \{0\},$$

have a common hypercyclic vector on $H(\mathbb{C})$. As before, by the León–Müller theorem, it suffices to consider real positive parameters λ. By Proposition 4.25 we may take $a = 1$.

We will show that condition (ii) of Theorem 11.5(b) is satisfied. Thus, let $K = [b, c]$, $0 < b < c$, and let U and V be nonempty open subsets of $H(\mathbb{C})$. Then there are functions $f \in U$ and $g \in V$, $\varepsilon > 0$, and some $N \in \mathbb{N}$ such that an entire function h belongs to U (or to V) whenever $\sup_{|z|\leq N} |f(z) - h(z)| < \varepsilon$ (or $\sup_{|z|\leq N} |g(z) - h(z)| < \varepsilon$, respectively).

Let $M = \max_{|z|\leq N} |g(z)|$ and $\delta = \frac{\varepsilon b}{2M}$. There is then some $J \geq 1$ such that

$$b + \sum_{\nu=1}^{J-1} \frac{\delta}{3\nu N} \leq c < b + \sum_{\nu=1}^{J} \frac{\delta}{3\nu N}.$$

We set

$$\lambda_0 = b, \quad \lambda_j = b + \sum_{\nu=1}^{j} \frac{\delta}{3\nu N} \; (0 < j < J), \quad \lambda_J = c$$

and $n_j = 3jN$ for $j = 1, \ldots, J$.

Applying Runge's theorem to the union of the closed balls of radius N around $0, n_1, \ldots, n_J$ (see Figure 11.1), we obtain an entire function h such that

$$\sup_{|z|\leq N} |f(z) - h(z)| < \varepsilon$$

and, for $j = 1, \ldots, J$,

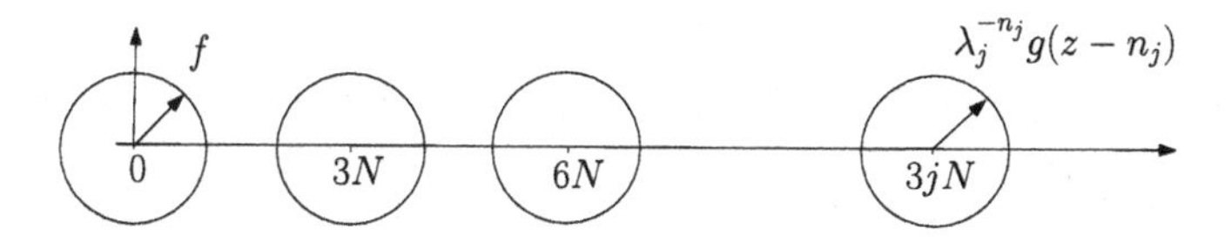

Fig. 11.1 Constructing common Birkhoff hypercyclic functions I

$$\sup_{|z-n_j|\leq N} |\lambda_j^{-n_j} g(z-n_j) - h(z)| < \frac{\varepsilon}{2}\lambda_j^{-n_j}.$$

By the first inequality we have that $h \in U$.

Moreover, let $\lambda \in K$. Then there is some j with $1 \leq j \leq J$ such that $\lambda_{j-1} \leq \lambda \leq \lambda_j$. Thus the second inequality implies that, for $|z| \leq N$,

$$\begin{aligned}|g(z) - \lambda^{n_j} h(z+n_j)| &\leq \left|1 - \left(\frac{\lambda}{\lambda_j}\right)^{n_j}\right| |g(z)| + \left(\frac{\lambda}{\lambda_j}\right)^{n_j} |g(z) - \lambda_j^{n_j} h(z+n_j)| \\ &< \left|1 - \left(\frac{\lambda}{\lambda_j}\right)^{n_j}\right| M + \frac{\varepsilon}{2}.\end{aligned}$$

Using the estimate (11.6) and the fact that $0 \leq \lambda_j - \lambda \leq \frac{\delta}{n_j}$ we obtain that

$$\sup_{|z|\leq N} |g(z) - (\lambda T_1)^{n_j} h(z)| < \frac{\delta}{b} M + \frac{\varepsilon}{2} = \varepsilon,$$

so that $(\lambda T_1)^{n_j} h \in V$, which proves the claim.

11.3 Further examples

As a first example of a family that is not made up of multiples of a given operator we turn to families of weighted shifts. Following the notation and terminology of Section 4.1, we will consider weighted shifts $B_{w(\lambda)}$, where the weight sequence $w(\lambda) = (w_n(\lambda))_n$ depends on a real parameter λ. Moreover, we recall that a function $f : [a,b] \to \mathbb{R}$ is called *Lipschitz continuous* if there is a constant $L \geq 0$ such that, for any $x, y \in [a,b]$,

$$|f(x) - f(y)| \leq L|x-y|.$$

The minimal constant L is called the *Lipschitz constant* of f.

Proposition 11.19. *Let X be a Fréchet sequence space in which $(e_n)_n$ is an unconditional basis. Let $\Lambda \subset \mathbb{R}$ be an interval and $w_n : \Lambda \to \mathbb{R}$, $n \geq 1$, be strictly positive functions such that, for each $\lambda \in \Lambda$, $B_{w(\lambda)}$ is an operator on X. Suppose that*

(i) *each function* w_n, $n \geq 1$, *is increasing;*
(ii) *for any compact subinterval* $K \subset \Lambda$, *the functions* $\log w_n$, $n \geq 1$, *are Lipschitz continuous on* K *with uniformly bounded Lipschitz constants;*
(iii) *for any* $\lambda \in \Lambda$ *the series*

$$\sum_{n=1}^{\infty} \Big(\prod_{\nu=1}^{n} w_\nu(\lambda) \Big)^{-1} e_n$$

converges in X.

Then the set of common hypercyclic vectors for the family $(B_{w(\lambda)})_{\lambda\in\Lambda}$ *is a dense* G_δ*-set, and in particular, nonempty.*

Proof. We apply the Common Hypercyclicity Criterion. Let $\|\cdot\|$ denote an F-norm defining the topology of X.

First, let $b \in \Lambda$ and $x \in X$. By assumption, the functions w_n, $n \geq 1$, are increasing and the series $\sum_{n=1}^{\infty} w_{n+1}(b)x_{n+1}e_n$ converges unconditionally. Hence, by Theorem A.16, for any $\varepsilon > 0$ there is some $N \geq 1$ such that $\|\sum_{n\geq N} w_{n+1}(\lambda)x_{n+1}e_n\| < \varepsilon$ for any $\lambda \in \Lambda$, $\lambda \leq b$. Now the continuity of the w_n implies that $(B_{w(\lambda)})_{\lambda\in\Lambda}$ is a continuous family.

In order to verify the main conditions of the criterion, let $K = [a, b] \subset \Lambda$. As usual for weighted shifts we take X_0 as the space of finite sequences, which is dense in X, and we define $S_n(\lambda) : X_0 \to X$ by

$$S_n(\lambda)e_k = \frac{1}{w_{k+1}(\lambda)\cdots w_{k+n}(\lambda)} e_{k+n}, \quad n \geq 0, k \geq 1,$$

and linear extension to X_0. In view of Remark 11.10(a) we need only verify conditions (i)–(iii) of the Common Hypercyclicity Criterion.

Condition (i) of the criterion holds because, for any $k \geq 1$ and $\lambda \in \Lambda$,

$$B^m_{w(\lambda)} S_{m-n}(\mu)e_k = 0 \quad \text{for } n \geq k.$$

As for condition (ii) we have for any $k \geq 1$ that

$$B^m_{w(\lambda)} S_{m+n}(\mu_n)e_k = \frac{w_{k+1+n}(\lambda)\cdots w_{k+m+n}(\lambda)}{w_{k+1}(\mu_n)\cdots w_{k+m+n}(\mu_n)} e_{k+n}. \tag{11.7}$$

Using the fact that each function w_n is increasing we obtain, whenever $\mu_n, \lambda \in K$ with $\mu_n \geq \lambda$,

$$0 \leq \frac{w_{k+1+n}(\lambda)\cdots w_{k+m+n}(\lambda)}{w_{k+1}(\mu_n)\cdots w_{k+m+n}(\mu_n)} \leq \frac{1}{w_{k+1}(a)\cdots w_{k+n}(a)}. \tag{11.8}$$

It follows from assumption (iii) and the unconditionality of the basis that

$$\sum_{n=1}^{\infty} \frac{1}{w_{k+1}(a)\cdots w_{k+n}(a)} e_{k+n}$$

converges unconditionally. By Example 11.8(iii), (11.7) and (11.8) then imply condition (ii) of the Common Hypercyclicity Criterion.

Next we find that, for any $k \geq 1$,

$$B_{w(\lambda)}^n S_n(\mu)e_k - e_k = \Big(\frac{w_{k+1}(\lambda)\cdots w_{k+n}(\lambda)}{w_{k+1}(\mu)\cdots w_{k+n}(\mu)} - 1\Big)e_k.$$

Choose $\delta > 0$ such that $\|(e^t - 1)e_k\| < \varepsilon$ whenever $|t| < \delta$. By assumption (ii) there is some $L > 0$ such that, for all $\nu \geq 1$ and $\lambda, \mu \in K$,

$$|\log w_\nu(\lambda) - \log w_\nu(\mu)| \leq L|\lambda - \mu|.$$

Thus, if $n \geq 1$ and $\lambda, \mu \in K$ with $0 \leq \mu - \lambda < \frac{\delta}{Ln}$ then

$$\begin{aligned}\Big|\log\Big(\frac{w_{k+1}(\lambda)\cdots w_{k+n}(\lambda)}{w_{k+1}(\mu)\cdots w_{k+n}(\mu)}\Big)\Big| &\leq \sum_{\nu=k+1}^{k+n} |\log w_\nu(\lambda) - \log w_\nu(\mu)| \\ &\leq nL(\mu - \lambda) < \delta,\end{aligned}$$

so that $\|B_{w(\lambda)}^n S_n(\mu)e_k - e_k\| < \varepsilon$. Hence also condition (iii) of the Common Hypercyclicity Criterion holds, which concludes the proof. □

We apply the proposition in some known and some new situations.

Example 11.20. (a) Let X be one of the spaces ℓ^p, $1 \leq p < \infty$, or c_0. Then the multiples λB, $\lambda > 1$, have a common hypercyclic vector. Indeed, the functions $w_n(\lambda) = \lambda$, $n \geq 1$, satisfy all the assumptions of the proposition on $\Lambda =]1, \infty[$. This confirms Example 11.11.

(b) On the same spaces as in (a) we consider the weighted shifts $B_{w(\lambda)}$ with $w_n(\lambda) = 1 + \frac{\lambda}{n}$, $\lambda > 0$, $n \geq 1$. Since $1 + \frac{\lambda}{n} \geq \frac{n+1}{n}$ for $\lambda \geq 1$ and $1 + \frac{\lambda}{n} \geq (\frac{n+1}{n})^\lambda$ for $0 < \lambda \leq 1$ we have that $w_1(\lambda)\cdots w_n(\lambda) \geq (n+1)^{\min(1,\lambda)}$ for $\lambda > 0$. Thus, assumption (iii) is satisfied for $\lambda > 1/p$ for $X = \ell^p$, and for all $\lambda > 0$ for $X = c_0$. Since the other assumptions hold as well, we have common hypercyclic vectors for the respective families of operators.

(c) Consider the multiples λD, $\lambda > 0$, of the differentiation operator D on the space $H(\mathbb{C})$ of entire functions. By Example 4.9(b), D can be regarded as a weighted shift with weights $w_n = n$, $n \geq 1$, on a suitable sequence space. The proposition then implies easily that the operators λD, $\lambda > 0$, have a common hypercyclic vector, confirming Example 11.17(a).

We return to Birkhoff's operators. Instead of the multiples of a single operator we consider the family of all hypercyclic Birkhoff operators.

Example 11.21. **(Birkhoff's operators, II)** We claim that the translation operators

$$T_a f(z) = f(z + a), \quad a \in \mathbb{C} \setminus \{0\},$$

have a common hypercyclic vector on $H(\mathbb{C})$. Since this is a continuous family of commuting operators, the set of common hypercyclic vectors is then a dense G_δ-set.

First, for every $\theta \in [0, 2\pi]$, the family $(T_{ae^{i\theta}})_{a \geq 0}$ is a C_0-semigroup. One may easily verify that the proof of the Conejero–Müller–Peris theorem remains valid on Fréchet spaces. Hence the operators $T_{ae^{i\theta}}$, $a > 0$, have the same hypercyclic vectors. Consequently it suffices to show that the operators $T_{e^{i\theta}}$, $\theta \in [0, 2\pi]$, have a common hypercyclic vector.

The proof is very similar to the one given in Example 11.18. We will show again that condition (ii) in Theorem 11.5(b) is satisfied, this time for $K = [0, 2\pi]$. Thus, let $U, V \subset H(\mathbb{C})$ be nonempty and open, let $f \in U$, $g \in V$, and let $N \in \mathbb{N}$ and $\varepsilon > 0$ be such that $h \in H(\mathbb{C})$ belongs to U (or to V) whenever $\sup_{|z| \leq N} |f(z) - h(z)| < \varepsilon$ (or $\sup_{|z| \leq N} |g(z) - h(z)| < \varepsilon$, respectively).

By the continuity of g there is some $\delta > 0$ such that $|g(z) - g(\zeta)| < \frac{\varepsilon}{2}$ whenever $|z| \leq N$, $|\zeta| \leq \frac{5}{4}N$ and $|z - \zeta| \leq \delta$. We can assume that $\delta \leq \frac{N}{4}$. Then there is some $J \geq 1$ such that

$$\sum_{\nu=1}^{J-1} \frac{\delta}{3\nu N} \leq 2\pi < \sum_{\nu=1}^{J} \frac{\delta}{3\nu N}.$$

We set

$$\theta_0 = 0, \quad \theta_j = \sum_{\nu=1}^{j} \frac{\delta}{3\nu N} \; (0 < j < J), \quad \theta_J = 2\pi$$

and $n_j = 3jN$ for $j = 1, \ldots, J$.

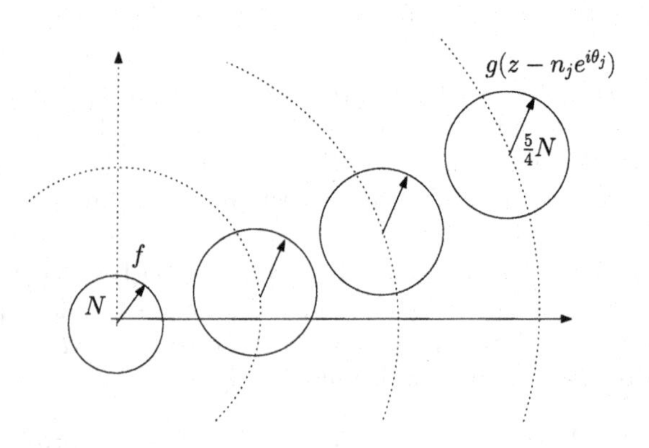

Fig. 11.2 Constructing common Birkhoff hypercyclic functions II

Applying Runge's theorem to the union of the closed balls of radius N around 0 and of radius $\frac{5}{4}N$ around $n_1 e^{i\theta_1}, \ldots, n_J e^{i\theta_J}$ (see Figure 11.2), we obtain an entire function h such that

$$\sup_{|z|\leq N} |f(z) - h(z)| < \varepsilon$$

and, for $j = 1, \ldots, J$,

$$\sup_{|z-n_je^{i\theta_j}|\leq \frac{5}{4}N} \left|g(z - n_je^{i\theta_j}) - h(z)\right| < \frac{\varepsilon}{2}.$$

By the first inequality we have that $h \in U$.

Moreover, let $\theta \in [0, 2\pi]$. Then there exists some j with $1 \leq j \leq J$ such that $\theta_{j-1} \leq \theta \leq \theta_j$. Hence $|\theta - \theta_j| \leq \frac{\delta}{3jN}$ and

$$\left|n_je^{i\theta} - n_je^{i\theta_j}\right| \leq n_j|\theta - \theta_j| \leq \delta \leq \frac{1}{4}N.$$

Now, if $|z - n_je^{i\theta}| \leq N$ then

$$\left|z - n_je^{i\theta_j}\right| \leq \left|z - n_je^{i\theta}\right| + \left|n_je^{i\theta} - n_je^{i\theta_j}\right| \leq N + \delta \leq \frac{5}{4}N.$$

Consequently we have that

$$\begin{aligned}
\sup_{|z|\leq N} \left|g(z) - h\left(z + n_je^{i\theta}\right)\right| &\leq \sup_{|z|\leq N} \left|g(z) - g\left(z + n_je^{i\theta} - n_je^{i\theta_j}\right)\right| \\
&\quad + \sup_{|z-n_je^{i\theta}|\leq N} \left|g\left(z - n_je^{i\theta_j}\right) - h(z)\right| \\
&\leq \sup_{|z|\leq N, |\zeta|\leq \frac{5}{4}N, |z-\zeta|\leq\delta} |g(z) - g(\zeta)| \\
&\quad + \sup_{|z-n_je^{i\theta_j}|\leq \frac{5}{4}N} \left|g\left(z - n_je^{i\theta_j}\right) - h(z)\right| \\
&< \frac{\varepsilon}{2} + \frac{\varepsilon}{2} = \varepsilon.
\end{aligned}$$

This shows that $(T_{e^{i\theta}})^{n_j}h \in V$, which implies the claim.

11.4 Common hypercyclic subspaces

In this section we combine the ideas of this chapter with those of Chapter 10. Having established criteria for the existence of common hypercyclic vectors one may wonder if a family of operators even has a common hypercyclic subspace. Our first result might already come as a surprise: not even two operators with hypercyclic subspaces need to share a hypercyclic subspace.

Example 11.22. We consider the weights $w_n = \frac{n+1}{n}$ and $v_n = 2$, $n \geq 1$, and the corresponding weighted shifts B_w and B_v on the Hilbert space ℓ^2. Then the direct sums $T_1 = B_w \oplus B_v$ and $T_2 = B_v \oplus B_w$ on the Hilbert space $\ell^2 \oplus \ell^2$ satisfy the conditions of Montes' theorem and therefore have hypercyclic subspaces; see Exercise 11.4.1.

Let us suppose that T_1 and T_2 have a common hypercyclic subspace $M \subset \ell^2 \oplus \ell^2$. We can construct inductively an orthonormal system $(e_n)_n$ of vectors $e_n = (x_n, y_n) \in M$, $n \geq 1$, such that $(x_n)_n$ and $(y_n)_n$ are orthogonal sequences in ℓ^2. Indeed, for e_1 we choose a unit vector in M. Now let $e_k = (x_k, y_k)$, $k = 1, \ldots, n$, be chosen. Then the linear map $M \to \mathbb{K}^{2n}$, $z = (x, y) \to (\langle x, x_1\rangle, \langle y, y_1\rangle, \ldots, \langle x, x_n\rangle, \langle y, y_n\rangle)$ must have nontrivial kernel because M is infinite dimensional. Hence there is a unit vector $e_{n+1} = (x_{n+1}, y_{n+1})$ in M such that $(x_k)_{1\leq k\leq n+1}$ and $(y_k)_{1\leq k\leq n+1}$ are orthogonal, so that $(e_k)_{1\leq k\leq n+1}$ is orthonormal.

The sequence $(e_n)_n$ then satisfies $\|x_n\|^2 + \|y_n\|^2 = 1$ for $n \geq 1$. By passing to a subsequence and interchanging x and y if necessary, we can assume that $\|x_n\|^2 \geq \frac{1}{2}$ for all $n \geq 1$. Then $(x_n/\|x_n\|)_n$ is an orthonormal basis for $L := \overline{\operatorname{span}}\{x_n \ ; \ n \geq 1\}$ in ℓ^2. Let $x = \sum_{n=1}^{\infty} a_n x_n/\|x_n\| \in L$, $x \neq 0$. Since $\sum_{n=1}^{\infty}(|a_n|/\|x_n\|)^2 \leq 2\sum_{n=1}^{\infty}|a_n|^2 < \infty$, then $z := \sum_{n=1}^{\infty} a_n/\|x_n\| \, e_n$ converges in M, and $z \neq 0$. Therefore z is hypercyclic for T_2, which implies that x is hypercyclic for B_v. But then L is a hypercyclic subspace for B_v, which contradicts Example 10.26.

Thus, even in the case of finitely or countably many operators, one is lead to look for a sufficient condition on a family of operators to share a hypercyclic subspace. The second approach to Montes' theorem via left-multiplication operators, paired with the Baire category theorem, yields the following quite natural result.

Theorem 11.23. *Let X be a separable Fréchet space with a continuous norm, and let T_j, $j \geq 1$, be operators on X. If they satisfy the hypotheses of Montes' theorem for the same infinite-dimensional closed subspace M_0 of X then they have a common hypercyclic subspace.*

Proof. Let $\mathcal{K}$ be the separable Fréchet space defined in Lemma 10.36. By assumption, for any $j \geq 1$, there exists an increasing sequence $(n_{j,k})_k$ of positive integers such that T_j satisfies the Hypercyclicity Criterion for $(n_{j,k})_k$ and $T_j^{n_{j,k}}x \to 0$ as $k \to \infty$, for all $x \in M_0$. It follows from Lemma 10.37 that the sequences $(L_{T_j^{n_{j,k}}})_k$, $j \geq 1$, are hypercyclic on $\mathcal{K}$ and therefore, by the Baire category theorem, have a common hypercyclic vector $S \in \mathcal{K}$. Then, by Lemma 10.38, there is some $\lambda \neq 0$ such that $(I + \lambda S)M_0$ is a hypercyclic subspace for each operator T_j, $j \geq 1$. □

Example 11.24. Let Ω be a domain in $\mathbb{C}$. By Proposition 10.33, every hypercyclic composition operator C_φ, $\varphi \in \operatorname{Aut}(\Omega)$, possesses a hypercyclic subspace. We want to show here that any two of them, C_φ and C_ψ, $\varphi, \psi \in \operatorname{Aut}(\Omega)$, possess a common hypercyclic subspace. In particular, every hypercyclic operator C_φ has a common hypercyclic subspace with its inverse operator $(C_\varphi)^{-1} = C_{\varphi^{-1}}$; note that the hypercyclicity of the inverse operator is automatic by Proposition 2.23.

In fact, we need only modify the argument of Proposition 10.33 slightly. As in that proof there are subsequences $(m_n)_n$ and $(m'_n)_n$ such that C_φ and C_ψ satisfy the Hypercyclicity Criterion for the sequence $(m_n)_n$ and $(m'_n)_n$, respectively, and such that $(\varphi^{m_n})_n$ and $(\psi^{m'_n})_n$ are run-away sequences. Assuming again that Ω contains $\overline{\mathbb{D}}$ and fixing an exhaustion $(K_n)_n$ of Ω by compact sets with $K_1 = \overline{\mathbb{D}}$, we can construct inductively Ω-convex compact subsets L_k, and a strictly increasing sequence $(n_k)_k$ of positive integers with $n_{2k-1} \in \{m_n \; ; \; n \geq 1\}$ and $n_{2k} \in \{m'_n \; ; \; n \geq 1\}$ such that, for $k \geq 1$,

$$\begin{aligned}
\varphi^{n_{2k-1}}(L_{2k-1}) \cap L_{2k-1} = \varnothing, \; \varphi^{n_{2k-1}}(L_{2k-1}) \cup L_{2k-1} \text{ is } \Omega\text{-convex},\\
\psi^{n_{2k}}(L_{2k}) \cap L_{2k} = \varnothing, \; \psi^{n_{2k}}(L_{2k}) \cup L_{2k} \text{ is } \Omega\text{-convex},\\
K_{2k-1} \cup L_{2k-2} \cup \psi^{n_{2k-2}}(L_{2k-2}) \subset L_{2k-1},\\
K_{2k} \cup L_{2k-1} \cup \varphi^{n_{2k-1}}(L_{2k-1}) \subset L_{2k},
\end{aligned}$$

where $L_0 = \varnothing$ and $n_0 = 0$.

Denoting by e_n the functions $e_n(z) = z^n$, $n \geq 1$, and applying successively Runge's theorem we obtain, for any $j \geq 1$, functions $f_{j,k} \in H(\Omega)$, $k \geq 1$, such that

$$\sup_{z \in L_1} |e_j(z) - f_{j,1}(z)| < \frac{1}{2^{j+2}}, \qquad \sup_{z \in L_{k+1}} |f_{j,k}(z) - f_{j,k+1}(z)| < \frac{1}{2^{j+k+2}},$$

$$\sup_{z \in \varphi^{n_{2k-1}}(L_{2k-1})} |f_{j,2k-1}(z)| < \frac{1}{2^{j+2k}}, \qquad \sup_{z \in \psi^{n_{2k}}(L_{2k})} |f_{j,2k}(z)| < \frac{1}{2^{j+2k+1}}.$$

It can now be shown exactly as in the proof of Proposition 10.33 that

$$f_j := f_{j,1} + \sum_{k=1}^{\infty} (f_{j,k+1} - f_{j,k}), \quad j \geq 1,$$

converges in $H(\Omega)$, that the closed linear span M_0 of the functions f_j, $j \geq 1$, is infinite dimensional and that $(C_\varphi)^{n_{2k-1}} g \to 0$ and $(C_\psi)^{n_{2k}} g \to 0$ in $H(\Omega)$ for any $g \in M_0$. Since C_φ and C_ψ satisfy the Hypercyclicity Criterion for $(n_{2k-1})_k$ and $(n_{2k})_k$, respectively, Theorem 11.23 implies that C_φ and C_ψ have a common hypercyclic subspace.

We now turn to uncountable families. Not surprisingly, in this case we have to impose stronger assumptions in order to obtain common hypercyclic subspaces. Again the approach using left-multiplication operators, this time paired with the Common Hypercyclicity Criterion, leads to the desired result.

Theorem 11.25. *Let $\Lambda \subset \mathbb{R}$ be an interval and $(T_\lambda)_{\lambda \in \Lambda}$ a continuous family of operators on a separable Fréchet space X with a continuous norm. Suppose that*

(i) *the family $(T_\lambda)_{\lambda \in \Lambda}$ satisfies the Common Hypercyclicity Criterion;*
(ii) *there exists an infinite-dimensional closed subspace M_0 of X such that $T_\lambda^n x \to 0$ for all $x \in M_0$, $\lambda \in \Lambda$.*

Then the family $(T_\lambda)_{\lambda\in\Lambda}$ has a common hypercyclic subspace.

Proof. We will use the terminology of Section 10.5. Let $|||\cdot|||$ be a continuous norm on X, let Φ be a countable set of continuous linear functionals on $X_{|||\cdot|||}$ that is weak-$*$-dense in $(X_{|||\cdot|||})^*$, and let $\mathcal{K}=\mathcal{K}_\Phi$ be the closed subspace of $L(X_{|||\cdot|||},X)$ defined in Lemma 10.36. Then $\mathcal{K}$ is a separable Fréchet space under the seminorms

$$\|S\|_k = \sup_{|||x|||\leq 1} p_k(Sx), \quad k\geq 1,$$

where $(p_k)_k$ is an increasing sequence of seminorms defining the topology of X. Also, for every $\lambda\in\Lambda$, the left-multiplication operator L_{T_λ} is an operator on $\mathcal{K}$. We claim that the family $(L_{T_\lambda})_{\lambda\in\Lambda}$ satisfies the Common Hypercyclicity Criterion on $\mathcal{K}$.

We leave the proof of continuity of the family to the reader; see Exercise 11.4.6. Now let $K=[a,b]\subset\Lambda$. By hypothesis there is a dense subset X_0 of X and maps $S_n(\lambda): X_0\to X$, $n\geq 0$, $\lambda\in K$, such that, for any $x\in X_0$, conditions (i)–(iv) of Theorem 11.9 are satisfied. By Lemma 10.36, $\mathcal{F}=\mathcal{F}_{\Phi,X_0}$ is dense in $\mathcal{K}$. For any $S\in\mathcal{F}$ we choose a representation

$$S=\sum_{j=1}^{l}\langle\cdot\,,y_j^*\rangle x_j,\quad y_j^*\in\Phi,\ x_j\in X_0,$$

and define, for any $n\geq 0$, $\lambda\in K$,

$$M_n(\lambda)S=\sum_{j=1}^{l}\langle\cdot\,,y_j^*\rangle S_n(\lambda)x_j\in\mathcal{K}.$$

We then have that, for $k\geq 1$, $N\geq 0$ and any finite set $F\subset\{N,N+1,\ldots\}$,

$$\begin{aligned}\Big\|\sum_{\substack{0\leq n\leq m\\ n\in F}} L_{T_\lambda}^m M_{m-n}(\mu_n)S\Big\|_k &= \Big\|\sum_{\substack{0\leq n\leq m\\ n\in F}} T_\lambda^m\sum_{j=1}^{l}\langle\cdot\,,y_j^*\rangle S_{m-n}(\mu_n)x_j\Big\|_k\\ &= \sup_{|||x|||\leq 1} p_k\Big(\sum_{j=1}^{l}\langle x,y_j^*\rangle \sum_{\substack{0\leq n\leq m\\ n\in F}} T_\lambda^m S_{m-n}(\mu_n)x_j\Big)\\ &\leq \sum_{j=1}^{l}\sup_{|||x|||\leq 1}|\langle x,y_j^*\rangle|\, p_k\Big(\sum_{\substack{0\leq n\leq m\\ n\in F}} T_\lambda^m S_{m-n}(\mu_n)x_j\Big),\end{aligned}$$

and by assumption the right-hand side can be made arbitrarily small if N is sufficiently large, uniformly for $m\geq 0$ and $\lambda\geq\mu_0\geq\ldots\geq\mu_m$ from K. In the same way one can show also that conditions (ii)–(iv) of the Common Hypercyclicity Criterion are satisfied for the family $(L_{T_\lambda})_{\lambda\in\Lambda}$. This implies that there is an operator $S \in \mathcal{K}$ that is hypercyclic for any operator L_{T_λ} on $\mathcal{K}$, $\lambda \in \Lambda$. Then, by Lemma 10.38, there is some $\mu \neq 0$ such that $(I + \mu S)M_0$ is a hypercyclic subspace for each operator T_λ, $\lambda \in \Lambda$. □

Combining this result with (the proof of) Theorem 11.13 we obtain the following special case.

Theorem 11.26. *Let X be a separable Fréchet space with a continuous norm, let T be an operator on X with dense generalized kernel X_0 and $\lambda_0 \geq 0$. Suppose that*

(i) *there is a map $S : X_0 \to X$ such that $TSx = x$ and $\frac{1}{\lambda^n}S^n x \to 0$ for all $x \in X_0$, $\lambda > \lambda_0$;*

(ii) *there exists an infinite-dimensional closed subspace M_0 of X such that $\lambda^n T^n x \to 0$ for all $x \in M_0$, $\lambda > \lambda_0$.*

Then the family $(\lambda T)_{|\lambda|>\lambda_0}$ has a common hypercyclic subspace.

Remark 11.27. Remark 11.14 applies here as well.

Example 11.28. Let X be one of the spaces ℓ^p, $1 \leq p < \infty$, or c_0, and T the operator on X defined by

$$T(x_1, x_2, x_3, \ldots) = (x_2, x_4, x_6, \ldots);$$

see also Proposition 8.5. Then the operators λT, $|\lambda| > 1$, have a common hypercyclic subspace. To this end consider for X_0 the set of finite sequences, for S_n the maps $S_n = S^n$ with $S(x_n)_n = (0, x_1, 0, x_2, 0, x_3, 0, \ldots)$, $x \in X$, and for M_0 the subspace of all sequences $x = (x_n)_n \in X$ whose even coordinates vanish. Then the claim follows from Theorem 11.26 and Remark 11.27.

We end this section by mentioning a related problem, that of the existence of a common dense subspace of hypercyclic vectors, excluding 0 as usual. By the Herrero–Bourdon theorem, every hypercyclic operator has such a subspace. This easily implies that if a commuting family of operators has a common hypercyclic vector then it has a dense subspace of hypercyclic vectors in common; see Exercise 11.1.5. In the noncommuting case we still get a positive answer when the family satisfies the Common Hypercyclicity Criterion and the underlying space has a continuous norm; see Exercise 11.4.10.

Exercises

Exercise 11.1.1. Consider the family of all weighted backward shifts B_w on ℓ^2 whose weight sequence $(w_n)_n$ only takes the values 1 and 2, with 2 appearing infinitely often. Show that each of these operators is hypercyclic but that they do not possess a common hypercyclic vector. (*Hint*: Construct, for any $x \in \ell^2$, a weight w such that $w_1 \cdots w_n |x_n| \leq 1$ for all sufficiently large n.)

Exercise 11.1.2. Let S and T be operators on a Banach space X and $\Lambda \subset]0,\infty[^2$. Show that if the family $(\lambda S \oplus \mu T)_{(\lambda,\mu)\in\Lambda}$ of operators on $X \oplus X$ has a common hypercyclic vector then Λ has two-dimensional Lebesgue measure 0. (*Hint*: First show as in Example 11.3 that the set $\{(\log\lambda, \log\mu)\ ;\ (\lambda,\mu)\in\Lambda\}$ has Lebesgue measure 0.)

Exercise 11.1.3. Let $1 \le p < \infty$, $\alpha > 0$, $v(t) = \frac{1}{1+|t|^\alpha}$ $(t\in\mathbb{R})$, and $b \in \mathbb{R}$. Show that the following define families of operators on $L^p_v(\mathbb{R})$ and can be extended to semigroups of operators:
(i) $T_\lambda f(x) = f(x+\lambda)$, $\lambda > 0$;
(ii) $T_\lambda f(x) = f(\lambda x)$, $\lambda > 1$;
(iii) $T_\lambda f(x) = f(\lambda(x-b)+b)$, $\lambda > 1$.
Deduce that in each family the operators have the same sets of hypercyclic vectors. (*Hint*: When suitable, replace λ with $\lambda > 1$ by e^t with $t > 0$.)

Exercise 11.1.4. Let B be the backward shift on one of the spaces ℓ^p, $1 \le p < \infty$, or c_0. Show that, although the operators λB, $\lambda > 1$, have common hypercyclic vectors, no two of them have the same hypercyclic vectors. More precisely, if $\lambda, \mu > 1$ with $\lambda \neq \mu$ then there is a vector that is λB-hypercyclic but not μB-hypercyclic. (*Hint*: Show that there exists $x \in HC(\lambda B)$ with $\inf_{n\ge1} \mu^n |x_n| > 0$ $(\lambda < \mu)$ or $\sup_{n\ge1}\mu^n|x_n| < \infty$ $(\lambda > \mu)$.)

Exercise 11.1.5. Let $(T_\lambda)_{\lambda\in\Lambda}$ be a family of hypercyclic operators on a separable Fréchet space X. Show that, if some T_{λ_0} commutes with each operator T_λ, $\lambda\in\Lambda$, then the set

$$\bigcap_{\lambda\in\Lambda} HC(T_\lambda)$$

of common hypercyclic vectors is either empty or it contains, except for 0, a dense subspace.

Exercise 11.1.6. Let X be a separable Fréchet space, Λ a σ-compact metric space and $(T_\lambda)_{\lambda\in\Lambda}$ a continuous family of operators on X. Show the following:
(i) if $x \in X$ is a common hypercyclic vector then, for any compact subset $K \subset \Lambda$ and for any nonempty open subset V of X there are $n_1,\ldots,n_J \ge 0$ such that for any $\lambda \in K$ there is some j such that $T_\lambda^{n_j} x \in V$;
(ii) the set of common hypercyclic vectors is a dense G_δ-set if and only if, for any compact set $K \subset \Lambda$ and for any nonempty open subsets U and V of X, there are $x \in U$ and $n_1,\ldots,n_J \ge 0$ such that, for any $\lambda \in K$ there is some j such that $T_\lambda^{n_j} x \in V$.
Moreover, show that the following condition is satisfied for *any* continuous family $(T_\lambda)_{\lambda\in\Lambda}$ of hypercyclic operators and therefore does not characterize common hypercyclicity:
(iii) for any compact set $K \subset \Lambda$ and for any nonempty open subsets U and V of X there are $n_1,\ldots,n_J \ge 0$ such that for any $\lambda\in K$ there is some j such that $T_\lambda^{n_j}(U) \cap V \neq \varnothing$.

Exercise 11.1.7. (a) Let $\Lambda \subset \mathbb{R}$ be an interval and $(T(\lambda))_{\lambda\in\Lambda}$ a continuous family of operators on a separable Fréchet space X. Suppose that, for any compact subinterval $K \subset \Lambda$, there are dense subsets X_0, Y_0 of X, an increasing sequence $(n_k)_k$ of positive integers, and maps $S_{n_k}(\lambda) : Y_0 \to X$, $k \ge 1$, $\lambda \in \Lambda$, such that, for any $x \in X_0$, $y \in Y_0$,
(i) $T^{n_k}(\lambda)x \to 0$, uniformly for $\lambda \in K$;
(ii) $S_{n_k}(\lambda)y \to 0$ for any $\lambda \in K$;
(iii) for any $\varepsilon > 0$ there is some $\delta > 0$ such that, for any $\mu \in K$ and $N \ge 1$ there is some $k \ge N$ such that, if $\lambda \in K$ with $|\lambda-\mu| < \delta$ then $\|T^{n_k}(\lambda)S_{n_k}(\mu)y - y\| < \varepsilon$.

Show that the set of common hypercyclic vectors of the family $(T_\lambda)_{\lambda\in\Lambda}$ is a dense G_δ-set, and in particular, nonempty. (*Hint*: Fix a partition $K_1,\ldots,K_m$ of K of intervals of length $< \delta$; obtain $U \supset U_1 \supset \ldots \supset U_m$ and $\nu_1,\ldots,\nu_m$ such that $T^{\nu_j}(U_j) \subset V$ for $\lambda \in K_j$.)

(b) Show that, for a singleton Λ, this criterion reduces to (a weak form of) the Hypercyclicity Criterion.

(c) Show that the criterion cannot be applied in the canonical way to the family of Rolewicz operators.

Exercise 11.2.1. Show that the family $(\lambda(B\oplus B))_{|\lambda|>1} = (\lambda B\oplus\lambda B)_{|\lambda|>1}$ of operators on $\ell^p \oplus \ell^p$, $1 \le p < \infty$, has a common hypercyclic vector. Compare the result with Exercise 11.1.2.

Exercise 11.2.2. Let B_w be a weighted shift on one of the spaces ℓ^p, $1 \le p < \infty$, or c_0, where $w = (w_n)_n$ is a bounded weight sequence; see Section 4.1. Furthermore, let $\lambda_0 > 0$ be such that λB_w is a mixing operator for any $\lambda > \lambda_0$. Show then that $(\lambda B_w)_{|\lambda|>\lambda_0}$ has a common hypercyclic vector. Formulate λ_0 in terms of the weights.

Exercise 11.2.3. Let D be the differentiation operator on $H(\mathbb{C})$ and p a nonconstant polynomial. By using the following steps, show that $(\lambda p(D))_{\lambda\neq 0}$ has a common hypercyclic vector on $H(\mathbb{C})$.

(a) Fix a root $\mu \in \mathbb{C}$ of p. Then the set X_0 of functions $f(z) = e^{\mu z}q(z)$, q a polynomial, is dense in $H(\mathbb{C})$.

(b) If $f(z) = e^{\mu z}q(z) = \sum_{n=0}^\infty a_n z^n$ belongs to X_0 then $\widetilde{f}(z) = \sum_{n=0}^\infty \frac{n!a_n}{z^{n+1}}$ is holomorphic for $|z| > |\mu|$. (*Hint*: Look at the proof of Lemma 4.18.)

(c) Fix $\rho > |\mu|$ such that p has no roots on $|z| = \rho$. Then, for any $f \in X_0$ and $n \ge 0$, $S_n f(z) = \frac{1}{2\pi i}\int_{|\zeta|=\rho} e^{z\zeta}\frac{\widetilde{f}(\zeta)}{p(\zeta)^n}\,d\zeta$ defines an entire function with $p(D)^n S_n f = f$.

(d) Prove the claim by letting $\rho \to \infty$.

Exercise 11.2.4. Let $\mu \in \mathbb{C}$ with $|\mu| < 1$. Consider the adjoint multiplier $T = M_\varphi^*$ on the Hardy space H^2 given by $\varphi(z) = z - \mu$; see Section 4.4.

(a) Let $g_n(z) = \frac{z^n}{(1-\overline{\mu}z)^{n+1}}$, $n \ge 0$. Show that $X_0 := \operatorname{span}\{g_n\ ;\ n \ge 0\}$ is dense in H^2 and that $Tg_n = g_{n-1}$ with $g_{-1} = 0$. (*Hint*: Show that $\langle f, g_n\rangle$ is a multiple of $f^{(n)}(\mu)$.)

(b) Deduce that the operators λM_φ^*, $|\lambda| > \frac{1}{1-|\mu|}$, have common hypercyclic vectors.

(c) Deduce also that the operators $\lambda(\mu I + B)$ on ℓ^2, $|\lambda| > \frac{1}{1-|\mu|}$, have common hypercyclic vectors, where B is the backward shift.

Exercise 11.2.5. Let φ be a nonconstant complex polynomial all of whose roots lie in $\mathbb{D}$. Show that the operators λM_φ^* on H^2, $|\lambda| > \max_{|z|=1}\frac{1}{|\varphi(z)|}$, have common hypercyclic vectors. (*Hint*: Fix a root μ of φ; consider the set X_0 of Exercise 11.2.4 and define $S_n : X_0 \to H^2$ by $S_n = M_{\psi_n}$, where $\psi_n(z) = (1/\varphi^*(1/z))^n$; see Proposition 4.41. Use Proposition 4.40.)

Exercise 11.2.6. Let φ be a nonconstant bounded holomorphic function on $\mathbb{D}$ and M_φ^* the corresponding adjoint multiplier on H^2. Using Theorem 4.42, show that λM_φ^* is hypercyclic if and only if $\inf_{z\in\mathbb{D}}|\varphi(z)|^{-1} < |\lambda| < \sup_{z\in\mathbb{D}}|\varphi(z)|^{-1}$, where $1/0 = \infty$. Deduce that, for a certain φ, Theorem 11.13 cannot be used to obtain common hypercyclic vectors.

Exercise 11.2.7. Let T be an operator on a Fréchet space X with a continuous norm, and set $T(\lambda) = \lambda T$, $\lambda > 0$. Let X_0 be a dense subset of X and $S_n : X_0 \to X$ be maps such that $T^n S_n x = x$ for all $x \in X_0$, $n \ge 0$. Set $S_n(\lambda) = \frac{1}{\lambda^n}S_n$. Show that if condition (i) of the Common Hypercyclicity Criterion is satisfied for some set $K = [a,b]$ with $0 < a < b$ then T has a dense generalized kernel.

Exercise 11.2.8. The proof of the existence of entire functions that are hypercyclic for all multiples λT_a, $\lambda > 0$, of the Birkhoff operator T_a (see Example 11.18), can easily be turned into a purely constructive proof. Explain.

Exercise 11.3.1. Show that Proposition 11.19 remains true when assumption (ii) is weakened as follows: for any compact subinterval $K \subset \Lambda$, the functions $\log w_n$ are Lipschitz continuous on K with Lipschitz constant L_n such that $\sum_{n=1}^{\infty}(\sum_{k=1}^{n} L_k)^{-1} = \infty$. Deduce that the weighted shifts $B_{w(\lambda)}$, $\lambda > 0$, with $w_n(\lambda) = n^\lambda$, $n \geq 1$, have a common hypercyclic vector on any of the spaces ℓ^p, $1 \leq p < \infty$, or c_0. (*Hint*: Use Remark 11.10(d).)

Exercise 11.3.2. Proposition 11.19 remains true if each function w_n, $n \geq 1$, is decreasing. More generally, it remains true if the interval I can be partitioned into countably many subintervals on which either all functions w_n are decreasing or all functions w_n are increasing. Explain.

Exercise 11.3.3. Let $\Lambda \subset \mathbb{R}$ be an interval. Let $w_n : \Lambda \to \mathbb{R}$, $n \geq 1$, be strictly positive functions such that, for any compact subinterval $K \subset \Lambda$, the functions $\log w_n$ are Lipschitz continuous on K with uniformly bounded Lipschitz constant. Show that the family $(B_{w(\lambda)})_{\lambda\in\Lambda}$ of weighted backward shifts has a common hypercyclic vector on $\omega = \mathbb{K}^{\mathbb{N}}$ and that the set of common hypercyclic vectors is a dense G_δ-set.

Exercise 11.3.4. Use the results of this section (and the León–Müller theorem) to confirm the claim of Exercise 11.2.2.

Exercise 11.4.1. Show that the operators T_1 and T_2 of Example 11.22 have hypercyclic subspaces.

Exercise 11.4.2. Explain how the construction in Example 11.24 can be modified in order to show that any countable family of hypercyclic composition operators C_{φ_j}, $\varphi_j \in \mathrm{Aut}(\Omega)$, $j \geq 1$, on a domain $\Omega \subset \mathbb{C}$ possesses a common hypercyclic subspace.

Exercise 11.4.3. Let $\mathcal{B}$ be the family of all weighted backward shifts B_w on $X = \ell^p$, $1 \leq p < \infty$, or on $X = c_0$, such that

$$\lim_{n\to\infty} \prod_{\nu=1}^{n} |w_\nu| = \infty \quad \text{and} \quad \sup_{n\geq 1} \limsup_{k\to\infty} \prod_{\nu=1}^{n} |w_{\nu+k}| < \infty.$$

Show that any countable family of operators in $\mathcal{B}$ has a common hypercyclic subspace.

Exercise 11.4.4. Let T_j, $j \geq 1$, be operators on a separable Banach space X. Suppose that there exist increasing sequences $(n_{j,k})_k$ of positive integers such that

(i) for $j \geq 1$, T_j satisfies the Hypercyclicity Criterion for $(n_{j,k})_k$,
(ii) there is a decreasing sequence $(M_k)_k$ of infinite-dimensional closed subspaces of X such that $\sup_{k\geq 1, j\leq k} \|T_j^{n_{j,k}}|_{M_k}\| < \infty$.

Then the operators T_j, $j \geq 1$, have a common hypercyclic subspace. (*Hint*: Modify the proof of Theorem 10.29 using Exercise 3.4.9.)

Exercise 11.4.5. Let T_j, $j \geq 1$, be weakly mixing operators on a separable Banach space that are of the form $T_j = U_j + K_j$, $\|U_j\| \leq 1$ and K_j compact, $j \geq 1$. Show that these operators have a common hypercyclic subspace. (*Hint*: Modify the proof of Corollary 10.31 using the previous exercise.)

Exercise 11.4.6. Show that, in the setting of Theorem 11.25 and its proof, $(L_{T_\lambda})_{\lambda\in\Lambda}$ is a continuous family on $\mathcal{K}$. (*Hint*: Restrict the λ to a compact subset; show that the corresponding T_λ are equicontinuous on X; use that $\mathcal{F}$ is dense in $\mathcal{K}$).

Exercise 11.4.7. Let X be one of the spaces $\ell^p(\mathbb{Z})$, $1 \le p < \infty$, or $c_0(\mathbb{Z})$. Show that the bilateral weighted backward shifts $B_{w(\lambda)}$, $\lambda > 1$, with weights

$$w(\lambda) = (\ldots, \tfrac{1}{\lambda}, \tfrac{1}{\lambda}, \tfrac{1}{\lambda}, 2, 2, 2, \ldots)$$

have a common hypercyclic subspace.

Exercise 11.4.8. Generalize the result of Example 11.28 to arbitrary operators D_w given by $D_w(x_n)_n = (w_2x_2, w_4x_4, w_6x_6, \ldots)$ such that $(w_{2n})_n$ is bounded and, for any $m \ge 1$ odd, $\liminf_{n\to\infty} \prod_{k=1}^n w_{m2^k} > 0$.

Exercise 11.4.9. Let $a > 0$, and let T be the operator on $C_0(\mathbb{R}_+)$ given by $Tf(x) = f(x+a)$; see Exercise 10.1.1. Show that the operators λT, $\lambda > 1$, have a common hypercyclic subspace.

Exercise 11.4.10. Let $\Lambda \subset \mathbb{R}$ be an interval and $(T_\lambda)_{\lambda\in\Lambda}$ a continuous family of operators on a separable Fréchet space X with a continuous norm. Show that if $(T_\lambda)_{\lambda\in\Lambda}$ satisfies the Common Hypercyclicity Criterion then its set of common hypercyclic vectors contains, except for 0, a dense subspace. (*Hint*: Use the ideas of Exercise 10.5.1.)

Sources and comments

Section 11.1. The problem of the existence of common hypercyclic vectors for uncountable families was first raised by Godefroy and Shapiro [165] when they asked if the differential operators $\varphi(D)$ on $H(\mathbb{C})$ with nonconstant φ (see Section 4.2) share a hypercyclic vector. However, this question was largely ignored, so that it was Salas [275] who initiated the present study into common hypercyclicity by asking if the Rolewicz operators λB, $|\lambda| > 1$, have a common hypercyclic vector. For real parameters, a positive answer was given by Abakumov and Gordon [1] and, independently, by Peris [252]; the general case is due to Costakis and Sambarino [123]; see Example 11.11.

Example 11.3 is due to Borichev (see [1, 42]). Theorem 11.5(a) is due to Saint Raymond (see [1, 33]); the proof given here is taken from Costakis and Sambarino [123, p. 304]. Assertion (b) was obtained by Shkarin [288], while its equivalent formulation given in Remark 11.6(b) is due to Chan and Sanders [102]. Remark 11.6(a) was observed by Bayart [33].

The Common Hypercyclicity Criterion, under stronger assumptions, is due to Costakis and Sambarino [123]. The present form was inspired by the Frequent Hypercyclicity Criterion. An alternative criterion was found by Bayart and Matheron [42]. They use it to show that, for the family of all automorphisms φ of $\mathbb{D}$ having 1 as an attractive fixed point, the corresponding composition operators C_φ on H^2 (see Section 4.5) share hypercyclic vectors; see also Bayart and Grivaux [39]. By Bayart [33], this result breaks down when the attractive fixed points are allowed to cover a set of positive Lebesgue measure on $\mathbb{T}$.

We have already commented on the difficulty of extending the criterion to more than one-dimensional families of operators. Shkarin [288] confirmed that there is an intrinsic obstruction to common hypercyclicity for higher-dimensional families by proving the following remarkably general result.

Theorem 11.29. *Let T be an operator on a complex Fréchet space and $\Lambda \subset \mathbb{C} \times \mathbb{R}_+$ be such that the family $(\lambda I + \mu T)_{(\lambda,\mu)\in\Lambda}$ has a common hypercyclic vector. Then Λ has three-dimensional Lebesgue measure* 0.

For example, if D denotes the differentiation operator on $H(\mathbb{C})$, then the operators $\lambda I + \mu D$, $\lambda \in \mathbb{C}$, $\mu > 0$, cannot have a common hypercyclic vector. This finally solved the Godefroy–Shapiro problem in the negative.

Section 11.2. Proposition 11.12 is due to Bayart [33]. Theorem 11.13, Corollary 11.15 and Example 11.16 are due to Bayart and Matheron [42]; see also Bayart [33]. Remark 11.14 was made by Costakis and Mavroudis [120] and Bernal [59].

Common hypercyclic vectors for the multiples of MacLane's operator (see Example 11.17) were found by Costakis and Sambarino [123]. This was extended to multiples of certain operators $\varphi(D)$ by Costakis and Mavroudis [120] and Bernal [59]. Finally, Shkarin [288] proved that for any nonconstant entire function φ of exponential type the differential operators $\lambda\varphi(D)$, $\lambda \neq 0$, share hypercyclic vectors. This then also includes Example 11.18.

Bayart [33] and Gallardo and Partington [160] have obtained common hypercyclic vectors for multiples λM_φ^* of certain adjoint multipliers on H^2. Shkarin [288] proved that, for any bounded holomorphic function φ, all hypercyclic multiples λM_φ^* share hypercyclic vectors.

The two mentioned results by Shkarin [288] are based on a powerful sufficient condition for common hypercyclicity for multiples of operators.

Section 11.3. Proposition 11.19 is a special case of a theorem by Bayart and Matheron [42]. The first result in this direction is due to Costakis and Sambarino [123]; see Example 11.20(b). These authors also found Example 11.21; in fact, they gave a direct proof that for parameters along rays starting from zero the sets of hypercyclic vectors are the same.

Chan and Sanders [102] showed that between any two hypercyclic weighted shifts on ℓ^2 there is a continuous path of weighted shifts that shares hypercyclic vectors; and there is another such path without any common hypercyclic vector.

Shkarin [288] studied common hypercyclicity for genuinely two- or higher-dimensional families, that is, where one cannot reduce the number of dimensions by the León–Müller theorem or the Conejero–Müller–Peris theorem. He was the first to obtain a genuinely higher-dimensional family with common hypercyclic vectors: the multiples of the Birkhoff operators $f \to \lambda f(z+a)$, $\lambda, a \in \mathbb{C}\setminus\{0\}$, on $H(\mathbb{C})$ (note that one can reduce the two complex dimensions to two real ones).

Section 11.4. Theorem 11.23 is due to Aron, Bès, León and Peris [14] where Example 11.22 can also be found. Theorem 11.25 improves on the corresponding result in Bayart [35], Theorem 11.26 seems to be new. Common hypercyclic subspaces for some sequence of operators on $\omega = \mathbb{K}^{\mathbb{N}}$ were studied by Bès and Conejero [69].

As for the problem mentioned at the end of the section we recall that by a remarkable result of Grivaux [169], any countable family of hypercyclic operators on a Banach space has a common dense subspace of hypercyclic vectors, except for 0. See also Exercise 3.4.10.

Exercises. Exercise 11.1.1 is from Bayart and Matheron [42], Exercise 11.1.2 is due to Borichev (see [42]), Exercise 11.1.5 is taken from Grivaux [169], and Exercises 11.1.6 and 11.1.7 from Chan and Sanders [102]. Exercise 11.2.1 is mentioned in Bayart and Matheron [42], while Exercise 11.2.3 follows Costakis and Mavroudis [120] and Bernal [59]. Exercises 11.3.1 and 11.3.2 are taken from Bayart and Matheron [42], Exercise 11.4.5 from Aron, Bès, León and Peris [14], and Exercise 11.4.10 partially improves a result by Bayart [34].

Chapter 12
Linear dynamics in topological vector spaces

So far, we have been working with operators on Banach or Fréchet spaces. One of the main reasons was that we then had the Baire category theorem at our disposal, which is a basic tool in hypercyclicity.

We have made one exception. In Chapter 10, the left-multiplication operators that we needed were defined on the space $L(X)$ with the strong operator topology, which is not a Fréchet space unless X is finite dimensional. But even there, in the final analysis, we worked in a separable Fréchet space $\mathcal{K}$ of operators on X.

Dealing with more general spaces in which Baire category arguments cannot be applied makes life certainly more difficult for hypercyclicity; but there are several dynamical properties, like mixing or weak mixing, where the previous arguments extend, essentially unchanged, to arbitrary topological vector spaces. Also, several interesting and natural operators are defined on non-Fréchet topological vector spaces, which is a good motivation to study linear dynamics in a wider context. This is the purpose of this chapter.

12.1 Topological vector spaces

A *topological vector space* is a vector space X over the scalar field $\mathbb{K} = \mathbb{R}$ or $\mathbb{C}$ endowed with a Hausdorff topology such that addition and scalar multiplication,

$$+ : X \times X \to X, \quad (x, y) \to x + y,$$
$$\cdot : \mathbb{K} \times X \to X, \quad (\lambda, x) \to \lambda x,$$

are continuous maps. We recall that a topology is Hausdorff if any two distinct points in the space have disjoint neighbourhoods.

K.-G. Grosse-Erdmann, A. Peris Manguillot, *Linear Chaos*, Universitext,
DOI 10.1007/978-1-4471-2170-1_12, © Springer-Verlag London Limited 2011

Many arguments in Banach and Fréchet spaces use the triangle inequality of the norm or the seminorms. In general topological vector spaces, such arguments are replaced by operations with 0-neighbourhoods.

A subset A of a vector space X is called *balanced* if $\lambda A \subset A$ whenever $\lambda \in \mathbb{K}$, $|\lambda| \leq 1$.

Proposition 12.1. *Let X be a topological vector space.*

(a) *A set U is a neighbourhood of a point $x \in X$ if and only if there is a 0-neighbourhood W such that*

$$x + W \subset U.$$

(b) *Let W be a 0-neighbourhood. For any $\lambda, \mu \in \mathbb{K}$ there is a 0-neighbourhood W_1 such that*

$$\lambda W_1 + \mu W_1 \subset W.$$

In particular, there is a 0-neighbourhood W_1 such that

$$W_1 + W_1 \subset W \quad \textit{and} \quad W_1 - W_1 \subset W.$$

(c) *If W is a 0-neighbourhood and $M > 0$, then there is a 0-neighbourhood $W_1 \subset W$ such that $\lambda W_1 \subset W$ for every $\lambda \in \mathbb{K}$ with $|\lambda| \leq M$. In particular, every 0-neighbourhood contains a balanced 0-neighbourhood.*

Proof. Properties (a) and (b) are easy consequences of the continuity of the vector operations. For property (c), given W and M, by the continuity of scalar multiplication we can find $\varepsilon > 0$ and a 0-neighbourhood W_0 such that $\lambda W_0 \subset W$ for every $\lambda \in \mathbb{K}$ with $|\lambda| < \varepsilon$. Let $\delta = \varepsilon/(M+1)$, and consider $W_1 = \delta W_0$, which is a 0-neighbourhood since multiplication by a fixed nonzero scalar is a homeomorphism of X. If $|\lambda| \leq M$ then $\lambda W_1 = (\lambda\delta) W_0 \subset W$. As a consequence, $\bigcap_{|\lambda|\geq 1} \lambda W$ is a balanced 0-neighbourhood contained in W. □

Let us apply the proposition to obtain some basic facts.

Proposition 12.2. *Let X be a topological vector space.*

(a) *If A is an arbitrary subset of X and U an open set then $A+U$ is open.*

(b) *For any 0-neighbourhood W there is a 0-neighbourhood W_1 such that*

$$\overline{W_1} \subset W;$$

in particular, every 0-neighbourhood contains a closed 0-neighbourhood.

Proof. (a) Let $x = y + z$, $y \in A$, $z \in U$. Since U is open there is a 0-neighbourhood W such that $z + W \subset U$. Then $x + W = y + (z + W) \subset A + U$, so that x is an interior point of $A + U$. This proves the claim.

(b) By Proposition 12.1(b) there is a 0-neighbourhood W_1 such that $W_1 - W_1 \subset W$. Let $x \in \overline{W_1}$. Then $(x + W_1) \cap W_1 \neq \varnothing$, hence $x \in W_1 - W_1 \subset W$. □

In view of Propositions 12.1(c) and 12.2(b), the set of all closed and balanced 0-neighbourhoods of a topological vector space X is a base of 0-neighbourhoods in X, which will be denoted by $\mathcal{U}_0(X)$.

There are some classes of topological vector spaces that deserve special consideration.

To start with, any finite-dimensional topological vector space X is isomorphic to $\mathbb{K}^N$, for some $N \geq 0$, where $\mathbb{K}$ is the scalar field of X; see Exercise 12.1.2.

If a topological vector space X admits a countable base $(W_n)_n$ of 0-neighbourhoods, then there is a translation-invariant metric d on X generating the topology of X. If, moreover, (X, d) is complete, then X is called an *F-space.* Metrizable topological vector spaces are, thus, exactly the topological vector spaces admitting a countable base of 0-neighbourhoods, and the completion $\widehat{X}$ of a metrizable topological vector space is an F-space. See also the related discussion in Section 2.1.

A topological vector space X whose topology is defined by a family of seminorms is called a *locally convex space*; that is, X is locally convex if there is a family $(p_\alpha)_{\alpha\in\mathcal{A}}$ of seminorms on X such that a subset W of X is a 0-neighbourhood if and only if there are $\alpha_1, \ldots, \alpha_n \in \mathcal{A}$, $n \geq 1$, and $\varepsilon > 0$ such that

$$\{x \in X\ ;\ p_{\alpha_k}(x) < \varepsilon \ \text{ for } k = 1, \ldots, n\} \subset W.$$

Fréchet spaces are, precisely, the locally convex F-spaces.

A subset A of a vector space X is called *absolutely convex* if, for any $x_1, x_2 \in A$ and $\lambda_1, \lambda_2 \in \mathbb{K}$ with $|\lambda_1| + |\lambda_2| \leq 1$, the absolutely convex combination $\lambda_1 x_1 + \lambda_2 x_2$ belongs to A. An easy observation is that, if $A \subset X$ is absolutely convex, $x_k \in A$, $\lambda_k \in \mathbb{K}$, $k = 1, \ldots, n$, with $\sum_{k=1}^n |\lambda_k| \leq 1$, then

$$\sum_{k=1}^{n} \lambda_k x_k \in A.$$

Also, if p is a seminorm on X and $M \geq 0$, then the set $A = \{x \in X\ ;\ p(x) \leq M\}$ is absolutely convex. Conversely, if $A \subset X$ is an absolutely convex set, then the associated *gauge* of A, also called its *Minkowski functional*, is defined as

$$p_A(x) = \inf\{\lambda > 0\ ;\ x \in \lambda A\}, \quad x \in \operatorname{span} A.$$

One can verify that p_A is a seminorm on $\operatorname{span} A$; see Exercise 12.1.5. Therefore, a topological vector space X is locally convex if, and only if, it has a base of 0-neighbourhoods $(W_\alpha)_{\alpha\in\mathcal{A}}$ consisting of absolutely convex sets.

Example 12.3. In Sections 8.3 and 10.2 we considered the space $L(X)$ of operators on a Fréchet space X, endowed with the strong operator topology (SOT). In this topology, a base of neighbourhoods of $T \in L(X)$ is given by

$$U_{x_1,\ldots,x_n}(T, \varepsilon) = \{S \in L(X)\ ;\ \|Tx_k - Sx_k\| < \varepsilon \ \text{ for } k = 1, \ldots, n\},$$

where $x_1, \ldots, x_n$, $n \geq 1$, is an arbitrary collection of linearly independent vectors of X, $\|\cdot\|$ is an F-norm defining the topology of X, and $\varepsilon > 0$. We immediately obtain that $L(X)$ with the strong operator topology is a locally convex space.

Example 12.4. Given $0 < p < 1$ and $a < b$, let

$$L^p[a,b] = \Big\{ f : [a,b] \to \mathbb{K} \ ; f \text{ is measurable and } \int_a^b |f(t)|^p \, dt < \infty \Big\}.$$

We set $W_n = \{f \in L^p[a,b] \ ; \ \int_a^b |f(t)|^p \, dt < 1/n\}$, $n \in \mathbb{N}$. Then the sequence $(W_n)_n$ defines a base of 0-neighbourhoods and, by translation, a topology on $L^p[a,b]$ that makes it an F-space that is not locally convex; see Exercise 12.1.3. Therefore, $L^p[a,b]$ is not a Fréchet space.

Example 12.5. Let $(X_n)_n$ be an increasing sequence of Banach spaces such that each inclusion map $i_n : X_n \to X_{n+1}$, $n \geq 1$, is continuous. We consider $X = \bigcup_{n=1}^\infty X_n$. For each sequence $\delta = (\delta_n)_n$ of strictly positive numbers, let

$$W_\delta = \bigcup_{n=1}^{\infty} \sum_{k=1}^{n} \delta_k B_k,$$

where B_k is the open unit ball of X_k, $k \in \mathbb{N}$. The family of absolutely convex sets $\{W_\delta \ ; \ \delta = (\delta_n)_n \in \,]0,\infty[^{\mathbb{N}}\}$ forms a base of 0-neighbourhoods for a locally convex topology on X, called the *inductive limit* of $(X_n)_n$; see Exercise 12.1.7.

12.2 Hypercyclicity, topological transitivity, and linear chaos

We are now in a position to study linear dynamics in its widest possible framework, that of operators on arbitrary topological vector spaces.

In the sequel we will not always define a notion when its generalization from the Fréchet space setting is evident. Still, we cannot help stating the following.

Definition 12.6. An operator T on a topological vector space X is called *hypercyclic* if there is some $x \in X$ whose orbit under T is dense in X. In such a case, x is called a *hypercyclic vector* for T. The set of hypercyclic vectors is denoted by $HC(T)$.

Clearly, separability of a space is again a necessary condition for the existence of a hypercyclic operator. Moreover, any finite-dimensional topological vector space is isomorphic to some $\mathbb{K}^N$ and therefore cannot support a hypercyclic operator. But unlike for the case of Fréchet spaces there are infinite-dimensional separable topological vector spaces that do not admit any hypercyclic operator.

Example 12.7. We consider the space φ of finite sequences,

$$\varphi = \{(x_n)_n \in \mathbb{K}^{\mathbb{N}} \; ; \text{ there is some } m \in \mathbb{N} \text{ such that } x_n = 0 \text{ for all } n > m\}.$$

The space φ has a natural locally convex topology, which is the strongest one that can be defined on it; it is generated by the family of norms

$$\|x\|_v = \sum_{n=1}^{\infty} |x_n| v_n, \quad x \in \varphi,$$

where $v = (v_n)_n$ is an arbitrary sequence of strictly positive numbers. Now let $x \in \varphi$ be a hypercyclic vector for an operator T on φ. We set $E_n = \{x \in \varphi \; ; \; x_k = 0 \text{ for } k > n\}$, $n \in \mathbb{N}$. Suppose that each E_n contains only a finite number of elements of the orbit of x. We define $F_1 = \operatorname{orb}(x,T) \cap (E_1 \setminus \{0\})$ and $F_n = \operatorname{orb}(x,T) \cap (E_n \setminus E_{n-1})$, $n > 1$. Then each F_n is finite, $\operatorname{orb}(x,T) = \bigcup_{n=1}^{\infty} F_n$, and every element $y \in F_n$ satisfies $y_n \neq 0$. We can therefore define

$$v_n = \frac{1}{\min\{|y_n| \; ; \; y \in F_n\}}, \qquad n \in \mathbb{N},$$

if F_n is nonempty, and $v_n = 1$ otherwise. Considering the sequence $v = (v_n)_n$ we then find that $\|y\|_v \geq 1$ for every $y \in \operatorname{orb}(x,T)$, which contradicts the hypercyclicity of x. Therefore, some E_n, $n \geq 1$, must contain an infinite number of elements of $\operatorname{orb}(x,T)$, which is impossible because E_n is finite dimensional and the vectors in a dense orbit are linearly independent; note that Proposition 2.60 continues to hold. Consequently, φ admits no hypercyclic operator.

By the Birkhoff transitivity theorem, an operator on a separable Fréchet space is hypercyclic if and only if it is topologically transitive. One implication remains true since no topological vector space has isolated points.

Observation 12.8. *Any hypercyclic operator on a topological vector space is topologically transitive.*

But the converse is no longer true, as the following example shows.

Example 12.9. We consider again the space $X = \varphi$ of finite sequences, but this time endowed with the topology inherited from ℓ^2. Consider the multiple of the backward shift operator $T = 2B : X \to X$, $B(x_n)_n = 2(x_{n+1})_n$. Then T is topologically transitive, and even mixing, because Rolewicz's operator is. On the other hand, T cannot be hypercyclic since the orbit of any vector in X is finite.

Example 12.10. For a separable Banach space X, let us consider the space $L(X)$ of operators on X, endowed with the strong operator topology. In Theorem 10.20 we proved that an operator T on X satisfies the Hypercyclicity Criterion if, and only if, the left-multiplication operator $L_T : L(X) \to L(X)$, $S \to TS$, is hypercyclic. This characterization provides a good collection of hypercyclic operators on the non-metrizable locally convex space $L(X)$.

Since hypercyclicity and topological transitivity no longer coincide, we adopt Devaney's original definition of chaos in the general setting.

Definition 12.11 (Linear chaos). An operator T on a topological vector space X is said to be *chaotic* if it satisfies the following conditions:
(i) T is topologically transitive;
(ii) T has a dense set of periodic points.

We recall the useful result, Proposition 2.33, that the set $\mathrm{Per}(T)$ of periodic points for an operator T on a complex space X is given by

$$\mathrm{Per}(T) = \mathrm{span}\{x \in X \ ; \ Tx = e^{\alpha\pi i}x \text{ for some } \alpha \in \mathbb{Q}\},$$

whose density can often be checked easily for a concrete operator T.

Example 12.12. Let T be a chaotic operator on a separable Banach space X. Then there is a countable dense set $E \subset X$ such that each element of E is a periodic point for T. In the proof of Proposition 10.14 we showed that, if Φ is a countable weak-$*$-dense subset of X^*, then the countable set

$$\mathcal{F} = \mathcal{F}_{\Phi,E} = \Big\{\sum_{j=1}^{m}\langle\,\cdot\,, y_j^*\rangle e_j \ ; \ y_j^* \in \Phi, e_j \in E, 1 \le j \le m\Big\}$$

of operators on X is SOT-dense in $L(X)$. We have that every element of $\mathcal{F}$ is periodic for the left-multiplication operator L_T on $L(X)$. On the other hand, hypercyclicity, and therefore topological transitivity, of L_T follows from Theorem 10.20 and the fact that chaotic operators satisfy the Hypercyclicity Criterion; see Theorem 3.18. We thus conclude that L_T is chaotic.

We next want to show that many fundamental results for hypercyclic operators on Fréchet spaces extend to arbitrary topological vector spaces. For this we need the notion of a quotient space.

Let X be a vector space and $L \subset X$ a subspace. Defining $x \sim y$ if $x-y \in L$, we obtain an equivalence relation on X. Let us denote by $[x] = x + L$ the equivalence class of $x \in X$, and by X/L the set of equivalence classes. Then X/L inherits in a natural way a vector space structure, and we denote by $q : X \to X/L$, $x \to [x]$, the *quotient map*, which is linear and surjective.

If, now, X is a topological vector space and $L \subset X$ is a closed subspace, then X/L becomes a topological vector space, called the *quotient space* of X modulo L, when endowed with the induced topology:

$$U \subset X/L \text{ is open if and only if there is an open set } \widetilde{U} \subset X \text{ with } q(\widetilde{U}) = U.$$

The quotient map q is then a continuous and open map. The requirement that L be closed is necessary for the Hausdorff property of X/L; see Exercise 12.2.5.

The following result, which generalizes Bourdon's theorem, is the key to the announced extensions.

Lemma 12.13 (Wengenroth). *Let T be an operator on topological vector space X. If either*

(i) T is topologically transitive, or

(ii) T has a somewhere dense orbit,

then, for any nonzero polynomial p, the operator $p(T)$ has dense range.

Proof. We will only show the complex case. The real case can be deduced in a similar way after some minor considerations; see Exercise 12.2.6.

As in the proof of Bourdon's theorem it suffices to show that $T - \lambda I$ has dense range for every $\lambda \in \mathbb{C}$. Let $L = \overline{(T - \lambda I)(X)}$, which is a closed subspace of X, and suppose that $L \neq X$. We then consider the quotient space X/L, which is nontrivial, and the quotient map $q : X \to X/L$. Since, for any $x \in X$, $q((T - \lambda I)x) = 0$ we have that $q(Tx) = \lambda q(x)$. Hence the operator S on X/L given by $S[x] = \lambda[x]$ is quasiconjugate to T via q and therefore inherits the stated properties from T; see the following section.

Under assumption (i), S is topologically transitive. On the other hand, let $[x] \in X/L$, $[x] \neq 0$. By the Hausdorff property there is an open neighbourhood U of $[x]$ and a balanced 0-neighbourhood W such that $U \cap W = \varnothing$. Now, if $|\lambda| \geq 1$, then $W \subset \lambda^n W$, and therefore $S^n(U) \cap W = \varnothing$ for all $n \in \mathbb{N}_0$. And if $|\lambda| < 1$, then $\lambda^n W \subset W$, and therefore $S^n(W) \cap U = \varnothing$ for all $n \in \mathbb{N}_0$. Thus, for any $\lambda \in \mathbb{C}$, S is not topologically transitive, a contradiction.

Under assumption (ii), S has a somewhere dense orbit $\{\lambda^n[x] \; ; \; n \in \mathbb{N}_0\}$. Then $\text{span}\{[x]\} = \overline{\text{span}}\{[x]\} = X/L$ (see Exercises 12.1.1(v) and 12.1.2), so that X/L is isomorphic to $\mathbb{C}$. But every orbit $\{\lambda^n z \; ; \; n \in \mathbb{N}_0\}$, $z \in \mathbb{C}$, is nowhere dense, a contradiction. □

Now, looking back at the proofs of the following fundamental results we see that they work unrestrictedly once one has Wengenroth's lemma at hand. They therefore hold for operators on all topological vector spaces.

Herrero–Bourdon theorem. *Any hypercyclic operator admits a dense invariant subspace consisting, except for zero, of hypercyclic vectors.*

Ansari's theorem. *Any power of a hypercyclic operator is hypercyclic.*

Costakis–Peris theorem. *Any multi-hypercyclic operator is hypercyclic.*

Bourdon–Feldman theorem. *Any somewhere dense orbit is (everywhere) dense.*

Indeed, each result holds in the more detailed form given in Chapters 2 and 6.

12.3 Dynamical transference principles

Working with general topological vector spaces instead of F-spaces sometimes requires some abstract considerations; for example, instead of sequences, balls, or the distance one needs to use notions like nets or neighbourhoods. In addition, of course, one has to do without the Baire category theorem. There do exist topological vector spaces beyond F-spaces in which Baire's theorem holds but they are rare.

In this section we want to discuss three techniques that allow us to transfer dynamical properties from operators on F-spaces to operators on general topological vector spaces.

The first technique is by now well known, that of quasiconjugacies. As before, if X and Y are topological vector spaces then an operator T on X is called *quasiconjugate* to an operator S on Y via a continuous map $\phi : Y \to X$ with dense range if $T \circ \phi = \phi \circ S$. Then the usual notions of linear dynamics are preserved under quasiconjugacy: hypercyclicity, topological transitivity, (weak) mixing, chaos, frequent hypercyclicity, etc. Moreover, if $y \in Y$ is a hypercyclic vector for S, then $x := \phi(y)$ is hypercyclic for T.

A particular case of quasiconjugacy that frequently occurs naturally is when one can find a T-invariant dense subspace $Y \subset X$ that carries its own, not necessarily the induced, vector space topology such that the restriction $T|_Y$ is an operator Y. If, in addition, the embedding $Y \to X$ is continuous, then T is quasiconjugate to $T|_Y$, so that T inherits dynamical properties from $T|_Y$. This is commonly known as the *hypercyclic comparison principle*; see Exercise 2.2.6.

Now, if Y is, in particular, an F-space, then the results of the previous chapters can be applied. We illustrate this by an example.

Example 12.14. Let $(X_n)_n$ be an increasing sequence of Banach spaces with continuous inclusions, and let X be the inductive limit of $(X_n)_n$; see Example 12.5. Suppose that T is an operator on X such that, for some $n \geq 1$, X_n is dense in X, $T(X_n) \subset X_n$ and $T|_{X_n}$ is continuous and hypercyclic. Then, by the comparison principle, T is hypercyclic.

As a particular case, let $1 < p \leq \infty$, and consider the space $\ell^{p-} := \bigcup_{q<p} \ell^q$. Obviously, $\ell^{p-} = \bigcup_{n=1}^{\infty} \ell^{p_n}$ for any strictly increasing sequence $(p_n)_n$ in $]1, p[$ tending to p. A natural topology on ℓ^{p-} is the corresponding inductive limit topology. If $\lambda \in \mathbb{K}$ is any scalar with $|\lambda| > 1$, then the multiple $T = \lambda B$ of the backward shift satisfies the above requirements and is therefore hypercyclic on ℓ^{p-}.

The second method is a kind of converse to the first technique. Of course, if, for a given operator T, all operators S that are quasiconjugate to T are hypercyclic then T itself must be hypercyclic; one may simply take $S = T$. It is, however, remarkable that the result remains true when we only admit operators S defined on F-spaces. In addition, the map ϕ defining the quasiconjugacy may be required to be linear.

Proposition 12.15. *Let T be an operator on a topological vector space X, and $x \in X$.*

(a) *If every operator S, defined on an F-space and quasiconjugate to T via a linear map, is hypercyclic (topologically transitive, weakly mixing, or mixing), then the same holds for T.*

(b) *If for any operator S, defined on an F-space and quasiconjugate to T via a linear map ϕ, $\phi(x)$ is (frequently) hypercyclic for S, then x is (frequently) hypercyclic for T.*

(c) *If X is a locally convex space then it suffices in* (a) *and* (b) *to allow operators S on Fréchet spaces.*

Proof. We will only show assertion (a) for the mixing property; the remaining cases follow similarly.

Let $U, V \subset X$ be arbitrary nonempty open subsets. Let $x_1 \in U$, $x_2 \in V$, and choose $W \in \mathcal{U}_0(X)$ such that $x_1 + W \subset U$ and $x_2 + W \subset V$. By continuity, we obtain a decreasing sequence $(W_n)_n$ of closed balanced 0-neighbourhoods such that $W_1 = W$, $W_{n+1} + W_{n+1} \subset W_n$, and $T(W_{n+1}) \subset W_n$, $n \in \mathbb{N}$. Let $L := \bigcap_{n=1}^{\infty} W_n$, which is easily seen to be a closed T-invariant subspace of X. We set $Y' = X/L$, and endow it with the topology τ generated by the family of neighbourhoods $\{y' + \widetilde{W}_n \ ; \ y' \in Y', \ n \in \mathbb{N}\}$, where $\widetilde{W}_n$ is the image of W_n under the quotient map $q : X \to Y'$, $n \in \mathbb{N}$. It is routine to verify that (Y', τ) is a topological vector space, which is metrizable since it has a countable base of 0-neighbourhoods, and that the operator T induces an operator $S' : Y' \to Y'$ that is quasiconjugate to T via q.

Now let Y be the completion of (Y', τ), which is an F-space, $S : Y \to Y$ the extension of S' to the completion, and $\phi : X \to Y$ the operator induced by q, which has dense range. It is clear that S is quasiconjugate to T via the linear map ϕ. It then follows from the assumption that S is a mixing operator. Therefore, also S' is mixing, so that there is some $N \in \mathbb{N}_0$ such that

$$q(T^n(x_1 + W_2)) \cap q(x_2 + W_2) = (S')^n \left(q(x_1) + \widetilde{W}_2\right) \cap \left(q(x_2) + \widetilde{W}_2\right) \neq \varnothing$$

for every $n \geq N$. This implies that

$$T^n(U) \cap V \supset T^n(x_1 + W) \cap (x_2 + W) \supset T^n(x_1 + W_2) \cap (x_2 + W_2 + L) \neq \varnothing$$

for every $n \geq N$, so that T is mixing.

In the case that X is a locally convex space, the 0-neighbourhoods W_n, $n \in \mathbb{N}$, can be chosen to be absolutely convex, and Y is a Fréchet space. □

An application of this result yields the generalization of the León–Müller theorem to arbitrary complex topological vector spaces.

Corollary 12.16. *Let T be an operator on a complex topological vector space X. Then, for any $\lambda \in \mathbb{C}$ with $|\lambda| = 1$, T and λT have the same hypercyclic vectors, that is, $HC(T) = HC(\lambda T)$.*

Proof. Let $x \in HC(T)$ and $\lambda \in \mathbb{C}$ with $|\lambda| = 1$. By Proposition 12.15 it suffices to show that every operator S, defined on an arbitrary F-space Y and quasiconjugate to λT via a linear map $\phi : X \to Y$, has $\phi(x)$ as a hypercyclic vector. But under these assumptions, $\lambda^{-1}S$ is quasiconjugate to T via ϕ, so that $\phi(x)$ is a hypercyclic vector for $\lambda^{-1}S$. Since Y is an F-space, we can apply Theorem 6.7; note that its proof also works in arbitrary F-spaces. We then obtain that $\phi(x)$ is hypercyclic for S, as demanded. □

As another application of Proposition 12.15 we show that the space φ of finite sequences supports a mixing operator, which is surprising in view of the fact that φ does not admit any hypercyclic operator.

Example 12.17. We claim that the operator $T = I + B$ on φ is mixing, where B is the backward shift. Thus, let $S : Y \to Y$ be an operator on an arbitrary Fréchet space Y that is quasiconjugate to T via a linear map $\phi : \varphi \to Y$. We fix an increasing sequence of seminorms $(p_n)_n$ on Y generating its topology; by the continuity of ϕ there are strictly positive sequences $v(n) = (v_{n,k})_k$, $n \in \mathbb{N}$, such that $p_n(\phi(x)) \leq \|x\|_{v(n)}$, for any $n \in \mathbb{N}$ and $x \in \varphi$; see Example 12.7. Defining $v = (v_k)_k$ by $v_k = \max_{n,m\leq k} v_{n,m}$, $k \in \mathbb{N}$, a simple calculation shows that there are constants $M_n > 0$ such that $p_n(\phi(x)) \leq M_n\|x\|_v$ for any $n \in \mathbb{N}$, $x \in \varphi$. Since Y is complete, there is a continuous extension $\phi : \ell^1(v) \to Y$, and S is quasiconjugate to $I + B : \ell^1(v) \to \ell^1(v)$ via ϕ. Since, by Theorem 8.2, $I + B$ is mixing on $\ell^1(v)$, so is S on Y. Proposition 12.15 then implies that T is mixing on φ.

We turn to the third transference principle. For this we need a new concept. A *projective spectrum* $\mathcal{X}$ of Fréchet spaces consists of a family $(X_\alpha)_{\alpha\in I}$ of Fréchet spaces, where I is a directed index set, and operators $\varrho^\alpha_\beta : X_\beta \to X_\alpha$ for $\alpha \leq \beta$, called the *spectral maps*, that satisfy $\varrho^\alpha_\beta \circ \varrho^\beta_\gamma = \varrho^\alpha_\gamma$ and $\varrho^\alpha_\alpha = I_{X_\alpha}$, the identity on X_α, for any $\alpha \leq \beta \leq \gamma$. The *projective limit* of $\mathcal{X}$ is defined as

$$\operatorname{proj} \mathcal{X} = \Big\{(x_\alpha)_{\alpha\in I} \in \prod_{\alpha\in I} X_\alpha \ ; \ \varrho^\alpha_\beta x_\beta = x_\alpha \text{ for all } \alpha \leq \beta\Big\},$$

endowed with the topology inherited from the product topology on $\prod_{\alpha\in I} X_\alpha$; in this way, $\operatorname{proj} \mathcal{X}$ is a locally convex space. We denote by $\varrho^\alpha : \operatorname{proj} \mathcal{X} \to X_\alpha$ the projection onto the component with index α. It is not difficult to see that the sets $(\varrho^\alpha)^{-1}(W_\alpha)$, $W_\alpha \in \mathcal{U}_0(X_\alpha)$, $\alpha \in I$, form a base of 0-neighbourhoods for the topology of $\operatorname{proj} \mathcal{X}$. We say that $\mathcal{X}$ is *strongly reduced* if for each α there is a larger β such that $\varrho^\alpha_\beta(X_\beta)$ is contained in the closure of $\varrho^\alpha(\operatorname{proj} \mathcal{X})$ in X_α.

Now, a family $(T_\alpha)_{\alpha\in I}$ of operators T_α on X_α is called an *endomorphism* of $\mathcal{X}$ if their elements commute with the spectral maps in the sense that, for any $\alpha \leq \beta$, $T_\alpha \circ \varrho^\alpha_\beta = \varrho^\alpha_\beta \circ T_\beta$. The *projective limit* of the endomorphism is the operator T on $\operatorname{proj} \mathcal{X}$ defined by $T(x_\alpha)_{\alpha\in I} = (T_\alpha x_\alpha)_{\alpha\in I}$.

Proposition 12.18. *Let $\mathcal{X}$ be a strongly reduced projective spectrum of Fréchet spaces, $(T_\alpha)_{\alpha\in I}$ an endomorphism of $\mathcal{X}$, and T its projective limit.*

(a) *If every T_α, $\alpha \in I$, is topologically transitive (mixing, weakly mixing) on X_α then T is topologically transitive (mixing, weakly mixing) on* $\operatorname{proj}\mathcal{X}$.

(b) *If $x \in \operatorname{proj}\mathcal{X}$ is such that, for every $\alpha \in I$, $\varrho^\alpha x \in X_\alpha$ is (frequently) hypercyclic for T_α then x is (frequently) hypercyclic for T.*

Proof. (a) We will only show the result for the mixing property. Let $x, y \in X := \operatorname{proj}\mathcal{X}$ and $W_0 \in \mathcal{U}_0(X)$ be given. Then there are $\alpha \in I$ and $W_1 \in \mathcal{U}_0(X_\alpha)$ with $W_0 \supset (\varrho^\alpha)^{-1}(W_1)$, and there is some $\beta \geq \alpha$ with $\varrho^\alpha_\beta X_\beta \subset \varrho^\alpha(X)$. For each $W \in \mathcal{U}_0(X_\alpha)$ we obtain that $\varrho^\alpha_\beta(X_\beta) \subset \varrho^\alpha(X) + W$, and thus $X_\beta \subset \varrho^\beta(X) + (\varrho^\alpha_\beta)^{-1}(W)$. This means that the image of ϱ^β is dense in X_β with respect to the vector space topology τ having $\{(\varrho^\alpha_\beta)^{-1}(W)\ ;\ W \in \mathcal{U}_0(X_\alpha)\}$ as a base of 0-neighbourhoods. Moreover, T_β is continuous on (X_β, τ) since $T_\beta^{-1}((\varrho^\alpha_\beta)^{-1}(W)) = (\varrho^\alpha_\beta)^{-1}(T_\alpha^{-1}(W)) \in \tau$ for every $W \in \mathcal{U}_0(X_\alpha)$, and T_β is mixing on (X_β, τ) since τ is coarser than the original topology on X_β.

Hence there is some $N \in \mathbb{N}_0$ such that

$$U_n := (\varrho^\beta x + (\varrho^\alpha_\beta)^{-1}(W_1)) \cap (T_\beta^{-n}(\varrho^\beta y + (\varrho^\alpha_\beta)^{-1}(W_1))) \neq \varnothing$$

for all $n \geq N$. Since U_n is open with respect to τ, there are $z_n \in X$ such that $\varrho^\beta z_n \in U_n$ for all $n \geq N$. Then we have that $\varrho^\alpha(z_n - x) = \varrho^\alpha_\beta(\varrho^\beta z_n - \varrho^\beta x) \in W_1$, that is, $z_n \in x + W_0$, and

$$\varrho^\alpha(T^n z_n) = \varrho^\alpha_\beta(\varrho^\beta(T^n z_n)) = \varrho^\alpha_\beta(T^n_\beta(\varrho^\beta z_n)) \in \varrho^\alpha y + W_1,$$

which gives that $T^n z_n \in y + W_0$ for every $n \geq N$. This proves that

$$(x + W_0) \cap (T^{-n}(y + W_0)) \neq \varnothing,$$

for each $n \geq N$, and therefore that T is mixing on X.

(b) Now let $x \in X$ be such that, for every $\alpha \in I$, $\varrho^\alpha x \in X_\alpha$ is hypercyclic for T_α. Let $y \in X$ and $W_0 \in \mathcal{U}_0(X)$ be given. Then there are $\alpha \in I$ and $W_1 \in \mathcal{U}_0(X_\alpha)$ with $W_0 \supset (\varrho^\alpha)^{-1}(W_1)$. It follows that there is some $n \in \mathbb{N}_0$ such that $\varrho^\alpha T^n x = T^n_\alpha \varrho^\alpha x \in \varrho^\alpha y + W_1$, hence $T^n x - y \in (\varrho^\alpha)^{-1}(W_1) \subset W_0$. This shows that x is hypercyclic for T. The same argument also shows the claim for frequent hypercyclicity. □

We single out a particular case of this result. Let $(X_n)_n$ be a decreasing sequence of Fréchet spaces such that each inclusion map $i_n : X_{n+1} \to X_n$, $n \geq 1$, is continuous. For any $n \geq 1$, let $(p_{n,k})_k$ be an increasing sequence of seminorms defining the topology of X_n. We then consider the space $X := \bigcap_{n=1}^\infty X_n$ with the locally convex topology induced by the seminorms $p_k(x) := \max_{n\leq k} p_{n,k}(x)$, $x \in X$. Then X is a Fréchet space, also called the *projective limit* of $(X_n)_n$.

Taking the inclusion maps as spectral maps, we see that $(X_n)_n$ is also a projective spectrum, and $\operatorname{proj}\mathcal{X}$ is the space of constant sequences with entries from $X = \bigcap_{n=1}^\infty X_n$. It is then clear that $\operatorname{proj}\mathcal{X}$ is isomorphic to X. Moreover, the projective spectrum is strongly reduced if, for example, X is dense in each X_n, $n \geq 1$.

Corollary 12.19. *Let $(X_n)_n$ be a decreasing sequence of Fréchet spaces with continuous inclusion maps such that $X = \bigcap_{n=1}^\infty X_n$ is dense in X_n for all $n \geq 1$. Let $T : X \to X$ be an operator that can be extended to an operator $T_n : X_n \to X_n$ for any $n \geq 1$.*

(a) *If every T_n, $n \geq 1$, is topologically transitive (mixing, weakly mixing) on X_n then T is topologically transitive (mixing, weakly mixing) on X.*

(b) *If $x \in X$ is (frequently) hypercyclic for every T_n, $n \geq 1$, then x is (frequently) hypercyclic for T.*

Proof. This follows directly from Proposition 12.18 because, by the assumptions, $(T_n)_n$ is an endomorphism of $\operatorname{proj}\mathcal{X}$ and the projective limit of $(T_n)_n$ on $\operatorname{proj}\mathcal{X}$ turns into T via the identification of $\operatorname{proj}\mathcal{X}$ with X. □

We note that, by quasiconjugacy, the conditions in the corollary are also necessary.

Example 12.20. Let $X = L^{\infty-}[0,1] := \bigcap_{p<\infty} L^p[0,1]$ be endowed with the Fréchet space topology induced by the increasing sequence of norms $(p_n)_n$, where p_n is the norm of $L^n[0,1]$, $n \in \mathbb{N}$. Let $C : L^{\infty-}[0,1] \to L^{\infty-}[0,1]$ be the Cesàro operator given by $Cf(t) = \frac{1}{t}\int_0^t f(s)\,ds$. By Exercise 3.1.4, C is mixing on $L^p[0,1]$ for any $1 < p < \infty$. Hence C is mixing on $L^{\infty-}[0,1]$.

12.4 Mixing and weakly mixing operators

In this section we will convince ourselves that the central results of Sections 2.4 and 2.5 remain true in general topological vector spaces when we replace the assumption of hypercyclicity there by topological transitivity. We will omit the proofs when the arguments given in those sections translate directly to the general situation.

Proposition 12.21. *An operator T on a topological vector space X is mixing if and only if, for any nonempty open set $U \subset X$ and any 0-neighbourhood W, the sets*

$$N(U,W) \text{ and } N(W,U)$$

are cofinite.

Example 12.22. We pointed out in Example 12.17 that there are mixing operators on φ. Another easy example of a mixing operator is $T = \lambda B$, $|\lambda| > 1$, on the inductive limit ℓ^{p-} for $1 < p \leq \infty$; see Example 12.14. Also, if X is an arbitrary topological vector space, then the product space $X^{\mathbb{N}}$, endowed with the product topology, is a topological vector space, and the backward shift $B : X^{\mathbb{N}} \to X^{\mathbb{N}}$, $(x_1, x_2, \ldots) \to (x_2, x_3, \ldots)$ is a mixing operator. Thus, if I is an infinite set then the space X^I, endowed with the product topology, is isomorphic to $(X^I)^{\mathbb{N}}$ and therefore admits a mixing operator.

We turn to the weak mixing property. As before, under mild additional assumptions, topologically transitive operators turn out to be weakly mixing. We start with a useful auxiliary result.

Lemma 12.23. *Let T be a topologically transitive operator on a topological vector space X. Then, for any nonempty open sets U and V in X and for any 0-neighbourhood W, there is a nonempty open set $U_1 \subset U$ and a 0-neighbourhood $W_1 \subset W$ such that*

$$N(U_1, W_1) \subset N(V, W) \quad \text{and} \quad N(W_1, U_1) \subset N(W, V).$$

From this we can deduce the main result of this section.

Theorem 12.24. *Let T be a topologically transitive operator on a topological vector space X. If, for any nonempty open set $U \subset X$ and any 0-neighbourhood W, there is a continuous map $S : X \to X$ commuting with T such that*

$$S(U) \cap W \neq \varnothing \text{ and } S(W) \cap U \neq \varnothing,$$

then T is weakly mixing.

Recall that an operator T is flip transitive if, for any pair $U, V \subset X$ of nonempty open sets, $N(U, V) \cap N(V, U) \neq \varnothing$. Thus we have in particular:

Corollary 12.25. *Every flip transitive operator is weakly mixing.*

Theorem 12.24 also implies that Theorem 2.47 extends to general topological vector spaces.

Theorem 12.26. *An operator T on a topological vector space X is weakly mixing if and only if, for any nonempty open sets $U, V \subset X$ and any 0-neighbourhood W,*

$$N(U, W) \cap N(W, V) \neq \varnothing.$$

Another application of Theorem 12.24 provides us with a useful sufficient condition for a topologically transitive operator to be weakly mixing; see Exercise 12.2.4 for the notion of a bounded set.

Theorem 12.27. *Let T be a topologically transitive operator on a topological vector space X. If there exists a dense subset X_0 of X such that the orbit of each $x \in X_0$ is bounded, then T is weakly mixing.*

Proof. In order to adapt the proof of Theorem 2.48 one need only note that, for any $x \in X_0$ and any 0-neighbourhood W, there is some $\varepsilon > 0$ such that $\varepsilon T^n x \in W$ for all $n \in \mathbb{N}_0$. □

Recall that the generalized kernel of an operator T is given by $\bigcup_{n=0}^{\infty} \ker T^n$.

Corollary 12.28. *Let T be a topologically transitive operator on a topological vector space X. If one of the following conditions is satisfied:*
(i) *T is chaotic;*
(ii) *T has a dense set of points for which the orbits converge;*
(iii) *T has dense generalized kernel;*
then T is weakly mixing.

As a final application of Theorem 12.24 we will characterize weakly mixing operators by the behaviour of multiples of iterates of T. For the notion of topological transitivity for sequences of operators we refer to Section 1.6.

Theorem 12.29. *Let T be an operator on a topological vector space X and $\lambda, \mu \in \mathbb{K} \setminus \{0\}$ with $\lambda \neq \mu$. Then the following assertions are equivalent:*
(i) *T is weakly mixing;*
(ii) *for any $M > \delta > 0$ and for any $(\lambda_n)_n$ with $\delta \leq |\lambda_n| \leq M$, $n \in \mathbb{N}_0$, the sequence $(\lambda_n T^n)_n$ is topologically transitive;*
(iii) *for any $(\lambda_n)_n$ with $\{\lambda_n \,;\, n \in \mathbb{N}_0\} \subset \{\lambda, \mu\}$, the sequence $(\lambda_n T^n)_n$ is topologically transitive.*

Proof. (i) $\Longrightarrow$ (ii). Given $(\lambda_n)_n$ with $\delta \leq |\lambda_n| \leq M$, $n \in \mathbb{N}_0$, we let $U, V \subset X$ be nonempty open sets. By Exercise 12.1.1 there are nonempty open sets U_1 and V_1 and a 0-neighbourhood W such that $U_1 + W \subset U$ and $V_1 + W \subset V$. By Proposition 12.1, if $L = \max(\frac{1}{\delta}, M)$, then there is a 0-neighbourhood W_1 such that $\lambda W_1 \subset W$, for any $\lambda \in \mathbb{K}$ with $|\lambda| \leq L$. Now, since T is weakly mixing, there are $n \in \mathbb{N}_0$, $u \in U_1$ and $w \in W_1$ such that $T^n u \in W_1$ and $T^n w \in V_1$. Thus $u + \lambda_n^{-1} w \in U_1 + W \subset U$ and $\lambda_n T^n(u + \lambda_n^{-1} w) = \lambda_n T^n u + T^n w \in W + V_1 \subset V$. Hence $(\lambda_n T^n)_n$ is topologically transitive.

(ii) $\Longrightarrow$ (iii) is trivial.

(iii) $\Longrightarrow$ (i). By taking $(\lambda_n)_n$ to be a constant sequence, T is easily seen to be topologically transitive. It therefore suffices to verify the hypothesis of Theorem 12.24. Thus let W be a 0-neighbourhood and $U \subset X$ a nonempty open set. Let $x \in U$. Using the properties of Proposition 12.1 we can find some $M > 0$ and an open neighbourhood $U_1 \subset U$ of x such that

$$U_1 \subset \frac{M(\lambda - \mu)}{\lambda} W, \quad U_1 - U_1 \subset M^{-1} W, \quad \text{and} \quad \frac{\lambda}{\lambda - \mu} U_1 - \frac{\mu}{\lambda - \mu} U_1 \subset U.$$

Let $\alpha = M(\lambda - \mu)$. The hypothesis implies that there is some $n \in \mathbb{N}_0$ such that $\lambda T^n(U_1) \cap \alpha U_1 \neq \varnothing$ and $\mu T^n(U_1) \cap \alpha U_1 \neq \varnothing$; otherwise there would exist a sequence $(\lambda_n)_n$ with entries λ or μ such that $\lambda_n T^n(U_1) \cap \alpha U_1 = \varnothing$ for all $n \in \mathbb{N}_0$. Thus there are $u_1, u_2 \in U_1$ with $T^n(\alpha^{-1}\lambda u_1) \in U_1$ and $T^n(\alpha^{-1}\mu u_2) \in U_1$. Then

$$\alpha^{-1}\lambda u_1 \in \frac{\lambda}{M(\lambda-\mu)}U_1 \subset W, \qquad T^n(\alpha^{-1}\lambda u_1) \in U,$$

and

$$M\alpha^{-1}(\lambda u_1 - \mu u_2) \in U, \quad T^n(M\alpha^{-1}(\lambda u_1 - \mu u_2)) \in M(U_1 - U_1) \subset W.$$

We then conclude with Theorem 12.24. □

12.5 Criteria for weak mixing, mixing and chaos

We extend here the criteria of Chapter 3 to general topological vector spaces. Since the arguments given there can be adapted directly to the general situation we omit the proofs.

Following the same order, we start with the criterion based on a large supply of eigenvectors.

Theorem 12.30 (Godefroy–Shapiro criterion). *Let T be an operator on a topological vector space X. Suppose that the subspaces*

$$X_0 := \operatorname{span}\{x \in X \ ; \ Tx = \lambda x \text{ for some } \lambda \in \mathbb{K} \text{ with } |\lambda| < 1\},$$

$$Y_0 := \operatorname{span}\{x \in X \ ; \ Tx = \lambda x \text{ for some } \lambda \in \mathbb{K} \text{ with } |\lambda| > 1\}$$

are dense in X. Then T is mixing.

If, moreover, X is a complex space and also the subspace

$$Z_0 := \operatorname{span}\{x \in X \ ; \ Tx = e^{\alpha\pi i}x \text{ for some } \alpha \in \mathbb{Q}\}$$

is dense in X, then T is chaotic.

Kitai's Criterion for mixing extends likewise.

Theorem 12.31 (Kitai's criterion). *Let T be an operator on a topological vector space X. If there are dense subsets $X_0, Y_0 \subset X$ and a map $S : Y_0 \to Y_0$ such that, for any $x \in X_0$, $y \in Y_0$,*
(i) $T^n x \to 0$,
(ii) $S^n y \to 0$,
(iii) $TSy = y$,
then T is mixing.

Example 12.32. Let $X = L^p[0,1]$, $0 < p < 1$, be the space of p-integrable functions on $[0,1]$; see Example 12.4. Let $\varphi : [0,1] \to [0,1]$ be the invertible function given by $\varphi(t) = t/2$ if $t \in [0,1/2]$, and $\varphi(t) = (3/2)t - 1/2$ if $t \in]1/2,1]$. We then consider the composition operator $C_\varphi : X \to X$, $C_\varphi f = f \circ \varphi$. The set

$$X_0 = Y_0 = \{f \in X \; ; \; f \text{ is continuous and } f(0) = f(1) = 0\},$$

is dense in X, and for $S : Y_0 \to Y_0$ we choose the composition operator $S = C_{\varphi^{-1}}$, so that $TSf = f$ for all $f \in Y_0$. Moreover, it is easy to check that, if $t \in [0,1[$, then $\lim_{n\to\infty} \varphi^n(t) = 0$, which implies that $C_\varphi^n f \to 0$ for every $f \in X_0$ by the dominated convergence theorem. Analogously, $\lim_{n\to\infty}(\varphi^{-1})^n(t) = 1$ for all $t \in]0,1]$, so that $C_{\varphi^{-1}}^n f \to 0$ for every $f \in Y_0$. An application of Kitai's criterion shows that T is mixing.

Finally, the Hypercyclicity Criterion turns out to be a weak mixing criterion within the general framework.

Theorem 12.33. *Let T be an operator on a topological vector space X. If there are dense subsets $X_0, Y_0 \subset X$, an increasing sequence $(n_k)_k$ of positive integers, and maps $S_{n_k} : Y_0 \to X$, $k \geq 1$, such that, for any $x \in X_0$, $y \in Y_0$,*
(i) $T^{n_k} x \to 0$,
(ii) $S_{n_k} y \to 0$,
(iii) $T^{n_k} S_{n_k} y \to y$,
then T is weakly mixing.

By Example 12.9 then, an operator satisfying this criterion need not be hypercyclic. With this realization that the Hypercyclicity Criterion is a misnomer we conclude the book.

Exercises

Exercise 12.1.1. Let X be a topological vector space. Prove the following assertions:
(i) if $x \in X$ and W is a 0-neighbourhood, then there exists some $M > 0$ and a neighbourhood U of x such that $U \subset MW$;
(ii) if U is a nonempty open set, then there is a 0-neighbourhood W and a nonempty open set $U_1 \subset U$ such that $U_1 + W \subset U$;
(iii) if U is a nonempty open set, $\lambda, \mu \in \mathbb{K}$ with $\lambda + \mu \neq 0$, and $x \in U$, then there is a neighbourhood $U_1 \subset U$ of x such that $\lambda U_1 + \mu U_1 \subset (\lambda + \mu) U$;
(iv) for any $\lambda \in \mathbb{K} \setminus \{0\}$ and $y \in X$, the operators $M_\lambda : X \to X$, $x \to \lambda x$, and $T_y : X \to X$, $x \to x + y$, are homeomorphisms;
(v) if $A \subset X$ is somewhere dense in X then $\operatorname{span} A$ is dense in X.

Exercise 12.1.2. Show that every finite-dimensional topological vector space X over the field $\mathbb{K} = \mathbb{R}$ or $\mathbb{C}$ is isomorphic to $\mathbb{K}^N$, where N is the dimension of X. Deduce that finite-dimensional subspaces of topological vector spaces are closed. Here, as usual, an isomorphism between two topological vector spaces is, by definition, a linear homeomorphism.

Exercise 12.1.3. Given $0 < p < 1$ and $a < b$, show that the vector space

$$X = L^p[a,b] = \left\{ f : [a,b] \to \mathbb{K} \; ; f \text{ is measurable and } \int_a^b |f(t)|^p \, dt < \infty \right\}$$

is an F-space if we endow it with the base of neighbourhoods $(g + W_n)_n$, $g \in X$, where $W_n = \{f \in X \; ; \; \int_a^b |f(t)|^p \, dt < 1/n\}$, $n \in \mathbb{N}$. Prove that X is not a Fréchet space.

Exercise 12.1.4. (a) Let X be a vector space. Show that a subset A of X is absolutely convex if and only if it is convex and balanced.

(b) Show that a topological vector space is locally convex if, and only if, it has a base of 0-neighbourhoods consisting of convex sets.

Exercise 12.1.5. Given a vector space X, prove that, if $A \subset X$ is an absolutely convex set, the associated gauge of A,

$$p_A(x) = \inf\{\lambda > 0 \; ; \; x \in \lambda A\}, \quad x \in \operatorname{span} A,$$

is a seminorm on span A.

Exercise 12.1.6. If X is an infinite-dimensional Fréchet space, then show that $L(X)$ endowed with the strong operator topology is not metrizable.

Exercise 12.1.7. Let $(X_n)_n$ be an increasing sequence of Banach spaces such that each inclusion map $i_n : X_n \to X_{n+1}$, $n \geq 1$, is continuous, and set $X = \bigcup_{n=1}^{\infty} X_n$. Show that the inductive limit topology defined in Example 12.5 is a locally convex topology on X. Prove that it is not metrizable unless there is some $m \in \mathbb{N}$ such that $X_n = X_m$ for all $n \geq m$.

Exercise 12.2.1. The *weak topology* on a Banach space X is the locally convex topology on X defined by the seminorms $x \to |\langle x, x^* \rangle|$, $x^* \in X^*$; that is, it is the topology of pointwise convergence on X^*. An operator T on a Banach space X is called *weakly hypercyclic* if it is hypercyclic on X endowed with the weak topology.

Let $X = \ell^p$, $1 \leq p < \infty$, or $X = c_0$. Show that a weighted backward shift B_w on X is hypercyclic if and only if it is weakly hypercyclic.

Exercise 12.2.2. Construct a chaotic and mixing operator that is not hypercyclic. (*Hint*: Enlarge the space in Example 12.9.)

Exercise 12.2.3. An operator T on a locally convex space X is called *compact* if there exists some $W \in \mathcal{U}_0(X)$ such that $\overline{T(W)}$ is compact. Show that no compact operator on a locally convex space is hypercyclic, and that no compact perturbation of a multiple of the identity is chaotic. (*Hint*: If $W \in \mathcal{U}_0(X)$ is absolutely convex and p_W is the gauge of W, then p_W induces a norm on $X/\ker p_W$. The completion of this normed space is called the *local Banach space* X_W. If $W \in \mathcal{U}_0(X)$ is absolutely convex such that $\overline{T(W)}$ is compact, then consider the operator T_W on X_W induced by T.)

Exercise 12.2.4. A subset B of a topological vector space X is called *bounded* if, for any $W \in \mathcal{U}_0(X)$, there is some $M > 0$ such that $B \subset MW$. If, in addition, B is absolutely convex then $X_B := \operatorname{span} B$ is a normed space when endowed with the gauge of B.

An operator T on X is called *bounded* if there is some $U \in \mathcal{U}_0(X)$ such that $T(U)$ is a bounded subset of X. Show that a bounded operator T with dense range is hypercyclic (or mixing, weakly mixing or chaotic) if and only if there is a bounded absolutely convex set $B \subset X$ such that X_B is a T-invariant dense subspace of X and the induced operator $T_B : X_B \to X_B$ is hypercyclic (or mixing, weakly mixing or chaotic, respectively).

Moreover, if T is a bounded operator and $\lambda \in \mathbb{K}$, show that there is some $M > 0$ such that $M(\lambda I + T)$ is not hypercyclic.

Exercise 12.2.5. Let X be a topological vector space and L a subspace of X. Show that X/L is Hausdorff if and only if L is closed. (*Hint*: Use some properties of Exercise 12.1.1.)

Exercise 12.2.6. Let X be a topological vector space over $\mathbb{R}$, T an operator on X that is topologically transitive or has a somewhere dense orbit, and p a nonzero polynomial over $\mathbb{R}$. Prove that $p(T)$ has dense range. (*Hint*: Define, as for Fréchet spaces, the *complexifications* $\widetilde{X}$ of X and $\widetilde{T}$ of T. As in the proof of Theorem 2.54 it suffices to show that $\widetilde{T}-\lambda I$ has dense range for all $\lambda \in \mathbb{C}$. *First case:* proceed as in the proof of Theorem 2.54, taking into account that $\widetilde{T}$ has the property that, for any $U, V \subset X$ open and nonempty, there is $n \in \mathbb{N}$ with $\widetilde{T}^n(U+iU) \cap (V+iV) \neq \varnothing$. *Second case:* there is $x \in X$ such that $\{T^n x + iT^m x\ ;\ n,m \geq 0\}$ is somewhere dense in $\widetilde{X}$. Applying the quotient map q, deduce that $\widetilde{X}$ can be identified with $\mathbb{C}$. By continuity and openness of $|q| : X \to \mathbb{R}_+$, $\{|q(T^n x)|\ ;\ n \geq 0\}$ is somewhere dense in $\mathbb{R}_+$, but $|q(T^n x)| = |q(\widetilde{T}^n x)| = |\lambda|^n |q(x)|$.)

Exercise 12.3.1. Given a compact subset $K \subset \mathbb{C}$, a *holomorphic germ* on K is a function that is defined and holomorphic on an open set U containing K. Let $H(K)$ be the space of holomorphic germs on K, and let $A(K)$ be the Banach space of continuous functions on K that are holomorphic on the interior of K, endowed with the sup-norm.

If $(K_n)_n$ is a decreasing sequence of compact sets such that the interior of each K_n contains K and $\bigcap_{n=1}^\infty K_n = K$, then $H(K)$ can be viewed as the inductive limit of the increasing sequence $(A(K_n))_n$ of Banach spaces. Prove that the differentiation operator D is a well-defined operator on $H(K)$, and that it is hypercyclic and chaotic if K is connected with connected complement $\mathbb{C} \setminus K$.

Exercise 12.3.2. Consider the weighted Banach space of holomorphic functions on the unit disk,

$$Hv_n(\mathbb{D}) := \Big\{ f \in H(\mathbb{D})\ ;\ \|f\| := \sup_{z\in\mathbb{D}} |f(z)| v_n(z) < \infty \Big\},$$

where $v_n(z) := (1-|z|)^n$, $n \in \mathbb{N}$; the inclusions $Hv_n(\mathbb{D}) \hookrightarrow Hv_{n+1}(\mathbb{D})$, $n \geq 1$, are continuous. The *Korenblum space* $A^{-\infty}$ is defined as the inductive limit of $(Hv_n(\mathbb{D}))_n$. Show that the differentiation operator D is a well-defined operator on $A^{-\infty}$, and prove that any finite-order differential operator on $A^{-\infty}$ that is not a multiple of the identity is chaotic. In contrast, observe that $D(Hv_n(\mathbb{D})) \not\subset Hv_n(\mathbb{D})$ for any $n \in \mathbb{N}$, so that the argument in Example 12.14 cannot be applied.

Exercise 12.3.3. Let $w = (w_n)_n$ be a weight sequence and B_w the corresponding weighted backward shift. Show that $T := I + B_w$ is mixing on φ. (*Hint*: Show that T is quasiconjugate to $I + B$ on φ via a suitable diagonal operator.)

Exercise 12.3.4. Let X be a *topological sequence space*, that is, a topological vector space X such that $X \subset \omega = \mathbb{K}^\mathbb{N}$ with continuous inclusion. Suppose that φ is contained and dense in X. If the weighted backward shift B_w is a well-defined operator on X, prove that $T := I + B_w$ is mixing on X.

Exercise 12.3.5. For this exercise we will need *Young's inequality:* given any $x \in \ell^p(\mathbb{Z})$ and $y \in \ell^q(\mathbb{Z})$, $1 \leq p,q < 2$, the *convolution product*

$$x * y := \Big(\sum_{k\in\mathbb{Z}} x_k y_{n-k} \Big)_{n\in\mathbb{Z}},$$

exists and belongs to $\ell^r(\mathbb{Z})$, where $1/p + 1/q = 1/r + 1$. Moreover, $\|x * y\|_r \leq \|x\|_p \|y\|_q$. We consider the Fréchet space $\ell^{1+}(\mathbb{Z}) = \bigcap_{p>1} \ell^p(\mathbb{Z})$. The space ℓ^{1+} is defined similarly.

(a) Show that, for any $y \in \ell^{1+}(\mathbb{Z})$, the map $y * \cdot : \ell^{1+}(\mathbb{Z}) \to \ell^{1+}(\mathbb{Z})$ given by $x \to y * x$ defines an operator on $\ell^{1+}(\mathbb{Z})$. Deduce that, for any function $f(z) = \sum_{n=0}^\infty \alpha_n z^n$ with $(\alpha_{n-1})_n \in \ell^{1+}$, $T = f(B) = \sum_{n=0}^\infty \alpha_n B^n$ defines an operator on ℓ^{1+}, where B is the (unweighted) backward shift.

(b) Let $\lambda \in \mathbb{K} \setminus \{0\}$, and suppose that $f(z) = \sum_{n=0}^{\infty} \alpha_n z^n$ with $(\alpha_{n-1})_n \in \ell^{1+}$ is a nonconstant function such that there exist $z_1, z_2 \in \mathbb{K}$ with $|z_1|, |z_2| < 1$ such that $|\lambda f(z_1)| < 1$ and $|\lambda f(z_2)| > 1$. Show that $\lambda f(B)$ is hypercyclic on ℓ^{1+}. If $\mathbb{K} = \mathbb{C}$ and, moreover, there exists $z \in \mathbb{C}$ with $|z| < 1$ such that $|\lambda f(z)| = 1$, show that $\lambda f(B)$ is chaotic on ℓ^{1+}. As a consequence obtain that λT is chaotic on ℓ^{1+} for every $\lambda \neq 0$, where $T := \sum_{n=1}^{\infty} \frac{1}{n} B^n$. (*Hint*: If $|\lambda| < 1$, then $e_\lambda := (\lambda^n)_n$ is an eigenvector of $f(B)$.)

Observe that $T(\ell^p) \not\subset \ell^p$ for any $p \geq 1$, which shows that the above result cannot be transferred from the Banach spaces defining the projective spectrum of ℓ^{1+}.

Exercise 12.4.1. Let $(T_n)_n$ be a topologically transitive commuting sequence of operators on a topological vector space X such that, for any nonempty open set $U \subset X$ and any 0-neighbourhood W, there is a continuous map $S : X \to X$ commuting with all T_n, $n \in \mathbb{N}_0$, such that

$$S(U) \cap U \neq \varnothing \quad \text{and} \quad S(U) \cap W \neq \varnothing.$$

Show that $(T_n)_n$ is weakly mixing. (*Hint*: Given U, W, find $U_1 \subset U$ and $W_1 \subset W$ with $U_1 - U_1 \subset W$, $U_1 - W_1 \subset U$. Apply the hypothesis and the 4-set trick to get that $N(U_1, U_1) \cap N(U_1, W_1) \neq \varnothing$. Then deduce the result from the analogue of Theorem 12.24 for sequences of operators.)

Exercise 12.4.2. Inspired by Theorems 1.54 and 12.29, prove that an operator T on a topological vector space X is weakly mixing if and only if, for any $M > \delta > 0$, for any $(\lambda_n)_n$ with $\delta \leq |\lambda_n| \leq M$, $n \in \mathbb{N}_0$, and for any syndetic sequence $(n_k)_k$, the sequence $(\lambda_k T^{n_k})_k$ is topologically transitive.

Exercise 12.5.1. Let $\varphi : [0,1] \to [0,1]$ be a continuous, surjective, and strictly increasing function such that $\varphi(t) \neq t$ for all $t \in]0,1[$. Show that the composition operator C_φ is mixing on $L^p[0,1]$ for any $p > 0$.

Exercise 12.5.2. Let T be an operator on a separable Banach space X satisfying the Godefroy–Shapiro criterion (or the Hypercyclicity Criterion with respect to $(n_k)_k$). Prove that the left-multiplication operator L_T on $L(X)$, endowed with the strong operator topology, satisfies the hypotheses of Theorem 12.30 (or the hypotheses of Theorem 12.33 with respect to $(n_k)_k$, respectively).

Exercise 12.5.3. Let X be a Banach space with separable dual X^* and T an operator on X whose adjoint $T^* : X^* \to X^*$ satisfies the Hypercyclicity Criterion (or is chaotic). Show that the right-multiplication operator $R_T : L(X) \to L(X)$, $S \to ST$, is hypercyclic and weakly mixing (or is hypercyclic and chaotic, respectively) on $L(X)$, endowed with the strong operator topology; see also Exercise 10.2.7.

Sources and comments

Section 12.1. All the basic results of this section can be found in the books by Meise and Vogt [237] and Rudin [271]. For F-spaces we also refer to Kalton, Peck and Roberts [212].

Section 12.2. Dynamical properties of linear operators on topological vector spaces beyond F-spaces were apparently first studied by Ansari [10]. The fact that the space φ admits no hypercyclic operators was obtained by Bonet and Peris [85] and Grosse-Erdmann [179]. The definition of chaos in general topological vector spaces was proposed by Bonet [78], where one also finds Example 12.9. The crucial Lemma 12.13 is due to Wengenroth [301].

We want to mention an interesting result on *weakly topologically transitive operators*, that is, operators that are topologically transitive with respect to the weak topology; it is due to Desch and Schappacher [130] (for Banach spaces) and Shkarin [286].

Theorem 12.34. *An operator T on a complex locally convex space is weakly topologically transitive if and only if T^* has no eigenvalues.*

Section 12.3. Proposition 12.15 is new, while Corollary 12.16 is due to Shkarin [284] and Bayart and Matheron [44].

Example 12.17 and Proposition 12.18 are due to Bonet, Frerick, Peris and Wengenroth [81]. Since φ is thus a topological vector space without hypercyclic operators, but that admits a mixing, and therefore topologically transitive, operator, one might wonder if every infinite-dimensional topological vector space necessarily admits a topologically transitive operator. This is not the case, as Bermúdez and Kalton [52] have shown: there are (non-separable) Banach spaces, like ℓ^∞, or $L(\ell^2)$ with the operator norm, without any topologically transitive operators. Further existence and nonexistence results concerning hypercyclic or topologically transitive operators on locally convex spaces beyond Fréchet spaces are due to Bonet and Peris [85], Bonet, Frerick, Peris and Wengenroth [81], and Shkarin [286, 288, 293].

Example 12.20 solves a problem from León, Piqueras and Seoane [224].

Section 12.4. For the results of this section we refer to Grosse-Erdmann and Peris [187]. Similar investigations can be found in Bayart and Matheron [44], [45] and Moothathu [245].

Section 12.5. Example 12.32 is from Grosse-Erdmann [179].

Exercises. Exercise 12.2.1 is taken from Chan and Sanders [101], where the authors also show that there are weakly hypercyclic bilateral shifts that are not hypercyclic. Exercise 12.2.2 is taken from Bonet [78], Exercise 12.2.3 from Bonet and Peris [85] and Martínez and Peris [229]. The first part of Exercise 12.2.4 is from Bonet and Peris [85]. The second part is extracted from Bonet [80]; this paper studies the open problem of the existence of non-normable Fréchet spaces X such that every operator on X is of the form $\lambda I + T$ with T a bounded operator. It is also asked whether, for every infinite-dimensional separable non-normable Fréchet space X, there exists an operator T on X such that λT is hypercyclic for any $\lambda \neq 0$. Exercise 12.2.6 is taken from Wengenroth [301]. For Exercise 12.3.2 we refer to Bonet [78], for Exercises 12.3.3 and 12.3.4 to Bonet, Frerick, Peris and Wengenroth [81], and for Exercise 12.3.5 to Frerick and Peris [155]. Exercise 12.5.3 is taken from Bonet, Martínez and Peris [84].

Appendix A – Prerequisites

Throughout this book we suppose that the reader is familiar with metric spaces, the basics of Hilbert and Banach space theory, and the fundamentals of complex analysis. An introduction to the theory of Fréchet spaces and their operators is given in Section 2.1. Some more advanced results of these theories will be provided here, with suitable references.

Metric spaces

A good understanding of metric spaces and their topology is essential for reading this book. A short introduction can be found in most texts on functional analysis; for a more thorough account we refer to Shirali and Vasudeva [282].

A point x in a metric space X is called *isolated* if some neighbourhood of x contains no other point from X. A G_δ*-set* is the intersection of countably many open sets. A set is of *first Baire category* if it is a countable union of nowhere dense sets, otherwise of *second Baire category*.

Theorem A.1 (Baire category theorem). *If X is a complete metric space then the intersection of countably many dense open sets in X is dense in X.*

For a proof see Rudin [270].

Banach spaces

We suppose that the reader has had a first introduction to Banach spaces, as can be found, for example, in Chapter 5 of Rudin [270], Chapter III of Conway [116], or Chapters 2–5 of Bollobás [77].

K.-G. Grosse-Erdmann, A. Peris Manguillot, *Linear Chaos*, Universitext, DOI 10.1007/978-1-4471-2170-1,

If X, Y are two normed spaces then $L(X, Y)$ denotes the space of (continuous, linear) operators $T : X \to Y$; under the operator norm $T \to \|T\|$ this space turns into a Banach space whenever Y is a Banach space.

As a special case, $L(X) = L(X, X)$ is the space of operators on X. The operator norm then has the property that

$$\|ST\| \leq \|S\|\|T\|$$

for $S, T \in L(X)$. The *identity operator* is denoted by I. Two operators S and T are said to *commute* if $ST = TS$.

If X is a normed space and Y is a Banach space then an operator $T \in L(X, Y)$ is called *compact* if the image of the closed unit ball is relatively compact. Recall that a subset is said to be *relatively compact* if its closure is compact. Thus, T is compact if and only if, for any sequence $(x_n)_n$ in X with $\|x_n\| \leq 1$, $n \geq 1$, the sequence $(Tx_n)_n$ has a convergent subsequence. One then writes $T \in K(X, Y)$, with $K(X) = K(X, X)$.

The *dual* $X^* = L(X, \mathbb{K})$ of a normed space X is the space of all continuous linear functionals on X. If $x^* \in X^*$ then we write

$$x^*(x) = \langle x, x^* \rangle, \quad x \in X.$$

The *adjoint* $T^* : X^* \to X^*$ of an operator T on X is defined by $T^*x^* = x^* \circ T$, that is,

$$\langle x, T^*x^* \rangle = \langle Tx, x^* \rangle, \quad x \in X, x^* \in X^*.$$

Proposition A.2. *Let S and T be operators on a Banach space X, and $\lambda \in \mathbb{K}$. Then*

(i) $I^* = I$;
(ii) $(S + T)^* = S^* + T^*$;
(iii) $(\lambda T)^* = \lambda T^*$;
(iv) $(ST)^* = T^*S^*$;
(v) *if T is invertible then* $(T^*)^{-1} = (T^{-1})^*$;
(vi) $\|T^*\| = \|T\|$.

Only the last assertion is not immediate; for a proof see [116, Chapter VI, Proposition 1.4].

Example A.3. (a) The spaces ℓ^p, $1 \leq p < \infty$, c_0 and $L^p[a, b]$ are recalled in Example 2.4. The dual of ℓ^p is given by ℓ^q, where $\frac{1}{p} + \frac{1}{q} = 1$, in the sense that the continuous linear functionals x^* on ℓ^p are precisely the maps of the form

$$x^*(x) = \langle x, x^* \rangle = \sum_{n=1}^{\infty} x_n y_n \tag{A.1}$$

with $y = (y_n)_n \in \ell^q$, and we have that $\|x^*\| = \|y\|$. In the same way the dual of c_0 is given by ℓ^1.

The dual of $L^p[a,b]$, $1 \le p < \infty$, is given by $L^q[a,b]$ with $\frac{1}{p} + \frac{1}{q} = 1$, in the sense that the continuous linear functionals x^* on $L^p[a,b]$ are precisely the maps of the form

$$x^*(f) = \langle f, x^* \rangle = \int_a^b f(t)g(t)\,dt \tag{A.2}$$

with $g \in L^q[a,b]$; moreover, $\|x^*\| = \|g\|$.

(b) Let $v = (v_n)_n$ be a strictly positive sequence. Then the weighted spaces $\ell^p(v)$, $1 \le p < \infty$, are defined as

$$\ell^p(v) = \Big\{(x_n)_n \; ; \; \sum_{n=1}^{\infty} |x_n|^p v_n < \infty\Big\}.$$

Under the representation (A.1), for $p > 1$, the dual of $\ell^p(v)$ is given by $\ell^q(w)$, where $\frac{1}{p} + \frac{1}{q} = 1$ and $w_n = v_n^{-q/p}$, $n \ge 1$; for $p = 1$, it is given by the space of sequences $(y_n)_n$ with $\sup_{n \ge 1} |y_n v_n^{-1}| < \infty$. This follows from the fact that $(x_n)_n \to (x_n v_n^{1/p})_n$ defines an isometric isomorphism from $\ell^p(v)$ to ℓ^p.

(c) The analogous representation, via (A.2), holds for the duals of the spaces $L^p_v(\mathbb{R}_+)$ introduced in Example 7.4; see also Exercise 2.1.6.

Of the four great principles in functional analysis, which will be stated in the larger context of Fréchet spaces below, only the Banach–Steinhaus theorem has a version that is specific to Banach spaces.

Theorem A.4 (Banach–Steinhaus theorem). *Let X, Y be Banach spaces and $T_j : X \to Y$, $j \in J$, operators. If, for every $x \in X$, $\sup_{j \in J} \|T_j x\| < \infty$, then*

$$\sup_{j \in J} \|T_j\| < \infty.$$

Further properties of Banach spaces will be treated in the context of Fréchet spaces below.

Hilbert spaces

We suppose that the reader is familiar with the basic properties of Hilbert spaces as can be found, for example, in Chapter 4 of Rudin [270], Chapter II of Conway [116], or Chapter 9 of Bollobás [77].

Let H be a Hilbert space with inner product $\langle \cdot\,, \cdot \rangle$. We formulate two consequences of the projection theorem.

Proposition A.5. *A subspace M of H is dense if and only if only the zero vector is orthogonal to M.*

Theorem A.6 (Riesz representation theorem). *Let $y \in H$. Then*

$$x^*(x) = \langle x, y\rangle, \quad x \in H$$

defines a continuous linear functional x^* on H with $\|x^*\| = \|y\|$. Conversely, any continuous linear functional on H can be represented in this way, and the vector y is uniquely determined by the functional.

Let T be an operator on H. As a consequence of the Riesz representation theorem there exists a unique operator $T^* : H \to H$ such that, for all $x, y \in H$,

$$\langle Tx, y\rangle = \langle x, T^*y\rangle.$$

It is called the *(Hilbert space) adjoint* of T.

Remark A.7. The Hilbert space adjoint $T^* = T^{hil*}$ is nothing but the (Banach space) adjoint $T^* = T^{ban*}$ of T as defined previously when we identify the dual H^* of H with H itself via the Riesz representation theorem. More precisely, if $J : H^* \to H$ denotes the Riesz representation of $x^* \in H^*$ by $y \in H$, then $T^{hil*} = JT^{ban*}J^{-1}$. In that sense it is justified to use the notation T^* for both adjoints, and it suggests using the notation $\langle \cdot\,, \cdot\rangle$ for the evaluation of continuous linear functionals in general Banach spaces.

We collect some useful properties.

Proposition A.8. *Let S and T be operators on a Hilbert space H, and $\lambda \in \mathbb{K}$. Then*

(i) $I^* = I$;
(ii) $(T^*)^* = T$;
(iii) $(S + T)^* = S^* + T^*$;
(iv) $(\lambda T)^* = \overline{\lambda} T^*$;
(v) $(ST)^* = T^*S^*$;
(vi) *if T is invertible then $(T^*)^{-1} = (T^{-1})^*$;*
(vii) $\|T^*\| = \|T\|$.

Fréchet spaces

Section 2.1 gives a short introduction to the theory of Fréchet spaces. For more detailed accounts we refer to Rudin [271] and Meise and Vogt [237]. We collect here some further definitions and results that are used in this book; their proofs can be found in the two books mentioned or in the references provided.

The *dual* X^* of a Fréchet space X and the *adjoint* T^* of an operator T on X are defined as in the case of Banach spaces. However we will not (need to) address the problem of topologizing X^*, which is more delicate than for Banach spaces; as a consequence we will consider T^* only as a linear map.

The compactness of an operator $T : X \to Y$ from a normed space X into a Fréchet space Y is defined as in the case of Banach spaces.

In a Banach space a set A is bounded if $\sup_{x\in A} \|x\| < \infty$. In a Fréchet space one cannot use the F-norm to the same effect because, for example, the F-norm (2.2) is always bounded.

Definition A.9. Let X be a Fréchet space with defining increasing sequence of seminorms $(p_n)_n$. Then a subset A of X is *bounded* if, for any $n \geq 1$,

$$\sup_{x\in A} p_n(x) < \infty.$$

If X and Y are Fréchet spaces, then a family $(T_j)_{j\in J}$ of operators $T_j : X \to Y$, $j \in J$, is called *equicontinuous* if for any neighbourhood W of 0 in Y there is a neighbourhood V of 0 in X such that $T_j(V) \subset W$ for all $j \in J$. In other terms, if $(p_n)_n$ and $(q_n)_n$ are defining increasing sequences of seminorms on X and Y, respectively, then $(T_j)_{j\in J}$ is equicontinuous if and only if, for any $m \geq 1$, there are $n \geq 1$ and $M > 0$ such that, for any $j \in J$,

$$q_m(T_j x) \leq M p_n(x), \quad x \in X.$$

We turn to the four great principles of functional analysis.

Theorem A.10 (Banach–Steinhaus theorem). *Let X and Y be Fréchet spaces and $T_j : X \to Y$, $j \in J$, operators.*

(a) *If, for every $x \in X$, the set $\{T_j x : j \in J\}$ is bounded in Y, then the family $(T_j)_{j\in J}$ is equicontinuous.*

(b) *If $J = \mathbb{N}$ and, for every $x \in X$, the sequence $(T_n x)_n$ converges in Y, then $Tx = \lim_{n\to\infty} T_n x$, $x \in X$, defines an operator $T : X \to Y$.*

Theorem A.11 (Open mapping theorem). *Let X, Y be Fréchet spaces. If $T : X \to Y$ is a surjective operator, then T is an open map, that is, for any open set $U \subset X$, $T(U)$ is open in Y.*

As a consequence, we have the following.

Corollary A.12 (Inverse mapping theorem). *Let X, Y be Fréchet spaces. If $T : X \to Y$ is a bijective operator, then T has a continuous linear inverse T^{-1}.*

Theorem A.13 (Closed graph theorem). *Let X, Y be Fréchet spaces. If a linear map $T : X \to Y$ has closed graph, that is, if $x_n \to x$ in X and $Tx_n \to y$ in Y implies that $Tx = y$, then T is continuous.*

Theorem A.14 (Hahn–Banach theorem). *Let X be a vector space, M a subspace of X, p a seminorm on X and $u : M \to \mathbb{K}$ a linear functional such that $|u(x)| \leq p(x)$ for all $x \in M$. Then u has a linear extension $\widetilde{u}$ to X such that $|\widetilde{u}(x)| \leq p(x)$ for all $x \in X$.*

We will mostly apply the Hahn–Banach theorem through one of the following corollaries:

(i) *if p is a seminorm on X and $x_0 \in X$ then there exists a linear functional u on X such that $u(x_0) = p(x_0)$ and $|u(x)| \leq p(x)$ for all $x \in X$;*

if X is a Fréchet space then

(ii) *every continuous linear functional on a subspace of X extends to a continuous linear functional on X (with preservation of norm, if X is a Banach space);*

(iii) *if M is a closed subspace of X and $x \notin M$ then there exists a continuous linear functional x^* on X that vanishes on M with $\langle x, x^* \rangle \neq 0$;*

(iv) *a subspace M of X is dense in X if and only if every continuous linear functional that vanishes on M also vanishes on X;*

(v) *for any $x \in X$, if $\langle x, x^* \rangle = 0$ for all $x^* \in X^*$ then $x = 0$.*

A sequence $(e_n)_n$ in a Fréchet space X is called a *basis* if every $x \in X$ has a unique representation

$$x = \sum_{n=1}^{\infty} a_n e_n$$

with scalars $a_n \in \mathbb{K}$, $n \geq 1$. The *coefficient functionals*

$$e_n^* : X \to \mathbb{K}, \quad x \to a_n,$$

are then continuous.

At several places in this book unconditional convergence plays a crucial role.

Definition A.15. A series $\sum_{n=1}^{\infty} x_n$ in a Fréchet space is called *unconditionally convergent* if for any bijection $\pi : \mathbb{N} \to \mathbb{N}$ the series

$$\sum_{n=1}^{\infty} x_{\pi(n)}$$

converges.

There are several useful equivalent formulations; in the following result, let $\|\cdot\|$ denote an F-norm that induces the topology of X.

Theorem A.16. *Let X be a Fréchet space. Then the following assertions are equivalent:*

(i) *$\sum_{n=1}^{\infty} x_n$ is unconditionally convergent;*

(ii) *for any 0-1-sequence $(\varepsilon_n)_n$, $\sum_{n=1}^{\infty} \varepsilon_n x_n$ converges;*

(iii) *for any bounded sequence $(\alpha_n)_n$ of scalars, $\sum_{n=1}^{\infty} \alpha_n x_n$ converges;*

(iv) *for any $\varepsilon > 0$ there is some $N \in \mathbb{N}$ such that for any finite set $F \subset \{N, N+1, N+2, \ldots\}$ we have that*

$$\Big\| \sum_{n \in F} x_n \Big\| < \varepsilon;$$

(v) *for any* $\varepsilon > 0$ *there is some* $N \in \mathbb{N}$ *such that for any* 0-1*-sequence* $(\varepsilon_n)_n$, $\sum_{n=1}^{\infty} \varepsilon_n x_n$ *converges and*

$$\Big\| \sum_{n \geq N} \varepsilon_n x_n \Big\| < \varepsilon;$$

(vi) *for any* $\varepsilon > 0$ *there is some* $N \in \mathbb{N}$ *such that whenever* $\sup_{n\geq 1} |\alpha_n| \leq 1$ *then* $\sum_{n=1}^{\infty} \alpha_n x_n$ *converges and*

$$\Big\| \sum_{n \geq N} \alpha_n x_n \Big\| < \varepsilon.$$

For a proof we refer to [269, 3.8.2 and p. 153] and [213, 3.3.8 and 3.3.9].

Unconditional convergence of series of the form $\sum_{n\in Z} x_n$ is defined similarly and has the corresponding properties.

A sequence $(e_n)_n$ in a Fréchet space X is called an *unconditional basis* if it is a basis such that, for every $x \in X$, the representation

$$x = \sum_{n=1}^{\infty} a_n e_n,$$

converges unconditionally.

In connection with eigenvalue criteria for hypercyclicity, but also for Appendix B on spectral theory, we need the Riemann integral for Fréchet space-valued continuous functions. Its definition poses no particular problems and follows exactly the same lines as in the real-valued case. Thus, if $f : [a,b] \to X$ is a continuous function with values in a Fréchet space X, then its *Riemann integral* is defined as

$$\int_a^b f(t)\,dt = \lim \sum_{k=0}^{N-1} f(t_k)(t_{k+1} - t_k),$$

where the limit is taken as $\max_{0\leq k\leq N-1} |t_{k+1} - t_k| \to 0$. Apart from the usual properties of linearity in f and additivity with respect to the domain of integration we have that, for any continuous seminorm p on X, for the norm $\|\cdot\|$ on a Banach space X, for any operator $T \in L(X,Y)$, where Y is another Fréchet space, and for any $x^* \in X^*$ that

$$p\Big(\int_a^b f(t)\,dt\Big) \leq \int_a^b p(f(t))\,dt,$$

$$\Big\|\int_a^b f(t)\,dt\Big\| \leq \int_a^b \|f(t)\|\,dt,$$

$$T\int_a^b f(t)\,dt = \int_a^b Tf(t)\,dt,$$

$$\left\langle \int_a^b f(t)\,dt, x^* \right\rangle = \int_a^b \langle f(t), x^* \rangle\,dt.$$

As usual, if X is a complex Fréchet space and Γ is a continuously differentiable curve in $\mathbb{C}$ then one defines

$$\int_\Gamma f(z)\,dz = \int_0^1 f(\gamma(t))\gamma'(t)\,dt,$$

where $\gamma : [0,1] \to \Gamma$ is a parametrization of the curve. This extends to integrals over finite collections of such curves.

Finite-dimensional Fréchet spaces

Finite-dimensional Fréchet spaces play a particular role in the theory, for several reasons.

First, every Fréchet space of dimension $N \geq 1$ is isomorphic to $\mathbb{K}^N$ under the Euclidean topology. In particular, all Fréchet space topologies on finite-dimensional vector spaces coincide.

Moreover, any subspace of finite dimension of a Fréchet space is closed and therefore a Fréchet space in the induced topology.

Finally, a Fréchet space with a relatively compact neighbourhood of 0 is finite dimensional.

Indeed, all these results hold in arbitrary topological vector spaces, which are studied in Chapter 12; see Rudin [271].

Complex analysis

Finally, we assume familiarity with the basic concepts of complex analysis as can be gained, for example, from Chapter 10 of Rudin [270] or Chapters II–IV of Conway [115]. We collect here some important results.

Theorem A.17 (Liouville). *Every bounded entire function is constant.*

Theorem A.18 (Casorati–Weierstrass). *For any punctured neighbourhood U of an essential singularity, possibly ∞, of a holomorphic function f, $f(U)$ is dense in $\mathbb{C}$.*

As a consequence, every nonconstant entire function has dense range.

Theorem A.19 (Open mapping theorem). *Any nonconstant holomorphic function f on a domain Ω is an open mapping, that is, for any open set $O \subset \Omega$, $f(O)$ is open.*

Theorem A.20 (Rouché). *Let f and g be holomorphic functions on a neighbourhood of $\{z \in \mathbb{C} \; ; \; |z-a| \leq r\}$, where $a \in \mathbb{C}$, $r > 0$, such that*

$$|f(z) - g(z)| < |g(z)| \quad \text{for } |z-a| = r.$$

Then f and g have the same number of zeros, counting multiplicity, for $|z-a| < r$.

Let now $\mathbb{D} = \{z \in \mathbb{C} \; ; \; |z| < 1\}$ be the open unit disk.

Theorem A.21 (Schwarz lemma). *Let $f : \mathbb{D} \to \overline{\mathbb{D}}$ be a holomorphic function with $f(0) = 0$. Then $|f(z)| \leq |z|$ for any $z \in \mathbb{D}$. Moreover, if there is some $z \in \mathbb{D}$, $z \neq 0$, with $|f(z)| = |z|$ then there is some $a \in \mathbb{C}$ with $|a| = 1$ such that $f(z) = az$, $z \in \mathbb{D}$.*

Theorem A.22 (Big Picard theorem). *Let f be an entire function that is not a polynomial. Then f takes every value in $\mathbb{C}$, with at most one exception, infinitely often.*

The following is a consequence of Jensen's formula; see Rudin [270, 15.20] or Conway [115, Chapter XI, § 1].

Theorem A.23. *Let f be an entire function, and let $N(r)$ denote the number of zeros of f in $|z| < r$, counting multiplicity. If $f(0) = 1$ and $M(r) = \max\{|f(z)| \; ; \; |z| = r\}$ then*

$$N(r) \log 2 \leq \log M(2r).$$

The Runge approximation theorem is a crucial tool for several of our examples. By $\widehat{\mathbb{C}}$ we denote the *extended complex plane* $\mathbb{C} \cup \{\infty\}$.

Theorem A.24 (Runge's theorem). *Let $K \subset \mathbb{C}$ be a compact set and A a set that contains at least one point from each connected component of $\widehat{\mathbb{C}} \setminus K$.*

Let f be a function that is holomorphic on some neighbourhood of K, and let $\varepsilon > 0$. Then there exists a rational function h with poles only at points from A such that

$$\sup_{z \in K} |f(z) - h(z)| < \varepsilon.$$

We note that the complement of a compact set is open and therefore has at most a countable number of connected components. Thus, A may always be chosen to be at most countable.

Moreover, if $K \subset \Omega$ is such that $\widehat{\mathbb{C}} \setminus K$ is connected then A may be taken to be $\{\infty\}$, and the function h will a polynomial.

Corollary A.25. *Let $\Omega \subset \mathbb{C}$ be a domain.*

(a) *Let $A \subset \mathbb{C}$ contain at least one point from each bounded component of $\widehat{\mathbb{C}} \setminus \Omega$. Then the rational functions with poles only in A form a dense set in $H(\Omega)$.*

(b) *If Ω is simply connected, then the polynomials form a dense set in* $H(\Omega)$.

Next we discuss an important family of maps. A *linear fractional transformation*, also known as a *Möbius transformation*, is a map of the form

$$f(z) = \frac{az+b}{cz+d},$$

with constants $a, b, c, d \in \mathbb{C}$. In addition, one demands that $ad - bc \neq 0$, which excludes the constant functions. Under obvious conventions for dealing with the point at infinity, each linear fractional transformation defines a bijection $f : \widehat{\mathbb{C}} \to \widehat{\mathbb{C}}$, where $\widehat{\mathbb{C}}$ is the extended complex plane, and its inverse is also a linear fractional transformation. Moreover, any composition of linear fractional transformations is a linear fractional transformation.

Every linear fractional transformation maps each circle and each line onto either a circle or a line. It has either one or two fixed points, unless it is the identity function. And given any distinct points $z_1, z_2, z_3 \in \widehat{\mathbb{C}}$ and distinct points $w_1, w_2, w_3 \in \widehat{\mathbb{C}}$ there is a (unique) linear fractional transformation that maps z_j to w_j, $j = 1, 2, 3$.

For more information on the topic we refer to Ahlfors [3].

Finally, the notion of homotopy is only used once in the book, but then in a crucial way; see Section 6.4. Let X and Y be metric spaces. Two maps $f, g : X \to Y$ are called *homotopic* if there is a continuous map $H : X \times [0, 1] \to Y$, called a *homotopy*, such that $H(x, 0) = f(x)$ and $H(x, 1) = g(x)$, $x \in X$. The map f is called *homotopically trivial* or *null-homotopic* if it is homotopic to a constant map.

In particular, let $\mathbb{T} = \{z \in \mathbb{C} \; ; \; |z| = 1\}$ be the unit circle in $\mathbb{C}$. Then every continuous map $f : \mathbb{T} \to \mathbb{T}$ has an *index* $n \in \mathbb{Z}$. Indeed, f can be identified with a closed curve that lies entirely in $\mathbb{T}$, and the index is nothing but the winding number of the curve with respect to the origin. Then f has index n if and only if it is homotopic to the map $z \to z^n$, and it is homotopically trivial if and only if it has index 0.

For more details we refer the reader to Chapter 10 of Rudin [270]; note also Exercise 28 there.

Appendix B – Spectral theory

Throughout this appendix, T denotes an operator on a complex Banach space X.

Basic spectral theory

In the finite-dimensional setting, $\lambda \in \mathbb{C}$ is an eigenvalue of an operator T if and only if $\lambda I - T$ is not injective, or equivalently, not invertible. For general complex Banach spaces, the latter is taken as the defining property of elements in the spectrum.

Definition B.1. Let T be an operator on a complex Banach space X. The *spectrum* $\sigma(T)$ of T is defined as

$$\sigma(T) = \{\lambda \in \mathbb{C} \; ; \; \lambda I - T \text{ is not invertible}\}.$$

The *point spectrum* $\sigma_p(T)$ of T is the set of eigenvalues of T.

We recall that an operator is said to be invertible if it is bijective and its inverse is also continuous. However, by the inverse mapping theorem (see Appendix A), the continuity of the inverse is automatic in our setting. Therefore, $\lambda \in \mathbb{C}$ belongs to the spectrum of T if and only if either $\lambda I - T$ fails to be injective, in which case λ is an eigenvalue, or if it fails to be surjective.

One obviously has the useful formula

$$\sigma(\lambda I + \mu T) = \lambda + \mu\sigma(T)$$

for any $\lambda, \mu \in \mathbb{C}$, which is a special case of the spectral mapping theorem; see below.

Proposition B.2. *The spectrum $\sigma(T)$ is a nonempty compact set. Moreover, $|\lambda| \leq \|T\|$ for any $\lambda \in \sigma(T)$.*

K.-G. Grosse-Erdmann, A. Peris Manguillot, *Linear Chaos*, Universitext,
DOI 10.1007/978-1-4471-2170-1, © Springer-Verlag London Limited 2011

The *spectral radius* of T is defined as $r(T) = \sup_{\lambda \in \sigma(T)} |\lambda|$.

Theorem B.3 (Spectral radius formula). *For the spectral radius we have that*

$$r(T) = \lim_{n \to \infty} \|T^n\|^{1/n}.$$

The existence of the limit is part of the result. In particular, we have that $\sigma(T) = \{0\}$ if and only if $\lim_{n \to \infty} \|T^n\|^{1/n} = 0$.

For the adjoint $T^* : X^* \to X^*$ of T we refer to Appendix A.

Proposition B.4. *We have that* $\sigma(T^*) = \sigma(T)$.

For a more detailed introduction to spectral theory we refer to Chapter 18 of Rudin [270] or Chapter 12 of Bollobás [77].

The Riesz–Dunford functional calculus

If $f(z) = \sum_{n=0}^{\infty} a_n z^n$ is a holomorphic function on some disk $\{z \in \mathbb{C}\,;\ |z| < r\}$ with $r > \|T\|$ then $\sum_{n=0}^{\infty} \|a_n T^n\| \leq \sum_{n=0}^{\infty} |a_n| \|T\|^n < \infty$, so that

$$f(T) = \sum_{n=0}^{\infty} a_n T^n$$

defines an operator on X; we apply this procedure, for example, in Section 4.4. If f is only holomorphic on a neighbourhood O of the spectrum $\sigma(T)$ of T then one can still define an operator $f(T)$ by

$$f(T) = \frac{1}{2\pi i} \int_{\Gamma} f(\lambda)(\lambda I - T)^{-1}\, d\lambda,$$

where Γ is the union of finitely many continuously differentiable Jordan curves in O that contains $\sigma(T)$ in its interior and $\mathbb{C} \setminus O$ in its exterior; see Figure B.1. The integral is to be understood as an operator-valued Riemann integral; see Appendix A. This definition of $f(T)$ goes back to Riesz and Dunford and is therefore referred to as the *Riesz–Dunford functional calculus*.

Theorem B.5. *Let f and g be holomorphic functions on a neighbourhood of $\sigma(T)$ and $\lambda \in \mathbb{C}$. Then we have the following:*

(i) $(\lambda f)(T) = \lambda(f(T))$;
(ii) $(f+g)(T) = f(T) + g(T)$;
(iii) $(fg)(T) = f(T)g(T)$; *in particular, $f(T)$ and $g(T)$ commute;*
(iv) $f(T)^* = f(T^*)$;
(v) *if $f(z) \neq 0$ for all $z \in \sigma(T)$, then $f(T)$ is invertible and* $f(T)^{-1} = (1/f)(T)$;

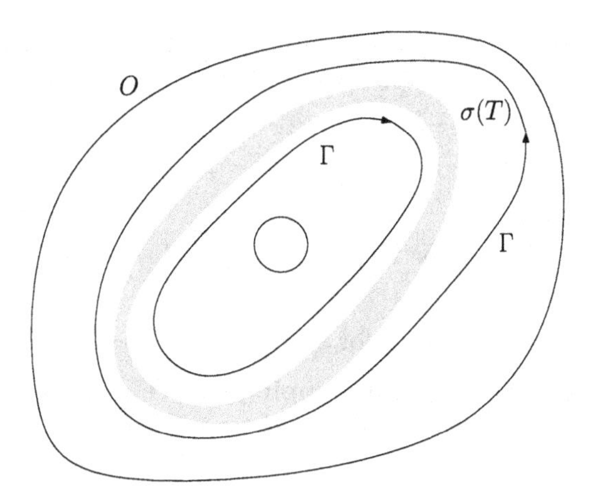

Fig. B.1 Jordan curves for the definition of $f(T)$

(vi) *if* $p(z) = \sum_{n=0}^{N} a_n z^n$ *is a polynomial then* $p(T) = \sum_{n=0}^{N} a_n T^n$*;*
(vii) *if* $f(z) = \sum_{n=0}^{\infty} a_n z^n$ *is holomorphic on* $\{z \in \mathbb{C}\ ;\ |z| < r\}$ *for some* $r > \|T\|$ *then* $f(T) = \sum_{n=0}^{\infty} a_n T^n$.

The following is a central result of spectral theory.

Theorem B.6 (Spectral mapping theorem). *Let f be a holomorphic function on a neighbourhood of $\sigma(T)$. Then*

$$\sigma(f(T)) = f(\sigma(T)).$$

There is also a version for the point spectrum. Since the result is more difficult to find in the literature we give a proof here.

Theorem B.7 (Point spectral mapping theorem). *Let f be a holomorphic function on an open neighbourhood O of $\sigma(T)$ that is not constant on any connected component of O. Then*

$$\sigma_p(f(T)) = f(\sigma_p(T)).$$

Proof. First suppose that $\lambda \in \sigma_p(T)$. Since the function $z \to f(\lambda) - f(z)$ has a zero at λ, it factorizes as $f(\lambda) - f(z) = (\lambda - z)g(z)$, where g is holomorphic on some neighbourhood of $\sigma(T)$. Hence

$$f(\lambda)I - f(T) = g(T)(\lambda I - T).$$

Since $\lambda I - T$ is non-injective, the same is true for $f(\lambda)I - f(T)$, so that $f(\lambda) \in \sigma_p(f(T))$.

Conversely, let $\mu \in \sigma_p(f(T))$. It follows from the assumption that, on $\sigma(T)$, the function $z \to \mu - f(z)$ only has a finite number of zeros $\lambda_1, \dots, \lambda_n$, so that it factorizes as $\mu - f(z) = \prod_{k=1}^{n}(\lambda_k - z)^{N_k} g(z)$, where $N_k \geq 1$ and g is holomorphic on O and zero-free on $\sigma(T)$. Thus $g(T)$ is invertible and

$$\mu I - f(T) = \prod_{k=1}^{n}(\lambda_k I - T)^{N_k} g(T).$$

Since $\mu I - f(T)$ is non-injective, so is some $\lambda_k I - T$. Hence $\mu = f(\lambda_k) \in f(\sigma_p(T))$. □

We state an analogous result for semigroups of operators; see Section 7.1.

Theorem B.8 (Point spectral mapping theorem for semigroups). *Let $(A, D(A))$ be the generator of a C_0-semigroup $(T_t)_{t\geq 0}$ defined on a complex Banach space X. Then we have the following identities:*

$$\sigma_p(T_t) \setminus \{0\} = e^{t\sigma_p(A)} \quad \textit{for } t \geq 0,$$

$$\ker(\lambda I - A) = \bigcap_{t\geq 0} \ker\left(e^{\lambda t} I - T_t\right) \quad \textit{for } \lambda \in \mathbb{C},$$

$$\ker(e^{\lambda t} I - T_t) = \overline{\operatorname{span}} \bigcup_{n\in\mathbb{Z}} \ker\left(\left(\lambda + \frac{2\pi n i}{t}\right) I - A\right) \quad \textit{for } t > 0.$$

The Riesz–Dunford functional calculus is treated in detail in Chapter VII of Conway [116] or in Radjavi and Rosenthal [264]. The point spectral mapping theorem is taken from Hille and Phillips [203, Theorem 5.12.2]. Its version for C_0-semigroups can be found in the book by Engel and Nagel [143].

The Riesz decomposition theorem

We turn to a result that is crucial for the application of spectral theory to linear dynamics. Since its proof is not so well known we will provide one here. We recall that an invariant closed subspace is called nontrivial if it is different from $\{0\}$ and the whole space.

Theorem B.9 (Riesz decomposition theorem). *Suppose that the spectrum $\sigma(T)$ splits into two disjoint nonempty closed subsets σ_1 and σ_2:*

$$\sigma(T) = \sigma_1 \cup \sigma_2.$$

Then there are nontrivial T-invariant closed subspaces M_1 and M_2 of X such that $X = M_1 \oplus M_2$,

$$\sigma(T|_{M_1}) = \sigma_1 \quad \textit{and} \quad \sigma(T|_{M_2}) = \sigma_2.$$

It should be noted that, the spectrum being closed, a subset of $\sigma(T)$ is closed in $\sigma(T)$ if and only if it is closed in $\mathbb{C}$.

Proof. Since $\sigma(T)$ is compact, the sets σ_1 and σ_2 are disjoint compact subsets of $\mathbb{C}$, so that there exist disjoint open neighbourhoods O_1 of σ_1 and O_2 of σ_2. The function f_1 that takes the value 1 on O_1 and the value 0 on O_2 is then holomorphic on a neighbourhood of $\sigma(T)$, as is the function f_2 that takes the value 0 on O_1 and the value 1 on O_2. We can therefore define two operators by

$$P_1 = f_1(T) \quad \text{and} \quad P_2 = f_2(T).$$

Since the functions f_1 and f_2 have the properties

$$f_1 + f_2 = 1, \quad f_1 f_2 = f_2 f_1 = 0, \quad f_1^2 = f_1, \quad f_2^2 = f_2,$$

an application of Theorem B.5 yields that

$$P_1 + P_2 = I, \quad P_1 P_2 = P_2 P_1 = 0, \quad P_1^2 = P_1, \quad P_2^2 = P_2. \tag{B.1}$$

We define

$$M_1 = \operatorname{ran} P_1 \quad \text{and} \quad M_2 = \operatorname{ran} P_2$$

and claim that these are the desired subspaces.

First, (B.1) implies that

$$x = P_1 x + P_2 x \quad \text{and} \quad P_1 P_2 x = P_2 P_1 x = 0 \quad \text{for all } x \in X, \tag{B.2}$$

and hence that $X = M_1 + M_2$, $M_1 \subset \ker P_2$ and $M_2 \subset \ker P_1$. Now if $x \in M_1 \cap M_2$ then $P_2 x = P_1 x = 0$ and hence $x = 0$ by (B.2); this shows that $X = M_1 \oplus M_2$. Moreover, if $x \in \ker P_2$ then $x = P_1 x \in M_1$ by (B.2); this shows that $M_1 = \ker P_2$ is a closed subspace and $x = P_1 x$ for all $x \in M_1$. The corresponding statements hold for M_2. Moreover, for $x \in M_1$, $Tx = TP_1 x = P_1 T x \in M_1$, so that M_1 is T-invariant; note that P_1 and T commute because P_1 is a function of T. As a consequence, $T|_{M_1}$ is an operator on M_1. By symmetry the same holds for M_2. Finally, $M_1 \neq \{0\}$ and hence $M_2 \neq X$. Otherwise, $\ker P_1 = M_2 = X$, so that $f_1 = 0$ on $\sigma(T)$ (see Exercise 5.0.8), which contradicts the fact that $\sigma_1 \neq \varnothing$. By symmetry we also have that $M_2 \neq \{0\}$ and hence $M_1 \neq X$.

It remains to identify the spectra of $T|_{M_1}$ and $T|_{M_2}$. Since the operator $\lambda I - T$ leaves the spaces M_1 and M_2 invariant, it is invertible if and only if $(\lambda I - T)|_{M_1}$ and $(\lambda I - T)|_{M_2}$ are; in other words,

$$\sigma(T) = \sigma(T|_{M_1}) \cup \sigma(T|_{M_2}). \tag{B.3}$$

Moreover, let $\lambda \notin \sigma_1$. Then the function

$$g(z) = (\lambda - z) f_1(z) + f_2(z)$$

is holomorphic on a neighbourhood of $\sigma(T)$ and nonzero on $\sigma(T)$. Hence the operator

$$g(T) = (\lambda I - T) P_1 + P_2$$

is invertible. It follows from the properties of P_1 and P_2 that $g(T)|_{M_1} = (\lambda I - T)|_{M_1}$ and $g(T)|_{M_2} = I|_{M_2}$. Since X is the direct sum of M_1 and M_2, $(\lambda I - T)|_{M_1}$ must then be invertible, hence $\lambda \notin \sigma(T|_{M_1})$.

Thus we have shown that $\sigma(T|_{M_1}) \subset \sigma_1$, and similarly $\sigma(T|_{M_2}) \subset \sigma_2$. But since σ_1 and σ_2 form a partition of $\sigma(T)$, (B.3) implies that these inclusions are in fact identities, which had to be shown. □

For the Riesz decomposition theorem we have followed Radjavi and Rosenthal [264].

References

1. E. Abakumov and J. Gordon, Common hypercyclic vectors for multiples of backward shift, *J. Funct. Anal.* 200 (2003), 494–504.
2. Y.A. Abramovich and C.D. Aliprantis, *An invitation to operator theory*, American Mathematical Society, Providence, RI, 2002.
3. L.V. Ahlfors, *Complex analysis*, third edition, McGraw-Hill, New York, 1978.
4. E. Akin, *Recurrence in topological dynamics. Furstenberg families and Ellis actions*, Plenum Press, New York, 1997.
5. A.A. Albanese, Construction of operators with prescribed orbits in Fréchet spaces with a continuous norm, *Math. Scand.*, to appear.
6. M.P. Aldred and D.H. Armitage, Harmonic analogues of G.R. MacLane's universal functions, *J. London Math. Soc. (2)* 57 (1998), 148–156.
7. M.P. Aldred and D.H. Armitage, Harmonic analogues of G.R. MacLane's universal functions. II, *J. Math. Anal. Appl.* 220 (1998), 382–395.
8. T. Ando, Operators with a norm condition, *Acta Sci. Math. (Szeged)* 33 (1972), 169–178.
9. S.I. Ansari, Hypercyclic and cyclic vectors, *J. Funct. Anal.* 128 (1995), 374–383.
10. S.I. Ansari, Existence of hypercyclic operators on topological vector spaces, *J. Funct. Anal.* 148 (1997), 384–390.
11. R. Arens, Dense inverse limit rings, *Michigan Math. J.* 5 (1958), 169–182.
12. S.A. Argyros and R.G. Haydon, A hereditarily indecomposable $\mathcal{L}_\infty$-space that solves the scalar-plus-compact problem, *Acta Math.* 206 (2011), 1–54.
13. D.H. Armitage, Permissible growth rates for Birkhoff type universal harmonic functions, *J. Approx. Theory* 136 (2005), 230–243.
14. R. Aron, J. Bès, F. León and A. Peris, Operators with common hypercyclic subspaces, *J. Operator Theory* 54 (2005), 251–260.
15. R. Aron and D. Markose, On universal functions, *J. Korean Math. Soc.* 41 (2004), 65–76.
16. R.M. Aron, J.B. Seoane-Sepúlveda and A. Weber, Chaos on function spaces, *Bull. Austral. Math. Soc.* 71 (2005), 411–415.
17. A. Atzmon, Power regular operators, *Trans. Amer. Math. Soc.* 347 (1995), 3101–3109.
18. J. Auslander and J.A. Yorke, Interval maps, factors of maps, and chaos, *Tôhoku Math. J. (2)* 32 (1980), 177–188.
19. C. Badea and S. Grivaux, Unimodular eigenvalues, uniformly distributed sequences and linear dynamics, *Adv. Math.* 211 (2007), 766–793.
20. C. Badea, S. Grivaux and V. Müller, Multiples of hypercyclic operators, *Proc. Amer. Math. Soc.* 137 (2009), 1397–1403.

K.-G. Grosse-Erdmann, A. Peris Manguillot, *Linear Chaos*, Universitext, DOI 10.1007/978-1-4471-2170-1, © Springer-Verlag London Limited 2011

21. C. Badea, S. Grivaux and V. Müller, Epsilon-hypercyclic operators, *Ergodic Theory Dynam. Systems* 30 (2010), 1597–1606.
22. J. Banasiak, Chaos in some linear kinetic models, *"WASCOM 2003"—12th Conference on Waves and Stability in Continuous Media* (Proc. Conf., Villasimius, 2003), 32–37, World Sci. Publishing, River Edge, NJ, 2004.
23. J. Banasiak, Birth-and-death type systems with parameter and chaotic dynamics of some linear kinetic models, *Z. Anal. Anwendungen* 24 (2005), 675–690.
24. J. Banasiak and M. Lachowicz, Chaos for a class of linear kinetic models, *C. R. Acad. Sci. Paris Sér. IIB Méc.* 329 (2001), 439–444.
25. J. Banasiak and M. Lachowicz, Topological chaos for birth-and-death-type models with proliferation, *Math. Models Methods Appl. Sci.* 12 (2002), 755–775.
26. J. Banasiak, M. Lachowicz and M. Moszyński, Semigroups for generalized birth-and-death equations in l^p spaces, *Semigroup Forum* 73 (2006), 175–193.
27. J. Banasiak, M. Lachowicz and M. Moszyński, Chaotic behavior of semigroups related to the process of gene amplification–deamplification with cell proliferation, *Math. Biosci.* 206 (2007), 200–215.
28. J. Banks, Regular periodic decompositions for topologically transitive maps, *Ergodic Theory Dynam. Systems* 17 (1997), 505–529.
29. J. Banks, Topological mapping properties defined by digraphs, *Discrete Contin. Dynam. Systems* 5 (1999), 83–92.
30. J. Banks, Chaos for induced hyperspace maps, *Chaos Solitons Fractals* 25 (2005), 681–685.
31. J. Banks, J. Brooks, G. Cairns, G. Davis and P. Stacey, On Devaney's definition of chaos, *Amer. Math. Monthly* 99 (1992), 332–334.
32. W. Bauer and K. Sigmund, Topological dynamics of transformations induced on the space of probability measures, *Monatsh. Math.* 79 (1975), 81–92.
33. F. Bayart, Common hypercyclic vectors for composition operators, *J. Operator Theory* 52 (2004), 353–370.
34. F. Bayart, Topological and algebraic genericity of divergence and universality, *Studia Math.* 167 (2005), 161–181.
35. F. Bayart, Common hypercyclic subspaces, *Integral Equations Operator Theory* 53 (2005), 467–476.
36. F. Bayart, Dynamics of holomorphic groups, *Semigroup Forum* 82 (2011), 229–241.
37. F. Bayart and T. Bermúdez, Semigroups of chaotic operators, *Bull. Lond. Math. Soc.* 41 (2009), 823–830.
38. F. Bayart and S. Grivaux, Hypercyclicité: le rôle du spectre ponctuel unimodulaire, *C. R. Math. Acad. Sci. Paris* 338 (2004), 703–708.
39. F. Bayart and S. Grivaux, Hypercyclicity and unimodular point spectrum, *J. Funct. Anal.* 226 (2005), 281–300.
40. F. Bayart and S. Grivaux, Frequently hypercyclic operators, *Trans. Amer. Math. Soc.* 358 (2006), 5083–5117.
41. F. Bayart and S. Grivaux, Invariant Gaussian measures for operators on Banach spaces and linear dynamics, *Proc. Lond. Math. Soc. (3)* 94 (2007), 181–210.
42. F. Bayart and É. Matheron, How to get common universal vectors, *Indiana Univ. Math. J.* 56 (2007), 553–580.
43. F. Bayart and É. Matheron, Hypercyclic operators failing the hypercyclicity criterion on classical Banach spaces, *J. Funct. Anal.* 250 (2007), 426–441.
44. F. Bayart and É. Matheron, *Dynamics of linear operators*, Cambridge University Press, Cambridge, 2009.
45. F. Bayart and É. Matheron, (Non-)weakly mixing operators and hypercyclicity sets, *Ann. Inst. Fourier (Grenoble)* 59 (2009), 1–35.
46. B. Beauzamy, Un opérateur, sur l'espace de Hilbert, dont tous les polynômes sont hypercycliques, *C. R. Acad. Sci. Paris Sér. I Math.* 303 (1986), 923–925.
47. B. Beauzamy, An operator on a separable Hilbert space with many hypercyclic vectors, *Studia Math.* 87 (1987), 71–78.

48. B. Beauzamy, *Introduction to operator theory and invariant subspaces*, North-Holland, Amsterdam, 1988.
49. T. Bermúdez, A. Bonilla, J.A. Conejero and A. Peris, Hypercyclic, topologically mixing and chaotic semigroups on Banach spaces, *Studia Math.* 170 (2005), 57–75.
50. T. Bermúdez, A. Bonilla and A. Martinón, On the existence of chaotic and hypercyclic semigroups on Banach spaces, *Proc. Amer. Math. Soc.* 131 (2003), 2435–2441.
51. T. Bermúdez, A. Bonilla and A. Peris, On hypercyclicity and supercyclicity criteria, *Bull. Austral. Math. Soc.* 70 (2004), 45–54.
52. T. Bermúdez and N.J. Kalton, The range of operators on von Neumann algebras, *Proc. Amer. Math. Soc.* 130 (2002), 1447–1455.
53. T. Bermúdez and V.G. Miller, On operators T such that $f(T)$ is hypercyclic, *Integral Equations Operator Theory* 37 (2000), 332–340.
54. L. Bernal-González, On hypercyclic operators on Banach spaces, *Proc. Amer. Math. Soc.* 127 (1999), 1003–1010.
55. L. Bernal-González, Hypercyclic sequences of differential and antidifferential operators, *J. Approx. Theory* 96 (1999), 323–337.
56. L. Bernal-González, Norms of hypercyclic sequences, *Math. Pannon.* 15 (2004), 221–230.
57. L. Bernal-González, Hypercyclic subspaces in Fréchet spaces, *Proc. Amer. Math. Soc.* 134 (2006), 1955–1961.
58. L. Bernal-González, Disjoint hypercyclic operators, *Studia Math.* 182 (2007), 113–131.
59. L. Bernal-González, Common hypercyclic functions for multiples of convolution and non-convolution operators, *Proc. Amer. Math. Soc.* 137 (2009), 3787–3795.
60. L. Bernal-González and A. Bonilla, Exponential type of hypercyclic entire functions, *Arch. Math. (Basel)* 78 (2002), 283–290.
61. L. Bernal-González and M.D.C. Calderón-Moreno, Dense linear manifolds of monsters, *J. Approx. Theory* 119 (2002), 156–180.
62. L. Bernal-González and K.-G. Grosse-Erdmann, The hypercyclicity criterion for sequences of operators, *Studia Math.* 157 (2003), 17–32.
63. L. Bernal-González and K.-G. Grosse-Erdmann, Existence and nonexistence of hypercyclic semigroups, *Proc Amer. Math. Soc.* 135 (2007), 755–766.
64. L. Bernal-González and A. Montes-Rodríguez, Universal functions for composition operators, *Complex Variables Theory Appl.* 27 (1995), 47–56.
65. L. Bernal González and A. Montes Rodríguez, Non-finite dimensional closed vector spaces of universal functions for composition operators, *J. Approx. Theory* 82 (1995), 375–391.
66. J.P. Bès, Invariant manifolds of hypercyclic vectors for the real scalar case, *Proc. Amer. Math. Soc.* 127 (1999), 1801–1804.
67. J. Bès and K.C. Chan, Approximation by chaotic operators and by conjugate classes, *J. Math. Anal. Appl.* 284 (2003), 206–212.
68. J. Bès and K.C. Chan, Denseness of hypercyclic operators on a Fréchet space, *Houston J. Math.* 29 (2003), 195–206.
69. J. Bès and J.A. Conejero, Hypercyclic subspaces in omega, *J. Math. Anal. Appl.* 316 (2006), 16–23.
70. J. Bès, Ö. Martin, A. Peris and S. Shkarin, Disjoint mixing operators, preprint.
71. J. Bès and A. Peris, Hereditarily hypercyclic operators, *J. Funct. Anal.* 167 (1999), 94–112.
72. J. Bès and A. Peris, Disjointness in hypercyclicity, *J. Math. Anal. Appl.* 336 (2007), 297–315.
73. J.J. Betancor and A. Bonilla, On a universality property of certain integral operators, *J. Math. Anal. Appl.* 250 (2000), 162–180.
74. G.D. Birkhoff, Surface transformations and their dynamical applications, *Acta Math.* 43 (1920), 1–119.

75. G.D. Birkhoff, Démonstration d'un théorème élémentaire sur les fonctions entières, *C. R. Acad. Sci. Paris* 189 (1929), 473–475.
76. O. Blasco, A. Bonilla and K.-G. Grosse-Erdmann, Rate of growth of frequently hypercyclic functions, *Proc. Edinb. Math. Soc. (2)* 53 (2010), 39–59.
77. B. Bollobás, *Linear analysis*, second edition, Cambridge University Press, Cambridge, 1999.
78. J. Bonet, Hypercyclic and chaotic convolution operators, *J. London Math. Soc. (2)* 62 (2000), 253–262.
79. J. Bonet, Dynamics of the differentiation operator on weighted spaces of entire functions, *Math. Z.* 261 (2009), 649–657.
80. J. Bonet, A problem on the structure of Fréchet spaces, *Rev. R. Acad. Cienc. Exactas Fís. Nat. Ser. A Mat. RACSAM* 104 (2010), 427–434.
81. J. Bonet, L. Frerick, A. Peris and J. Wengenroth, Transitive and hypercyclic operators on locally convex spaces, *Bull. London Math. Soc.* 37 (2005), 254–264.
82. J. Bonet, F. Martínez-Giménez and A. Peris, A Banach space which admits no chaotic operator, *Bull. London Math. Soc.* 33 (2001), 196–198.
83. J. Bonet, F. Martínez-Giménez and A. Peris, Linear chaos on Fréchet spaces, *Internat. J. Bifur. Chaos Appl. Sci. Engrg.* 13 (2003), 1649–1655.
84. J. Bonet, F. Martínez-Giménez and A. Peris, Universal and chaotic multipliers on spaces of operators, *J. Math. Anal. Appl.* 297 (2004), 599–611.
85. J. Bonet and A. Peris, Hypercyclic operators on non-normable Fréchet spaces, *J. Funct. Anal.* 159 (1998), 587–595.
86. A. Bonilla and K.-G. Grosse-Erdmann, On a theorem of Godefroy and Shapiro, *Integral Equations Operator Theory* 56 (2006), 151–162.
87. A. Bonilla and K.-G. Grosse-Erdmann, Frequently hypercyclic operators and vectors, *Ergodic Theory Dynam. Systems* 27 (2007), 383–404. Erratum: *Ergodic Theory Dynam. Systems* 29 (2009), 1993–1994.
88. A. Bonilla and K.-G. Grosse-Erdmann, Frequently hypercyclic subspaces, preprint.
89. A. Bonilla and P. Miana, Hypercyclic and topologically mixing cosine functions on Banach spaces, *Proc. Amer. Math. Soc.* 136 (2008), 519–528.
90. P.S. Bourdon, Invariant manifolds of hypercyclic vectors, *Proc. Amer. Math. Soc.* 118 (1993), 845–847.
91. P.S. Bourdon, The second iterate of a map with dense orbit, *Proc. Amer. Math. Soc.* 124 (1996), 1577–1581.
92. P.S. Bourdon, Orbits of hyponormal operators, *Michigan Math. J.* 44 (1997), 345–353.
93. P.S. Bourdon and N.S. Feldman, Somewhere dense orbits are everywhere dense, *Indiana Univ. Math. J.* 52 (2003), 811–819.
94. P.S. Bourdon and J.H. Shapiro, Cyclic composition operators on H^2, *Operator Theory: Operator Algebras and Applications, Part 2* (Proc. Summer Res. Inst., Durham, NH, 1988), 43–53, Amer. Math. Soc., Providence, RI, 1990.
95. P.S. Bourdon and J.H. Shapiro, Cyclic phenomena for composition operators, *Mem. Amer. Math. Soc.* 125 (1997), no. 596.
96. P.S. Bourdon and J.H. Shapiro, Hypercyclic operators that commute with the Bergman backward shift, *Trans. Amer. Math. Soc.* 352 (2000), 5293–5316.
97. M. Brin and G. Stuck, *Introduction to dynamical systems*, Cambridge University Press, Cambridge, 2002.
98. L. Carleson and T.W. Gamelin, *Complex dynamics*, Springer, New York, 1993.
99. K.C. Chan, Hypercyclicity of the operator algebra for a separable Hilbert space, *J. Operator Theory* 42 (1999), 231–244.
100. K.C. Chan, The density of hypercyclic operators on a Hilbert space, *J. Operator Theory* 47 (2002), 131–143.
101. K.C. Chan and R. Sanders, A weakly hypercyclic operator that is not norm hypercyclic, *J. Operator Theory* 52 (2004), 39–59.

102. K.C. Chan and R. Sanders, Two criteria for a path of operators to have common hypercyclic vectors, *J. Operator Theory* 61 (2009), 191–223.
103. K.C. Chan and R. Sanders, Common hypercyclic vectors for the conjugate class of a hypercyclic operator, *J. Math. Anal. Appl.* 375 (2011), 139–148.
104. K.C. Chan and I. Seceleanu, Orbital limit points and hypercyclicity of operators on analytic function spaces, *Math. Proc. R. Ir. Acad.* 110A (2010), 99–109.
105. K.C. Chan and I. Seceleanu, Hypercyclicity of shifts as a zero-one law of orbital limit points, *J. Operator Theory*, to appear.
106. K.C. Chan and J.H. Shapiro, The cyclic behavior of translation operators on Hilbert spaces of entire functions, *Indiana Univ. Math. J.* 40 (1991), 1421–1449.
107. K.C. Chan and R.D. Taylor, Jr., Hypercyclic subspaces of a Banach space, *Integral Equations Operator Theory* 41 (2001), 381–388.
108. J.A. Conejero, *Operadores y semigrupos de operadores en espacios de Fréchet y espacios localmente convexos*, Thesis, Univ. Politécnica de Valencia, Valencia, 2004.
109. J.A. Conejero and E.M. Mangino, Hypercyclic semigroups generated by Ornstein–Uhlenbeck operators, *Mediterr. J. Math.* 7 (2010), 101–109.
110. J.A. Conejero, V. Müller and A. Peris, Hypercyclic behaviour of operators in a hypercyclic C_0-semigroup, *J. Funct. Anal.* 244 (2007), 342–348.
111. J.A. Conejero and A. Peris, Linear transitivity criteria, *Topology Appl.* 153 (2005), 767–773.
112. J.A. Conejero and A. Peris, Chaotic translation semigroups, *Discrete Contin. Dyn. Syst.* 2007, Dynamical Systems and Differential Equations (Proc. Conf., Poitiers, 2006), suppl., 269–276.
113. J.A. Conejero and A. Peris, Hypercyclic translation C_0-semigroups on complex sectors, *Discrete Contin. Dyn. Syst.* 25 (2009), 1195–1208.
114. J.A. Conejero, A. Peris and M. Trujillo, Chaotic asymptotic behaviour of the hyperbolic heat transfer equation solutions, *Internat. J. Bifur. Chaos Appl. Sci. Engrg.* 20 (2010), 2943–2947.
115. J.B. Conway, *Functions of one complex variable*, second edition, Springer, New York-Berlin, 1978.
116. J.B. Conway, *A course in functional analysis*, second edition, Springer, New York-Berlin, 1990.
117. G. Costakis, On a conjecture of D. Herrero concerning hypercyclic operators, *C. R. Acad. Sci. Paris Sér. I Math.* 330 (2000), 179–182.
118. G. Costakis and A. Manoussos, J-class weighted shifts on the space of bounded sequences of complex numbers, *Integral Equations Operator Theory* 62 (2008), 149–158.
119. G. Costakis and A. Manoussos, J-class operators and hypercyclicity, *J. Operator Theory*, to appear.
120. G. Costakis and P. Mavroudis, Common hypercyclic entire functions for multiples of differential operators, *Colloq. Math.* 111 (2008), 199–203.
121. G. Costakis and A. Peris, Hypercyclic semigroups and somewhere dense orbits, *C. R. Math. Acad. Sci. Paris* 335 (2002), 895–898.
122. G. Costakis and I.Z. Ruzsa, Frequently Cesàro hypercylic operators are hypercyclic, preprint.
123. G. Costakis and M. Sambarino, Genericity of wild holomorphic functions and common hypercyclic vectors, *Adv. Math.* 182 (2004), 278–306.
124. G. Costakis and M. Sambarino, Topologically mixing hypercyclic operators, *Proc. Amer. Math. Soc.* 132 (2004), 385–389.
125. C.C. Cowen and B.D. MacCluer, *Composition operators on spaces of analytic functions*, CRC Press, Boca Raton, FL, 1995.
126. M. De la Rosa and C. Read, A hypercyclic operator whose direct sum $T \oplus T$ is not hypercyclic, *J. Operator Th.* 61 (2009), 369–380.
127. M. De la Rosa, L. Frerick, S. Grivaux and A. Peris, Frequent hypercyclicity, chaos, and unconditional Schauder decompositions, *Israel J. Math.*, to appear.

128. R. deLaubenfels and H. Emamirad, Chaos for functions of discrete and continuous weighted shift operators, *Ergodic Theory Dynam. Systems* 21 (2001), 1411–1427.
129. W. Desch and W. Schappacher, On products of hypercyclic semigroups, *Semigroup Forum* 71 (2005), 301–311.
130. W. Desch and W. Schappacher, Spectral characterization of weak topological transitivity, *Rev. R. Acad. Cienc. Exactas Fís. Nat. Ser. A Mat. RACSAM*, to appear.
131. W. Desch, W. Schappacher and G.F. Webb, Hypercyclic and chaotic semigroups of linear operators, *Ergodic Theory Dynam. Systems* 17 (1997), 793–819.
132. L.R. Devaney, *An introduction to chaotic dynamical systems*, Benjamin/Cummings, Menlo Park, CA, 1986; second edition, Addison-Wesley, Redwood City, CA, 1989.
133. J. Diestel, *Sequences and series in Banach spaces*, Springer, New York, 1984.
134. J. Diestel, H. Jarchow and A. Tonge, *Absolutely summing operators*, Cambridge University Press, Cambridge, 1995.
135. S.J. Dilworth and V.G. Troitsky, Spectrum of a weakly hypercyclic operator meets the unit circle, *Trends in Banach spaces and operator theory* (Proc. Conf., Memphis, TN, 2001), 67–69, Amer. Math. Soc., Providence, RI, 2003.
136. P.L. Duren, *Theory of H^p spaces*, Academic Press, New York-London, 1970.
137. S.M. Duyos-Ruiz, On the existence of universal functions, *Soviet Math. Dokl.* 27 (1983), 9–13.
138. O.P. Dzagnidze, The universal harmonic function in the space E_n (Russian), *Soobshch. Akad. Nauk Gruzin. SSR* 55 (1969), 41–44.
139. S. El Mourchid, On a hypercyclicity criterion for strongly continuous semigroups, *Discrete Contin. Dyn. Syst.* 13 (2005), 271–275.
140. S. El Mourchid, The imaginary point spectrum and hypercyclicity, *Semigroup Forum* 73 (2006), 313–316.
141. H. Emamirad, G.R. Goldstein and J.A. Goldstein, Chaotic solution for the Black–Scholes equation, preprint.
142. P. Enflo, On the invariant subspace problem for Banach spaces, *Acta Math.* 158 (1987), 213–313.
143. K.-J. Engel and R. Nagel, *One-parameter semigroups for linear evolution equations*, Springer, New York-Berlin, 2000.
144. K.-J. Engel and R. Nagel, *A short course on operator semigroups*, Springer, New York, 2006.
145. J. Esterle, Mittag-Leffler methods in the theory of Banach algebras and a new approach to Michael's problem, *Banach algebras and several complex variables* (Proc. Conf., New Haven, CT, 1983), 107–129, American Mathematical Society, Providence, RI, 1984.
146. A. Fathi, Existence de systèmes dynamiques minimaux sur l'espace de Hilbert séparable, *Topology* 22 (1983), 165–167.
147. N.S. Feldman, *Linear chaos?*, <http://home.wlu.edu/~feldmann/research.html>, 2001.
148. N.S. Feldman, Perturbations of hypercyclic vectors, *J. Math. Anal. Appl.* 273 (2002), 67–74.
149. N.S. Feldman, Countably hypercyclic operators, *J. Operator Theory* 50 (2003), 107–117.
150. N.S. Feldman, Hypercyclicity and supercyclicity for invertible bilateral weighted shifts, *Proc. Amer. Math. Soc.* 131 (2003), 479–485.
151. N.S. Feldman, V.G. Miller and T.L. Miller, Hypercyclic and supercyclic cohyponormal operators, *Acta Sci. Math. (Szeged)* 68 (2002), 965–990.
152. E. Flytzanis, Mixing properties of linear operators in Hilbert spaces, *Séminaire d'Initiation à l'Analyse* 34ème année (1994/1995), Exposé no. 6.
153. E. Flytzanis, Unimodular eigenvalues and linear chaos in Hilbert spaces, *Geom. Funct. Anal.* 5 (1995), 1–13.

154. G.L. Forti, Various notions of chaos for discrete dynamical systems. A brief survey, *Aequationes Math.* 70 (2005), 1–13.
155. L. Frerick and A. Peris, Hypercyclic operators for which every non-zero multiple is also hypercyclic, preprint.
156. H. Furstenberg, The structure of distal flows, *Amer. J. Math.* 85 (1963) 477–515.
157. H. Furstenberg, Disjointness in ergodic theory, minimal sets, and a problem in Diophantine approximation, *Math. Systems Theory* 1 (1967), 1–49.
158. E.A. Gallardo-Gutiérrez and A. Montes-Rodríguez, The role of the spectrum in the cyclic behavior of composition operators, *Mem. Amer. Math. Soc.* 167 (2004), no. 791.
159. E.A. Gallardo-Gutiérrez and A. Montes-Rodríguez, The Volterra operator is not supercyclic, *Integral Equations Operator Theory* 50 (2004), 211–216.
160. E.A. Gallardo-Gutiérrez and J.R. Partington, Common hypercyclic vectors for families of operators, *Proc. Amer. Math. Soc.* 136 (2008), 119–126.
161. R.M. Gethner and J.H. Shapiro, Universal vectors for operators on spaces of holomorphic functions, *Proc. Amer. Math. Soc.* 100 (1987), 281–288.
162. E. Glasner, *Ergodic theory via joinings*, American Mathematical Society, Providence, RI, 2003.
163. E. Glasner and B. Weiss, Sensitive dependence on initial conditions, *Nonlinearity* 6 (1993), 1067–1075.
164. G. Godefroy, Linear dynamics, *Advanced courses of mathematical analysis. II* (Proc. 2nd Int. School, Granada, 2004), 57–75, World Sci. Publ., Hackensack, NJ, 2007.
165. G. Godefroy and J.H. Shapiro, Operators with dense, invariant, cyclic vector manifolds, *J. Funct. Anal.* 98 (1991), 229–269.
166. M.C. Gómez-Collado, F. Martínez-Giménez, A. Peris and F. Rodenas, Slow growth for universal harmonic functions, *J. Inequal. Appl.* 2010, Art. ID 253690, 6 pp.
167. M. González, F. León-Saavedra and A. Montes-Rodríguez, Semi-Fredholm theory: hypercyclic and supercyclic subspaces, *Proc. London Math. Soc. (3)* 81 (2000), 169–189.
168. P. Gorkin, F. León-Saavedra and R. Mortini, Bounded universal functions in one and several complex variables, *Math. Z.* 258 (2008), 745–762.
169. S. Grivaux, Construction of operators with prescribed behaviour, *Arch. Math. (Basel)* 81 (2003), 291–299.
170. S. Grivaux, Sums of hypercyclic operators, *J. Funct. Anal.* 202 (2003), 486–503.
171. S. Grivaux, Hypercyclic operators with an infinite dimensional closed subspace of periodic points, *Rev. Mat. Complut.* 16 (2003), 383–390.
172. S. Grivaux, Hypercyclic operators, mixing operators, and the bounded steps problem, *J. Operator Theory* 54 (2005), 147–168.
173. S. Grivaux, A probabilistic version of the frequent hypercyclicity criterion, *Studia Math.* 176 (2006), 279–290.
174. S. Grivaux, A new class of frequently hypercyclic operators, with applications, *Indiana Univ. Math. J.*, to appear.
175. S. Grivaux, A hypercyclic rank one perturbation of a unitary operator, preprint.
176. S. Grivaux and S. Shkarin, Non-mixing hypercyclic operators, unpublished (2007).
177. K.-G. Grosse-Erdmann, Holomorphe Monster und universelle Funktionen, *Mitt. Math. Sem. Giessen* 176 (1987).
178. K.-G. Grosse-Erdmann, On the universal functions of G.R. MacLane, *Complex Variables Theory Appl.* 15 (1990), 193–196.
179. K.-G. Grosse-Erdmann, Universal families and hypercyclic operators, *Bull. Amer. Math. Soc. (N.S.)* 36 (1999), 345–381.
180. K.-G. Grosse-Erdmann, Hypercyclic and chaotic weighted shifts, *Studia Math.* 139 (2000), 47–68.
181. K.-G. Grosse-Erdmann, Rate of growth of hypercyclic entire functions, *Indag. Math. (N.S.)* 11 (2000), 561–571.

182. K.-G. Grosse-Erdmann, Dynamics of linear operators, *Topics in complex analysis and operator theory* (Proc. Winter School, Málaga, 2006), 41–84, Univ. Málaga, Málaga, 2007.
183. K.-G. Grosse-Erdmann, F. León-Saavedra and A. Piqueras-Lerena, The iterates of a map with dense orbit, *Acta Sci. Math. (Szeged)* 74 (2008), 245–257.
184. K.-G. Grosse-Erdmann and R. Mortini, Universal functions for composition operators with non-automorphic symbol, *J. Anal. Math.* 107 (2009), 355–376.
185. K.-G. Grosse-Erdmann and A. Peris, Frequently dense orbits, *C. R. Math. Acad. Sci. Paris* 341 (2005), 123–128.
186. K.-G. Grosse-Erdmann and A. Peris, Corrigendum to the note "Frequently dense orbits" (2008), not accepted for publication in *C. R. Math. Acad. Sci. Paris.*
187. K.-G. Grosse-Erdmann and A. Peris, Weakly mixing operators on topological vector spaces, *Rev. R. Acad. Cienc. Exactas Fís. Nat. Ser. A Mat. RACSAM* 104 (2010), 413–426.
188. D. Gulick, *Encounters with chaos*, McGraw-Hill, New York, 1992.
189. D.W. Hadwin, E.A. Nordgren, H. Radjavi and P. Rosenthal, Most similarity orbits are strongly dense, *Proc. Amer. Math. Soc.* 76 (1979), 250–252.
190. P. Hájek and P. Vivi, On ω-limit sets of ordinary differential equations in Banach spaces, *J. Math. Anal. Appl.* 371 (2010), 793–812.
191. P.R. Halmos, *A Hilbert space problem book*, second edition, Springer, New York-Berlin, 1982.
192. I. Halperin, C. Kitai and P. Rosenthal, On orbits of linear operators, *J. London Math. Soc. (2)* 31 (1985), 561–565.
193. G.H. Hardy, J.E. Littlewood and G. Pólya, *Inequalities*, second edition, Cambridge University Press, Cambridge, 1952.
194. D.A. Herrero, Limits of hypercyclic and supercyclic operators, *J. Funct. Anal.* 99 (1991), 179–190.
195. D.A. Herrero, Hypercyclic operators and chaos, *J. Operator Theory* 28 (1992), 93–103.
196. D.A. Herrero and C. Kitai, On invertible hypercyclic operators, *Proc. Amer. Math. Soc.* 116 (1992), 873–875.
197. G. Herzog, On linear operators having supercyclic vectors, *Studia Math.* 103 (1992), 295–298.
198. G. Herzog, On a universality of the heat equation, *Math. Nachr.* 188 (1997), 169–171.
199. G. Herzog and R. Lemmert, Über Endomorphismen mit dichten Bahnen, *Math. Z.* 213 (1993), 473–477.
200. G. Herzog and C. Schmoeger, On operators T such that $f(T)$ is hypercyclic, *Studia Math.* 108 (1994), 209–216.
201. G. Herzog and A. Weber, A class of hypercyclic Volterra composition operators, *Demonstratio Math.* 39 (2006), 465–468.
202. H.M. Hilden and L.J. Wallen, Some cyclic and non-cyclic vectors of certain operators, *Indiana Univ. Math. J.* 23 (1973/74), 557–565.
203. E. Hille and R.S. Phillips, *Functional analysis and semi-groups*, American Mathematical Society, Providence, R.I., 1957.
204. T. Hosokawa, Chaotic behavior of composition operators on the Hardy space, *Acta Sci. Math. (Szeged)* 69 (2003), 801–811.
205. P.R. Hurst, Relating composition operators on different weighted Hardy spaces, *Arch. Math. (Basel)* 68 (1997), 503–513.
206. V. Istrăţescu, T. Saitô and T. Yoshino, On a class of operators, *Tôhoku Math. J. (2)* 18 (1966), 410–413.
207. L. Ji and A. Weber, L^p spectral theory and heat dynamics of locally symmetric spaces, *J. Funct. Anal.* 258 (2010), 1121–1139.
208. L. Ji and A. Weber, Dynamics of the heat semigroup on symmetric spaces, *Ergodic Theory Dynam. Systems* 30 (2010), 457–468.

209. T. Kalmes, Hypercyclic, mixing, and chaotic C_0-semigroups induced by semiflows, *Ergodic Theory Dynam. Systems* 27 (2007), 1599–1631.
210. T. Kalmes, Hypercyclic C_0-semigroups and evolution families generated by first order differential operators, *Proc. Amer. Math. Soc.* 137 (2009), 3833–3848.
211. T. Kalmes, Hypercyclicity and mixing for cosine operator functions generated by second order partial differential operators, *J. Math. Anal. Appl.* 365 (2010), 363–375.
212. N.J. Kalton, N.T. Peck and J.W. Roberts, *An F-space sampler*, Cambridge University Press, Cambridge, 1984.
213. P.K. Kamthan and M. Gupta, *Sequence spaces and series*, Marcel Dekker, New York, 1981.
214. K.-T. Kim and S.G. Krantz, The automorphism groups of domains, *Amer. Math. Monthly* 112 (2005), 585–601.
215. C. Kitai, *Invariant closed sets for linear operators*, Thesis, University of Toronto, Toronto, 1982.
216. S.F. Kolyada, Li-Yorke sensitivity and other concepts of chaos, *Ukrainian Math. J.* 56 (2004), 1242–1257.
217. S. Kolyada and L. Snoha, Some aspects of topological transitivity—a survey, *Iteration theory* (Proc. Conf., Opava, 1994) 3–35, Grazer Math. Ber. 334, Graz, 1997.
218. C.S. Kubrusly, *Hilbert space operators. A problem solving approach*, Birkhäuser, Boston, MA, 2003.
219. F. León-Saavedra, Notes about the hypercyclicity criterion, *Math. Slovaca* 53 (2003), 313–319.
220. F. León-Saavedra and A. Montes-Rodríguez, Linear structure of hypercyclic vectors, *J. Funct. Anal.* 148 (1997), 524–545.
221. F. León-Saavedra and A. Montes-Rodríguez, Spectral theory and hypercyclic subspaces, *Trans. Amer. Math. Soc.* 353 (2001), 247–267.
222. F. León-Saavedra and V. Müller, Rotations of hypercyclic and supercyclic operators, *Integral Equations Operator Theory* 50 (2004), 385–391.
223. F. León-Saavedra and V. Müller, Hypercyclic sequences of operators, *Studia Math.* 175 (2006), 1–18.
224. F. León-Saavedra, A. Piqueras-Lerena and J.B. Seoane-Sepúlveda, Orbits of Cesàro type operators, *Math. Nachr.* 282 (2009), 764–773.
225. G.R. MacLane, Sequences of derivatives and normal families, *J. Analyse Math.* 2 (1952/53), 72–87.
226. M. Marano and H.N. Salas, Maps with dense orbits: Ansari's theorem revisited and the infinite torus, *Bull. Belg. Math. Soc. Simon Stevin* 16 (2009), 481–492.
227. F. Martínez-Giménez, Operadores hipercíclicos en espacios de Fréchet, *Rev. Colombiana Mat.* 33 (1999), 51–76.
228. F. Martínez-Giménez, Chaos for power series of backward shift operators, *Proc. Amer. Math. Soc.* 135 (2007), 1741–1752.
229. F. Martínez-Giménez and A. Peris, Chaos for backward shift operators, *Internat. J. Bifur. Chaos Appl. Sci. Engrg.* 12 (2002), 1703–1715.
230. F. Martínez-Giménez and A. Peris, Hypercyclic differential operators, *Analysis and operators* (3rd Conf. EU Network, Tenerife, 2003).
231. F. Martínez-Giménez and A. Peris, Universality and chaos for tensor products of operators, *J. Approx. Theory* 124 (2003), 7–24.
232. F. Martínez-Giménez and A. Peris, Hypercyclic differential operators on spaces of entire and harmonic functions, preprint.
233. V. Matache, Notes on hypercyclic operators, *Acta Sci. Math. (Szeged)* 58 (1993), 401–410.
234. V. Matache, Spectral properties of operators having dense orbits, *Topics in Operator Theory, Operator Algebras and Applications* (Proc. Conf., Timişoara, 1994), 221–237, Inst. Math. Roman. Acad., Bucharest, 1995.

235. É. Matheron, Subsemigroups of transitive semigroups, *Ergodic Theory Dynam. Systems*, to appear.
236. M. Matsui, M. Yamada and F. Takeo, Supercyclic and chaotic translation semigroups, *Proc. Amer. Math. Soc.* 131 (2003), 3535–3546. Erratum: *Proc. Amer. Math. Soc.* 132 (2004), 3751–3752.
237. R. Meise and D. Vogt, *Introduction to functional analysis*, Oxford University Press, New York, 1997.
238. G. Metafune and V.B. Moscatelli, Dense subspaces with continuous norm in Fréchet spaces, *Bull. Polish Acad. Sci. Math.* 37 (1989), 477–479.
239. T.L. Miller and V.G. Miller, Local spectral theory and orbits of operators, *Proc. Amer. Math. Soc.* 127 (1999), 1029–1037.
240. A. Montes-Rodríguez, Banach spaces of hypercyclic vectors, *Michigan Math. J.* 43 (1996), 419–436.
241. A. Montes-Rodríguez, A Birkhoff theorem for Riemann surfaces, *Rocky Mountain J. Math.* 28 (1998), 663–693.
242. A. Montes-Rodríguez and M.C. Romero-Moreno, Supercyclicity in the operator algebra, *Studia Math.* 150 (2002), 201–213.
243. A. Montes-Rodríguez and H.N. Salas, Supercyclic subspaces, *Bull. London Math. Soc.* 35 (2003), 721–737.
244. T.K.S. Moothathu, Quantitative views of recurrence and proximality, *Nonlinearity* 21 (2008), 2981–2992.
245. T.K.S. Moothathu, Weak mixing and mixing of a single transformation of a topological (semi)group, *Aequationes Math.* 78 (2009), 147–155.
246. V. Müller, Local behaviour of the polynomial calculus of operators, *J. Reine Angew. Math.* 430 (1992), 61–68.
247. V. Müller, *Spectral theory of linear operators and spectral systems in Banach algebras*, second edition, Birkhäuser, Basel, 2007.
248. V. Müller, On the Salas theorem and hypercyclicity of $f(T)$, *Integral Equations Operator Theory* 67 (2010), 439–448.
249. V. Nestoridis, Universal Taylor series, *Ann. Inst. Fourier (Grenoble)* 46 (1996), 1293–1306.
250. M.H.A. Newman, *Elements of the topology of plane sets of points*, second edition, Cambridge University Press, Cambridge, 1951.
251. J.C. Oxtoby and S.M. Ulam, Measure-preserving homeomorphisms and metrical transitivity, *Ann. of Math. (2)* 42 (1941), 874–920.
252. A. Peris, Common hypercyclic vectors for backward shifts, *Operator Theory Seminar*, Michigan State University, 2000–2001.
253. A. Peris, Hypercyclicity criteria and the Mittag-Leffler theorem, *Bull. Soc. Roy. Sci. Liège* 70 (2001), 365–371.
254. A. Peris, Multi-hypercyclic operators are hypercyclic, *Math. Z.* 236 (2001), 779–786.
255. A. Peris, Chaotic polynomials on Banach spaces, *J. Math. Anal. Appl.* 287 (2003), 487–493.
256. A. Peris, Set-valued discrete chaos, *Chaos Solitons Fractals* 26 (2005), 19–23.
257. A. Peris and L. Saldivia, Syndetically hypercyclic operators, *Integral Equations Operator Theory* 51 (2005), 275–281.
258. H. Petersson, Hypercyclic subspaces for Fréchet space operators, *J. Math. Anal. Appl.* 319 (2006), 764–782.
259. A. Pietsch, *Operator ideals*, North-Holland, Amsterdam-New York, 1980.
260. G.T. Prajitura, The density of the hypercyclic operators in the strong operator topology, *Integral Equations Operator Theory* 49 (2004), 559–560.
261. G. Prăjitură, Irregular vectors of Hilbert space operators, *J. Math. Anal. Appl.* 354 (2009), 689–697.
262. V. Protopopescu, Linear vs nonlinear and infinite vs finite: An interpretation of chaos, *Oak Ridge National Laboratory Report* TM-11667, Oak Ridge, TN, 1990.

263. V. Protopopescu and Y.Y. Azmy, Topological chaos for a class of linear models, *Math. Models Methods Appl. Sci.* 2 (1992), 79–90.
264. H. Radjavi and P. Rosenthal, *Invariant subspaces*, second edition, Dover Publications, Mineola, NY, 2003.
265. C.J. Read, A solution to the invariant subspace problem, *Bull. London Math. Soc.* 16 (1984), 337–401.
266. C.J. Read, The invariant subspace problem for a class of Banach spaces. II. Hypercyclic operators, *Israel J. Math.* 63 (1988), 1–40.
267. C. Robinson, *Dynamical systems. Stability, symbolic dynamics, and chaos*, second edition, CRC Press, Boca Raton, FL, 1999.
268. S. Rolewicz, On orbits of elements, *Studia Math.* 32 (1969), 17–22.
269. S. Rolewicz, *Metric linear spaces*, second edition, D. Reidel Publishing Co., Dordrecht, 1985.
270. W. Rudin, *Real and complex analysis*, third edition, McGraw-Hill, New York, 1987.
271. W. Rudin, *Functional analysis*, second edition, McGraw-Hill, New York, 1991.
272. R. Rudnicki, Gaussian measure-preserving linear transformations, *Univ. Iagel. Acta Math.* 30 (1993), 105–112.
273. H. Salas, A hypercyclic operator whose adjoint is also hypercyclic, *Proc. Amer. Math. Soc.* 112 (1991), 765–770.
274. H.N. Salas, Hypercyclic weighted shifts, *Trans. Amer. Math. Soc.* 347 (1995), 993–1004.
275. H.N. Salas, Supercyclicity and weighted shifts, *Studia Math.* 135 (1999), 55–74.
276. H.N. Salas, Banach spaces with separable duals support dual hypercyclic operators, *Glasg. Math. J.* 49 (2007), 281–290.
277. H.N. Salas, Dual disjoint hypercyclic operators, *J. Math. Anal. Appl.* 374 (2011), 106–117.
278. W. Seidel and J.L. Walsh, On approximation by Euclidean and non-Euclidean translations of an analytic function, *Bull. Amer. Math. Soc.* 47 (1941), 916–920.
279. J.H. Shapiro, *Composition operators and classical function theory*, Springer, New York, 1993.
280. J.H. Shapiro, Simple connectivity and linear chaos, *Rend. Circ. Mat. Palermo (2) Suppl.* (1998), no. 56, 27–48.
281. J.H. Shapiro, *Notes on the dynamics of linear operators*, <http://www.mth.msu.edu/~shapiro>, 2001.
282. S. Shirali and H.L. Vasudeva, *Metric spaces*, Springer, London, 2006.
283. S.A. Shkarin, On the growth of D-universal functions, *Moscow Univ. Math. Bull.* 48 (1993), no. 6, 49–51.
284. S. Shkarin, Universal elements for non-linear operators and their applications, *J. Math. Anal. Appl.* 348 (2008), 193–210.
285. S. Shkarin, The Kitai criterion and backward shifts, *Proc. Amer. Math. Soc.* 136 (2008), 1659–1670.
286. S. Shkarin, Existence theorems in linear chaos, arXiv:0810.1192v2 (2008).
287. S. Shkarin, On the spectrum of frequently hypercyclic operators, *Proc. Amer. Math. Soc.* 137 (2009), 123–134.
288. S. Shkarin, Remarks on common hypercyclic vectors, *J. Funct. Anal.* 258 (2010), 132–160.
289. S. Shkarin, A hypercyclic finite rank perturbation of a unitary operator, *Math. Ann.* 348 (2010), 379–393.
290. S. Shkarin, On the set of hypercyclic vectors for the differentiation operator, *Israel J. Math.* 180 (2010), 271–283.
291. S. Shkarin, A short proof of existence of disjoint hypercyclic operators, *J. Math. Anal. Appl.* 367 (2010), 713–715.
292. S. Shkarin, Chaotic Banach algebras, preprint.
293. S. Shkarin, Hypercyclic operators on topological vector spaces, preprint.

294. S. Silverman, On maps with dense orbits and the definition of chaos, *Rocky Mountain J. Math.* 22 (1992), 353–375.
295. L.A. Smith, *Chaos: A very short introduction*, Oxford University Press, Oxford, 2007.
296. C.L. Stewart and R. Tijdeman, On infinite-difference sets, *Canad. J. Math.* 31 (1979), 897–910.
297. F. Takeo, Chaos and hypercyclicity for solution semigroups to some partial differential equations, *Nonlinear Anal.* 63 (2005), e1943–e1953.
298. M. Taniguchi, Chaotic composition operators on the classical holomorphic spaces, *Complex Var. Theory Appl.* 49 (2004), 529–538.
299. P. Touhey, Yet another definition of chaos, *Amer. Math. Monthly* 104 (1997), 411–414.
300. P. Walters, *An introduction to ergodic theory*, Springer, New York-Berlin, 1982.
301. J. Wengenroth, Hypercyclic operators on non-locally convex spaces, *Proc. Amer. Math. Soc.* 131 (2003), 1759–1761.
302. P.Y. Wu, Sums and products of cyclic operators, *Proc. Amer. Math. Soc.* 122 (1994), 1053–1063.
303. B. Yousefi and H. Rezaei, Conditions for hypercyclicity criterion, *Int. J. Contemp. Math. Sci.* 1 (2006), 99–107.
304. N. Zorboska, Cyclic composition operators on smooth weighted Hardy spaces, *Rocky Mountain J. Math.* 29 (1999), 725–740.

Index

K.-G. Grosse-Erdmann, A. Peris Manguillot, *Linear Chaos*, Universitext,
DOI 10.1007/978-1-4471-2170-1,